HVAC
Troubleshooting
Guide

697

M

HVAC
Troubleshooting
Guide

Rex Miller

New York   Chicago   San Francisco   Lisbon   London   Madrid
Mexico City   Milan   New Delhi   San Juan   Seoul
Singapore   Sydney   Toronto

The McGraw·Hill Companies

Library of Congress Cataloging-in-Publication Data

Miller, Rex, date.
    HVAC troubleshooting guide / Rex Miller.
        p.   cm.
    Includes index.
    ISBN 978-0-07-160499-4 (alk. paper)
    1. Heating—Equipment and supplies—Maintenance and repair.
    2. Air conditioning—Equipment and supplies—Maintenance and repair.
    3. Ventilation—Equipment and supplies—Maintenance and repair.
    I. Title.
    TH7012.M528   2009
    697.00028′8—dc22   2008040449

1 2 3 4 5 6 7 8 9 0   DOC/DOC   0 1 5 4 3 2 1 0 9

ISBN 978-0-07-160499-4
MHID 0-07-160499-5

Sponsoring Editor
    Larry S. Hager

Production Supervisor
    Richard C. Ruzycka

Editing Supervisor
    Stephen M. Smith

Project Manager
    Harleen Chopra, International
    Typesetting and Composition

Copy Editor
    Surendra Nath Shivam, International
    Typesetting and Composition

Proofreader
    Ragini Pandey, International Typesetting
    and Composition

Art Director, Cover
    Jeff Weeks

Composition
    International Typesetting and Composition

Printed and bound by RR Donnelley.

McGraw-Hill books are available at special discounts to use as premium and sales promotions, or for use in corporate training programs. To contact a special sales representative, please visit the Contact Us page at www.mhprofessional.com.

This book is printed on acid-free paper.

ABOUT THE AUTHOR

Rex Miller is professor emeritus of industrial technology at State University of New York college at Buffalo and has taught technical curriculum at the college and high school levels for more than 40 years. He is the coauthor of McGraw-Hill's best-selling *Carpentry & Construction*, now in its Fourth Edition, and the author of more than 100 texts for vocational and industrial arts programs. He lives in Round Rock, Texas.

Contents

Preface

The value of a troubleshooting guide increases when it is available and needed. The format of this book will allow you to locate information quickly. The material reflects recent changes in the field of refrigeration and air-conditioning.

The purpose of the book is to aid you in your everyday tasks and keep you current with some of the latest facts, figures, and devices in this important trade. It is truly a portable library for use on the job, on the reference shelf in the shop, in your pocket, or in the toolbox.

Many tables, charts, and illustrations are included. They show a variety of parts and techniques often found in present-day practice in the field. Obviously, not all related problems can be presented here, since there is a great deal of ingenuity required by the worker on the job. For standard procedures, however, the *ASHRAE Handbook* gives the numbers and sizes of various types of equipment that is safe to install in a given location.

This guide should be kept within each reach and as part of your tool kit. An effort has been made to make the book useable for classroom work as well as on the job.

Rex Miller

Acknowledgments

I would like to thank the following manufacturers and organizations for their generous help. They furnished photographs, drawings, and technical assistance. Without their valuable time and effort this book would not have been possible.

Air Conditioning and
  Refrigeration Institute
Air Temp Division of Chrysler
  Corp.
Americold Compressor
  Corporation
Amprobe Instruments
Arkla Industries, Inc.
Bryant Manufacturing Co.
Carrier Corporation
Environmental Protection Agency
General Controls
General Electric Co.
Honeywell, Inc.
Hussman Refrigeration, Inc.
Lennox Industries
Lima Register Co.

Marley Company
Mueller Brass Co.
National Refrigerants
National Safety Council
Rheem Manufacturing Co.
Robertshaw Controls Co.
Sporlan Valve Co.
Tecumseh Products Co.
Tyler Refrigeration Corp.
Virginia Chemicals, Inc.
Wadsworth Electric Co.
Wagner Electric Co.
Wagner Electric Motors
Westinghouse Electric Co.
Weston Electrical Instruments Co.
Worthington Compressors

I hope this acknowledgment of these contributors will let you know that the field you are in or are about to enter is one of the best. Individuals, too numerous to mention, have also played a role in this book. I would like to take this opportunity to thank them too.

HVAC
Troubleshooting
Guide

Tools and Instruments

Tools and Equipment

The air-conditioning technician must work with electricity. Equipment that has been wired may have to be replaced or rewired. In any case, it is necessary to identify and use safely various tools and pieces of equipment. Special tools are needed to install and maintain electrical service to air-conditioning units. Wires and wiring should be installed according to the National Electrical Code (NEC). However, it is possible that this will not have been done. In such a case, the electrician will have to be called to update the wiring to carry the extra load of the installation of new air-conditioning or refrigeration equipment.

This section deals only with interior wiring. Following is a brief discussion of the more important tools used by the electrician in the installation of air-conditioning and refrigeration equipment.

Pliers and clippers

Pliers come in a number of sizes and shapes designed for special applications. Pliers are available with either insulated or uninsulated handles. Although pliers with insulated handles are always used when working on or near "hot" wires, they must not be considered sufficient protection alone. Other precautions must be taken. Long-nose pliers are used for close work in panels or boxes. Slip-joint, or gas, pliers are used to tighten locknuts or small nuts (see Fig. 1-1). Wire cutters are used to cut wire to size.

Figure 1-1   Pliers.

Figure 1-2   A fuse puller.

Fuse puller

The fuse puller is designed to eliminate the danger of pulling and replacing cartridge fuses by hand (see Fig. 1-2). It is also used for bending fuse clips, adjusting loose cutout clips, and handling live electrical parts. It is made of a phenolic material, which is an insulator. Both ends of the puller are used. Keep in mind that one end, is for large-diameter fuses; the other is for small-diameter fuses.

Screwdrivers

Screwdrivers come in many sizes and tip shapes. Those used by electricians and refrigeration technicians should have insulated handles. One variation of the screwdriver is the screwdriver bit. It is held in a brace and used for heavy-duty work. For safe and efficient use, screwdriver tips should be kept square and sharp. They should be selected to match the screw slot (see Fig. 1-3).

The Phillips-head screwdriver has a tip pointed like a star and is used with a Phillips screw. These screws are commonly found in production equipment. The presence of four slots, rather than two, ensures that the screwdriver will not slip in the head of the screw. There are number of sizes of Phillips-head screwdrivers. They are designated as No. 1, No. 2, and so on. The proper point size must be used to prevent damage to the slot in the head of the screw (see Fig. 1-4).

Wrenches

Three types of wrenches used by the air-conditioning and refrigeration trade are shown in Fig. 1-5. The *adjustable open-end wrenches* are commonly called *crescent wrenches*.

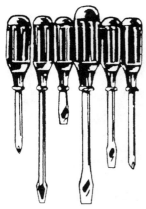

Figure 1-3 Screwdrivers.

Figure 1-4 A Phillips-head screwdriver.

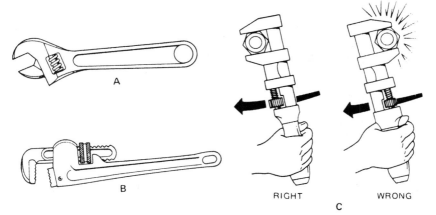

RIGHT   WRONG

C

Figure 1-5 Wrenches. (A) Crescent wrench. (B) Pipe wrench. (C) Using a monkey wrench.

- **Monkey wrenches** are used on hexagonal and square fittings such as machine bolts, hexagonal nuts, or conduit unions.
- **Pipe wrenches** are used for pipe and conduit work. They should not be used where crescent or monkey wrenches can be used. Their construction will not permit the application of heavy pressure on square or hexagonal material. Continued misuse of the tool in this manner will deform the teeth on the jaw face and mar the surfaces of the material being worked.

Soldering equipment

The standard soldering kit used by electricians consists of the same equipment that the refrigeration mechanics use (see Fig. 1-6). It consists of a

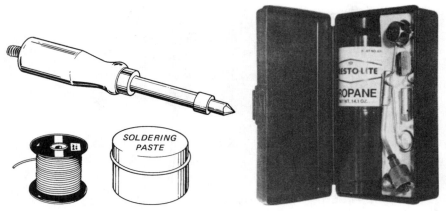

Figure 1-6 Soldering equipment.

nonelectric soldering device—in the form of a torch with propane fuel cylinder or an electric soldering iron, or both. The torch can be used for heating the solid-copper soldering iron or for making solder joints in copper tubing. A spool of solid tin-lead wire solder or flux-core solder is used. Flux-core solder with a rosin core is used for electrical soldering.

Solid-core solder is used for soldering metals. It is strongly recommended that acid-core solder not be used with electrical equipment. Soldering paste is used with the solid-wire solder for soldering joints on copper pipe or solid material. It is usually applied with a small stiff-haired brush.

Drilling equipment

Drilling equipment consists of a brace, a joint-drilling fixture, an extension bit to allow for drilling into and through thick material, an adjustable bit, and a standard wood bit. These are required in electrical work to drill holes in building structures for the passage of conduit or wire in new or modified construction. Similar equipment is required for drilling holes in sheet-metal cabinets and boxes. In this case, high-speed or carbide-tipped drills should be used in place of the carbon-steel drills that are used in wood drilling. Electric power drills are also used (see Fig. 1-7).

Woodworking tools.   Crosscut saws, keyhole saws, and wood chisels are used by electricians and refrigeration and air-conditioning technicians (see Fig. 1-8). They are used to remove wooden structural members obstructing a wire or conduit run and to notch studs and joists to take conduit, cable, box-mounting brackets, or tubing.

They are also used in the construction of wood-panel mounting brackets. The keyhole saw will again be used when cutting an opening in a wall of existing buildings where boxes are to be added or tubing is to be inserted for a refrigeration unit.

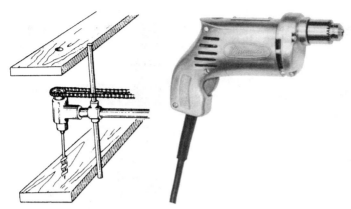

Figure 1-7 Drilling equipment.

Figure 1-8 Woodworking tools.

Metalworking tools.   The cold chisel and center punch are used when working on steel panels (see Fig. 1-9). The knockout punch is used either in making or in enlarging a hole in a steel cabinet or outlet box.

The hacksaw is usually used when cutting conduit, cable, or wire that is too large for wire cutters. It is also a handy device for cutting copper tubing or pipe. The mill file is used to file the sharp ends of such cutoffs. This is a precaution against short circuits or poor connections in tubing.

Masonry working tools.   The air-conditioning technician should have several sizes of masonry drills in the tool kit. These drills normally are carbide-tipped. They are used to drill holes in brick or concrete walls. These holes are used for anchoring apparatus with expansion screws or to allow the

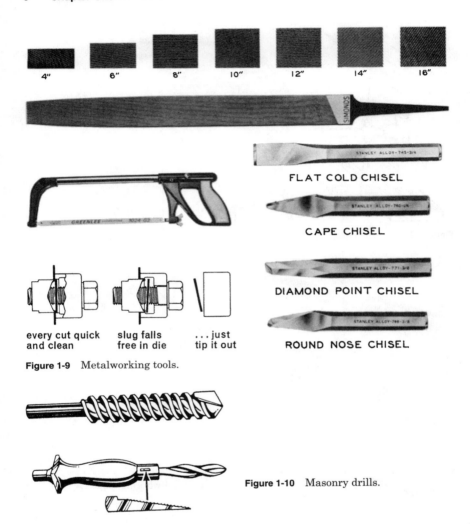

4"   6"   8"   10"   12"   14"   16"

FLAT COLD CHISEL

CAPE CHISEL

DIAMOND POINT CHISEL

every cut quick   slug falls   ...just
and clean   free in die   tip it out

ROUND NOSE CHISEL

Figure 1-9 Metalworking tools.

Figure 1-10 Masonry drills.

passage of conduit, cable, or tubing. Figure 1-10 shows the carbide-tipped bit used with a power drill and a hand-operated masonry drill.

Knives and other insulation-stripping tools

The stripping or removing of wire and cable insulation is accomplished by the use of tools shown in Fig. 1-11. The knives and patented wire strippers are used to bare the wire of insulation before making connections. The scissors shown are used to cut insulation and tape.

The armored cable cutter may be used instead of a hacksaw to remove the armor from the electrical conductors at box entry or when cutting the cable to length.

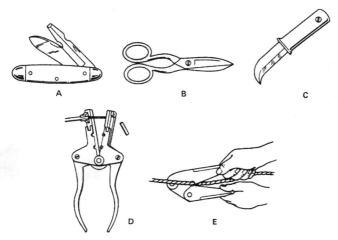

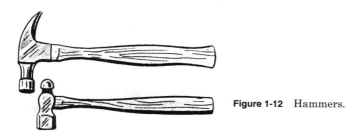

Figure 1-11 Tools for cutting and stripping. (A) Electrician's knife. (B) Electrician's scissors. (C) Skinning knife. (D) Stripper. (E) Cable cutter.

Figure 1-12 Hammers.

Hammers. Hammers are used either in combination with other tools such as chisels or in nailing equipment to building supports (see Fig. 1-12). The figure shows a carpenter's claw hammer and a machinist's ball-peen hammer.

Tape. Various tapes are available. They are used for replacing removed insulation and wire coverings. Friction tape is a cotton tape impregnated with an insulating adhesive compound. It provides weather resistance and limited mechanical protection to a splice already insulated.

Rubber tape or varnished cambric tape may be used as an insulator when replacing wire covering.

Plastic electrical tape is made of a plastic material with an adhesive on one side of the tape. It has replaced friction and rubber tape in the field for 120- and 208-V circuits. It serves a dual purpose in taping joints. It is preferred over the former tapes.

Ruler and measuring tape. The technician should have a folding rule and a steel tape. Both of these are aids for cutting to exact size.

Figure 1-13   Extension light.

Extension cord and light.   The extension light shown in Fig. 1-13 normally is supplied with a long extension cord. It is used by the technician whenever normal building lighting has not been installed and where the lighting system is not functioning.

Wire code markers.   Tapes with identifying numbers or nomenclature are available for permanently identifying wires and equipment. The wire code markers are particularly valuable for identifying wires in complicated wiring circuits in fuse boxes, circuit breaker panels, and in junction boxes (see Fig. 1-14).

Meters and test prods

An indicating voltmeter or test lamp is used when determining the system voltage. It is also used in locating the ground lead and for testing circuit continuity through the power source. They both have a light that glows in the presence of voltage (see Fig. 1-15).

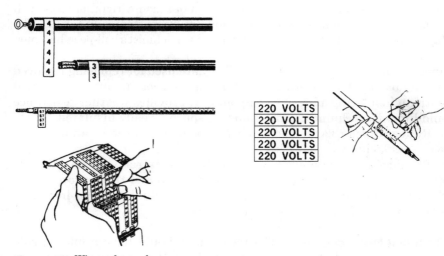

Figure 1-14   Wire code markers.

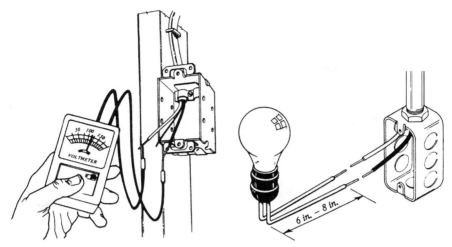

Figure 1-15   Test devices.

A modern method of measuring current flow in a circuit uses the hook-on voltammeter (see Fig. 1-16). This instrument does not have to be hooked into the circuit. It can be operated with comparative ease. Just remember that it measures only one wire. Do not clamp it over a cord running from the consuming device to the power source. In addition, this meter is used only on alternating current (AC) circuits. The AC current will cancel the reading if two wires are covered by the clamping circle. Note how the clamp-on part of the meter is used on the one wire of the motor.

To make a measurement, the hook-on section is opened by hand and the meter is placed against the conductor. A slight push on the handle snaps the section shut. A slight pull on the handle springs open the tool on the C-shaped current transformer and releases a conductor. Applications of this meter are shown in Fig. 1-16. Figure 1-16*b* shows current being measured by using the hook-on section. Figure 1-16*c* shows the voltage being measured using the meter leads. An ohmmeter is included in some of the newer models. However, power in the circuit must be off when the ohmmeter is used. The ohmmeter uses leads to complete the circuit to the device under test.

Use of the voltammeter is a quick way of testing the air-conditioning or refrigeration unit motor that is drawing too much current. A motor that is drawing too much current will overheat and burn out.

Tool kits

Some tool manufacturers make up tool kits for the refrigeration and appliance trade. See Fig. 1-17 for a good example. In the Snap-on® tool kit, the leak detector is part of the kit. The gages are also included. An

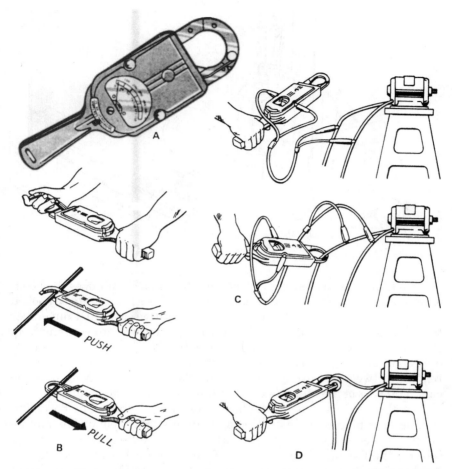

Figure 1-16 Hook-on volt-ammeter. (A) The volt-ammeter. (B) Correct operation. (C) Measuring alternating current and voltage with a single setup. (D) Looping conductor to extend current range of transformer.

adjustable wrench, tubing cutter, hacksaw, flaring tool, and ball-peen hammer can be hung on the wall and replaced when not in use. One of the problems for any repairperson is keeping track of tools. Markings on a board will help locate at a glance when one is missing.

Figure 1-18 shows a portable tool kit. Figure 1-18*j* shows a pulley puller. This tool is used to remove the pulley if necessary to get to the seals. A cart (Fig. 1-18*a*) is included so that the refrigerant and vacuum pump can be easily handled in large quantities. The goggles (Fig. 1-18*q*) protect the eyes from escaping refrigerant.

Figure 1-19 shows a voltmeter probe. It detects the presence of 115 to 750 V. The handheld meter is used to find whether the voltage

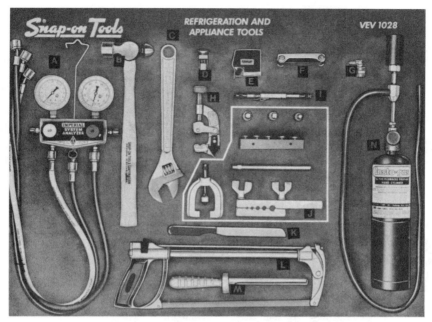

Figure 1-17   Refrigeration and appliance tools. (A) Servicing manifold. (B) Ball peen hammer. (C) Adjustable wrench. (D) Tubing tapper. (E) Tape measure. (F) Allen wrench set. (G) 90° adapter service part. (H) Tubing cutter. (I) Thermometer. (J) Flaring tool kit. (K) Knife. (L) Hacksaw. (M) Jab saw. (N) Halide leak detector. (*Snap-on Tools.*)

is AC or DC and what the potential difference is. It is rugged and easy to handle. This meter is useful when working around unknown power sources in refrigeration units.

Figure 1-20 shows a voltage and current recorder. It can be left hooked to the line for an extended period. Use of this instrument can be used to determine the exact cause of a problem, since voltage and current changes can affect the operation of air-conditioning and refrigeration units.

Gages and Instruments

It is impossible to install or service air-conditioning and refrigeration units and systems without using gages and instrument.

A number of values must be measured accurately if air-conditioning and refrigeration equipment is to be operated properly. Refrigeration and air-conditioning units must be properly serviced and monitored if they are to give the maximum efficiency for the energy expended. Here, the use of gages and instruments becomes important. It is not possible to analyze a system's operation without proper equipments and procedures. In some cases, it takes thousands of dollars worth of

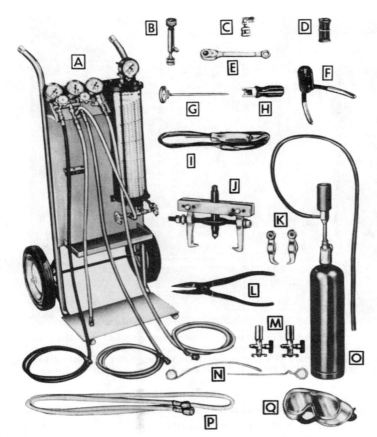

Figure 1-18 Air conditioning and refrigeration portable tool kit. (A) Air conditioning charging station. (B) Excavating/charging valve. (C) 90 adapter service port. (D) O-ring installer. (E) Refrigeration ratchet. (F) Snap-ring pliers. (G) Stem thermometer. (H) Seal remover and installer. (I) Test light. (J) Puller. (K) Puller jaws. (L) Retainer ring pliers. (M) Refrigerant can tapper. (N) Dipsticks for checking oil level. (O) Halide leak detector. (P) Flexible charging hose. (Q) Goggles. (*Snap-on Tools.*)

equipment to troubleshoot or maintain modern refrigeration and air-conditioning system. Instruments are used to measure and record such values as temperature, humidity, pressure, airflow, electrical quantities, and weight. Instruments and monitoring tools can be used to detect incorrectly operating equipment. They can also be used to check efficiency. Instruments can be used on a job, in the shop, or in the laboratory.

If properly cared for and correctly used, modern instruments are highly accurate.

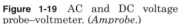

Figure 1-19 AC and DC voltage probe–voltmeter. (*Amprobe.*)

Figure 1-20 Voltage and current recorder. (*Amprobe.*)

Pressure gages

Pressure gages are relatively simple in function (see Fig. 1-21). They read positive pressure or negative pressure, or both (see Fig. 1-22). Gage components are relatively few. However, different combinations of gage components can produce literally millions of design variations (see Fig. 1-23). One gage buyer may use a gage with 0 to 250 psi range, while another person with the same basic measurement requirements will order a gage with a range of 0 to 300 psi. High-pressure gages can be purchased with scales of 0 to 1000, 2000, 3000, 4000, or 5000 psi.

There are, of course, many applications that will continue to require custom instruments, specially designed and manufactured. Most gage manufacturers have both stock items and specially manufactured gages.

Figure 1-21   Pressure gage. (*Weksler.*)

Figure 1-22   This gage measures up to 150 psi pressure and also reads from 0 to 30 for vacuum. The temperature scaled runs from −40° to 115°F (−40° to 46.1°C).

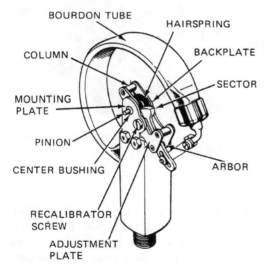

Figure 1-23   Bourdon tube arrangement and parts of a gage. (*Marsh.*)

Gage selection

Since 1939, gages used for pressure measurements have been standardized by the American National Standards Institute. Most gage manufacturers are consistent in face patterns, scale ranges, and grades of accuracy. Industry specifications are revised and updated periodically.

Gage accuracy is stated as the limit that error must not exceed when the gage is used within any combination of rated operating conditions. It is expressed as a percentage of the total pressure (dial) span.

Classification of gages by ANSI standards has significant bearing on other phases of gage design and specification. As an example, a test gage with ±0.25 percent accuracy would not be offered in a 2-in. dial size. Readability of smaller dials is not sufficient to permit the precision indication necessary for this degree of accuracy. Most gages with accuracy of ±0.5 percent and better have dials that are at least 4.5 in. Readability can be improved still further by increasing the dial size.

Accuracy.   How much accuracy is enough? That is a question only the application engineer can answer. However, from the gage manufacturer's point of view, increased accuracy represents a proportionate increase in the cost of building a gage. Tolerances of every component must be more exacting as gage accuracy increases.

Time is needed for technicians to calibrate the gage correctly. A broad selection of precision instruments is available and grades A (±1 percent), 2A (±0.5 percent), and 3A (±0.25 percent) are examples of tolerances available.

Medium.   In every gage selection, the medium to be measured must be evaluated for potential corrosiveness to the Bourdon tube of the gage.

There is no ideal material for Bourdon tubes. No one material adapts to all applications. Bourdon tube materials are chosen for their elasticity, repeatability, and ability to resist "set" and corrosion resistance to the fluid mediums.

Ammonia refrigerants are commonly used in refrigeration. All-steel internal construction is required. Ammonia gages have corresponding temperature scales. A restriction screw protects the gage against sudden impact, shock, or pulsating pressure. A heavy-duty movement of stainless steel and Monel steel prevents corrosion and gives extra-long life. The inner arc on the dial shows pressure. The other arc shows the corresponding temperature (see Fig. 1-24).

Line pressure

The important consideration regarding line pressures is to determine whether the pressure reading will be constant or whether it will fluctuate. The maximum pressure at which a gage is continuously operated should not exceed 75 percent of the full-scale range. For the best performance, gages should be graduated to twice the normal system-operating pressure.

This extra margin provides a safety factor in preventing overpressure damage. It also helps avoid a permanent set of the Bourdon tube. For

Figure 1-24 Ammonia gage. (*Marsh.*)

applications with substantial pressure fluctuations, this extra margin is especially important. In general, the lower the Bourdon tube pressure, the greater the overpressure percentage it will absorb without damage. The higher the Bourdon tube pressure, the less overpressure it will safely absorb.

Pulsation causes pointer flutter, which makes gage reading difficult. Pulsation also can drastically shorten gage life by causing excessive wear of the movement gear teeth. A pulsating pressure is defined as a pressure variation of more than 0.1 percent full scale per second. Following are conditions often encountered and suggested means of handling them.

The restrictor is a low-cost means of combating pulsation problems. This device reduces the pressure opening. The reduction of the opening allows less of the pressure change to reach the Bourdon tube in a given time interval. This dampening device protects the Bourdon tube by the retarding overpressure surges. It also improves gage readability by reducing pointer flutter. When specifying gages with restrictors, indicate whether the pressure medium is liquid or gas. The medium determines the size of the orifice. In addition, restrictors are not recommended for dirty line fluids. Dirty materials in the line can easily clog the orifice. For such conditions, diaphragm seals should be specified.

The needle valve is another means of handling pulsation if used between the line and the gage (see Fig. 1-25). The valve is throttled down to a point where pulsation ceases to register on the gage.

In addition, to the advantage of precise throttling, needle valves also offer complete shutoff, an important safety factor in many applications. Use of a needle valve can greatly extend the life of the gage by allowing it to be used only when a reading is needed.

SOFT SEAT NEEDLE VALVES

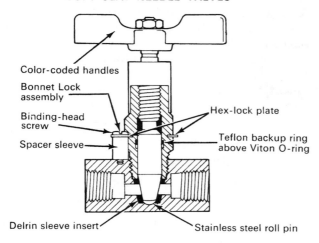

Color-coded handles

Bonnet Lock assembly

Binding-head screw

Spacer sleeve

Hex-lock plate

Teflon backup ring above Viton O-ring

Delrin sleeve insert

Stainless steel roll pin

STAINLESS STEEL NEEDLE VALVES

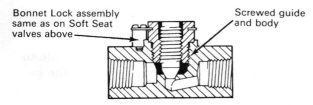

Bonnet Lock assembly same as on Soft Seat valves above

Screwed guide and body

ALLOY STEEL NEEDLE VALVES

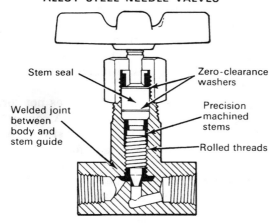

Stem seal

Welded joint between body and stem guide

Zero-clearance washers

Precision machined stems

Rolled threads

Figure 1-25   Different types of needle valves. (*Marsh.*)

Liquid-filled gages are another very effective way to handle line pulsation problems. Because the movement is constantly submerged in lubricating fluid, reaction to pulsating pressure is dampened and the pointer flutter is practically eliminated.

Silicone-oil-treated movements dampen oscillations caused by line pressure pulsations and/or mechanical oscillation. The silicone oil, applied to the movement, bearings, and gears, acts as a shock absorber. This extends the gage life while helping to maintain accuracy and readability.

Effects of temperature on gage performance

Because of the effects of temperature on the elasticity of the tube material, the accuracy may change. Gages calibrated at 75°F (23.9°C) may change by more than 2 percent at

- Full scale (FS) below −30°F (−34°C)
- Above 150°F (65.6°C)

Care of gages

The pressure gage is one of the service person's most valuable tool. Thus, the quality of the work depends on the accuracy of the gages used. Most are precision-made instruments that will give many years of dependable service if properly treated.

The test gage set should be used primarily to check pressures at the low and high side of the compressor. The ammonia gage should be used with a steel Bourdon tube tip and socket to prevent damage.

Once you become familiar with the construction of your gages, you will be able to handle them more efficiently. The internal mechanism of a typical gage is shown in Fig. 1-23. The internal parts of a vapor tension thermometer are very similar.

Drawn brass is usually used for case material. It does not corrode. However, some gages now use high-impact plastics. A copper alloy Bourdon tube with a brass tip and socket is used for most refrigerants. Stainless steel is used for ammonia. Engineers have found that moving parts involved in rolling contact will last longer if made of unlike metals. That is why many top-grade refrigeration gages have bronze-bushed movements with a stainless steel pinion and arbor.

The socket is the only support for the entire gage. It extends beyond the case. The extension is long enough to provide a wrench flat enough for use in attaching the gage to the pressure source. Never twist the case when threading the gage into the outlet. This could cause misalignment or permanent damage to the mechanism.

NOTE: Keep gages and thermometers separate from other tools in your service kit. They can be knocked out of alignment by a jolt from a heavy tool.

Most pressure gages for refrigeration testing have a small orifice restriction screw. The screw is placed in the pressure inlet hole of the socket. It reduces the effects of pulsations without throwing off pressure readings. If the orifice becomes clogged, the screw can be easily removed for cleaning.

Gage recalibration

Most gages retain a good degree of accuracy in spite of daily usage and constant handling. Since they are precision instruments, however, you should set up a regular program for checking them. If you have a regular program, you can be sure that you are working with accurate instruments.

Gages will develop reading errors if they are dropped or subjected to excessive pulsation, vibration, or a violent surge of overpressure. You can restore a gage to accuracy by adjusting the recalibration screw (see Fig. 1-26). If the gage does not have a recalibration screw, remove the ring and glass. Connect the gage you are testing and a gage of known accuracy to the same pressure source. Compare readings at mid-scale. If the gage under test is not reading the same as the test gage, remove the pointer and reset.

This type of adjustments on the pointer acts merely as a pointer setting device. It does not reestablish the original even increment (linearity) of pointer travel. This becomes more apparent as the correction requirement becomes greater.

If your gage has a recalibrator screw on the face of the dial as in Fig. 1-26, remove the ring and glass. Relieve all pressure to the gage. Turn the recalibration screw until the pointer rests at zero.

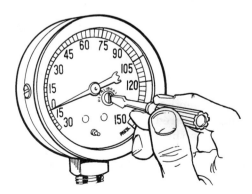

Figure 1-26   Recalibrating a gage. (*Marsh.*)

The gage will be as accurate as when it left the factory if it has a screw recalibration adjustment. Resetting the dial to zero restores accuracy throughout the entire range of dial readings. If you cannot calibrate the gage by either of these methods, take it to a qualified specialist for repair.

Thermometers

Thermometers are used to measure heat. A thermometer should be chosen according to its application. Consider first the kind of installation—direct mounting or remote reading.

If remote readings are necessary, then the vapor tension thermometer is best. It has a closed, filled Bourdon tube. A bulb is at one end for temperature sensing. Changes in the temperature at the bulb result in pressure changes in the fill medium. Remote reading thermometers are equipped with 6 ft of capillary tubing as standard. Other lengths are available on special order.

The location of direct or remote is important when choosing a thermometer. Four common types of thermometers are used to measure temperature:

- Pocket thermometer
- Bimetallic thermometer
- Thermocouple thermometer
- Resistance thermometer

Pocket thermometer

The pocket thermometer depends upon the even expansion of a liquid. The liquid may be mercury or colored alcohol. This type of thermometer is versatile. It can be used to measure temperatures of liquids, air, gas, and solids. It can be strapped to the suction line during a superheat measurement. For practical purposes, it can operate wet or dry. This type of thermometer can withstand extremely corrosive solutions and atmospheres.

When the glass thermometer is read in place, temperatures are accurate if proper contact is made between the stem and the medium being measured. Refrigeration service persons are familiar with the need to attach the thermometer firmly to the suction line when taking superheat readings (see Fig. 1-27a and b). Clamps are available for this purpose. One thing should be kept in mind. That is, the depth at which the thermometer is to be immersed in the medium being measured. Most instruction sheets point out that for liquid measurements the thermometer should be immersed so many inches. When used in a duct, a specified length of stem should be in the airflow. Dipping only the bulb into a glass of water does not give the same readings as immersing to the prescribed length.

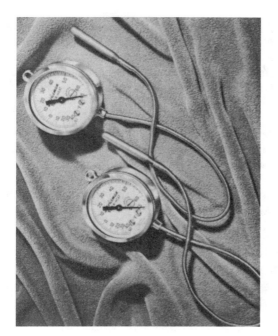

(A)

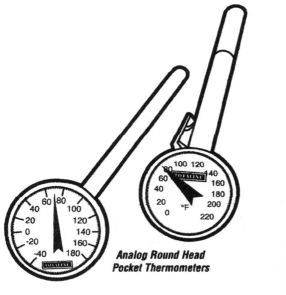

**Analog Round Head
Pocket Thermometers**

**Swivel Head Digital
Pocket Thermometer**

(B)

Figure 1-27   Thermometers used to measure superheat. (*Marsh.*)

Shielding is frequently overlooked in the application of the simple glass thermometer. The instrument should be shielded from radiated heat. Heating repairpersons often measure air temperature in the furnace bonnet. Do not place the thermometer in a position where it receives direct radiation from the heat exchanger surfaces. This causes erroneous readings.

The greatest error in the use of the glass thermometer is that it is often not read in place. It is removed from the outlet grille of a packaged air conditioner. Then it is carried to eye level in the room at ambient temperatures. Here it is read a few seconds to a minute later. It is read in a temperature different from that which it was measuring.

A liquid bath temperature reading is taken with the bulb in the bath. It is then left for a few minutes, immersed, and raised so it can be read.

A simple rule helps eliminate incorrect readings:

- Read glass thermometers while they are actually in contact with the medium being measured.

- If a thermometer must be handled, do so with as little hand contact as possible. Read the thermometer immediately!

A recurring problem with mercury-filled glass thermometers is separation of the mercury column (see Fig. 1-28). This results in what is frequently termed a "split thermometer." The cause of the column's splitting is always rough handling. Such handling cannot be avoided at all times in service work. Splitting does not occur in thermometers that do not have a gas atmosphere over the mercury. Such thermometers allow the mercury to move back and forth by gravity, as well as temperature change. Such thermometers may not be used in other than vertical positions.

A split thermometer can be repaired. Most service thermometers have the mercury reservoir at the bottom of the tube. In this case, cool the thermometer bulb in shaved ice. This draws the mercury to the lower part of the reservoir. Add more ice or salt to lower the temperature, if necessary. With the thermometer in an upright position, tap the bottom of the bulb on a padded piece of paper or cloth. The entrapped gas causing the split column should then rise to the top of the mercury. After the column has been joined, test the service thermometer against a standard thermometer. Do this at several service temperatures.

Bimetallic thermometer

Dial thermometers are actuated by bimetallic coils, by mercury, by vapor pressure, or gas. They are available in varied forms that allow the dial to be used in a number of locations (see Fig. 1-29). The sensing portion of the instrument may be located somewhere else. The dial can be read in a convenient location.

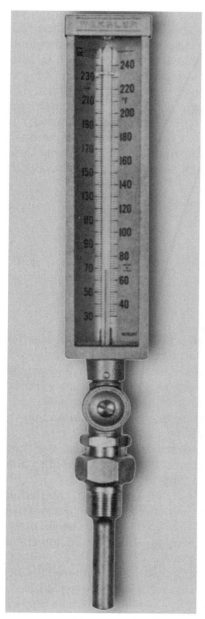

Figure 1-28 Mercury thermometer. (*Weksler.*)

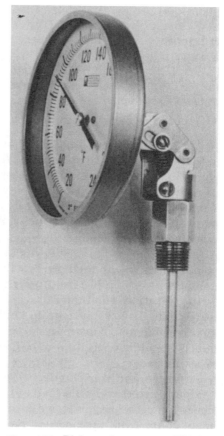

Figure 1-29 Dial-type thermometer. (*Weksler.*)

Bimetallic thermometers have a linear dial face. There are equal increments throughout any given dial ranges. Dial ranges are also available to meet higher temperature measuring needs. Ranges are available to 1000°F (537.8°C). In four selected ranges, dials giving both Celsius and Fahrenheit readings are available. Bimetallic thermometers are economical. There is no need for a machined movement or gearing. The temperature-sensitive bimetallic element is connected directly to the pointer. This type of thermometry is well adapted to measuring the temperature of a surface. Dome-mounted thermal protectors actually react to the surface temperature of the compressor skin. These thermometers are used where direct readings need to be taken, such as on

- Pipelines
- Tanks
- Ovens
- Ducts
- Sterilizers
- Heat exchangers
- Laboratory temperature baths

The simplest type of dial thermometer is a stem. The stem is inserted into the medium to be measured. With the stem immersed 2 in. in liquids and 4 in. in gases, this thermometer gives reasonably accurate readings.

Although dial thermometers have many uses, there are some limitations. They are not as universally applicable as the simple glass thermometer. When ordering a dial thermometer, specify the stem length, scale range, and medium in which it will be used.

One of the advantages of bimetallic thermometry is that the thermometer can be applied directly to surfaces. It can be designed to take temperatures of pipes from 1/2 through 2 in.

In operation, the bimetallic spiral is closely coupled to the heated surface that is to be measured. The thermometer is held fast by two permanent magnets. One manufacturer claims their type of thermometer reaches stability within 3 min. Its accuracy is said to be ±2 percent in working ranges.

A simple and inexpensive type of bimetallic thermometer scribes temperature travel on a load of food in transit. It can be used also to check temperature variations in controlled industrial areas. The replacement chart gives a permanent record of temperature variations during the test period.

Bimetallic drives are also used in control devices. For example, thermal overload sensors for motors and other electrical devices use bimetallic elements. Other examples will be discussed later.

Thermocouple thermometer

Thermocouples are made of two dissimilar metals. Once the metals are heated, they give off an EMF (electro-motive force or voltage). This electrical energy can be measured with a standard type of meter designed to measure small amounts of current. The meter can be calibrated in degrees, instead of amperes, milliamperes, or microamperes.

In use, the thermocouples are placed in the medium that is to be measured. Extension wires run from the thermocouple to the meter. The meter then gives the temperature reading at the remote location.

The extension wires may be run outside closed chests and rooms. There is no difficulty in closing a door, and the wires will not be pinches. In air-conditioning work, one thermocouple may be placed in the supply grille and another in the return grille. Readings can be taken seconds apart without handling a thermometer.

Thermocouples are easily taped onto the surface of pipes to check the inside temperature. It is a good idea to insulate the thermocouple from ambient and radiated heat.

Although this type of thermometer is rugged, it should be handled with care. It should not be handled roughly.

Thermocouples should be protected from corrosive chemicals and fumes. Manufacturer's instructions for protection and use are supplied with the instrument.

Resistance thermometer

One of the newer ways to check temperature is with a thermometer that uses a resistance-sending element. An electrical sensing unit may be made of a thermistor. A thermistor is a piece of material that changes resistance rapidly when subjected to temperature changes. When heated, the thermistor lowers its resistance. This decrease in resistance makes a circuit increase its current. A meter can be inserted in the circuit. The change in current can be calibrated against a standard thermometer. The scale can be marked to read temperature in degrees Celsius or degrees Fahrenheit.

Another type of resistance thermometer indicates the temperature by an indicating light. The resistance-sensing bulb is placed in the medium to be measured. The bridge circuit is adjusted until the light comes on. The knob that adjusts the bridge circuit is calibrated in degrees Celsius or degrees Fahrenheit.

The knob then shows the temperature. The sensing element is just one of the resistors in the bridge circuit. The bridge circuit is described in detail in Chap. 3.

There is the possibility of having practical precision of ±1°F (0.5°C). In this type of measurement, the range covered is −325 to 250°F

(–198 to 121°C). A unit may be used for deep freezer testing, for air-conditioning units, and for other work. Response is rapid. Special bulbs are available for use in rooms, outdoors, immersion, surfaces, and ducts.

Superheat thermometer

The superheat thermometer is used to check for correct temperature differential of the refrigerator gas. The inlet and outlet side of the evaporator coil have to be measured to obtain the two temperatures. The difference is obtained by subtracting.

Test thermometers are available in boxes (see Fig. 1-30). The box protects the thermometer. It is important to keep the thermometer in operating condition. Several guidelines must be followed. Figure 1-31 illustrates how to keep the test thermometer in good working condition. Preventing kinks in the capillary is important. Keep the capillary clean by removing grease and oil. Clean the case and crystal with a mild detergent.

Figure 1-30   Test thermometer (*Marsh.*)

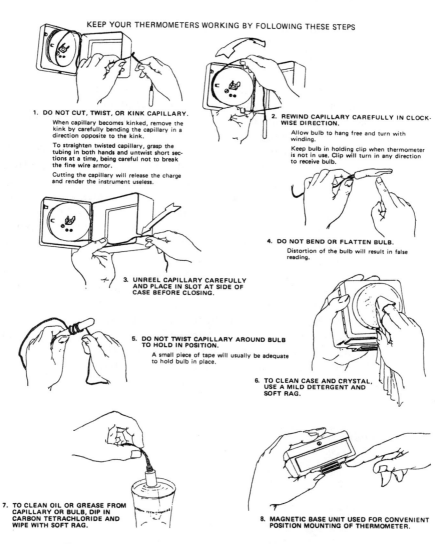

KEEP YOUR THERMOMETERS WORKING BY FOLLOWING THESE STEPS

1. **DO NOT CUT, TWIST, OR KINK CAPILLARY.**

   When capillary becomes kinked, remove the kink by carefully bending the capillary in a direction opposite to the kink.

   To straighten twisted capillary, grasp the tubing in both hands and untwist short sections at a time, being careful not to break the fine wire armor.

   Cutting the capillary will release the charge and render the instrument useless.

2. **REWIND CAPILLARY CAREFULLY IN CLOCK-WISE DIRECTION.**

   Allow bulb to hang free and turn with winding.

   Keep bulb in holding clip when thermometer is not in use. Clip will turn in any direction to receive bulb.

3. **UNREEL CAPILLARY CAREFULLY AND PLACE IN SLOT AT SIDE OF CASE BEFORE CLOSING.**

4. **DO NOT BEND OR FLATTEN BULB.**

   Distortion of the bulb will result in false reading.

5. **DO NOT TWIST CAPILLARY AROUND BULB TO HOLD IN POSITION.**

   A small piece of tape will usually be adequate to hold bulb in place.

6. **TO CLEAN CASE AND CRYSTAL, USE A MILD DETERGENT AND SOFT RAG.**

7. **TO CLEAN OIL OR GREASE FROM CAPILLARY OR BULB, DIP IN CARBON TETRACHLORIDE AND WIPE WITH SOFT RAG.**

8. **MAGNETIC BASE UNIT USED FOR CONVENIENT POSITION MOUNTING OF THERMOMETER.**

Figure 1-31   How to take care of the thermometer. (*Marsh.*)

Superheat Measurement Instruments

Superheat plays an important role in refrigeration and air-conditioning service. For example, the thermostatic expansion valve operates on the principle of superheat. In charging capillary tube systems, the superheat measurement must be carefully watched. The suction line superheat is an indication of whether the liquid refrigerant is flooding the compressor from the suction side. A measurement of zero superheat is a definite indicator that liquid is reaching the compressor. A measurement of 6 to 10°F (−14.4 to −12.2°C) for the expansion valve system and 20°F (−6.7°C) for capillary tube system indicates that all refrigerant is vaporized before entering the compressor.

The superheat at any point in a refrigeration system is found by first measuring the actual refrigerant temperature at that point using an electronic thermometer. Then the boiling point temperature of the refrigerant is found by connecting a compound pressure gage to the system and reading the boiling temperature from the center of the pressure gage. The difference between the actual temperature and the boiling point temperature is *superheat*. If the superheat is zero, the refrigerant must be boiling inside. Then, there is a good chance that some of the refrigerant is still liquid. If the superheat is greater than zero, by at least 5°F or better, then the refrigerant is probably past the boiling point stage and is all vapor.

The method of measuring superheat described here has obvious faults. If there is no attachment for a pressure gage at the point in the system where you are measuring superheat, the hypothetical boiling temperature cannot be found. To determine the superheat at such a point, the following method can be used. This method is particularly useful for measuring the refrigerant superheat in the suction line.

Instead of using a pressure gage, the boiling point of the refrigerant in the evaporator can be determined by measuring the temperature in the line just after the expansion valve where the boiling is vigorous. This can be done with any electronic thermometer (see Fig. 1-32). As the

Figure 1-32 Handheld electronic thermometer. (*Amprobe.*)

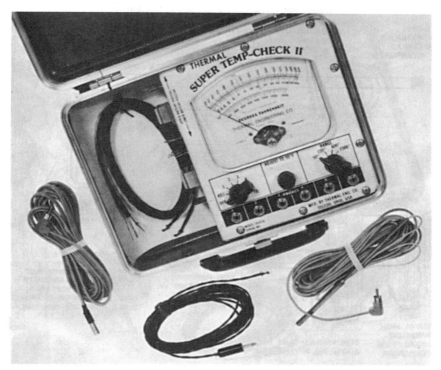

Figure 1-33 Electronic thermometer for measuring superheat. The probes are made of thermo-couple wire. They can be strapped on anywhere with total contact with the surface. This thermometer covers temperatures from −50°F to 1500°F on four scales. The temperature difference between any two points directly means it can read superheat directly. It is battery operated and has a ±2% accuracy on all ranges. Celsius scales are available. (*Thermal Engineering.*)

refrigerant heats up through the evaporator and the suction line, the actual temperature of the refrigerant can be measured at any point along the suction line. Comparison of these two temperatures gives a superheat measurement sufficient for field service unless a distributor-metering device is used or the evaporator is very large with a great amount of pressure drop across the evaporator.

By using the meter shown in Fig. 1-33, it is possible to read superheat directly, using the temperature differential feature. Strap one end of the differential probe to the outlet of the metering device. Strap the other end to the point on the suction line where the superheat measure is to be taken. Turn the meter to temperature differential and the superheat will be directly read on the meter.

Figure 1-34 illustrates the way superheat works. The bulb "opening" force (F-l) is caused by bulb temperature. This force is balanced against

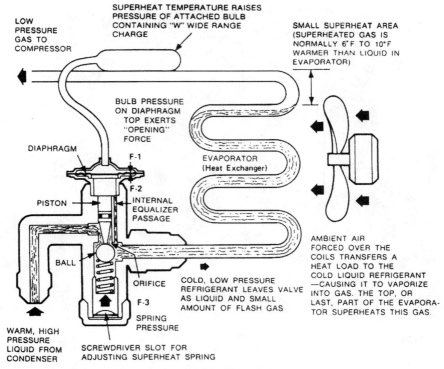

Figure 1-34   How superheat works. (*Parker-Hannefin.*)

the system backpressure (F-2) and the valve spring force (F-3). The force holds the evaporator pressure within a range that will vaporize the entire refrigerant just before it reaches the upper part or end of the evaporator.

The method of checking superheat is shown in Fig. 1-35. The procedure is as follows:

1. Measure the temperature of the suction line at the bulb location. In the example, the temperature is 37°F.

2. Measure the suction line pressure. In the example, the suction line pressure is 27 psi.

3. Convert the suction line pressure to the equivalent saturated (or liquid) evaporator temperature by using a standard temperature-pressure chart (27 psi = 28°F).

4. Subtract the two temperatures. The difference is superheat. In this case, superheat is found by the formula: 37°F − 28°F = 9°F

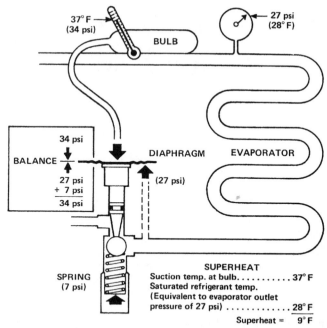

Figure 1-35 Where and how to check superheat. (*Parker-Hannefin.*)

Suction pressure at the bulb may be obtained by either of the following methods:

- If the valve has an external equalizer line, the gage in this line may be read directly.

- If the valve is internally equalized, take a pressure gage reading at the compressor base valve. Add to this the estimated pressure drop between the gage and the bulb location. The sum will approximate the pressure at the bulb.

The system should be operating normally when the superheat is between 6 and 10°F (−14.4 and −12.2°C).

Halide Leak Detectors

Not too long ago leaks were detected by using soap bubbles and water. If possible, the area of the suspected leak was submerged in soap water. Bubbles pinpointed the leak area. If the unit or suspected area was not easily submerged in water then it was coated with soap solution. In addition, where the leak was covered with soap, bubbles would

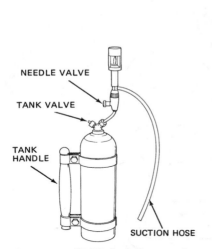

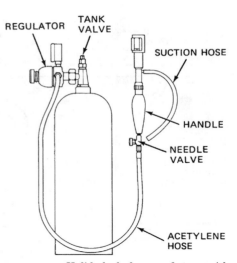

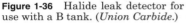

Figure 1-36 Halide leak detector for use with a B tank. (*Union Carbide.*)

Figure 1-37 Halide leak detector for use with an MC tank. (*Union Carbide.*)

be produced. These indicated the location of the leak. These methods are still used today in some cases. However, it is now possible to obtain better indications of leaks with electronic equipment with halide leak detectors.

Halide leak detectors are used in the refrigeration and air-conditioning industry. They are designed for locating leaks and noncombustible halide refrigerant gases (see Figs. 1-36 and 1-37).

The supersensitive detector will detect the presence of as little as 20 parts per million of refrigerant gases (see Fig. 1-38). Another model will detect 100 parts of halide gas per million parts of air.

Setting up

The leak detector is normally used with a standard torch handle. The torch handle has a shut-off valve. Acetylene can be supplied by a B tank (40 ft^3) or MC tank (10 ft^3). In either case, the tank must be equipped with a pressure-reducing regulator; the torch handle is connected to the regulator by a suitable length of fitted acetylene hose (see Fig. 1-36).

An alternate setup uses an adapter to connect the leak detector stem to an MC tank. No regulator is required. The tank must be fitted with a handle (see Fig. 1-37).

In making either setup, be sure all seating sources are clean before assembling. Tighten all connections securely. Use a wrench to tighten hose and regulator connections. If you use the B tank setup, be sure to follow the instructions supplied with the torch handle and regulator.

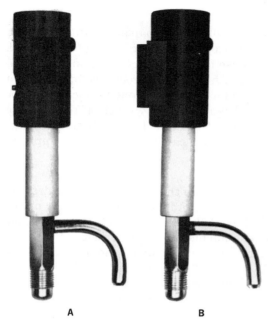

Figure 1-38 Detectors. (A) Supersensitive detector of refrigerant gases. This detects 20 parts per million. (B) Standard model detector torch. This detects 100 parts per million. (*Union Carbide.*)

Lighting

Setup with tank, regulator, and torch handle. Refer to Fig. 1-36.

- Open the tank valve one-quarter turn, using a P-O-L tank key.

- Be sure the shut-off valve on the torch handle is closed. Then, adjust the regulator to deliver 10 psi. Do this by turning in the pressure-adjusting screw until the "C" marking on the flat surfaces of the screw is opposite the face of the front cap. Test for leaks.

- Open the torch handle, shut-off valve, and light the gas above the reaction plate. Use a match or taper.

- Adjust the torch until a steady flame is obtained.

Setup with MC tank and adaptor. Refer to Fig. 1-37.

- With the needle valve on the adaptor closed tightly, just barely open the tank valve, suing a P-O-L tank key. Test for leaks.

- Open the adapter needle valve about one-quarter turn. Light the gas above the reaction plate. Use a match or taper.

Leak testing the setup

Using a small brush, apply a thick solution of soap and water to test for leaks. Check for leaks at the regulator and any connection point. Check the hose to handle connection, hose to regulator connection, and regulator or adaptor connection. If you find a leak, correct it before you light the gas. A leak at the valve stem of a small acetylene tank can often be corrected by tightening the packing nut with a wrench. If this will not stop a leak, remove the tank. Tag it to indicate valve stem leakage. Place it outdoors in a safe spot until you can return it to the supplier.

Adjusting the flame

Place the inlet end of the suction hose so that it is unlikely to draw in air to contaminate the refrigerant vapor. Adjust the needle valve on the adapter or torch handle until the pale blue outer envelope of the flame extends about 1 in. above the reaction plate. The inner cone of the flame, which should also be visible above the reaction plate, should be clear and sharply defined.

If the outer envelope of the flame, when of proper length, is yellow, not pale blue, the hose is picking up refrigerant vapors. There may also be some obstruction in the suction hose. Make sure the suction tube is not clogged or bent sharply. If the suction tube is clear, shut off the flame. Close the tank valve. Disconnect the leak detector from the handle or adaptor. Check for dirt in the filter screw or mixer disk (see Fig. 1-39). Use a 1/8-in. socket key (Allen wrench) to remove or replace the filter screw. This screw retains the mixer disk.

Detecting leaks

To explore the leaks, move the end of the suction hose around all points where there might be leaks. Be careful not to kink the suction hose.

Watch for color changes in the flame as you move the end of the suction hose:

- With the model that has a large opening in the flame shield (wings on each side), a small leak will change the color of the outer flame to a yellow or an orange-yellow hue. As the concentration of halide gas increases, the yellow will disappear. The lower part of the flame will become a bright, light blue. The top of the flame will become a vivid purplish blue.

- With the model that has no wings alongside the flame shield opening, small concentrations of halide gas will change the color. A bright blue-green outer flame indicates a leak. As the concentration of the halide gas increases, the lower part of the flame will lose its greenish tinge. The upper portion will become a vivid purplish blue.

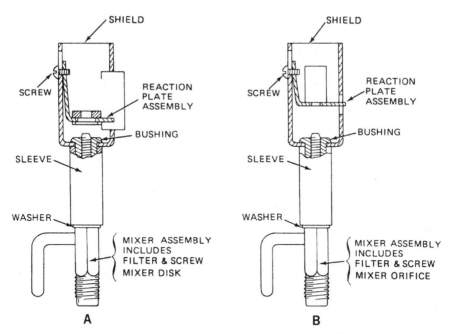

Figure 1-39   Position of filter screw and mixer disc on Prest-O-Lite® halide leak detector. (A) Standard model; (B) supersensitive model.

■ Watch for color intensity changes. The location of small color leaks can be pinpointed rapidly. Color in the flame will disappear almost instantly after the intake end of the hose has passed the point of leakage. With larger leaks, you will have to judge the point of leakage. Note the color change from yellow to purple-blue or blue-green to blue-purple, depending upon the model used.

Maintenance

With intensive usage, an oxide scale may form on the surface of the reaction plate. Thus, sensitivity is reduced. Usually this scale can be easily broken away from the late surface. If you suspect a loss in sensitivity, remove the reaction plate. Scrape its surface with a knife or screwdriver blade, or install a new plate.

Electrical Instruments

Several electrical instruments are used by the air-conditioning serviceperson to see if the equipment is working properly. Studies show that the most trouble calls on heating and cooling equipment are electrical in nature.

The most frequently measured quantities are volts, amperes, and ohms. In some cases, wattage is measured to check for shorts and other malfunctions. A wattage meter is available. However, it must be used to measure volt-amperes instead of watts. To measure watts, it is necessary to use DC only or convert the volt-amperes (VA) to watts by using the power factor. The power factor times the volt-amperes produces the actual power consumed in watts. Since most cooling equipment uses AC, it is necessary to convert to watts by this method.

A number of factors can be checked with electrical instruments. For example, electrical instruments can be used to check the flow rate from a centrifugal water pump, the condition of a capacitor, or the character of a start or run winding of an electric motor.

Ammeter

The ammeter is used to measure current. It can measure the amount of current flowing in a circuit. It may use one of the number of different basic meter movements to accomplish this. The most frequently used of the basic meter movements is the D'Arsonval type (see Fig. 1-40). It uses a permanent magnet and an electromagnet to determine circuit current. The permanent magnet is used as a standard basic source of magnetism. As the current flows through the coil of wire, it creates a magnetic

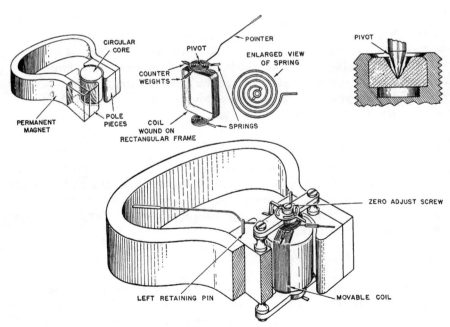

Figure 1-40   Moving coil (D'Arsonval) meter movement.

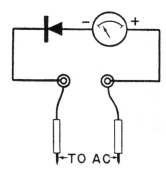

Figure 1-41 Diode inserted in the circuit with a D'Arsonval movement to produce an AC ammeter.

field around it. This magnetic field is strong or weak, depending upon the amount of current flowing through it. The stronger the magnetic field created by the moving coil, the more it is repelled by the permanent magnet. This repelling motion is calibrated to read amperes, milliamperes (0.001 A), or microamperes (0.000001 A).

The D'Arsonval meter movement may also be used on alternating current when a diode is placed in series with the moving coil winding. The diode changes the AC to DC and the meter works as on DC (see Fig. 1-41). The dial or face of the instrument is calibrated to indicate the AC readings.

There are other types of AC ammeters. They are not always as accurate as the D'Arsonval, but they are effective. In some moving magnet meters, the coil is stationary and the magnet moves. Although rugged, this type is not as accurate as the D'Arsonval type meter.

The moving vane meter is useful in measuring current when AC is used (see Fig. 1-42).

The clamp-on ammeter has already been discussed. It has some limitations. However, it does have one advantage in that, it can be used without having to break the line to insert it. Most ammeters must be connected in series with the consuming device. That means one line has to be broken or disconnected to insert the meter into the circuit.

The ampere reading can be used to determine if the unit is drawing too much current or insufficient current. The correct current amount is usually stamped on the nameplate of the motor or the compressor.

Starting and running amperes may be checked to see if the motor is operating with too much load or it is shorted. The flow rate of some pumps can be determined by reading the current the motor pulls. The load on the entire line can be checked by inserting the ammeter in the line. This is done by taking out the fuse and completing the circuit with the meter. Be careful.

If the ammeter has more than one range, it is best to start on the highest range and work down. The reading should be in or near the center of the meter scale for a more accurate reading.

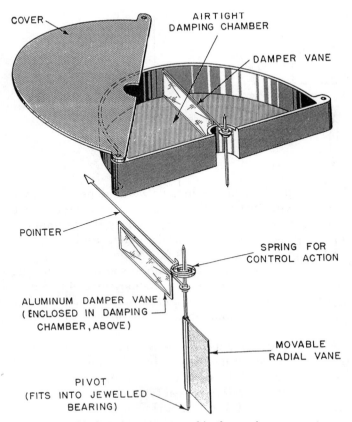

COVER

AIRTIGHT
DAMPING CHAMBER

DAMPER VANE

POINTER

SPRING FOR
CONTROL ACTION

ALUMINUM DAMPER VANE
(ENCLOSED IN DAMPING
CHAMBER, ABOVE)

MOVABLE
RADIAL VANE

PIVOT
(FITS INTO JEWELLED
BEARING)

Figure 1-42   Air damping system used in the moving-vane meter.

Make sure you have some idea what the current in the circuit should be before inserting the meter. Thus, the correct range—or, in some instances, the correct meter—can be selected.

Voltmeter

The voltmeter is used to measure voltage. Voltage is the electrical pressure needed to cause current to flow. The voltmeter is used across the line or across a motor or whatever is being used as a consuming device.

Voltmeters are nothing more than ammeters that are calibrated to read volts. There is, however, an important difference. The voltmeter has a very high internal resistance. That means very small amounts of current flow through its coil (see Fig. 1-43). This high resistance is produced by multipliers. Each range on the voltmeter has a different resistor to increase the resistance so the line current will not be diverted through it (see Fig. 1-44). The voltmeter is placed across the line, whereas the

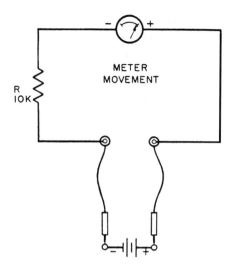

Figure 1-43 A voltmeter circuit with high resistance in series with the meter movement allows it to measure voltage.

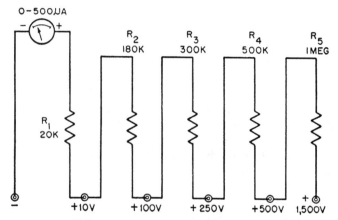

Figure 1-44 Different types of multirange voltmeters. This view shows the interior of the meter box or unit.

ammeter is placed in series. You do not have to break the line to use the voltmeter. The voltmeter has two leads. If you are measuring DC, you have to observe polarity. The red lead is the positive "+" and the black lead is the negative "–". However, when AC is used, it does not matter which lead is placed on which terminal. Using a D'Arsonval meter movement, voltmeters can be made with the proper diode to change AC to DC. Voltmeters can be made with a stationary coil and a moving magnet. Other types of voltmeters are available. They use various means of registering voltage.

If the voltage is not known, use the highest scale on the meter. Turn the range switch to appoint where the reading is in the midrange of the meter movement.

Normal line voltage in most locations is 120 V. When the line voltage is lower than normal, it is possible for the equipment to draw excessive current. This will cause overheating and eventual failure and/or burnout. The correct voltage is needed for the equipment to operate according to its designed specifications. The voltage range is usually stamped on the nameplate of the device. Some will state 208 V. This voltage is obtained from a three-phase connection. Most home or residential voltage is supplied at 120 or 230 V. The range is 220 to 240 V for normal residential service. The size of the wire used to connect the equipment to the line is important. If the wire is too small, it will drop voltage. There will be low voltage at the consuming device. For this reason a certified electrician with knowledge of the National Electrical Code should wire a new installation.

Ohmmeter

The ohmmeter measures resistance. The basic unit of resistance is the ohm "Ω". Every device has resistance. That is why it is necessary to know the proper resistance before trying to troubleshoot a device by using an ohmmeter. The ohmmeter has its own power supply (see Fig. 1-45). Do not use an ohmmeter on a line that is energized or connected to a power source of any voltage.

An ohmmeter can read the resistance of the windings of a motor. If the correct reading has been given by the manufacturer, it is then possible to see if the reading has changed. If the reading is much lower, it may indicate a shorted winding. If the reading is infinite "∞", it may mean there is a loose connection or an open circuit.

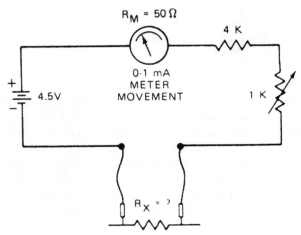

Figure 1-45   Internal circuit of an ohmmeter.

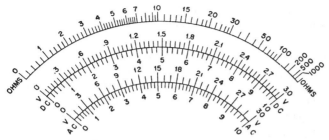

Figure 1-46 A multimeter scale. Note the ohms and volt scales.

Ohmmeters have ranges. Figure 1-46 shows a meter scale. The $R \times 1$ range means the scale is read as is. If the $R \times 10$ range is used, it means that the scale reading must be multiplied by 10. If the $R \times 1000$ range is selected, then the scale reading must be multiplied by 1000. If the meter has a $R \times 1$ megohm range, the scale reading must be multiplied by one million. A megohm is one million ohms.

Multimeter

The multimeter is a combination of meters (see Fig. 1-47). It may have a voltmeter, ammeter, and ohmmeter in the same case. This is the usual

Figure 1-47 Two types of multimeters.

arrangement for fieldwork. This way it is possible to have all three meters in one portable combination. It should be checked for each of the functions.

The snap-around meter uses its scale for a number of applications. It can read current by snapping around the current-carrying wire. If the leads are used, it can be used as a voltmeter or an ohmmeter. Remember that the power must be off to use the ohmmeter. This meter is mounted in its own case. It should be protected from shock and vibration just as any other sensitive instrument.

Wattmeter

The wattmeter is used to measure watts. However, when used on an AC line it measures volt-amperes. If watts are to be measured, the reading must be converted to watts mathematically. Multiply the reading on the wattmeter by the power factor (usually available on the nameplate) to obtain the reading in watts.

Wattmeters use the current and the voltage connections as with individual meters (see Fig. 1-48). One coil is heavy wire and is connected in series. It measures the current. The other connection is made in the

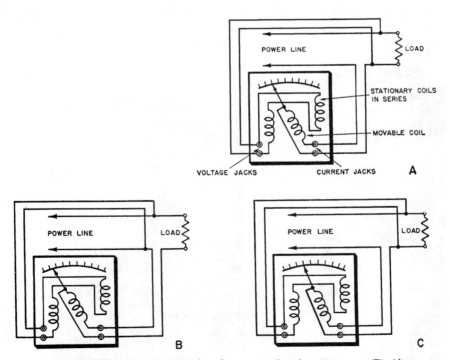

Figure 1-48 (A) Wattmeter connection for measuring input power. (B) Alternate wattmeter connection. (C) With load disconnected, uncompensated wattmeter measures its own power loss.

same way as with the voltmeter and connected across the line. This coil is made of many turns of fine wire. It measures the voltage. By the action of the two magnetic fields, the current is multiplied by the voltage. Wattage is read on the meter scale.

The volt-ampere is the unit used to measure volts time amperes in an AC circuit. If a device has inductance (as in a motor) or capacitance (some motors have run-capacitors), the true wattage is not given on a wattmeter. The reading is in volt-amperes instead of watts. It is converted to watts by multiplying the reading by the power factor. A wattmeter reads watts only when it is connected to a DC circuit or to an AC circuit with resistance only.

The power factor is the ratio of true power to apparent power. Apparent power is what is read on a wattmeter on an AC line. True power is the wattage reading of DC. The two can be used to find the power factor. The power factor is the cosine of the phase angle. The power factor can be found by using a mathematical computation or a very delicate meter designed for the purpose. However, the power factor of equipment using alternating current is usually stamped on the nameplate of the compressor, the motor, or the unit itself.

Wattmeters are also used to test capacitors. Some companies provide charts to convert wattage ratings to microfarad ratings. The wattmeter can test the actual connection of the capacitor. The ohmmeter tells if the capacitor is good or bad. However, it is hard to indicate how a capacitor will function in a circuit with the voltage applied. This is why testing with the wattmeter is preferred.

Other Instruments

Many types of meters and gages are available to test almost any quantity or condition, for example,

- Air-filter efficiency gages
- Air-measurement gages
- Humidity-measuring devices
- Moisture analyzers
- Btu meters

Vibration and sound meters and recorders are also available.

Air-filter efficiency gages

Air measurements are taken in an air-distribution system. They often reveal the existence and location of unintentionally closed or open dampers and obstructions. Leaks in the ductwork and sharp bends are located this way.

Air measurements frequently show the existence of a blocked filter. Dirty and blocked filters can upset the balance of either a heating or cooling system. This is important whether it is in the home or in a large building.

Certain indicators and gages can be mounted in air plenums. They can be used to show the filter has reached the point that it is restricting the airflow. An air plenum is a large space above the furnace heating or cooling unit.

Air-measurement gages

The volume and velocity of air are important measurements in the temperature control industries. Proper amounts of air are indispensable to the best functioning of refrigeration cycles, regardless of the size of the system. Air-conditioning units and systems also rely upon volume and velocity for proper distribution of conditioned air.

Only a small number of contractors are equipped to measure volume and velocity correctly. The companies that are doing the job properly are in great demand. Professional handling of air volume and velocity ensure the efficient use of equipment. Large buildings are very much in need of the skills of air balancing teams.

Some people attempt to obtain proper airflow by measuring air temperature. They adjust damper and blower speeds. However, they usually fail in their attempts to balance the airflow properly.

There are instruments available to measure air velocity and volume. Such instruments can accurately measure the low pressures and differentials involved in air distribution.

Draft gages do measure pressure. However, their specific application to air control makes it more appropriate to discuss them here, rather than under pressure gages. They measure pressure in inches of water. They come in several styles. The most familiar is the slanted type. It may be used either in the field or in the shop.

Meter-type draft gages are better for fieldwork. They can be carried easily. They can sample air at various locations, with the meter box in one location.

Besides air pressure, it is frequently necessary to measure air volume. Air volume is measured in cubic feet per minute or cfm. Air velocity is measured in feet per minute or fpm.

The measure of airflow is still somewhat difficult. However, newer instruments are making accurate measurements possible.

Humidity-measurement instruments

Many hygroscopic (moisture absorbing) materials can be used as relative-humidity sensors. Such materials absorb or lose moisture until a balance is reached with the surrounding air. A change in material moisture content causes a dimensional change, and this change can be used as an input signal to a controller. Commonly used materials include

- Human hair
- Wood
- Biwood combinations similar in action to a bimetallic temperature sensor
- Organic films
- Some fabrics, especially certain synthetic fabrics

All these have the drawbacks of slow response and large hysteresis effects. Accuracy also tends to be questionable unless they are frequently calibrated. Field calibration of humidity sensors is difficulty.

Humidity is read in rh (relative humidity). To obtain the rh it is necessary to use two thermometers. One thermometer is a dry bulb, the other is a wet bulb. The device used to measure relative humidity is the sling psychrometer. It has two glass-stem thermometers. The wet bulb thermometer is moistened by a wick attached to the bulb. As the dual thermometers are whirled, air passes over them. The dry and wet bulb temperatures are recorded. Relative humidity is determined by

- Graphs
- Slide rules
- Similar devices

Thin-film sensors are now available which use an absorbent deposited on a silicon substrate such that the resistance or capacitance varies with relative humidity. They are quite accurate in the range of ±3 to 5 percent. They also have low maintenance requirements.

Stationary psychrometers

Stationary psychrometers take the same measurements as sling psychrometers. They do not move, however. They use a blower or fan to move the air over the thermometer bulbs.

For approximate rh readings, there are metered devices. They are used on desks and walls. They are not accurate enough to be used in engineering work.

Humidistats, which are humidity controls, are used to control humidifiers. They operate in the same way as thermometers in closing contacts to complete a circuit. They do not use the same sensing element, however.

Moisture analyzers

It is sometimes necessary to know the percentage of water in a refrigerant. The water vapor or moisture is measured in parts per million. The necessary measuring instrument is still used primarily in the laboratory. Instruments for measuring humidity are not used here.

Btu meters

The British thermal unit (Btu) is used to indicate the amount of heat present. Meters are especially designed to indicate the Btu in a chilled water line, a hot water line, or a natural gas line. Specially designed, they are used by skilled laboratory personnel at present.

Vibration and sound meters

More cities are now prohibiting conditioning units that make too much noise. In most cases, vibration is the main problem. However, it is not an easy task to locate the source of vibration. However, special meters have been designed to aid in the search for vibration noise.

Portable noise meters are available. Decibel (dB) is the unit for the measurement of sound. There are a couple of bands on the noise meters. The dB-A scale corresponds roughly to the human hearing range. Other scales are available for special applications.

More emphasis is now being placed on noise levels in factories, offices, and schools. The Occupational Safety and Hazards Act (OSHA) lays down strict guidelines regarding noise levels. There are penalties for noncompliance. Thus, it will be necessary for all new and previously installed units to be checked for noise.

High-velocity air systems—used in large buildings—are engineered to reduce noise to levels set by OSHA. For example, there are chambers to lower the noise in the ducts. Air engineers are constantly working on high-velocity systems to try to solve some of the problems associated with them.

Service Tools

Service personnel use some special devices to help them with repair jobs in the field. One of them is the chaser kit (see Fig. 1-49). It is used for cleaning partially plugged capillary tubes. The unit includes 10 spools of lead alloy wire. These wires can be used as chasers for the 10 most popular sizes of capillary tubes. In addition to the wire, a cap tube gage, a set of sizing tools, and a combination file/reamer are included in the metal case. This kit is used in conjunction with the Cap-Check™. The Cap-Check is a portable, self-contained hydraulic power unit with auxiliary equipment especially adapted to cleansing refrigeration capillary tubes (see Fig. 1-50). A small plug of wire from the chaser kit is inserted into

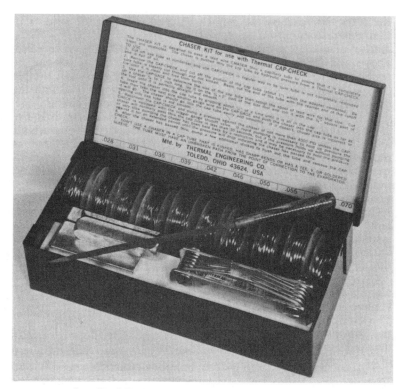

Figure 1-49   Cap-Check™ chaser kit. This is a means to clean partially plugged capillary tubes. It has 10 spools of lead alloy wire. These wires can be used as a chaser for the 10 most popular sizes of cap tubes. A cap tube gage, set of sizing tools, and a combination file/reamer are included in the kit.

Figure 1-50   Cap-Check™ is a portable self-contained hydraulic power unit with auxiliary Equipment that is especially adapted to cleaning refrigeration capillary tubes. It is hand operated.

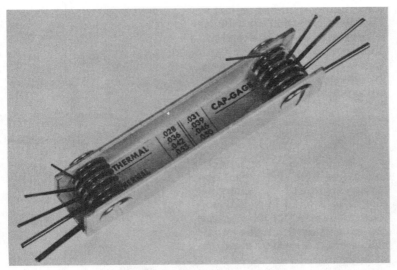

Figure 1-51 The Cap-Gage™ is a pocketknife-type cap tube gage with 10 stainless steel gages to measure the most popular sizes of cap tubes. (*Thermal Engineering.*)

the capillary tube. The wire is a few thousandths of an inch smaller than the internal diameter of the capillary tube. This wire is pushed like a piston through the capillary tube with hydraulic pressure from the Cap-Check. A 0 to 5000-psi gage shows pressure buildup if the capillary tube is restricted. It also shows when the chaser has passed through the tube. A trigger-operated gage shutoff is provided so the gage will not be damaged if pressures greater than 5000 psi are desired. When the piston stops against a partial restriction, high-velocity oil is directed around the piston and against the wall, washing the restriction away and allowing the wire to move through the tube. The lead wire eventually ends up in the bottom of the evaporator, where it remains. The capillary tube is then as clean as when it was originally installed.

A 30-in. high-pressure hydraulic hose with a 1/4-in. SAE male flare outlet connects the cap tube to the Cap-Check for simple handling. An adapter comes with the Cap-Check for simple handling. Another adapter comes with the unit to connect the cap tube directly to the hose outlet without a flared fitting.

The Cap-Gage™ is a capillary tube gage. It has 10 stainless steel gages to measure the most popular sizes of capillary tubes (see Fig. 1-51).

Special Tools

Eventually, almost every refrigerant charging job turns into a vapor-charging job. Unless the compressor is turned on, liquid can be charged into the high side only so long before the system and cylinder pressures become unfavorable. When this happens, all refrigerant must be taken in the low side in the form of vapor.

Vapor charging is much slower than liquid charging. To create a vapor inside the refrigerant cylinder, the liquid refrigerant must be boiling. Boiling refrigerant absorbs heat. This is the principle on which refrigeration operates.

The boiling refrigerant absorbs heat from the refrigerant surrounding it in the cylinder. The net effect is that the cylinder temperature begins to drop soon after you begin charging with vapor. As the temperature drops, the remaining refrigerant will not vaporize as readily. Charging will slow.

To speed charging, service personnel add heat to the cylinder by immersing part of it in hot water. The cylinder temperature rises. The boiling refrigerant becomes vigorous and charging returns to a rapid rate. It is not long, though before all the heat has been taken from the water and more hot water must be added.

The Vizi-Vapr™ is an example of how a device can remove liquid from a cylinder and apply it to the system in the form of a vapor (see Fig. 1-52). No heat is required. This eliminates the hazards of using a torch and hot water. The change from a liquid to a gas or vapor

Figure 1-52   The Vizi-Vapr™ is a device that allows rapid charging of a compressor without heating the cylinder of refrigerant. (*Thermal Engineering.*)

takes place in the Vizi-Vapr. It restricts the charging line between the cylinder and compressor. This restriction is much like an expansion valve in that it maintains high cylinder pressure behind it to hold the refrigerant as a liquid.

However, it has a large pressure drop across it to start evaporation. Heat required to vaporize refrigerate is taken from the air surrounding the unit, not from the remaining refrigerant. This produces a dense, saturated vapor.

The amount of restriction in the unit is very critical. Too much restriction will slow charging considerably. It also will allow liquid to go through and cause liquid slugging in the compressor. The restriction setting is different for each size system, for different types of refrigerants and even for different ambient temperatures.

Vacuum Pumps

Use of the vacuum pump may be the single most important development in refrigeration and air-conditioning servicing. The purpose of a vacuum pump is to remove the undesirable materials that create pressure in a refrigeration system. These include

- Moisture
- Air (oxygen)
- Hydrochloric acid

In addition, other materials that will vaporize at low micrometer range. These, along with a wide variety of solid materials, are pulled into the vacuum pump in the same way a vacuum cleaner sucks up dirt. Evacuation is being routinely performed on almost every service call on which recharging is required.

NOTE: It is no longer permitted to simply add refrigerant to the system with one end open for evacuation into the atmosphere. This shortcut was a favorite of many service technicians over the year since it was quick and the refrigerant was inexpensive.

Vacuum levels formerly unheard of for field evacuations are being accomplished daily by service persons who are knowledgeable regarding vacuum equipment. These service persons have found through experience that the two-stage pump is much better than the single-stage pump for deep evacuations (see Figs. 1-53 and 1-54). It was devised as a laboratory instrument and with minor alterations; it has been adapted to the refrigeration field. It is the proper tool for vacuum evacuations in the field.

Figure 1-53   Single-stage portable vacuum pump. (*Thermal Engineering.*)

Figure 1-54   Dual or two-stage portable vacuum pump. (*Thermal Engineering.*)

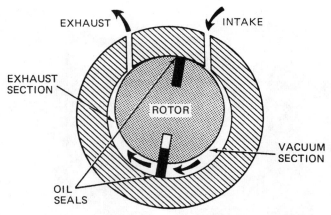

Figure 1-55 Two-stage vacuum pump showing seals and intake, exhaust, and vacuum section. (*Thermal Engineering.*)

To understand the advantages of a two-stage pump over a single-stage pump, refer to Fig. 1-55. This shows the interior of a two-stage vacuum pump. This is a simplified version of a vacuum stage. It is built on the principle of a Wankel engine. There is a stationary chamber with an eccentric rotor revolving inside. The sliding vanes pull gases through the intake. They compress them and force them into the atmosphere through the exhaust. The vanes create a vacuum section and a pressure section inside the pump. The seal between the vacuum and the pressure sections is made by the vacuum pump oil. These seals are the critical factor in the depth that a vacuum pump can pull. If the seals leak, the pump will not be able to draw a deep vacuum. Consequently, less gas can be processed. A pump with high leakage across the seal will be able to pull a deep vacuum on a small system, but the leakage will decrease the pumping speed (cfm) in the deep vacuum region. Long pull-down times will result.

There are three oil seals in a single-stage vacuum pump. Each seal must hold against a high pressure on one side and a deep vacuum on the other side. This places a great deal of strain on the oil seal. A two-stage vacuum pump cuts the pressure strain on the oil seal in half. Such a pump uses two chambers instead of one to evacuate a system. The first chamber is called the deep vacuum chamber. It pulls in the vacuum gases from the deep vacuum and exhausts them into the second chamber at a moderate vacuum. The second chamber, or stage, brings in these gases at a moderate vacuum and exhausts them into the atmosphere. By doing this, the work of a single chamber is split between two chambers. This, in turn, cuts in half the strain on each oil seal, which reduces the leakage up to 90 percent.

A two-stage vacuum pump is more effective than a single-stage vacuum pump. For example, a single-stage vacuum pump rated for 1.5 cfm capacity will take 1 and 1.5 h to evacuate one drop of water. A two-stage vacuum pump with the same rating will evacuate the drop in 12 min.

For evacuation of a 5-ton system saturated with moisture, a minimum of 15 h evacuation time is required in using a single-stage vacuum pump. A two-stage pump with the same cfm rating could do the job in as little as 2 h.

Another advantage of the two-stage pump is reliability. As you can see, if the oil seal is to be effective, the tolerances in these vacuum pumps must be very close between rotor and stator. If the tolerances are not correct, the oil seal will not be effective. Slippage of tolerance due to wear is the major cause of vacuum pump failure. With a single-stage pump, when the tolerance is in the stage slips, the pump loses effectiveness. With a two-stage pump, if one stage loses tolerance, the other one will still pull the vacuum of a single-stage pump.

Larger cfm, two-stage vacuum pumps are preferred to the one-stage vacuum pumps. The cost difference between the two is not great. In addition, the time saved by using the two-stage pump is evident on the first evacuation.

Vacuum pump maintenance

The purpose of vacuum pump oil is to lubricate the pump and act as a seal. To perform this function the oil must have

- A low vapor pressure that does not materially increase up to 125°F (51.7°C)
- A viscosity sufficiently low for use at 60°F (15.6°C) yet constant up to 125°F (51.7°C)

These requirements are easily met by using a low vapor pressure, paraffin-based oil having a viscosity of approximately 300 SSU (shearing stress units) at 100°F (37.8°C) and a viscosity index in the range of 95 to 100. This type of uninhibited oil is readily obtainable. It is the material provided by virtually all sellers of vacuum pump oil to the refrigeration trade.

Vacuum pump oil problems

The oils used in vacuum pumps are designed to lubricate and seal. Many of the oils available for other jobs are not designed to clean as they lubricate. Neither are they designed to keep in suspension the solids freed by the cleaning action of the oil. In addition, the oil is not usually heavily inhibited against the action of oxygen. Therefore, the vacuum pump must be run with flushing oil periodically to clean it. Otherwise, its efficiency will be reduced. The use of flushing oils is recommended by pump manufacturers.

If hydrochloric acid has been pulled into the pump, water, solids, and oil will bond together to form sludge or slime that may be acidic. The oil also may deteriorate due to oxidation (action on the oil by oxygen in air pulled through the pump). This results in a pump that will not pull a proper vacuum, may wear excessively, seriously corrode, or rust internally.

Operating instructions

Use vacuum pump oil in the pump when new. After 5 to 10 h of running time, change the oil. Make sure all of the original oil is removed from the pump. Thereafter, change the oil after every 30 h of operation when the oil becomes dark due to suspended solids drawn into the pump. Such maintenance will ensure peak efficiency in the pump operation.

If the pump has been operated for a considerable time on regular pump oil, drain the oil and replace with dual-purpose vacuum-pump oil. Drain the oil and replace with dual-purpose after 10 h of operation. The oil will probably be quite dark due to sludge removed from the pump. Operate the second charge of oil for 10 h and drain again. The second charge of oil may still be dark. However, it will probably be lighter in color than the oil drained after the first 10 h.

Change the oil at 30-h intervals after that. Change the oil before such intervals if it becomes dark due to suspended solids pulled into the pump. Be sure to change the oil every 30 h thereafter to keep the vacuum pump in peak condition.

Evacuating a system

How long should it take? Some techniques of evacuation will clean refrigeration and air-conditioning system to a degree never before reached. Properly used, a good vacuum pump will eliminate 99.99 percent of the air and virtually all of the moisture in a system. There is no firm answer regarding the time it will take a pump to accomplish this level of cleanliness. The time required for evacuation depends on many things. Some factors that must be considered are as follows:

- The size of the vacuum pump
- The type of vacuum pump—single or two stage
- The size of the hose connections
- The size of the system
- The contamination in the system
- The application for the system

Figure 1-56 This vacuum check gage is designed to be as handy as a charging manifold. (*Thermal Eng.*)

Evacuating down to 29 in. eliminates 97 percent of all air. Moisture removal, however, does not begin until a vacuum below 29 in. is reached. This is the micrometer level of vacuum. It can be measured only with an electronic vacuum gage. Dehydration of system does not certainly begin until the vacuum gage reads below 5000 μm. If the system will not pump down to this level, something is wrong. There may be a leak in the vacuum connections. The vacuum pump oil may be contaminated. There may be a leak in the system. Vacuum gage readings between 5000 and 1000 μm ensure that dehydration is proceeding. When all moisture is removed, the micron gage will pull down below 1000 μm (see Figs. 1-56 and 1-57).

Pulling a system down below 1000 μm is not a perfect test for cleanliness. If the vacuum pump is too large for the system it may pull down this level before all of the moisture is removed. Another test is preferred. Once the system is pulled down below 1000 μm it will not go any further. The system should be valved-off from the vacuum pump and the pump turned off. If the vacuum in the system does not rise over 2000 μm in the next 5 min, evacuation has been completed. If it goes over this level, either the moisture is not completely removed or the system has a slight leak. To find which, reevacuate the system to its lowest level. Valve it off again and shutoff the vacuum pump. If the vacuum leaks back to the same level as before, there is a leak in the system. If, the rise is much slower than before, small amounts of moisture are probably left in the system. Reevacuate until the vacuum will hold.

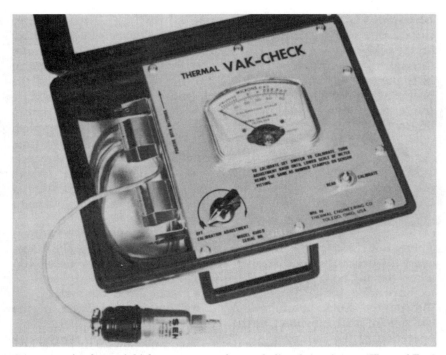

Figure 1-57   An electronic high-vacuum gage that reads directly in microns. (*Thermal Eng.*)

Charging Cylinder

The charging cylinder lets you charge with heat to speed up the charging process. This unit, with its heater assembly, allows up to 50 W of heat to be used in charging. Refrigerant is removed rapidly from the cylinder as liquid, but injected into the system as a gas with the Vizi-Vapr. It requires no heat during the charging process. The Extracta-Charge™ device allows the serviceperson to carry small amounts of refrigerant to the job. The refrigerant can be bought in large drums and stored at the shop. The Extracta-Charge comes in a rugged, steel carrying case to protect it from tough use. It provides a method for draining refrigerant even from capillary tube, sealed systems.

It is now mandatory to capture the escaping refrigerant. The Extracta-Charge is the instrument to use. When systems are overcharged, the excess can be transported back to the drum. The amount removed can be measured also. A leak found after the charging operation usually means the loss of the full charge. Using this device, the serviceperson can extract the charge and save it for use after the leak has been found and repaired.

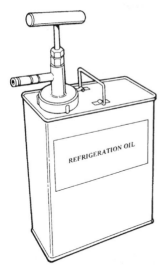

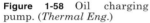

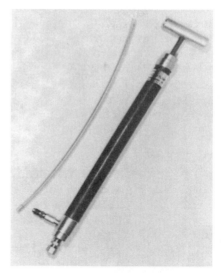

Figure 1-58 Oil charging pump. (*Thermal Eng.*)

Figure 1-59 Oil changing pump. (*Thermal Eng.*)

Charging Oil

In charging a compressor with oil, there is danger of drawing air and moisture into the refrigeration system. Use of the pump shown in Fig. 1-58 eliminates this danger. This pump reduces charging time by over 70 percent without pumping down the compressor. The pump fits the can with a cap seal so the pump need not be removed until the can is empty. It is a piston-type high-pressure pump designed to operate at pressures up to 250 psi. It pumps 1 qt in 20 full strokes of the piston. The pump can be connected to the compressor by a refrigerant charging line or copper tubing from a 1/2-in. male flare fitting.

Changing Oil

Whenever it is impossible to drain oil in the conventional manner, it becomes necessary to hook up a pump. Removing oil from refrigeration compressors before dehydrating with a vacuum is a necessity. The pump shown in Fig. 1-59 has the ability to remove 1 quart of oil with about 10 strokes. It is designed for use in pumping oil from refrigeration compressors, marine engines, and other equipment.

Mobile Charging Stations

Mobile charging stations can be easily loaded into a pickup truck, van, or station wagon. They take little space (see Fig. 1-60). Stations come

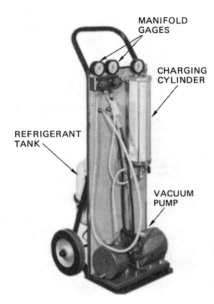

MANIFOLD
GAGES

CHARGING
CYLINDER

REFRIGERANT
TANK

VACUUM
PUMP

Figure 1-60   Mobile charging station. (*Thermal Engineering.*)

complete with manifold gage set, charging cylinder, instrument and tool sack, and vacuum pump. The refrigerant tank can also be mounted on the mobile charging station.

Tubing

Several types of tubing are used in plumbing, refrigeration, and air-conditioning work. Air-conditioning and refrigeration, however, use special tubing types. Copper, aluminum, and stainless steel are used for tubing materials. They ensure that refrigerants do not react with the tubing. Each type of tubing has a special application. Most of the tubing used in refrigeration and air-conditioning is made of copper. This tubing is especially processed to make sure it is clean and dry inside. It is sealed at the ends to make sure the cleanliness is maintained.

- Stainless steel tubing is used with R-717 or ammonia refrigerant.

- Brass or copper tubing should *not* be used in ammonia refrigerant systems.

- Aluminum tubing is used in condensers in air-conditioning systems for the home and automobile.

This calls for a special type of treatment for soldering or welding. Copper tubing is the type most often used in refrigeration systems. There are two types of copper tubing—hard-drawn and soft copper tubing. Each has a particular use in refrigeration.

TABLE 1-1   Inside and Outside Diameter of Small Capillary Tubing*

Inside diameter (ID), in.	Outside diameter (OD), in.
0.026	0.072
0.31	0.083
0.036	0.087
0.044	0.109
0.050	0.114
0.055	0.125
0.064	0.125
0.070	0.125
0.075	0.125
0.080	0.145
0.085	0.145

*Reducing bushing fits in 3/8 in. OD solder fitting and takes 3/8 in. OD tubing.

Soft copper tubing

Some commercial refrigeration systems use soft copper tubing. However, such tubing is most commonly found in domestic systems. Soft copper is annealed. *Annealing* is the process whereby the copper is heated to a blue surface color and allowed to cool gradually to room temperature. If copper is hammered or bent repeatedly, it will become hard. Hard copper tubing is subject to cracks and breaking.

Soft copper comes in rolls and is usually under 1/2 in. in outside diameter (OD). Small-diameter copper tubing is made for capillary use. It is soft drawn and flexible. It comes in random lengths of 90 to 140 ft. Table 1-1 gives the available inside and outside diameters. This type of tubing usually fits in a 1/4-in. OD solder fitting that takes a 1/8-in. OD tubing.

There are three types of copper tubing—types K, L, and M.

- **Type-K** tubing is heavy duty. It is used for refrigeration, general plumbing, and heating. It can also be used for underground applications.

- **Type-L** tubing is used for interior plumbing and heating.

- **Type-M** tubing is used for light-duty waste vents, water, and drainage purposes.

Type-K soft copper tubing that comes in 60-ft rolls is available in outside diameters of 5/8, 3/4, 7/8, and 1-1/8 in. It is used for underground water lines. Wall thickness and weight per foot are the same as for hard copper tubing.

Copper tubing used for air-conditioning and refrigeration purposes is marked *ACR*. It is deoxidized and dehydrated to ensure that there is no

TABLE 1-2   Dehydrated and Sealed Copper Tubing Outside Diameters, Wall Thicknesses, and Weights*

	50-Ft Coils	
Outside diameter (in.)	Wall thickness (in.)	Approximate weight (lb)
$\frac{1}{8}$	0.030	1.74
$\frac{3}{16}$	0.030	2.88
$\frac{1}{4}$	0.030	4.02
$\frac{5}{16}$	0.032	5.45
$\frac{3}{8}$	0.032	6.70
$\frac{1}{2}$	0.032	9.10
$\frac{5}{8}$	0.035	12.55
$\frac{3}{4}$	0.035	15.20
$\frac{7}{8}$	0.045	22.75
$1\frac{1}{8}$	0.050	44.20
$1\frac{3}{8}$	0.055	44.20

*The standard soft dehydrated copper tubing is made in the wall thickness recommended by the Copper and Brass Research Association to the National Bureau of Standards. Each size has ample strength for its capacity.

moisture in it. In most cases, the copper tubing is capped after it is cleaned and filled with nitrogen. Nitrogen keeps it dry and helps prevent oxides from forming inside when it is heated during soldering.

Refrigeration dehydrated and sealed soft copper tubing must meet standard sizes for wall thickness and outside diameter. These sizes are shown in Table 1-2.

Hard and soft copper tubing are available in two wall thicknesses—K and L. The L thickness is used most frequently in air-conditioning and refrigeration systems.

Hard-drawn copper tubing

Hard-drawn copper tubing is most frequently used in refrigeration and air-conditioning systems. Since it is hard and stiff, it does not need the supports required by soft copper tubing. This type of tubing is not easily bent. In fact, it should not be bent for refrigeration work. That is why there are several tubing fittings available for this type of tubing.

Hard-drawn tubing comes in 10 or 20 ft lengths (see Table 1-3). Remember, there is a difference between hard copper sizes and nominal pipe sizes. Table 1-4 shows the differences. Nominal sizes are used in water lines, home plumbing, and drains. They are *never* used in refrigeration systems. Keep in mind that Type K is heavy-wall tubing, Type L is medium-wall tubing, and Type M is thin-wall tubing. The thickness determines the pressure the tubing will safely handle.

TABLE 1-3   Outside Diameter, Wall Thickness, and Weight per Foot of Hard Copper Refrigeration Tubing

Outside diameter (in.)	Wall thickness	Weight per foot
Type-K tubing		
$3/8$	0.035	0.145
$1/2$	0.049	0.269
$5/8$	0.049	0.344
$3/4$	0.049	0.418
$7/8$	0.065	0.641
$1 1/8$	0.065	0.839
$1 3/8$	0.065	1.040
$1 5/8$	0.072	1.360
$2 1/8$	0.083	2.060
$2 5/8$	0.095	2.930
$3 1/8$	0.109	4.000
$4 1/8$	0.134	6.510
Type-L tubing		
$3/8$	0.030	0.126
$1/2$	0.035	0.198
$5/8$	0.040	0.285
$3/4$	0.042	0.362
$7/8$	0.045	0.445
$1 1/8$	0.050	0.655
$1 3/8$	0.055	0.884
$1 5/8$	0.060	1.114
$2 1/8$	0.070	1.750
$2 5/8$	0.080	2.480
Type-M tubing		
$1/2$	0.025	0.145
$5/8$	0.028	0.204
$7/8$	0.032	0.328
$1 1/8$	0.035	0.465
$1 3/8$	0.042	0.682
$1 5/8$	0.049	0.940

TABLE 1-4   Comparison of Outside Diameter and Nominal Pipe Size

Outside diameter (in.)	Nominal pipe size (in.)
$3/8$	$1/4$
$1/2$	$3/8$
$5/8$	$1/2$
$3/4$	—
$7/8$	$3/4$
$1 1/8$	1

Cutting copper tubing

Copper tubing can be cut with a copper tube cutter or a hacksaw. ACR tubing is cleaned, degreased, and dried before the end is sealed at the factory. The sealing plugs are reusable.

To provide further dryness and cleanliness, nitrogen, an inert gas, is used to fill the tube. It materially reduces the oxide formation during brazing. The remaining nitrogen limits excess oxides during succeeding brazing operations. Where tubing will be exposed inside food compartments, tinned copper is recommended.

To uncoil the tube without kinks, hold one free end against the floor or on a bench. Uncoil along the floor or bench to the desired length. The tube may be cut to length with a hacksaw or a tube cutter. In either case, deburr the end before flaring. Bending is accomplished by use of an internal or external bending spring. Lever-type bending tools may also be used. These tools will be shown and explained later.

The hacksaw should have a 32-tooth blade. The blade should have a wave set. No filings or chips can be allowed to enter the tubing. Hold the tubing so that when it is cut the scraps will fall out of the usable end.

Figure 1-61 shows some of the tubing cutters available. The tubing cutter is moved over the spot to be cut. The cutting wheel is adjusted so it touches the copper. A slight pressure is applied to the tightening knob

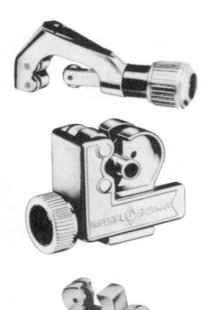

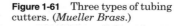

Figure 1-61   Three types of tubing cutters. (*Mueller Brass.*)

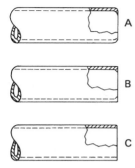

Figure 1-62   The three steps in removing a burr after the tubing has been cut with a tubing cutter. (A) The end of the cut tubing. (B) Squaring with a file produces a flat end. (C) The tube has been filed and reamed. It can now be flared.

on the cutter to penetrate the copper slightly. Then the knob is rotated around the tubing. Once around, it is tightened again to make a deeper cut. Rotate again to make a deeper cut. Do this by degrees so that the tubing is not crushed during the cutting operation.

After the tubing is cut through, it will have a crushed end. The crushed end is prepared for flaring by filing and reaming (see Fig. 1-62). A file and the deburring attachment on the cutting tool can also be used. After the tubing is cut to length, it probably will require flaring or soldering.

Flaring copper tubing

A flaring tool is used to spread the end of the cut copper tubing outward. Two types of tools are designed for this operation (see Fig. 1-63). The flaring process is shown in Fig. 1-64. Note that the flaring is done by holding the end of the tubing rigid at a point slightly below the protruding part of the tube. This protruding part allows for the stretching of the copper.

A flare is important for a strong, solid, leakproof joint. The flares shown in Fig. 1-64 are single flares. These are used in most refrigeration systems. The other type of flare is the double flare.

Here the metal is doubled over to make a stronger joint. They are used in commercial refrigeration and automobile air conditioners. Figure 1-65 shows how the double flare is made. The tool used is called a block-and-punch. Adapters can be used with a single-flare tool to produce a double flare (see Fig. 1-66).

Figure 1-67 shows joints that use the flare. The flared tubing fits over the beveled ends. The flare tee uses the flare connection on all three ends. The half-union elbow uses the flare at one end and a male pipe thread (MPT) on the other end. A female pipe thread is designated by the abbreviation FPT.

Double flaring is recommended for copper tubing 5/16 in. and over. Double flares are not easily formed on smaller sizes of tubing.

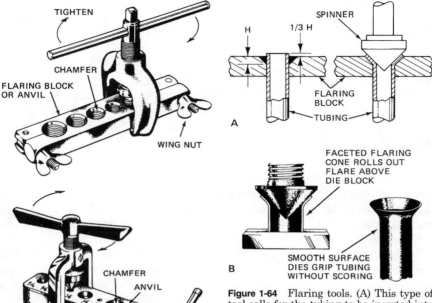

Figure 1-63   Two types of flaring tools for soft copper tubing.

Figure 1-64   Flaring tools. (A) This type of tool calls for the tubing to be inserted into the proper size hole with a small amount of the tube sticking above the flaring block. (B) This type of tool calls for the tubing to stick well above the flaring block. This type is able to maintain the original wall thickness at the base of the flare. The faceted flaring cone smoothes out any surface imperfections.

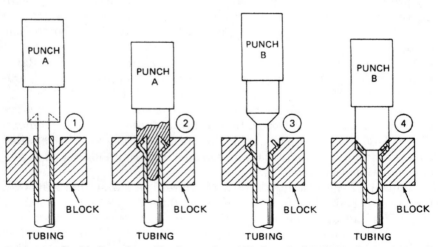

Figure 1-65   Double flares formed by the punch-and-block method. (1) Tubing is clamped into the block opening of the proper size. The female punch, Punch A, is inserted into the tubing. (2) Punch A is tapped to bend the tubing inward. (3) The male punch, Punch B, is tapped to bend the tubing inward. (4) The male punch is tapped to create the final double flare.

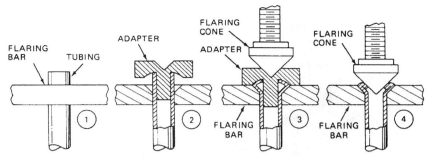

Figure 1-66   Making a double flare with an adapter for the single-flare tool. (1) Insert the tubing into the proper size hole in the flaring bar. (2) Place the adapter over the tubing. (3) Place the adapter inside the tubing. Apply pressure with the flaring cone to push the tubing into a doubled-over configuration. (4) Remove the adapter and use the flaring cone to form a double-thickness flare.

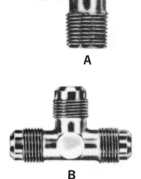

Figure 1-67   A half-union elbow (A) and a flare tee (B). Note the 45-degree angle on the end of the half-union elbow fitted for a flare. Also, note the 45-degree angles on both ends of the flare tee. Note that the flared end does not have threads to the end of the fitting.

Constricting tubing

A tubing cutter adapted with a roller wheel is used to constrict a tubing joint. Two tubes are placed so that one is inserted inside the other. They should be within 0.003 in. when inserted. This space is then constricted by a special wheel on the tube cutter (see Fig. 1-68). The one shown is a combination tube cutter and constrictor. The wheel tightens the outside tube around the inside tube. The space between the two is then filled with solder. Of course, proper cleanliness for the solder joint must be observed before attempting to fill the space with solder.

Both pieces of tubing must be hot enough to melt the solder. Flux must be used to prevent oxidation during the heating cycle. Place flux only on the tube to be inserted. No flux should be allowed to penetrate the inside of the tubing. It can clog filters and restrict refrigerant flow.

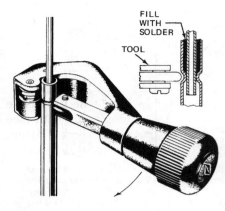

Figure 1-68 Tubing cutter adapter with a roller wheel to work as a tubing constrictor.

Swaging copper tubing

Swaging joins two pieces of copper without a coupling. This makes only one joint, instead of the two that would be formed if a coupling were used. With fewer joints, there are fewer chances of leaks. Punch-type swaging tools and screw-type swaging tools are used in refrigeration work. The screw-type swaging tool works the same as the flaring tool.

Tubing is swaged so that one piece of tubing is enlarged to the outside diameter of the other tube. The two pieces of soft copper are arranged so that the inserted end of the tubing is inside the enlarged end by the same amount as the diameter of the tubing used (see Fig. 1-69). Once the

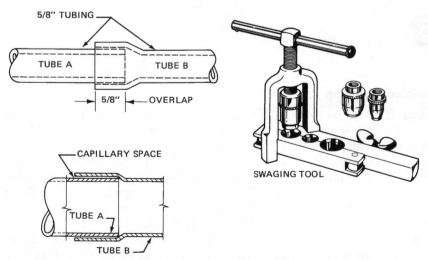

Figure 1-69 Swaging tool and swaging techniques. The swaging punches screw into the yoke and are changed for each size of tubing. Swages are available in $\frac{1}{2}''$, $\frac{5}{8}''$, and $\frac{7}{8}''$ OD, or $\frac{3}{8}''$, $\frac{1}{2}''$, and $\frac{3}{4}''$ nominal copper and aluminum tubing sizes.

areas have been properly prepared for soldering, the connection is soldered. Today, most mechanics use fittings, rather than take the time to prepare the swaged end.

Forming refrigerant tubing

There are two types of bending tools made of springs. One fits inside the tubing. The other fits outside and over the tubing being bent (see Fig. 1-70). Tubing must be bent so that it does not collapse and flatten. To prevent this, it is necessary to place some device over the tubing to make sure that the bending pressure is applied evenly. A tube bending spring may be fitted either inside or outside the copper tube while it is being bent (see Fig. 1-71). Keep in mind that the minimum safe distance for bending small tubing is 5 times its diameter. On larger tubing, the minimum safe distance is 10 times the diameter. This prevents the tubing from flattening or buckling.

Make sure the bending is done slowly and carefully. Make a large radius bend first, then go on to the smaller bends. Do not try to make the whole bend at one time. A number of small bends will equalize the applied pressure and prevent tubing collapse. When using the internal bending spring, make sure part of it is outside the tubing. This gives you a handle on it when it is time to remove it after the bending. You may have to twist the spring to release it after the bend. By bending it so the spring compresses, it will become smaller in diameter and pull out easily. The external spring is usually used in bending tubing along the midpoint. It is best to use the internal spring when a bend comes near the end of the tubing or close to a flared end.

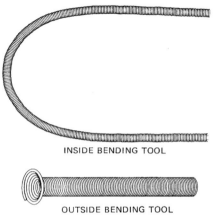

INSIDE BENDING TOOL

OUTSIDE BENDING TOOL

Figure 1-70   Bending tools for soft copper tubing.

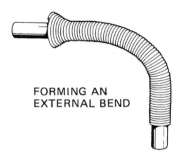

FORMING AN
EXTERNAL BEND

Figure 1-71   Using a spring-type tool to bend tubing.

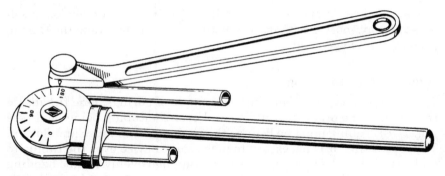

Figure 1-72   A tube bender.

The lever-type tube bender is also used for bending copper tubing (see Fig. 1-72). This one-piece open-side bender makes a neat, accurate bend since it is calibrated in degrees. It can be used to make bends up to 180°. A 180° bend is U shaped. This tool is to be used when working with hard-drawn copper or steel tubing. It can also be used to bend soft copper tubing. The springs are used only for soft copper, since the hard-drawn copper would be difficult to bend by hand. Hard-drawn copper tubing can be bent, if necessary, using tools that electricians use to bend conduit.

Fitting copper tubing by compression

Making leakproof and vibrationproof connections can be difficult. A capillary tube connection can be used (see Fig. 1-73). This compression fitting is used with a capillary tube. The tube extends through the nut and into the connector fitting. The nose section is forced tightly against the connector fitting as the nut is tightened. The tip of the nose is squeezed against the tubing.

If you service this type of fitting, you must cut back the tubing at the end and replace the soft nose nut. If the nut is reused, it will probably cause a leaky connection.

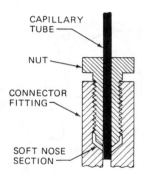

Figure 1-73   A capillary tube connection.

Soldering

Much refrigeration work requires soldering. Brass parts, copper tubing, and fittings are soldered. The cooling unit is also soldered. Thus, the air-conditioning and refrigeration mechanic should be able to solder properly.

Two types of solder are used in refrigeration and air-conditioning work. Soft solder and silver solder are most commonly used for making good joints. Brazing is actually silver soldering.

Brazing requires careful preparation of the products prior to heating for brazing or soldering. This preparation must include steps to prevent contaminants such as dirt, chips, flux residue, and oxides from entering and remaining in an installation. A general-purpose solder for water lines and temperatures below 250°F (121.1°C) is 50-50. It is made of 50 percent tin and 50 percent lead. The 50-50 solder flows at 414°F (212.2°C).

Another low-temperature solder is 95-5. It flows at 465°F (240.5°C). It has a higher resistance to corrosion. It will result in a joint shear strength approximately 2-1/2 times that of a 50-50 joint at 250°F (121.1°C).

A higher temperature solder is No. 122. It is 45 percent silver brazing alloy. This solder flows at 1145°F (618.2°C). It provides a joining material that is suitable for a joint strength greater than the other two solders. It is recommended for use on ACR copper tubing.

Number 50 solder is 50-50 lead and tin. Number 95 solder is 95 percent tin and 5 percent antimony. Silver solder is really a brazing rod, instead of solder. The higher temperature requires a torch to melt it.

Soft soldering

Soldering calls for a very clean surface. Sand-cloth is used to clean the copper surfaces. Flux must be added to prevent oxidation of the copper during the heating process. A no-corrode solder is necessary (see Fig. 1-6). Acid-core solder must not be used. The acid in the solder will corrode the copper and cause leaks.

Soldering is nothing more than applying a molten metal to join two pieces of tubing or a tubing end and a fitting. It is important that both pieces of metal being joined are at the flow point of the solder being used. Never use the torch to melt the solder. The torch is used to heat the tubing or fitting until it is hot enough to melt the solder.

The steps in making a good solder joint are shown in Fig. 1-74. Cleanliness is essential. Flux can damage any system. It is very important to keep flux out of the lines being soldered. The use of excessive amounts of solder paste affects the operation of a refrigeration system. This is especially true of R-22 systems. Solder paste will dissolve in the refrigerant at the high liquid line temperature. It is then carried through a drier or strainer and separated out at the colder expansion valve

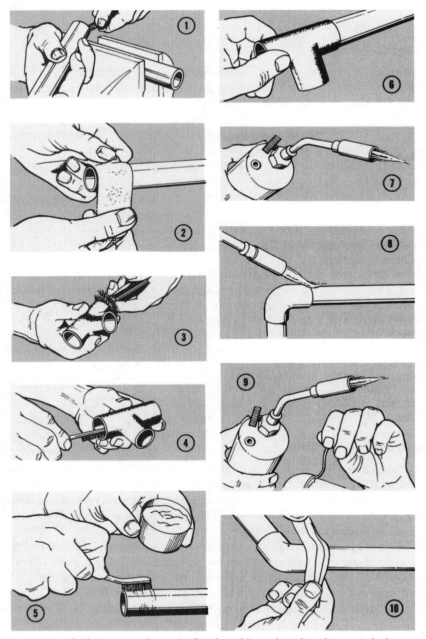

Figure 1-74 Soldering procedures. (1) Cut the tubing to length and remove the burrs. (2) Clean the joint area with sandpaper or sand-cloth. (3) Clean inside the fitting. Use sandpaper, sand-cloth, or wire brush. (4) Apply flux to the inside of the fitting. (5) Apply flux to the outside of the tubing. (6) Assemble the fitting onto the tubing. (7) Obtain proper tip for the torch and light it. Adjust the flame for the soldering being done. (8) Apply heat to the joint. (9) When solder can be melted by the heat of the copper (not the torch), simply apply solder so it flows around the joint. (10) Clean the joint of excess solder and cool it quickly with a damp rag.

temperature. Generally, R-22 systems will be more seriously affected than those carrying R-12. This is because the solid materials separate out at a higher temperature. Sound practice would indicate the use of only enough solder paste to secure a good joint. The paste should be applied according to directions specified by the manufacturer.

Silver soldering or brazing

Silver solder melts at about 1120°F (604.4°C) and flows at 1145°F (618°C). An acetylene torch is needed for the high heat. It is used primarily on hard-drawn copper tubing.

CAUTION: Before using silver solder, make sure it does not contain cadmium. Cadmium fumes are very poisonous. Make sure you work in a very well-ventilated room. The fumes should not contact your skin or eyes. Do not breathe the fumes from the cadmium type of silver solder. Most manufacturers will list the contents on the container.

Silver soldering also calls for a clean joint area. Use the same procedures as shown previously for soldering (see Fig. 1-74). Figure 1-75 shows good and poor design characteristics. No flux should enter the

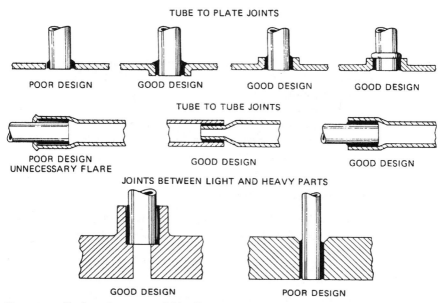

Figure 1-75 Designs that are useful in silver soldering copper tubing. Here, the clearances between the copper tubing are exaggerated for the sake of illustration. They should be much less than shown here. (*Handy and Harmon.*)

system being soldered. Make your plans carefully to prevent any flux entering the tubing being soldered.

Nitrogen or carbon dioxide can be used to fill the refrigeration system during brazing. This will prevent any explosion or the creation of phosgene when the joint has been cleaned with carbon tetrachloride.

In silver soldering, you need a tip that is several sizes larger than the one used for soft soldering. The pieces should be heated sufficiently to have the silver solder adhere to them. Never hold the torch in one place. Keep it moving. Use a slight feather on the inner cone of the flame to make sure you have the proper heat. A large soft flame may be used to make sure the tip does not burn through the fitting or the tubing being soldered.

It is necessary to disassemble sweat-type valves when soldering to the connecting lines. In soldering sweat-type valves where they connect to a line, make sure the torch flame is directed away from the valve. Avoid excessive heat on the valve diaphragm. As an extra precaution, a damp cloth may be wrapped around the diaphragm during the soldering operation. The same is true for soldering thermostatic expansion valves to the distributor.

Either soft or hard solder or silver brazing is acceptable in soldering thermostatic expansion valves. Keep the flame at the fittings and away from the valve body and distributor tube joints. *Do not overheat.* Always solder the outside diameter (OD) of the distributor, never the inside diameter (ID).

Testing for Leaks

Never use oxygen to test a joint for leaks. Any oil in contact with oxygen under pressure will form an explosive mixture.

Do not use emery cloth to clean a copper joint. Emery cloth contains oil. This may hinder the making of a good soldering joint. Emery cloth is made of silicon carbide, which is a very hard substance. Any grains of this abrasive in the refrigeration mechanism or lines can damage a compressor. Use a brush to help clean the area after sanding.

Cleaning and Degreasing Solvents

Solvents, including carbon tetrachloride (CCl_4), are frequently used in the refrigeration industry for cleaning and degreasing equipment. No solvent is absolutely safe. There are several that may be used with relative safety. Carbon tetrachloride is *not* one of them. Use of one of the safer solvents will reduce the likelihood of serious illness developing in the course of daily use. Some of these solvents are stabilized methyl chloroform, methylene chloride, trichloroethylene, and perchloroethylene. Some petroleum solvents are available. These are flammable in varying degrees.

Most solvents may be used safely if certain rules are followed:

- Use no more solvent than the job requires. This helps keep solvent vapor concentrations low in the work area.

- Use the solvent in a well-ventilated area and avoid breathing the vapors as much as possible. If the solvents are used in shop degreasing, it is wise to have a ventilated degreasing unit to keep the level of solvent vapors as low as possible.

- Keep the solvents off the skin as much as possible. All solvents are capable of removing the oils and waxes that keep the skin soft and moist. When these oils and waxes are removed, the skin becomes irritated, dry, and cracked. A skin rash may develop more easily.

A word of caution is in order concerning the commonly used solvent, carbon tetrachloride. While this material has many virtues as a solvent, it has caused much illness among those who use it. Each year several deaths result from its use. Usually, these occur in the small shop or the home. Most large industries have discontinued its use. It is used only with extreme caution. A measure of its harmful nature is indicated by the fact that it bears a *poison* label. It should never be placed in a container that is not labeled "poison." It is for industrial use only.

While occasional deaths result from swallowing carbon tetrachloride, the vast majority of deaths are caused by breathing its vapors. When exposure is very great, the symptoms will be headache, dizziness, nausea, vomiting, and abdominal cramping. The person may lose consciousness. While the person seems to recover from breathing too much of the vapor, a day or two later he or she again becomes ill. Now there is evidence of severe injury to the liver and kidneys. In many cases, this delayed injury may develop after repeated small exposures or after a single exposure not sufficient to cause illness at the time of exposure. The delayed illness is much more common and more severe among those who drink alcoholic beverages. In some episodes where several persons were equally exposed to carbon tetrachloride, the only one who became ill or the one who became most seriously ill was the person who stopped for a drink or two on the way home. When overexposure to carbon tetrachloride results in liver and kidney damage, the patient begins a fight for life without the benefit of an antidote. The only sure protection against such serious illness is not to breathe the vapors or allow contact with the skin.

Human response to carbon tetrachloride is not predictable. A person may occasionally use carbon tetrachloride in the same job in the same way without apparent harm. Then, one day severe illness may result. This unpredictability of response is one factor that makes the use of "carbon tet" so dangerous.

Other solvents will do a good job of cleaning and degreasing. It is much safer to select one of those solvents for regular use rather than to expose yourself to the potential dangers of carbon tetrachloride.

New and Old Tools

Refer to Fig. 1-76 for the following tools and supplies.

The Mastercool Company is indicative of the supply house supplies provided for those working in the refrigeration and air-conditioning field.

Some of the equipment you should be aware of as you continue to work in the field are shown in their catalog. A few of them are shown here as an example of some of the latest devices available to make your workday more efficient.

A convenient way to categorize the tools you work with is shown in the following listing of available tools. This listing may change in time as the requirements for handling new refrigerants are brought about by accrediting agencies and standards' writers.

- *Leak detection* relies on electronic detectors as well as the older types that have been around for years. Ultraviolet rays have now been utilized

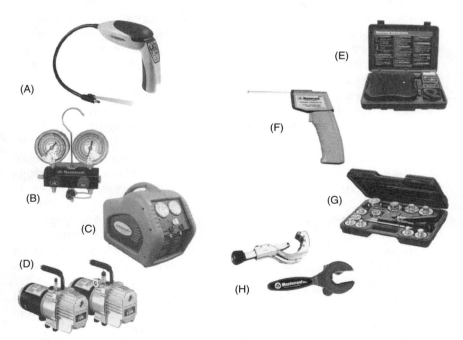

Figure 1-76 (*a*) Electronic leak detector, (*b*) manifold gages, (*c*) recovery unit, (*d*) vacuum pumps, (*e*) refrigerant scales, (*f*) laser thermometer, (*g*) hydraulic tools, (*h*) tube cutters. (*Mastercool; available at www.yellowjacket.com.*)

to more accurately identify and locate leaks. There are various dyes and injectors that need examining for keeping up. The combustible gas leak detector should also be examined as gases other than refrigerants are encountered on the job.

- *Manifold gages and hoses* is another category most often recognized as essential to the technician working in the field and in house. Hoses can stand some examination since they have been constantly improving through the years. And, there is always the chance a hose might be ruptured, leaked, or deteriorated. Newer hoses are usually designed for a longer life than previous ones.

- Another category for classifying devices, tools, and other equipment is the *recovery equipment*, now so necessary to keep within the letter of the law and protect the environment.

- *Vacuum pumps* now have the rotary vane to produce deep vacuums. There are a number of pumps, oils, and accessories that fall into this classification process.

- *Refrigerant scales* have certified scales and programmable scales. The charging program allows the user to program desired quantities, and before the charge is complete, an alarm will sound allowing ample time to turn off the refrigerant supply. There are new features such as pause/charge or empty/full tank which allows the user to know the amount of refrigerant left in the tank at any time. There is a repeat function that allows the user to charge to the previously stored amount. The scales are multilingual and have memory that allows programming for any number of vehicles or refrigerant applications.

- *Specialty hydraulic tools*, such as the tube expanding tool kit and the hydraulic flaring and swaging tool, are also updated. The new features are a handheld hydraulic press that accurately flares and swages copper tubing. Once the die and adapter are secured in the fixture, a few pumps of the handle and you are done. This tool really takes the work out of swaging and flaring, especially on larger tube sizes. The kit includes dies and adapters for flaring and swaging copper tubing sizes from 1/4 to 7/8 in.

- *Tube cutters* have carbide steel cutting wheels for cutting hard and soft copper, aluminum, brass, and thin-walled steel as well as stainless steel.

- *Charging station,* a lightweight durable steel frame cart, contains all the necessary tools to quickly and conveniently charge the AC system. No need for different charging cylinders with units that have a rugged die cast electronic scale. Simply place the refrigerant cylinder on the scale and charge.

- *Electronic tank heater blanket* speeds up recharge time. It also ensures total discharge of refrigerant from 30- and 50-lb tanks of 125°F (52°C) and maximum pressure of 185 psi (R-134a) and 170 psi for R-12. They are available for use with 120 or 240 V.

- *Air content analyzer*—when an AC system leaks, refrigerant is lost and air enters the system. Your refrigerant recycler cannot tell the difference between refrigerant and air—it cycles both from partially filled systems. You end up with an unknown quantity of efficiency robbing air in your supply tank. Excess actual pressure in your supply tank indicates the pressure of air, also called *noncondensible gases* (NCGs). When you release the excess pressure, you are also releasing air. This results in purer refrigerant which will work more efficiently. This one can be left on the supply tank for regular monitoring or it can be removed to check all your tanks.

- *Thermometers, valve core tools, and accessories*—valve core remover/installer controls refrigerant flow 1/4 turn of the valve lever. Lever position also gives visual indication of whether valve is opened or closed. The infrared thermometer with laser has a back-kit LCD display and an expanded temperature range of −20 to 500°C or −4 to 932°F. An alkaline battery furnishes power for up to 15 h.

2

Heat Pumps and Hot-Air Furnaces

Hot-Air Furnaces

Hot-air furnaces are self-contained and self-enclosed units. They are usually centrally located within a building or house. Their purpose is to make sure the temperature of the interior of the structure is maintained at a comfortable level throughout. The design of the furnace is determined by the type of fuel used to fire it. Cool air enters the furnace and is heated as it comes in contact with the hot, metal-heating surfaces. As the air becomes warmer, it also becomes lighter, which causes it to rise. The warmer, lighter air continues to rise until it is either discharged directly into a room, as in the pipeless gravity system, or carried through a duct system to warm-air outlets located at some distance from the furnace.

After the hot air loses its heat, it becomes cooler and heavier. Its increased weight causes it to fall back to the furnace, where it is reheated and repeats the cycle. This is a very simplified description of the operating principles involved in hot-air heating, and it is especially typical of those involved in gravity heating systems. The forced-air system relies on a blower to make sure the air is delivered to its intended location. The blower also causes the return air to move back to the furnace faster than with the gravity system.

With the addition of a blower to the system, there must be some way of turning the blower on when needed to move the air and to turn it off when the room has reached the desired temperature. Thus, electrical controls are needed to control the blower action.

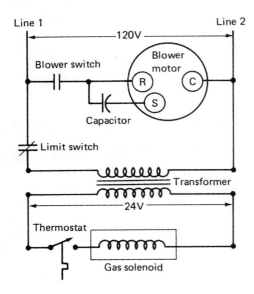

Figure 2-1 Simple one-stage furnace control system.

Basic Gas Furnace Operation

The gas furnace is the simplest to operate and understand. Therefore, we will use it here to look at a typical heating system. This type of natural-gas furnace is used to heat millions of homes in the United States.

Figure 2-1 is a simple circuit needed to control the furnace with a blower. Note the location of the blower switch and the limit switch. The transformer provides low voltage for control of the gas solenoid. If the limit switch opens (it is shown in a closed position), there is no power to the transformer and the gas solenoid cannot energize. This is a safety precaution because the limit switch will open if the furnace gets too hot. When the thermostat closes, it provides 24 V to the gas solenoid, which energizes and turns on the gas. The gas is ignited by the pilot light and provides heat to the plenum of the furnace. When the air in the plenum reaches 120°F (49°C), the fan switch closes and the fan starts. The fan switch provides the necessary 120 V to the fan motor for it to operate.

Once the room has heated up to the desired thermostat setting, the thermostat opens. When it opens, the gas solenoid is deenergized, and the spring action of the solenoid causes it to close off the gas supply, thereby turning off the source of heat. When the plenum on top of the furnace reaches 90°F (32°C), the blower switch opens and turns off the blower. As the room cools down, causing the thermostat to close once again, the cycle starts over again. The gas solenoid opens to let in the gas and the pilot light ignites it. The heat causes the temperature to rise in the plenum above the limit switch's setting and the switch closes to start the blower. Once the thermostat has been satisfied, it opens, and causes the gas solenoid to turn off the gas supply. The blower continues to run until the temperature in the plenum reaches 90°F (32°C) and it turns off the blower by opening. This cycle is repeated over and over again to keep the room or house at a desired temperature.

Basic Electric Heating System

Electric-fired heat is the only heat produced almost as fast as the thermostat calls for it. It is almost instantaneous. There are no heat exchangers to warm up. The heating elements start producing heat the moment the thermostat calls for it. Various types of electric-fired furnaces are available. They can be bought in 5- to 35-kW sizes. The outside looks almost the same as the gas-fired furnace. The heating elements are located where the heat exchangers would normally be located. Since they draw high amperage, they need electrical controls that can take the high currents.

The operating principle is simple. The temperature selector on the thermostat is set for the desired temperature. When the temperature in the room falls below this setting, the thermostat calls for heat and causes the first heating circuit in the furnace to be turned on. There is generally a delay of about 15 s before the furnace blower starts. This prevents the blower from circulating cool air in the winter. After about 30 s, the second heating circuit is turned on. The other circuits are turned on one by one in a timed sequence.

When the temperature reaches the desired level, the thermostat opens. After a short time, the first heating circuit is shut off. The others are shut off one by one in a timed sequence. The blower continues to operate until the air temperature in the furnace drops below a specified temperature.

Basic operation

In Fig. 2-2, the electrical heating system has a few more controls than the basic gas-fired furnace. The low-resistance element used for heating draws a lot of current, so the main contacts have to be of sufficient size to handle the current.

The thermostat closes and completes the circuit to the heating sequencer coil. The sequencer coil heats the bimetal strip that causes the main contacts to close. Once the main contacts are closed, the heating element is in the circuit and across the 240-V line. The auxiliary contacts will also close at the same time as the main contacts. When the auxiliary contacts close, they complete the low-voltage circuit to the fan relay. The furnace fan will be turned on at this time.

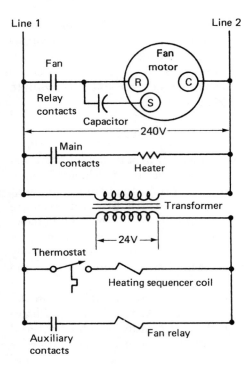

Figure 2-2   Ladder diagram for a hot-air furnace.

Once the thermostat has been satisfied, it opens. This allows the heating sequencer coil to cool down slowly. Thus, the main contacts do not open immediately to remove the heating element from the line. So the furnace continues to produce heat after the thermostat has been satisfied. The bimetal cools down in about 2 min. Once it cools, it opens the main and auxiliary contacts, which removes the heating element from the line and also stops the fan motor. After the room cools down below the thermostat setting, the thermostat closes and starts the sequence all over again.

Ladder Diagrams

Electrical schematics are used to make it simple to trace the circuits of various devices. Some of these can appear complicated, but they are usually very simple when you start at the beginning and wind up at the end. The beginning is one side of the power line and the end is the other side of the line. What happens in between is that a number of switches are used to make sure the device turns on or off when it is supposed to cool, freeze, or heat.

The ladder diagram makes it easier to see how these devices are wired. It consists of two wires drawn parallel and representing the main power source. Along each side you find connections. By simply looking from left to right, you are able to trace the required power for the device. Symbols are used to represent the devices. There is usually a legend on the side of the diagram to tell you, for example, that CC means compressor contactor, EFR means evaporator fan relay, and HR means heating relay (see Fig. 2-3).

Take a look at the thermostat in Fig. 2-3. The location of the switch determines whether the evaporator fan relay coil is energized, the compressor contactor coil is energized, or the heating relay coil is energized. Once the coil of the EFR is energized by having the thermostat turned

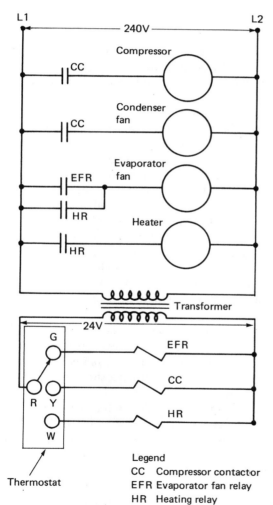

Legend
CC Compressor contactor
EFR Evaporator fan relay
HR Heating relay

Figure 2-3 Ladder diagram for a heat and cool installation.

to make contact with the desired point G, it closes the points in the relay and the evaporator fan motor starts to move. This means that the low voltage (24 V) has energized the relay. The relay energizes and closes the EFR contacts located in the high-voltage (240 V) circuit. If the thermostat is turned to W or the heating position, it will cause the heating relay coil to be energized when the thermostat switch closes and demands heat. The energized heating relay coil causes the HR contacts to close, which in turn places the heating element across the 240-V line and it begins to heat up. Note that the HR contacts are in parallel with the evaporator fan relay contacts. Thus, the evaporator fan will operate when either the heating relay or the evaporator fan relay is energized.

Manufacturer's Diagrams

Figure 2-4 shows how the manufacturer represents the location of the various furnace devices. The solid lines indicate the line voltage to be installed. The dotted lines are the low voltage to be installed when the furnace is put into service.

The motor is four speed. It has different colored leads to represent the speeds. You may have to change the speed of the motor to move the air to a given location. Most motors come from the factory with a medium-high speed selected. The speed is usually easily changed by removing a lead from one point and placing it on another, where the proper color is located. In the schematic of Fig. 2-5, the fan motor has white lead connected to one side of the 120-V line (neutral) and the red and black are switched by the indoor blower relay to black for the cooling speed and red for the heating speed. It takes a faster fan motor to push the cold air than for hot air because cold air is heavier than hot air.

In Fig. 2-5, the contacts on the thermostat are labeled R, W, Y, and G. R and W are used to place the thermostat in the circuit. It can be switched from W to Y manually by moving the heat-cool switch on the thermostat to the cool position.

Notice in Fig. 2-5 that the indoor blower relay coil is in the circuit all the time when the auto-on switch on the thermostat is located at the on position. The schematic also shows the cool position has been selected manually, and the thermostat contacts will complete the circuit when it moves from W1 to Y1.

In Fig. 2-4, note that the low-voltage terminal strip has a T on it. This is the common side of the low voltage from the transformer. In Fig. 2-5, the T is the common side of the low-voltage transformer secondary. In Fig. 2-4, the T terminal is connected to the compressor contactor by a wire run from the terminal to the contactor. Note that the other wire to the contactor runs from Y on the terminal strip. Now go back to Fig. 2-5, where the Y and T terminals are shown as connection points for the

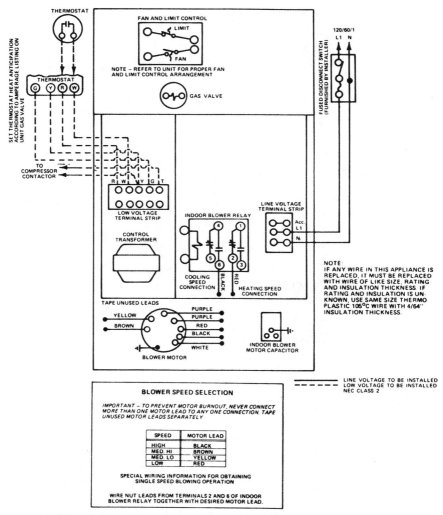

Figure 2-4   Manufacturer's diagram for a hot-air installation.

compressor contactor. Are you able to relate the schematic to the actual device? The gas valve is wired by having wire T of the terminal strip attached to one side of the solenoid and a wire run from the limit switch to the other side of the solenoid.

Figure 2-6 shows how the wiring diagram comes from the factory. It is usually located inside the cover for the cold-air return. In most instances, it is glued to the cover so that it is handy for the person working on the furnace whenever there is a problem after installation.

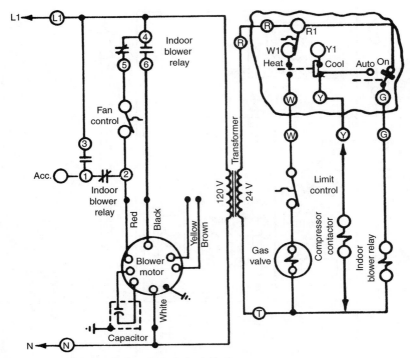

Figure 2-5   Schematic for a hot-air installation.

Field Wiring

The installation of a new furnace requires you to follow a factory diagram furnished in a booklet that accompanies the unit. The wiring to be done in the field is represented by the dotted lines in Fig. 2-7. All electrical connections should be made in accordance with the National Electrical Code and any local codes or ordinances that might apply.

> WARNING: The unit cabinet must have an uninterrupted or unbroken electrical ground to minimize personal injury if an electrical fault should occur. This may consist of electrical wire or approved conduit when installed in accordance with existing electrical codes.

Low-Voltage Wiring

Make the field low-voltage connections at the low-voltage terminal strip shown in Fig. 2-7. Set the thermostat heat anticipator at 0.60 A (or whatever is called for by the manufacturer). If additional controls are

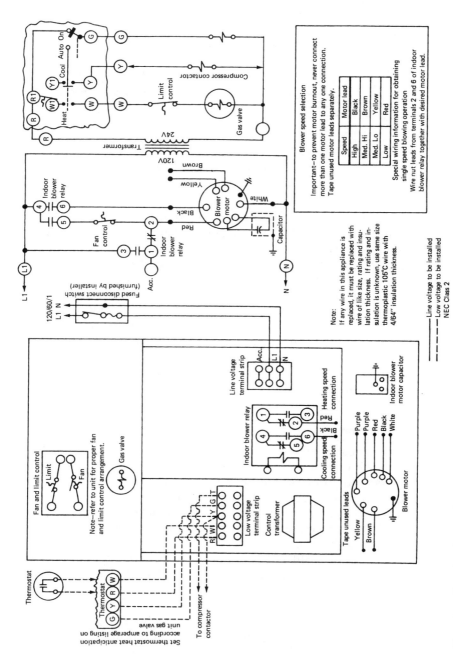

Figure 2-6 Complete instruction page packaged with a hot-air furnace.

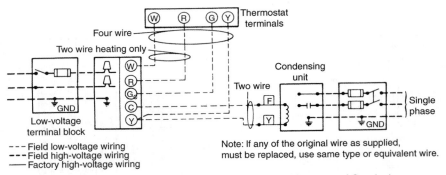

Figure 2-7   Heating and cool application wiring diagram. (*Courtesy of Carrier.*)

connected in the thermostat circuit, their amperage draw must be added to this setting. Failure to make the setting will result in improper operation of the thermostat.

With the addition of an automatic vent damper, the anticipator setting would then be 0.12 A. As you can see from this and the schematic (see Fig. 2-16), the anticipator resistor is in series with whatever is in the circuit and is to be controlled by the thermostat. The more devices controlled by the thermostat, the more current will be drawn from the transformer to energize them. As the current demand increases, the current through the anticipator is also increased. As you remember from previous chapters, a series circuit has the same current through each component in the circuit.

Thermostat location

The room thermostat should be located where it will be in the natural circulating path of room air. Avoid locations where the thermostat is exposed to cold-air infiltration, drafts from windows, doors, or other openings leading to the outside, or air currents from warm- or cold-air registers, or to exposure where the natural circulation of the air is cut off, such as behind doors and above or below mantels or shelves. Also keep the thermostat out of direct sunlight.

The thermostat should not be exposed to heat from nearby fireplaces, radios, televisions, lamps, or rays from the sun. Nor should the thermostat be mounted on a wall containing pipes, warm-air ducts, or a flue or vent that could affect its operation and prevent it from properly controlling the room temperature. Any hole in the plaster or panel through which the wires pass from the thermostat should be adequately sealed with suitable material to prevent drafts from affecting the thermostat.

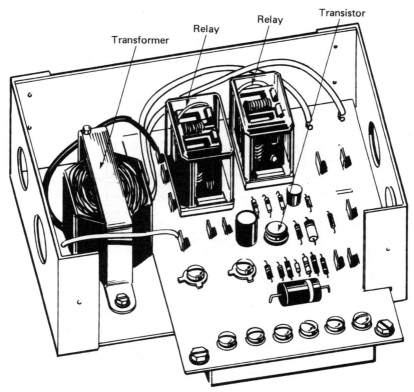

Figure 2-8   Printed circuit control center: heat and cool models. (*Courtesy of Carrier.*)

Printed circuit board control center

Newer hot-air furnaces feature printed circuit control. The board shown in Fig. 2-8 is such that it is easy for the technician installing the furnace to hook it up properly for the first time. The markings are designed for making it easy to connect the furnace for accessories, if needed. Figures 2-9 and 2-10 show the factory-furnished schematic. See if you can trace the schematic and locate the various points on the printed circuit boards.

Heat Pumps

The heat pump is a heat multiplier. It takes warm air and makes it hot air. This is done by compressing the air and increasing its temperature. Heat pumps received more attention during the fuel embargo of 1974. Energy conservation became a more important concern for everyone at that time. If a device can be made to take heat from the air and heat a home or commercial building, it is very useful.

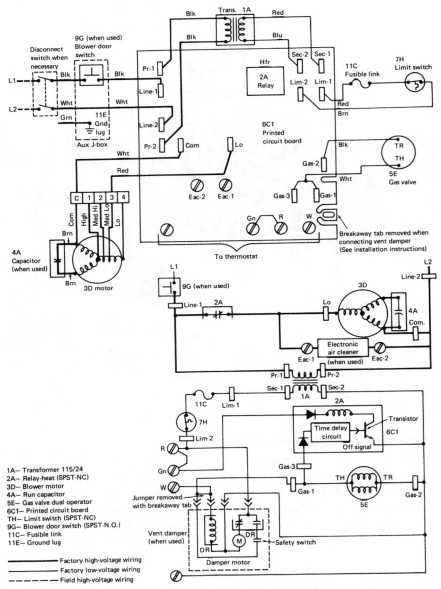

Figure 2-9 Wiring diagram. Heating only. (*Courtesy of Carrier.*)

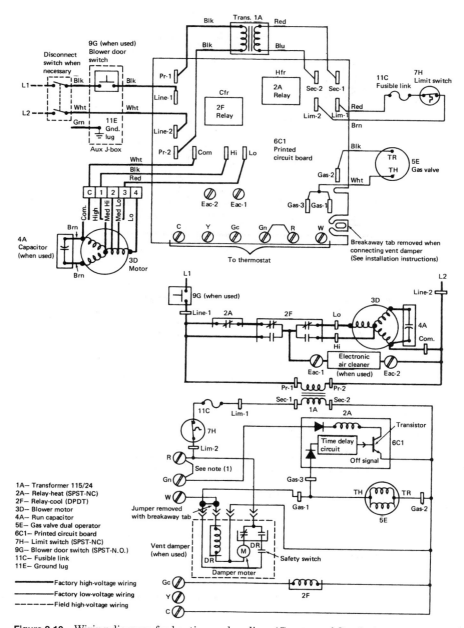

Figure 2-10 Wiring diagram for heating and cooling. (*Courtesy of Carrier.*)

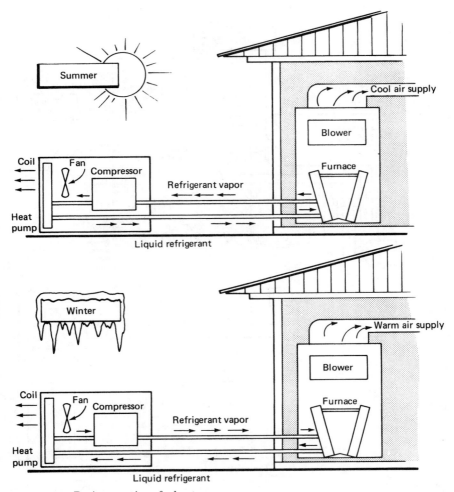

Figure 2-11   Basic operation of a heat pump.

The heat pump can take the heat generated by a refrigeration unit and use it to heat a house or room. Most of them take the heat from outside the home and move it indoors (see Fig. 2-11). This unit can be used to air condition the house in the summer and heat it in the winter by taking the heat from the outside air and moving it inside.

Operation

On mild temperature heating days, the heat pump handles all heating needs. When the outdoor temperature reaches the balance point of the

Figure 2-12   Single package heat pump.

home, that is, when the heat loss is equal to the heat-pump heating capacity, the two-stage indoor thermostat activates the furnace (a secondary heat source, in most cases electric heating elements). As soon as the furnace is turned on, a heat relay deenergizes the heat pump. When the second-stage (furnace) need is satisfied and the plenum temperature has cooled to below 90 to 100°F (32 to 38°C), the heat-pump relay turns the heat pump back on and controls the conditioned space, until the second-stage operation is required again. Figure 2-12 shows the heat-pump unit. The optional electric heat unit shown in Fig. 2-13 is added in geographic locations where needed. This particular unit can provide 23,000 to 56,000 Btu/h and up to 112,700 Btu/h with the addition of electric heat.

Figure 2-13 Optional electric heat for a heat pump.

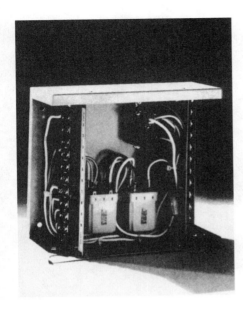

Figure 2-14 Control box for an add-on type heat pump.

If the outdoor temperature drops below the setting of the low-temperature compressor monitor, the control shuts off the heat pump completely and the furnace handles all the heating needs.

During the defrost cycle, the heat pump switches from heating to cooling. To prevent cool air from being circulated in the house when heating is needed, the control automatically turns on the furnace to compensate for the heat-pump defrost cycle (see Fig. 2-14). When supply air temperature climbs above 110 to 120°F (43 to 49°C), the defrost limit control turns off one furnace and keeps indoor air from getting too warm.

If, after a defrost cycle, the air downstream of the coil gets above 115°F (65°C), the closing point of the heat-pump relay, the compressor will stop until the heat exchanger has cooled down to 90 to 100°F (32 to 38°C) as it does during normal cycling operation between furnace and heat pump.

During summer cooling, the heat pump works as a normal split system, using the furnace blower as the primary air mover (see Fig. 2-15).

In a straight heat pump/supplementary electric heater application, at least one outdoor thermostat is required to cycle the heaters as the outdoor temperature drops. In the system shown here, the indoor thermostat controls the supplemental heat source (furnace). The outdoor thermostat is not required.

Since the furnace is serving as the secondary heat source, the system does not require the home rewiring usually associated with supplemental electric strip heating.

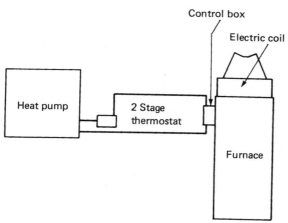

Figure 2-15 Heat pump with a two stage thermostat and control box mounted on the furnace.

Special requirements of heat-pump systems

The installation, maintenance, and operating efficiency of the heat-pump system are like those of no other comfort system. A heat-pump system requires the same air quantity for heating and cooling. Because of this, the air-moving capability of an existing furnace is extremely important. It should be carefully checked before a heat pump is added. Heating and load calculations must be accurate. System design and installation must be precise and according to the manufacturer's suggestions.

The air-distribution system and diffuser location are equally important. Supply ducts must be properly sized and insulated. Adequate return air is also required. Heating supply air is cooler than with other systems. This is quite noticeable to homeowners accustomed to gas or oil heat. This makes diffuser location and system balancing critical.

Heat-pump combinations

There are four ways to describe the heat-pump methods of transporting heat into the house:

1. *Air to air*: This is the most common method. It is the type of system previously described.

2. *Air to water*: This type uses two different types of heat exchangers. Warmed refrigerant flows through pipes to a heat exchanger in the boiler. Heated water flows into radiators located within the heated space.

3. *Water to water*: This type uses two water-to-refrigerant heat exchangers. Heat is taken from the water source (well water, lakes, or the sea) and is passed on by the refrigerant to the water used for heating. The reverse takes place in the cooling system.

4. *Water to air*: Well water furnishes the heat. This warms the refrigerant in the heat-exchanger coil. The refrigerant, compressed, flows to the top of the unit, where a fan blows air past the heat exchanger.

Each type of heat pump has its advantages and disadvantages. Each needs to be properly controlled. This is where the electrical connections and controls are used to do the job properly. Before attempting to work on this type of equipment, make sure you have a complete schematic of the electrical wiring and know all the component parts of the system.

High-Efficiency Furnaces

Furnaces have been designed (since 1981) with efficiencies of up to 97 percent, as compared to older types with efficiencies in the 60 percent range. The Lennox Pulse is one example of the types available. The G14 series pulse combustion up-flow gas furnace provides efficiency of up to 97 percent. Eight models for natural gas and LPG are available with input capacities of 40,000, 60,000, 80,000, and 100,000 Btu/h. The units operate on the pulse-combustion principle and do not require a pilot burner, main burners, conventional flue, or chimney. Compact, standard-sized cabinet design, with side or bottom return air entry, permits installation in a basement, utility room, or closet. Evaporator coils may be added, as well as electronic air cleaners and power humidifiers (see Fig. 2-16).

Operation

The high-efficiency furnaces achieve that level of fuel conversion by using a unique heat-exchanger design. It features a finned cast-iron combustion chamber, temperature-resistant steel tailpipe, aluminized steel exhaust decoupler section, and a finned stainless-steel tube condenser coil similar to an air-conditioner coil. Moisture, in the products of combustion, is condensed in the coil, thus wringing almost every usable Btu out of the gas. Since most of the combustion heat is utilized in the heat transfer from the coil, flue vent temperatures are as low as 100 to 130°F (38 to 54°C), allowing for the use of 2-in.-diameter polyvinyl chloride (PVC) pipe. The furnace is vented through a side wall or roof or to the top of an existing chimney with up to 25 ft of PVC pipe and four 90° elbows. Condensate created in the coil may be disposed of in an indoor drain (see Fig. 2-17). The condensate is not harmful to standard

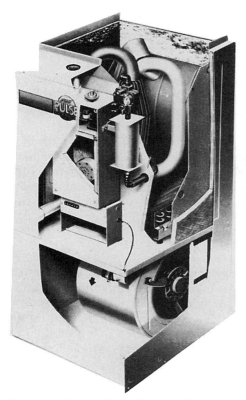

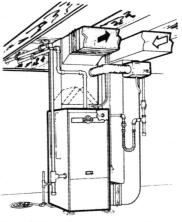

Figure 2-16   Lennox Pulse furnace. (*Courtesy of Lennox.*)

Figure 2-17   Basement installation of the Pulse furnace with cooling coil and automatic humidifier. Note the floor drain for condensate. (*Courtesy of Lennox.*)

household plumbing and can be drained into city sewers and septic tanks without damage.

The furnace has no pilot light or burners. An automotive-type spark plug is used for ignition on the initial cycle only, saving gas and electrical energy. Due to the pulse-combustion principle, the use of atmospheric gas burners is eliminated, with the combustion process confined to the heat-exchanger combustion chamber. The sealed combustion system virtually eliminates the loss of conditioned air due to combustion and stack dilution. Combustion air from the outside is piped to the furnace with the same type of PVC pipe used for exhaust gases.

Electrical controls

The furnace is equipped with a standard-type redundant gas valve in series with a gas expansion tank, gas intake flapper valve, and air intake flapper valve. Also factory installed are a purge blower, spark plug igniter and flame sensor with solid-state control circuit board. The standard equipment includes a fan and limit control, a 30-VA transformer, blower cooling relay, flexible gas line connector, and four isolation mounting pads, as well as a base insulation pad, condensate drip leg, and cleanable air filter. Flue vent/air intake line, roof or wall termination installation kits, LPG conversion kits, and thermostat are available as accessories and must be ordered extra, or you can use the existing one when replacing a unit.

The printed circuit board is replaceable as a unit when there is a malfunction of one of the components. It uses a multivibrator transistorized circuit to generate the high voltages needed for the spark plug. The spark plug gets very little use except to start the combustion process. It has a long life expectancy. Spark gap is 0.115 in. and the ground electrode is adjusted to 45° (see Fig. 2-20).

Sequence of operation

On a demand for heat, the room thermostat initiates the purge blower operation for a prepurge cycle of 34 s, followed by energizing of the ignition and opening of the gas valve. As ignition occurs the flame sensor senses proof of ignition and deenergizes the spark igniter and purge blower (see Fig. 2-18). The furnace blower operation is initiated 30 to 45 s after combustion ignition. When the thermostat is satisfied, the gas valve closes and the purge blower is reenergized for a postpurge cycle of 34 s. The furnace blower remains in operation until the preset temperature setting [90°F (32°C)] of the fan control is reached. Should the loss of flame occur before the thermostat is satisfied, flame sensor controls will initiate three to five attempts at reignition before locking out the unit operation. In addition, loss of either combustion intake air or flue exhaust will automatically shut down the system.

Combustion process

The process of pulse combustion begins as gas and air are introduced into the sealed combustion chamber with the spark plug igniter. Spark from the plug ignites the gas-air mixture, which in turn causes a positive pressure buildup that closes the gas and air inlets. This pressure relieves itself by forcing the products of combustion out of the combustion chamber through the tailpipe into the heat-exchanger exhaust decoupler and into the heat-exchanger coil. As the combustion chamber empties, its pressure becomes negative, drawing in air and gas for ignition of the next pulse. At the same instant, part of the pressure pulse is reflected back from the tailpipe at the top of the combustion chamber. The flame remnants of the previous pulse of combustion ignite the new gas-air mixture in the chamber, continuing the cycle.

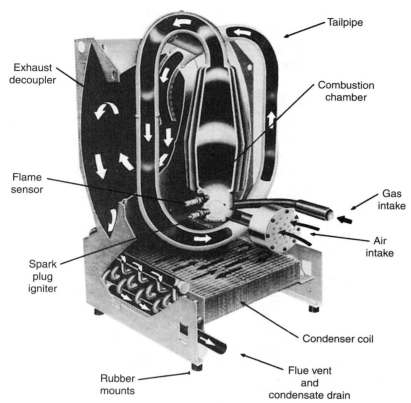

Figure 2-18   Cutaway view of the Pulse furnace combustion chamber. (*Courtesy of Lennox.*)

Once combustion is started, it feeds on itself, allowing the purge blower and spark igniter to be turned off. Each pulse of gas-air mixture is ignited at the rate of 60 to 70 times per second, producing 1/4 to 1/2 Btu per pulse of combustion. Almost complete combustion occurs with each pulse. The force of these series of ignitions creates great turbulence, which forces the products of combustion through the entire heat-exchanger assembly, resulting in maximum heat transfer (see Fig. 2-18).

Start-up procedures for the GSR-14Q series of Lennox Pulse furnaces, as well as maintenance and repair parts, are shown in Fig. 2-19.

Troubleshooting the Lennox Pulse Furnace

Troubleshooting procedures for the Lennox Pulse furnaces are shown in Fig. 2-20. Figure 2-21 shows the circuitry for the G-14Q series of furnaces. Note the difference in the electrical circuitry for the G-14

ELECTRICAL

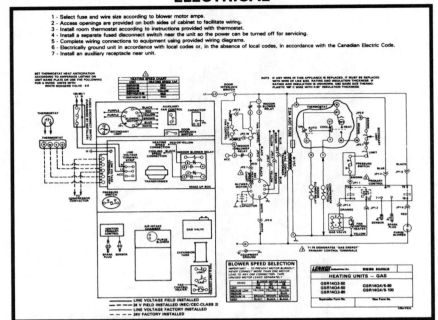

1 - Select fuse and wire size according to blower motor amps.
2 - Access openings are provided on both sides of cabinet to facilitate wiring.
3 - Install room thermostat according to instructions provided with thermostat.
4 - Install a separate fused disconnect switch near the unit so the power can be turned off for servicing.
5 - Complete wiring connections to equipment using provided wiring diagrams.
6 - Electrically ground unit in accordance with local codes or, in the absence of local codes, in accordance with the Canadian Electric Code.
7 - Install an auxiliary receptacle near unit.

START-UP/ADJUSTMENTS

START-UP

This unit is equipped with a direct spark ignition system with flame rectification. Once combustion has started, the purge blower and spark ignitor are turned off. To place furnace in operation:

1 - With thermostat set below room temperature make sure gas valve knob is in off position and wait 5 minutes.
2 - Turn manual knob of gas valve counterclockwise to ON position. Turn power on and set thermostat above room temperature. Unit will go into prepurge for approximately 30 seconds and then ignite.
3 - If the unit does not light on the first attempt, it will attempt four more ignitions before locking out.
4 - If lockout occurs, turn thermostat off and then back on.

To shut off furnace:

1 - Set thermostat to lowest temperature and turn power supply to furnace off.
2 - Turn manual knob of gas valve off.

FAILURE TO OPERATE

If unit fails to operate, check the following:

1 - Is thermostat calling for heat?
2 - Is main disconnect switch closed?
3 - Is there a blown fuse?
4 - Is filter dirty or plugged? Dirty or plugged filters will cause unit to go off on limit control.
5 - Is gas turned on at meter?
6 - Is manual main shut-off valve open?
7 - Is internal manual shut-off open?
8 - Are intake and exhaust pipes clogged?
9 - Is primary control locked out? (Turn thermostat off and then back on.)
10 - Is unit locked out on secondary limit? (Secondary limit is manually reset.)

GSR14 SERIES UNITS

Figure 2-19   Electrical start-up adjustments, maintenance, and repair list for the GSR-14Q. (*Courtesy of Lennox Industries Inc.*)

ADJUSTMENTS

GAS FLOW

To check proper gas flow to combustion chamber, determine BTU input from the appliance rating plate. Divide this input rating by the BTU per cubic foot of available gas. Result is the number of cubic feet per hour required. Determine the flow of gas through gas meter for 2 minutes and multiply by 30 to get the hourly flow of gas to burner.

GAS PRESSURE

1 - Check gas line pressure with unit firing at maximum rate. Normal natural gas inlet line pressure should be 7.0 in. (178 mm) w.c. Normal line pressure for LP gas is 11 in. (280 mm) w.c.

   IMPORTANT - Minimum gas supply pressure is listed on unit rating plate for normal input. Operation below minimum pressure may cause nuisance lockouts.

2 - After line pressure is checked and adjusted, check regulator pressure. Correct manifold pressure (unit running) is specified on nameplate. To measure, connect gauge to pressure tap in elbow below expansion tank.

HEAT ANTICIPATOR SETTINGS

Units with White Rodgers gas valves — 0.9

FAN/LIMIT CONTROL

Limit Control — Factory set: No adjustment necessary

Fan Control — Factory set: ON — No adjustment necessary
OFF — 90°F (32°C)

TEMPERATURE RISE AND EXTERNAL STATIC PRESSURE

Check temperature rise and external static pressure. If necessary, adjust blower speed to maintain temperature rise and external static pressure within range shown on unit rating plate.

ELECTRICAL

1 - Check all wiring for loose connections.

2 - Check for correct voltage at unit (unit operating).

3 - Check amp-draw on blower motor.

   Motor Nameplate_____Actual_____

NOTE - Do not secure electrical conduit directly to duct work or structure.

BLOWER SPEEDS

Multi-tap drive motors are wired for different heating and cooling speeds. Speed may be changed by simply interchanging motor connections at indoor blower relay and fan control. Refer to speed selection chart on unit wiring diagram.

CAUTION - To prevent motor burnout, never connect more than one (1) motor lead to any one connection. Tape unused motor leads separately.

MAINTENANCE

NOTE - Disconnect power before servicing.

ANNUAL MAINTENANCE

At the beginning of each heating season, system should be checked by qualified serviceman as follows:

A - Blower

1 - Check and clean blower wheel.

2 - Check motor lubrication.

   Always relubricate motor according to manufacturer's lubrication instructions on each motor. If no instructions are provided, use the following as a guide:

   a - Motors Without Oiling Ports — Prelubricated and sealed. No further lubrication required.

   b - Direct Drive Motors with Oiling Ports — Prelubricated for an extended period of operation. For extended bearing life, relubricate with a few drops of SAE No. 10 non-detergent oil once every two years. It may be necessary to remove blower assembly for access to oiling ports.

B - Electrical

1 - Check all wiring for loose connections.

2 - check for correct voltage at unit (unit operating).

3 - Check amp-draw on blower motor.

   Motor nameplate _____ Actual _____

4 - Check to see that heat tape (if applicable) is operating.

C - Filters

1 - Filters must be cleaned or replaced when dirty to assure proper furnace operation.

2 - Reusable foam filters supplied with GSR14 can be washed with water and mild detergent. They should be sprayed with filter handicoater when dry prior to reinstallation. Filter handicoater is RP products coating No. 481 and is available as Lennox part No. P-8-5069.

3 - If replacement is necessary, order Lennox part No. P-8-7831 for 20 X 25 inch (508 X 635 mm) filter.

D - Intake and Exhaust Lines

Check intake and exhaust PVC lines and all connections for tightness and make sure there is no blockage. Also check condensate line for free flow during operation.

E - Insulation

Outdoor piping insulation should be inspected yearly for deterioration. If necessary, replace with same materials.

Figure 2-19 *(Continued)*

REPAIR PARTS LIST

The following repair parts are available through independent Lennox dealers. When ordering parts, include the complete furnace model number listed on unit rating plate. Example: GSR14Q3-50-1.

CABINET PARTS
Blower access panel
Control access panel
Upper vestibule panel
Lower vestibule panel
Control box cover

CONTROL PANEL PARTS
Transformer
Indoor blower relay
Low voltage terminal strip
High voltage terminal strip

BLOWER PARTS
Blower wheel
Motor
Motor mounting frame
Motor capacitor
Blower housing cut-off plate
Blower housing

HEATING PARTS
Heat exchanger assembly
Gas orifice
Gas valve
Gas decoupler
Gas flapper valve
Purge blower
Air intake flapper valve
Primary control board
Ignition lead
Flame sensor lead
Flame sensor
Primary fan and limit control
Secondary limit control
Auxiliary fan control
Differential pressure switch
Door interlock switch
Air filter

Figure 2-19 Electrical start-up adjustments, maintenance, and repair list for the GSR-14Q. (*Courtesy of Lennox Industries Inc.*)

TROUBLESHOOTING

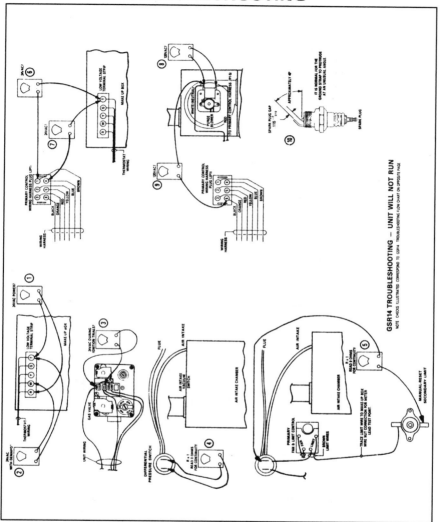

Figure 2-20   Troubleshooting the GSR-14Q furnace with a meter. (*Courtesy of Lennox.*)

TROUBLESHOOTING

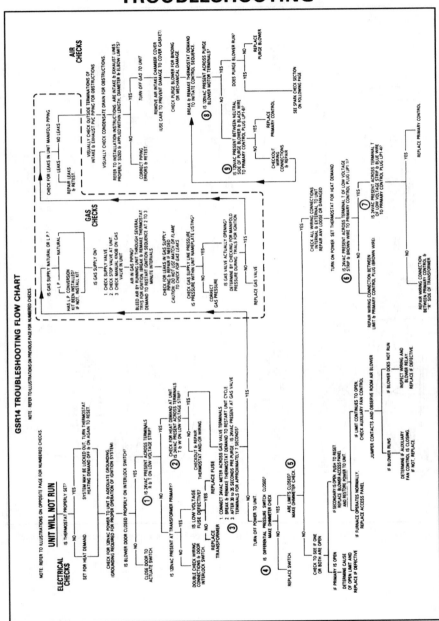

Figure 2-20

TROUBLESHOOTING (CONT.)

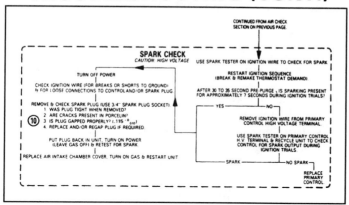

Figure 2-20

ELECTRICAL

1 - Select fuse and wire size according to blower motor amps.
2 - Access holes are provided on both sides of cabinet to facilitate wiring.
3 - Install room thermostat according to instruction provided with thermostat.
4 - Install a separate fused disconnect switch near the unit so the power can be turned off for servicing.

5 - Complete wiring connections to equipment using provided wiring diagrams.
6 - Electrically ground unit in accordance with local codes, or in the absence of local codes in accordance with the CSA Standards.
7 - Seal unused electrical openings with snap-plugs provided.

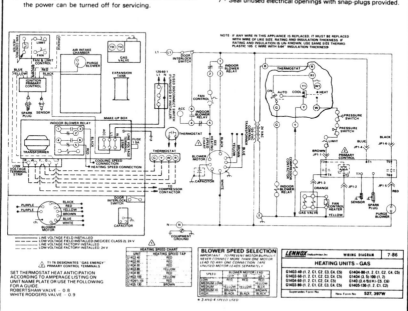

Figure 2-21   Electrical wiring for the G14Q series furnaces. (*Courtesy of Lennox.*)

and GSR-14. Blower speed color-coded wires are also indicated for the different units. The 40, 60, 80, and 100 after the G-14Q indicates whether it is a 40,000, 60,000, 80,000, or 100,000 Btu/h unit. Thermostat heat anticipation is also given for the Robertshaw valve and the Rodgers valve. This type of electrical diagram is usually glued to the cabinet so that it is with the unit whenever there is need for troubleshooting.

The troubleshooting flow chart is typical of those furnished with newer equipment in the technical manuals furnished the dealers who provide the service. After locating the exact symptoms, check with the other part of Fig. 2-20 to find how to use the multimeter to check out all the circuitry to see if the exact cause of the problem can be determined.

Ventilation Requirements

Ventilation is produced by two basic methods: natural and mechanical. Natural ventilation is obtained by open windows, vents, or drafts, whereas mechanical ventilation is produced by the use of fans.

Thermal effect is possibly better known as flue effect. Flue effect is the draft in a stack or chimney that is produced within a building when the outdoor temperature is lower than the indoor temperature. This is caused by the difference in weight of the warm column of air within the building and the cooler air outside.

Air may be filtered by two ways: dry filtering and wet filtering. Various air-cleaning equipments (such as filtering, washing, or combined filtering and washing devices) are used to purify the air. When designing the duct network, ample filter area must be included so that the air velocity passing through the filters is sufficient. Accuracy in estimating the resistance to the flow of air through the duct system is important in the selection of blower motors. Resistance should be kept as low as possible in the interest of economy. Ducts should be installed as short as possible.

The effect of dust on health has been properly emphasized by competent medical authorities. Air-conditioning apparatus removes these contaminants from the air and further provides the correct amount of moisture so that the respiratory tracts are not dehydrated, but are kept properly moist. Dust is more than just dry dirt. It is a complex, variable mixture of materials and, as a whole, is rather uninviting, especially the type found in and around human habitation. Dust contains fine particles of sand, soot, earth, rust, fiber, animal and vegetable refuse, hair, and chemicals.

Humidifiers add moisture to dry air. The types of humidifiers used in air-conditioning systems are spray-type air washers, pan evaporative, electrically operated, and air operated.

The function of an air washer is to cool the air and to control humidity. An air washer usually consists of a row of spray nozzles and a chamber or tank at the bottom that collects the water as it falls through the air contacting many baffles. Air passing over the baffles picks up the required amount of humidity.

Dehumidification is the removal of moisture from the air and is accomplished by two methods: cooling and adsorption. Cooling-type dehumidification operates on the refrigeration principle. It removes moisture from the air by passing the air over a cooling coil. The moisture in the air condenses to form water, which then runs off the coil into a collecting tray or bucket.

Ventilation is the process of supplying or removing air to or from any building or space. Such air may or may not have been conditioned. Methods of supplying or removing the air are accomplished by natural ventilation and mechanical methods. The treatment of air, both inside and out is very important in any ventilation system.

Air Leakage

Air leakage caused by cold air outside and warm air inside takes place when the building contains cracks or openings at different levels. This results in the cold and heavy air entering at low levels and pushing the warm and light air out at high levels; the same draft takes place in a chimney. When storm sashes are applied to well-fitted windows, very little redirection in infiltration is secured, but the application of the sash does give an air space that reduces heat transmission and helps prevent window frosting. By applying storm sashes to poorly fitted windows, a reduction in leakage of up to 50 percent may be obtained. The effect, insofar as air leakage is concerned, will be roughly equivalent to that obtained by the installation of weather stripping.

Natural Ventilation

The two natural forces available for moving air through and out of buildings are wind forces and temperature difference between the inside and the outside. Air movement may be caused by either of these forces acting alone, or by a combination of the two, depending on atmospheric conditions, building design, and location.

Wind forces

In considering the use of natural forces for producing ventilation, the conditions that must be considered are as follows:

- Average wind velocity
- Prevailing wind direction

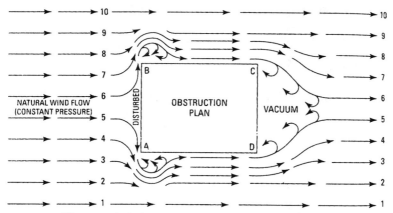

Figure 3-1 Diagram of wind action where the airstream meets an obstruction such as a building or ventilator.

- Seasonal and daily variations in wind velocity and direction
- Local wind interference by buildings and other obstructions

When the wind blows without encountering any obstructions to change its direction, the movements of the airstream (as well as the pressure) remain constant. If, on the other hand, the airstream meets an obstruction of any kind (such as a house or ventilator), the airstream will be pushed aside as illustrated in Fig. 3-1. In the case of a simple ventilator (see Fig. 3-2), the closed end forms an obstruction that changes the direction of the wind. The ventilator expands at the closed end and converges at the open end, thus producing a vacuum inside the head, which induces an upward flow of air through the flue and through the head.

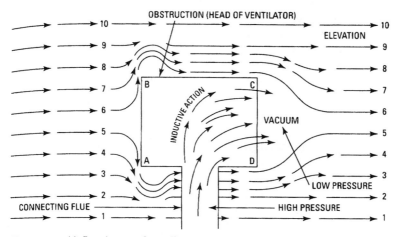

Figure 3-2 Airflow in a roof ventilator.

Temperature-difference forces

Perhaps the best example of the thermal effect is the draft in a stack or chimney known as the *flue effect*. The flue effect of a stack is produced within a building when the outdoor temperature is lower than the indoor temperature. It is caused by the difference in weight of the warm column of air within the building and the cooler air outside. The flow caused by flue effect is proportional to the square root of the draft head, or approximately.

$$Q = 9.4A\sqrt{h(t - t_o)}$$

where Q = air flow, ft^3/min
       A = free area of inlets or outlets (assumed equal)
       h = height from inlets to outlets, ft
       t = average temperature of indoor air, °F
      t_o = temperature of outdoor air, °F
    9.4 = constant of proportionality (including a value of 65 percent for effectiveness of openings)

The constant of proportionality should be reduced by 50 percent (constant = 7.2) if conditions are not favorable.

Combined wind and temperature forces

Note that when wind and temperature forces are acting together, even without interference, the resulting air flow is not equal to the sum of the two estimated quantities. However, the flow through any opening is proportional to the square root of the sum of the heads acting on that opening. When the two heads are equal in value, and the ventilating openings are operated so as to coordinate them, the total airflow through the building is about 10 percent greater than that produced by either head acting independently under ideal conditions. This percentage decreases rapidly as one head increases over the other. The larger head will predominate.

Roof ventilators

The function of a roof ventilator is to provide a storm and weather-proof air outlet. For maximum flow by induction, the ventilator should be located on that part of the roof where it will receive full wind without interference. Roof ventilators are made in a variety of shapes and styles (Fig. 3-3). Depending on their construction and application, they may be termed as stationary, revolving, turbine, ridge, or syphonage.

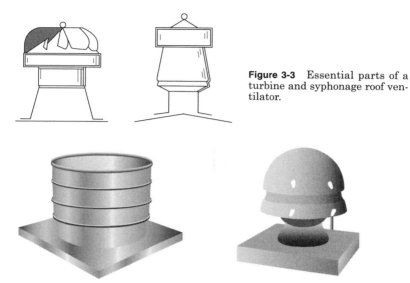

Figure 3-3 Essential parts of a turbine and syphonage roof ventilator.

Figure 3-4 Typical ventilator arrangement.

Ventilator capacity

Several factors must be taken into consideration in selecting the proper ventilator for any specific problem:

- Mean temperature difference
- Stack height (chimney effect)
- Induction effect of the wind
- Area of ventilator opening

Of several other minor factors, only one requires close attention. This is the area of the inlet air openings (Fig. 3-4). The action of a roof ventilator is to let air escape from the top of a building, which naturally means that a like amount of air must be admitted to the building to take the place of the air exhausted. The nature, size, and location of these inlet openings are of importance in determining the effectiveness of the ventilating system.

Fresh-Air Requirements

Table 3-1 is a guide to the amount of air required for efficient ventilation. This information should be used in connection with the ventilator-capacity tables provided by fan manufacturers to assist in the proper selection of the number and size of the units required.

TABLE 3-1   Fresh-Air Requirements

Type of building or room	Minimum air changes per hour	Cubic feet of air per minute per occupant
Attic spaces (for cooling)	12–15	
Boiler room	15–20	
Churches, auditoriums	8	20–30
College classrooms		25–30
Dining rooms (hotel)	5	
Engine rooms	4– 6	
Factory buildings:		
Ordinary manufacturing	2– 4	
Extreme fumes or moisture	10–15	
Foundries	15–20	
Galvanizing plants	20–30*	
Garages:		
Repair	20–30	
Storage	4–6	
Homes (night cooling)	9–17	
Hospitals:		
General		40–50
Children's		35–40
Contagious diseases		89–90
Kitchens:		
Hotel	10–20	
Restaurant	10–20	
Libraries (public)	4	
Laundries	10–15	
Mills:		
Paper	15–20*	
Textile-general buildings	4	
Textile-dye houses	15–20*	
Offices:		
Public	3	
Private	4	
Pickling plants	10–15[†]	
Pump rooms	5	
Restaurants	8–12	
Schools:		
Grade		15–25
High		30–35
Shops:		
Machine	5	
Paint	15–20*	
Railroad	5	
Woodworking	5	
Substations (electric)	5–10	
Theaters		10–15
Turbine rooms (electric)	5–10	
Warehouses	2	
Waiting rooms (public)	4	

*Hoods should be installed over vats or machines.
[†]Unit heaters should be directed on vats to keep fumes superheated.

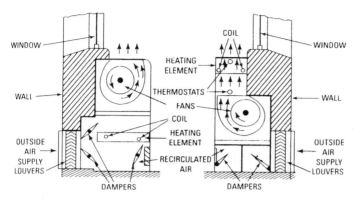

Figure 3-5 Typical mechanical ventilators for residential use showing placement of fans and other details.

Mechanical Ventilation

Mechanical ventilation differs from natural ventilation mainly in that the air circulation is performed by mechanical means (such as fans or blowers). In natural ventilation, the air is caused to move by natural forces. In mechanical ventilation, the required air changes are affected partly by diffusion, but chiefly by positive currents put in motion by electrically operated fans or blowers, as shown in Fig. 3-5. Fresh air is usually circulated through registers connected with the outside and warmed as it passes over and through the intervening radiators.

Volume of air required

The volume of air required is determined by the size of the space to be ventilated and the number of times per hour that the air in the space is to be changed. In many cases, existing local regulations or codes will govern the ventilating requirements. Some of these codes are based on a specified amount of air per person, and others on the air required per square foot of floor area.

Duct-system resistance

Air ducts may be designed with either a round or a rectangular cross section. The radius of elbows should preferably be at least 1.5 times the pipe diameter for round pipes, or the equivalent round pipe size in the case of rectangular ducts. Accuracy in estimating the resistance to the flow of air through the duct system is important in the selection of blowers for application in duct systems (Fig. 3-6).

Resistance should be kept as low as possible in the interest of economy, since underestimating the resistance will result in failure of the blower to deliver the required volume of air. You should carefully study the building drawings with consideration paid to duct locations and

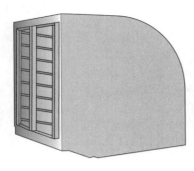

Figure 3-6 Central-station air-handling unit. (*Courtesy of Buffalo Forge Company.*)

clearances. Keep all duct runs as short as possible, bearing in mind that the airflow should be conducted as directly as possible from the source to the delivery points. Select the locations of the duct outlets to ensure proper air distribution. The ducts should be provided with cross-sectional areas that will permit air to flow at suitable velocities. Moderate velocities should be used in all ventilating work to avoid waste of power and to reduce noise. Lower velocities are more frequently used in schools, churches, theaters, and so on, instead of factories and other places where noise caused by airflow is not objectionable.

Air Filtration

The function of air filters as installed in air-conditioning systems is to remove the airborne dust that tends to settle in the air of the ventilated space and may become a menace to human health if not properly removed.

Effect of dust on health

The effect of dust on health has been properly emphasized by competent medical authorities. The normal human breathes about 17 times a minute. The air taken into the lungs may contain large quantities of dust, soot, germs, bacteria, and other deleterious matter. Most of this solid matter is removed in the nose and air passages of the normal person. However, if these passages are dry and permit the passage of these materials, colds and respiratory diseases result. Air-conditioning apparatus remove these contaminants from the air and further provide the correct amount of moisture so that the respiratory tracts are not dehydrated, but are kept properly moist. Among the airborne diseases are mumps, measles, scarlet fever, pneumonia, colds, tuberculosis, hay fever, grippe, influenza, and diphtheria.

Dust is more than just dry dirt. It is exceedingly complex, with variable mixtures of materials, and, as a whole, is rather uninviting, especially the type found in and around human habitation. Dust contains fine particles of sand, soot, earth, rust, fiber, animal and vegetable refuse, hair, chemicals, and compounds, all of which are abrasive, irritating, or both. The U.S. Weather Bureau estimates that there are 115,000 particles of dust per cubic inch of ordinary city air, and that each grain of this dust at breathing level contains from 85,000 to 125,000 g. The proximity of factories using oil or coal-burning equipment and the presence of considerable street traffic will aggravate this condition and increase the dust and germ content.

The Mellon Institute of Industrial Research conducted a series of experiments to determine dust precipitation in three large cities. The measurements published were the average amounts measured in more than 10 stations in each of the cities and were for a period of 1 year. It was observed that precipitation was considerably greater in the industrial districts. It is certainly plain that filtered air is almost a necessity, especially in large buildings where number of individuals are gathered. Hotels, theaters, schools, stores, hospitals, factories, and museums require the removal of dust from the atmosphere admitted to their interiors, not only because a purer product is made available for human health and consumption, but also so that the fittings, clothing, furniture, and equipment will not be damaged by the dust particles borne by the fresh airstream.

The air filter is one part of the air-conditioning system that should be operated the year around. There are times when the air washer (the cooling and humidifying apparatus) is not needed, but the filter is one part of the system that should be kept in continuous operation, purifying the air.

Various dust sources

The dust removed in the filters may be classified as poisonous, infective, obstructive, or irritating and may originate from an animal, vegetable, mineral, or metallic source. Industry produces many deleterious fumes and dusts, for example, those emanating from grinding, polishing, dyeing, gilding, painting, spraying, cleaning, pulverizing, baking, mixing of poisons, sawing, and sand blasting. This dust must be filtered out of the air because of its effect on health and efficiency, to recover valuable materials, to save heat by reusing filtered air, and to prevent damage to equipment and products in the course of manufacturing.

Air-Filter Classification

Air-cleaning equipment may be classified according to the following methods used:

- Filtering
- Washing
- Combined filtering and washing

There are several types of filters divided into two main classes: dry and wet. The wet filters may again be subdivided into what is known as the manual and automatic self-cleaning types.

Figure 3-7 Typical air-filter panel absorber containing activated charcoal for heavy-duty control of odors and vapors. (*Courtesy of Barneby Cheney Corporation.*)

Dry filters

Dry filters are usually made in standard units of definite rated capacity and resistance to the passage of air. At the time the installation is designed, ample filter area must be included so that the air velocity passing through the filters is not excessive; otherwise, proper filtration will not be secured. The proper area is secured by mounting the proper number of filters on an iron frame. A dry filter (Fig. 3-7) may be provided with permanent filtering material from which the dirt and dust must be removed pneumatically or by vibration. Or, it may be provided with an inexpensive filter medium that can be dispensed with when loaded with dirt and replaced with a new slab.

A dry-type filter is known as a *dry plate.* The cells are composed of a number of perforated aluminum plates, arranged in series and coated with a fireproof filament. Air drawn through the cell is split into thin streams by the perforations so that the dust particles are thrown against the filament by the change in the direction of the air between the plates. The filament tentacles hold the dust particles and serve to purify the air. Dust particles are deposited on the intervening flat surfaces. When the plate is loaded with dirt and requires cleaning, this may be accomplished in two ways.

In moderate-sized installations, the operator removes the cells from the sectional frame, rests one edge on the table or floor, and gently raps the opposite side with the open hand, causing the collected dirt to drop out and restoring the efficiency of the filter. Larger installations (Fig. 3-8), where a great number of cells are used, or where an abnormally high dust load is encountered, use a mechanical vibrator in which the cell is placed and rapped for about a minute to free it of dust. Under average dust conditions, cells must be cleaned or replaced about twice a year.

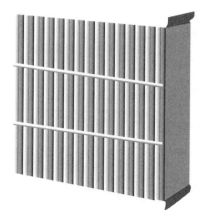

Figure 3-8 Activated-charcoal room-purification filter panel absorber. Filter absorbers of this type are recommended for central air-conditioning-system purification of recirculated air in occupied spaces, purification of exhaust to prevent atmospheric pollution, etc. They are suitable for light-, medium-, and heavy-duty requirements. (*Courtesy of Barneby Cheney Corporation.*)

Packed-type filters contain a fibrous mat so that the slow movement and irregular path taken by the air permits it to leave behind all particles of dirt in the mazes of the fibrous mat (Fig. 3-8). This type of unit must be removed every 12 to 15 weeks under average conditions.

Wet (or viscous) filters

The wet (or viscous) filter makes use of an adhesive in which the dust particles are caught and held upon impingement. It also makes use of densely packed layers of viscous-coated metal baffles, screens, sinuous passages, crimped wire, and glass wool inserted in the path of the air. The air divides into small streams that constantly change their direction and force the heavier dirt particles against the viscous-coated surfaces, where they are held.

The three major types of wet (or viscous) air filters are replaceable, manually cleaned, and automatically cleaned.

Wet replacement type. The glass-wool air filter is of the viscous-type replacement cell. Glass wool, being noncorrosive and nonabsorbing, maintains its density and leaves all the viscous adhesive free to collect and hold dirt and dust. This type of filter lends itself to both small and large installations. A carton of 12 filters weighs about 37 lb. After the filter replacement has been completed, the dirty filters can be put into the original shipping container and removed from the premises.

Manually cleaned type. The manually cleaned type is usually made in the form of a standard cell with a steel frame containing the filter media. The cell is fitted into the cell frame (see Fig. 3-9), with the use of a felt gasket to prevent leakage of air past the cell. As a usual practice, cleaning is necessary about once every eight weeks. This is made a simple matter by the use of automatic latches so that the filter sections can be

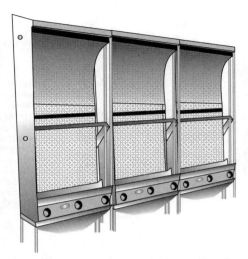

Figure 3-10 Typical automatic viscous filter. In a self-cleaning filter of this type, brushes on an endless carrier chain periodically sweep collected materials from the upstream side of the panels. Used in textile mills, laundries, and other ventilation systems involving a high volume of coarse, bulky contaminants. (*Courtesy of Rockwell-Standard Corporation.*)

Figure 3-9 Typical open-frame industrial air purifier. Each filter unit contains one or more activated charcoal absorbers, including blower and motor. (*Courtesy of Barneby Cheney Corporation.*)

pulled out of the frame, immersed in a solution of water and washing soda or cleaning compound, allowed to dry, then dipped into charging oil, drained to remove excess oil, and then replaced in the filter.

Automatic viscous filters.   The automatic filter is of the self-cleaning type and utilizes the same principle of adhesive dust impingement as the manual and replacement types, but the removal of the accumulated dirt from the filter medium is entirely automatic. The self-cleaning type may be divided into two distinct systems: immersion variety and flushing type.

The immersion-type air filter (Fig. 3-10) makes use of an endless belt that rotates the filter medium and passes it through an oil bath, washing the dirt off. The dirt, being heavier than the oil, settles to the bottom into special containers, which are removed and cleaned from time to time. The usual speed of the filter is from 1.5 to 3 in. every 12 min. Such an apparatus is particularly adapted for continuous operation. The sediment can be removed and the oil changed or added without stopping the apparatus.

The flushing type is constructed of cells in the form of a unit laid on shelves and connected by metal aprons. The dirt caught by the cells is flushed down into a sediment tank by flooding pipes, which travel back and forth over the clean-air side of the cells. Aprons catch the heavier dust particles before they get to the cells so that the cells are more efficient. This type does not flush the cells while the system is in operation. In many

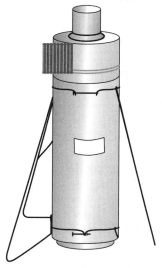

Figure 3-11 Cylindrical-type canister air purifier. (*Courtesy of Barneby Cheney Corporation.*)

Figure 3-12 Wall-mounted air purifier. (*Courtesy of Barneby Cheney Corporation.*)

cases, duplicate units are provided so that one unit can be flushed while the other bears the full load. Some of the flushing mechanisms are interlocked with the fan circuit, especially where a single unit is used, so that flushing cannot take place unless the fan is shut down. If air is permitted to pass through the filter while it is being flushed, oil will be carried along with the air.

A cylindrical absorber (shown in Fig. 3-11) can be used for the control of odor problems by forcing air through two perforated metal walls enclosing an annular bed of activated charcoal. A motor-operated blower draws contaminated air through the charcoal-filled container and recirculates pure odor-free air through the room or space to be purified.

The wall-mounted air purifier shown in Fig. 3-12 is equipped with a washable dust filter and centrifugal blower. Air purifiers of this type are ideal for toilets, utility rooms, public rest rooms, and similar odor-problem areas.

In residential-type air conditioners, however, unit filters provide the necessary dust protection, particularly since they are manufactured in a large variety of types and sizes. Where lint in a dry state predominates, a dry filter is preferable because of its lint-holding capacity. Throwaway filters are used increasingly where the cleaning process needs to be eliminated.

Filter Installation

Air filters are commonly installed in the outdoor air intake ducts of the air-conditioning system and often in the recirculating air ducts as well.

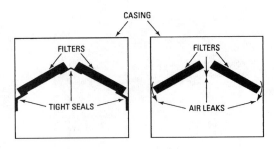

Figure 3-13 Good (left) and poor (right) air-filter installations. They should fit tightly to prevent air from bypassing the filter.

Air filters are logically mounted ahead of heating or cooling coils and other air-conditioning equipment in the system to prevent dust from entering. Filters should be installed so that the face area is at right angles to the airflow wherever possible (see Fig. 3-13). In most cases, failure of an air-filter installation can be traced to faulty installation, improper maintenance, or both.

The American Society of Heating and Ventilating Engineers gives the most important requirements for a satisfactory and efficiently functioning air-filter installation as follows:

- The filter must be of ample size for air it is expected to handle. An overload of 10 to 15 percent is regarded as the maximum allowable. When air volume is subject to increase, a larger filter should be installed.

- The filter must be suited to the operating conditions, such as degree of air cleanliness required, amount of dust in the entering air, type of duty, allowable pressure drop, operating temperatures, and maintenance facilities.

- The filter type should be the most economical for the specific application. The first cost of the installation should be balanced against depreciation as well as expense and convenience of maintenance.

The following recommendations apply to filters and washers installed with central fan systems:

- Duct connections to and from the filter should change their size or shape gradually to ensure even air distribution over the entire filter area.

- Sufficient space should be provided in front as well as behind the filter to make it accessible for inspection and service. A distance of 2 ft may be regarded as the minimum.

- Access doors of convenient size should be provided in the sheet-metal connections leading to and from the filters.

- All doors on the clean-air side should be lined with felt to prevent infiltration of unclean air. All connections and seams of the sheet-metal ducts on the clean-air side should be airtight.

- Electric lights should be installed in the chamber in front of and behind the air filter.

- Air washers should be installed between the tempering and heating coils to protect them from extreme cold in winter.

- Filters installed close to air inlets should be protected from the weather by suitable louvers, in front of which a large mesh wire screen should be provided.

- Filters should have permanent indicators to give a warning when the filter resistance reaches too high a value.

Humidity-Control Methods

Humidifiers, by definition, are devices for adding moisture to the air. Thus, to humidify is to increase the density of water vapor within a given space or room. Air humidification is affected by vaporization of water and always requires heat for its proper functioning. Thus, devices that function to add moisture to the air are termed *humidifiers,* whereas devices that function to remove moisture from the air are termed *dehumidifiers.*

Humidifiers

As previously noted, air humidification consists of adding moisture. Following are the types of humidifiers used in air-conditioning systems:

- Spray-type air washers
- Pan evaporative humidifiers
- Electrically operated humidifiers
- Air-operated humidifiers

Air-washer method

An air washer essentially consists of a row of spray nozzles inside a chamber or casing. A tank at the bottom of the chamber provides for collection of water as it falls through the air and comes into intimate contact with the wet surface of the chamber baffles. The water is generally circulated by means of a pump, the warm water being passed over refrigerating coils or blocks of ice to cool it before being passed to the spray chamber. The water lost in evaporation is usually replaced automatically by the use of a float arrangement, which admits water from the main tank as required. In many locations, the water is sufficiently cool to use as it is drawn from the source. In other places, the water is not cool enough and must be cooled by means of ice or with a refrigerating machine.

The principal functions of the air washer are to cool the air passed through the spray chamber and to control humidity. In many cases, the

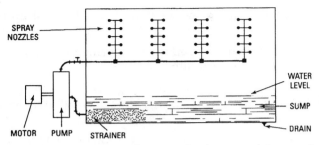

Figure 3-14   Elementary diagrams showing essential parts of air-washer unit with identification of parts.

cooling coils are located in the bottom of the spray chamber so that as the warm spray descends, it is cooled and ready to be again sprayed by the pump. In some cases, the water is passed through a double-pipe arrangement and is cooled on the counter-current principle.

Figure 3-14 shows a sketch of an air washer. In this case, the spray pipes are mounted vertically. In some instances, the spray pipes are horizontal so that the sprays are directed downward. As some of the finer water particles tend to be carried along with the air current, a series of curved plates or baffles is generally used, which forces the cooled and humidified air to change the direction of flow, throwing out or eliminating the water particles in the process.

Pan humidifiers

Figure 3-15 shows the essential parts of the pan-type humidifier. The main part is a tank of water heated by low-pressure steam or forced hot

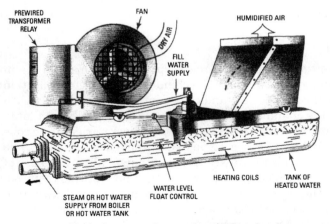

Figure 3-15   Typical evaporative pan humidifier showing operative components. (*Courtesy of Armstrong Machine Works.*)

water where a water temperature of 200°F (93°C) or higher is maintained. The evaporative-type humidifier is fully automatic, the water level being controlled by means of a float control. In operation, when the relative humidity drops below the humidity-control setting, the humidifier fan blows air over the surface of the heated water in the tank. The air picks up moisture. The air is blown to the space to be humidified. When the humidity control is satisfied, the humidifier fan stops.

Electrically operated humidifiers

Dry-steam electrically operated humidifiers operate by means of a solenoid valve, which is energized by a humidistat. When the relative humidity drops slightly below the desired level set by the humidistat (see Fig. 3-16), a solenoid valve actuated by the humidistat admits steam from the separating chamber to the reevaporating chamber. Steam passes from this chamber through the muffler directly to the atmosphere. The fan (which

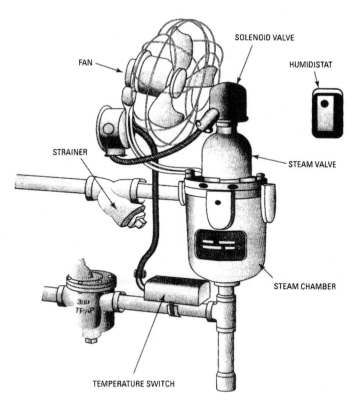

Figure 3-16   Armstrong dry-steam humidifier for direct discharge of steam into the atmosphere of the area to be humidified. (*Courtesy of Armstrong Machine Works.*)

is energized when the solenoid valve opens) assists in dispersing the steam into the area to be humidified. When the relative humidity reaches the desired level, the humidistat closes the solenoid valve and stops the fan.

Air-operated humidifiers

Air-operated humidifying units operate in the same manner as electrical units, except that they utilize a pneumatic hygrostat as a humidity controller and an air operator to open or close the steam valve (see Fig. 3-17). A decrease in relative humidity increases the air pressure under a spring-loaded diaphragm to open the steam valve wider. An increase in relative humidity reduces the pressure under the diaphragm and allows the valve to restrict the steam flow. In a humidifier operation of this type, the steam supply is taken off the top of the header (see Fig. 3-18). Any condensate formed in the supply line is knocked down to the humidifier drain by a baffle inside the inlet of the humidifier-separating chamber.

Any droplets of condensation picked up by the stream as it flows through the humidifier cap when the steam valve opens will be thrown to the bottom of the reevaporating chamber. Pressure in this chamber is essentially atmospheric. Since it is surrounded by steam at supply pressure and temperature, any water is reevaporated to provide dry steam at the outlet. The humidifier outlet is also surrounded by steam at supply pressure to ensure that there will be no condensation or drip at this point. A clamp-on temperature switch is attached to the condensate drain line to prevent the electric or pneumatic operator from opening the steam valve until the humidifier is up to steam temperature.

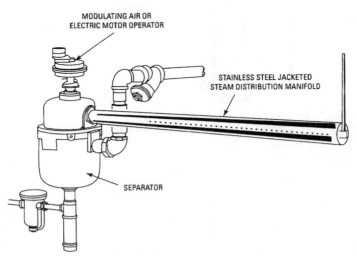

Figure 3-17   Armstrong dry-steam humidifier with steam-jacket distribution.

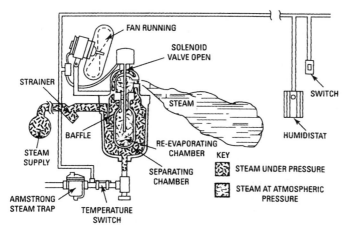

Figure 3-18   Operation of Armstrong dry-steam humidifier for area humidification.

Dehumidifiers

The removal of moisture from the air is termed *dehumidification*. Air dehumidification is accomplished by one of two methods: cooling and adsorption. Dehumidification can be accomplished by an air washer, providing the temperature of the spray is lower than the dew point of the air passing through the unit. If the temperature of the spray is higher than the dew point, condensation will not take place. Air washers having refrigerated sprays usually have their own recirculating pump.

Electric dehumidification

An electric dehumidifier operates on the refrigeration principle. It removes moisture from the air by passing the air over a cooling coil. The moisture in the air condenses to form water, which then runs off the coil into a collecting tray or bucket. The amount of water removed from the air varies, depending on the relative humidity and volume of the area to be dehumidified. In locations with high temperature and humidity conditions, 3 to 4 gal of water per day can usually be extracted from the air in an average-size home.

When the dehumidifier is first put into operation, it will remove relatively large amounts of moisture until the relative humidity in the area to be dried is reduced to the value where moisture damage will not occur. After this point has been reached, the amount of moisture removed from the air will be considerably less. This reduction in moisture removal indicates that the dehumidifier is operating normally and that it has reduced the relative humidity in the room or area to a safe value.

The performance of the dehumidifier should be judged by the elimination of dampness and accompanying odors rather than by the amount of moisture that is removed and deposited in the bucket. A dehumidifier cannot act as an air conditioner to cool the room or area to be dehumidified. In operation, the air that is dried when passed over the coil is slightly compressed, raising the temperature of the surrounding air, which further reduces the relative humidity of the air.

Controls

As mentioned previously, the dehumidifier (see Fig. 3-19) operates on the principles of the conventional household refrigerator. It contains a motor-operated compressor, a condenser, and a receiver. In a dehumidifier, the cooling coil takes the place of the evaporator, or chilling unit in a refrigerator. The refrigerant is circulated through the dehumidifier in the same manner as in a refrigerator. The refrigerant flow is controlled by a capillary tube. The moisture-laden air is drawn over the refrigerated coil by means of a motor-operated fan or blower.

The dehumidifier operates by means of a humidistat (see Fig. 3-20), which starts and stops the unit to maintain a selected humidity level. In a typical dehumidifier, the control settings range from *dry* to *extra dry* to *continuous* to *off*. For best operation, the humidistat control knob is normally set at *extra dry* for initial operation over a period of 3 to 4 weeks. After this period, careful consideration should be given to the dampness in the area being dried. If sweating on cold surfaces has discontinued and the damp odors are gone, the humidistat control should be reset to *dry*. At this setting, more economical operation is obtained,

Figure 3-19 Automatic electric dehumidifier. (*Courtesy of Westinghouse Electric Corporation.*)

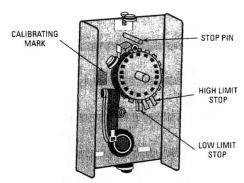

Figure 3-20 Typical humidistat designed to control humidifying or dehumidifying equipment or both with one instrument. (*Courtesy of Penn Controls, Inc.*)

but the relative humidity probably will be higher than at the *extra dry* setting.

After 3 or 4 weeks of operation at the *dry* setting, if the moisture condition in the area being dried is still satisfactory, the operation of the dehumidifier should be continued with the control set at this position. However, if at the setting the dampness condition is not completely corrected, the control should be returned to the *extra dry* setting. Minor adjustments will usually be required from time to time. Remember that the control must be set near *extra dry* to correct the dampness conditions but as close to *dry* as possible to obtain the most economical operation.

Adsorption-type dehumidifiers

Adsorption-type dehumidifiers operate on the use of sorbent materials for adsorption of moisture from the air. Sorbents are substances that contain a vast amount of microscopic pores. These pores afford a great internal surface to which water adheres or is adsorbed. A typical dehumidifier based on the honeycomb desiccant wheel principle is shown schematically in Fig. 3-21. The wheel is formed from thin corrugated and laminated asbestos sheets rolled to form wheels of various desired diameters and thickness. The wheels are impregnated with a desiccant cured and reinforced with a heat-resistant binder. The corrugations in the honeycomb wheel form narrow flutes perpendicular to the wheel diameter. Approximately 75 percent of the wheel face area is available for the adsorption or dehumidifying flow circuit, and 25 percent is available for the reactivation circuit. In the smaller units, the reactivated air is heated electrically; in the larger units, it is heated by electric, steam, or gas heaters.

Figure 3-22 shows another industrial adsorbent dehumidifier of the stationary bed type. It has two sets of stationary adsorbing beds arranged so that one set is dehumidifying the air while the other set is drying. With

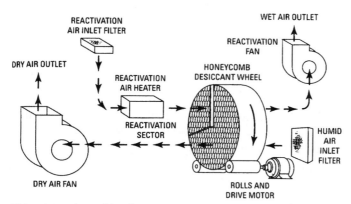

Figure 3-21 Assembly of components operating on the honeycomb method of dehumidification. (*Courtesy of Cargocaire Engineering Corporation.*)

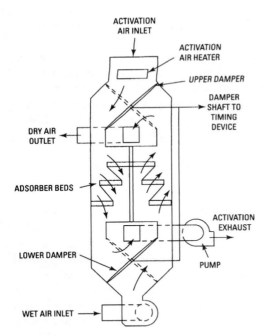

Figure 3-22 Stationary bed-type solid-absorbent dehumidifier.

the dampers in the position shown, air to be dried flows through one set of beds and is dehumidified while the drying air is heated and circulated through the other set. After completion of drying, the beds are cooled by shutting off the drying air heaters and allowing unheated air to circulate through them. An automatic timer controller is provided to allow the dampers to rotate to the opposite side when the beds have adsorbed moisture to a degree that begins to impair performance.

Air-Duct Systems

Air ducts for transmission of air in a forced-air heating, ventilation, or air-conditioner system must be carefully designed from the standpoint of economy, as well as for proper functioning. When designing air ducts, the following methods may be used:

- Compute the total amount of air to be handled per minute by the fan, as well as the fractional volumes composing the total, which are to be supplied to or withdrawn from different parts of the building.
- Locate the supply unit in the most convenient place and as close as possible to the center of distribution.
- Divide the building into zones, and proportion the air volumes per minute in accordance with the requirements of the different zones.

- Locate the air inlets or outlets for supply and recirculation, respectively. At the positions so located on the building plans, indicate the air volumes to be dealt with. The position of the outlets and inlets should be such as to produce a thorough diffusion of the conditioned air throughout the space supplied.

- Determine the size of each outlet or inlet based on passing the required amount of air per minute at a suitable velocity.

- Calculate the areas, and select suitable dimensions for all branch and main ducts. Do this based on creating equal frictional losses per foot of length. This involves reducing the velocities in smaller ducts.

- Ascertain the resistance of the ducts that sets up the greatest friction. In most cases, this will be the longest run, although not so invariably. This will be the resistance offered by the duct system as a whole to the flow of the required amount of air.

- Revise the dimensions and areas of the shorter runs so that the ducts themselves will create resistances equal to the longest run. This will cut down the cost of the sheet metal, and the result will be just the same as if dampers were used. Too high a velocity, however, must be avoided.

- To compensate for unforeseen contingencies, volume dampers should be provided for each branch.

Heat gains in ducts

In any air-conditioning installation involving a duct system, invariably there is an accession of heat by the moving air in the ducts between the coils and supply grilles when air is supplied below room temperature. If the ducts are located through much of their length in the conditioned space, then, of course, this heat absorption has no effect on the total load and frequently may be disregarded.

More frequently, however, the supply ducts must pass through spaces that are not air- conditioned. Under these conditions, the heat absorbed by the air in the ducts can be regarded as an additional load on the cooling equipment. The temperature rise in a duct system of a cooling installation depends on the following factors:

- Temperature of the space through which the duct passes
- Air velocity through the duct
- Type and thickness of insulation, if any

The first factor establishes the temperature differential between the air on either side of the duct walls. The dew point of the air surrounding the duct may also have some effect on the heat pickup, as condensation on the duct surface gives up the heat of vaporization to the air passing through the system. The highest air velocities consistent with the acoustic requirements of the installation should be used, not only for economy in the sheet metal material used, but also to reduce the heat pickup in the ducts.

The amount of heat absorbed by a unit area of sheet-metal duct conveying chilled air is almost directly proportional to the temperature difference between the atmosphere surrounding the duct and the chilled air, irrespective of the velocity of the latter. The heat pickup rate will be influenced somewhat by the outside finish of the duct and by the air motion, if any, in the space through which the duct passes.

Heat leakage in Btu per hour per square foot per degree difference in temperature for uncovered galvanized iron ductwork will be between 0.5 and 1.0, with an average value of 0.73. The rate of leakage, of course, will be greatest at the start of the duct run and will gradually diminish as the air temperature rises. Covering the duct with the equivalent of 0.5-in. rigid insulation board and sealing cracks with tape will reduce the average rate of heat pickup per square foot of surface per hour to 0.23 Btu per degree difference.

To summarize, the designer of a central unit system should observe the following:

■ Locate the equipment as close to the conditioned space as possible.

■ Use duct velocities as high as practical, considering the acoustic level of the space and operating characteristics of the fans.

■ Insulate all supply ducts with covering equivalent to at least 0.5 in. of rigid insulation, and seal cracks with tape.

Suitable duct velocities.   Table 3-2 gives the recommended air velocities in feet per minute for different requirements. These air velocities are in accordance with good practice. It should be understood that the fan-outlet velocities depend somewhat on the static pressure, as static pressures and fan-housing velocity pressures are interdependent for good operating conditions. Fan-outlet velocities are also affected by the particular type of fan installed.

Air-duct calculations.   For the sake of convenience, this section outlines a simplified air-duct sizing procedure, which eliminates the usual complicated engineering calculation required when designing a duct system. Refer to Figs. 3-23 to 3-25, which show the plans of a typical family residence having a total cubic content of approximately 19,000 ft^3. It is desired to provide humidification, ventilation, filtration, and air movement to all rooms on both the first and second floors.

The air conditioner located as shown in Fig. 3-23 had a total air-handling capacity of 1000 ft^3/min. If the rooms are to have individual

TABLE 3-2 Recommended Air Velocities

Designation	Residences, broadcasting studios, and so on	Schools, theaters, public buildings	Industrial applications
Initial air intake	750	800	1,000
Air washers	500	500	500
Extended surface heaters or coolers (face velocity)	450	500	500
Suction connections	750	800	1,000
Through fan outlet:			
For 1-1/2-in. static pressure	–	2,200	2,400
For 1-1/4-in. static pressure	–	2000	2,200
For 1-in. static pressure	1,700	1,800	2,000
For 3/4-in. static pressure	1,400	1,550	1,800
For 1/2-in. static pressure	1,200	1,300	1,600
Horizontal ducts	700	900	1,000–2,000
Branch ducts and risers	550	600	1,000–1,600
Supply grilles and openings	300	300 grille	400 opening
Exhaust grilles and openings	350	400 grille	500 opening
Duct outlets at high elevation	–	1,000	–

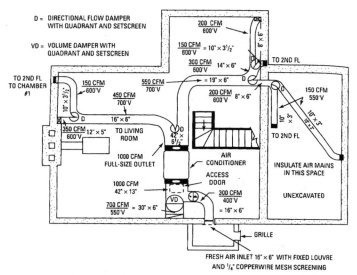

Figure 3-23 Basement floor of a typical family residence showing air-conditioner and heat-duct layout.

air mains, Table 3-3 shows the methods of computing the amount of air to be supplied to each room.

Second column in the table expresses the capacity of the individual room as a percentage of the total volume. For example, 3000 ft^3 is 30 percent

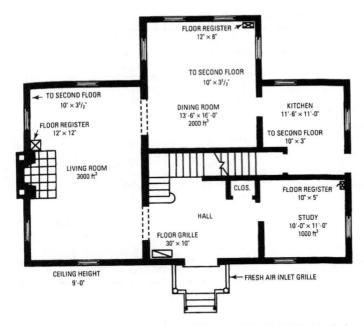

Figure 3-24 First-floor layout of residence of Fig. 3-23. Vertical air discharges are located in the floor of the living room, dining room, and study. Air is returned to the unit through the floor grilles in the entrance hall. Sizes for air mains are shown on the basement plan.

Figure 3-25 Second-floor layout of residence of Figs. 3-23 and 3-24. Air discharges are located in the basement of the three chambers.

TABLE 3-3 **Air-Duct Calculation**

	Rooms	Room volume, ft³	% of total volume	Supply, ft³/min
First floor	Living room	3,000	30	300
	Dining room	2,000	20	200
	Study	1,000	10	100
Second floor	Chamber No. 1	1,500	15	150
	Chamber No. 2	1,500	15	150
	Chamber No. 3	1,000	10	100
	Total volume	10,000	100	1,000

of the total space (10,000 ft³) to which air mains will lead. Third column indicates the cubic feet of air per minute to be supplied to the individual rooms. These figures are attained as follows. The air conditioner will handle 1000 ft³ of air per minute; 30 percent of this is 300 ft³/min. Similarly, 10 percent is 100 ft³/min, which would indicate the quantity of air to be supplied to the living room and chamber No. 3, respectively. Having thus established the air quantity to be delivered to each of the rooms, the design of the ducts can now be considered.

For duct sizes, consider the branch duct to both the living room and chamber No. 1 (see Table 3-3). Note that the branch leading into chamber No. 1 handles 150 ft³/min. The duct to the living room handles 300 ft³/min. Obviously, the connecting air main will handle 300 + 150, or 450, ft³/min, in accordance with the foregoing recommendation, allowing a velocity of 600 ft/min for the branches and 700 ft/min for the supply air main. Therefore, the necessary duct areas can be calculated using the following formula:

$$\frac{a \times 144}{6b} \, (\text{in.}^2/\text{ft}^2)$$

where a = necessary air supply to room, ft³/min
b = recommended velocity for a main or branch

Living-room (300 ft³/m) branch:

$$\frac{300 \times 144}{600} = 72 \, \text{in.}^2 = 12 \times 6 \, \text{in. branch}$$

Chamber No. 1 (150 ft³/min) branch:

$$\frac{150 \times 144}{600} = 36 \, \text{in.}^2 = 10 \times 3.5 \, \text{in. branch}$$

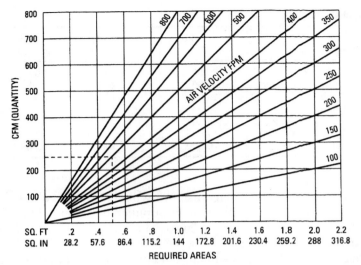

Figure 3-26 Air-duct sizing method.

Total (450 ft³/min) main:

$$\frac{450 \times 144}{700} = 93 \text{ in.}^2 = 16 \times 6 \text{ in. main}$$

The remaining ductwork may be calculated similarly. It is recommended that the main air supply leaving the unit be the same size as that of the outlet of the unit up to the first branch takeoff. The return main to the unit should run the same size as the inlet of the unit (that is for a distance of approximately 24 in. It should be provided with a large access door in the bottom of this length of the full size duct. Figure 3-26 is helpful as a further simplification in sizing the air ducts.

Example 3-1 It is desired to size a main duct for 250 ft³/min at 500 ft/min velocity. What cross-sectional area is required?

solution Locate 250 ft³/min on the left-hand side of Fig. 3-26. With a ruler or straightedge, carry a line across horizontally to the 500 velocity line and read off on the base line 72 in.², or 1/2 ft², which is the area required. All branches, risers, or grilles may be sized in the same manner.

Resistance losses in duct systems

In general, it can be said that duct sizes and the depth of a duct in particular are affected by the available space in the building. For this reason, although the round duct is the most economical shape from the standpoint of friction per unit area and from the standpoint of metal required for construction per unit of area, it is rarely possible, except in industrial buildings, to use round ducts largely. Square duct is the preferable shape among those of rectangular cross section. Headroom limitations usually require that the duct be flattened.

To illustrate the use of charts when sizing a duct system, consider Example 3-2.

Example 3-2 Assume a system requiring the delivery of 5000 ft^3/min. The distribution requirement is the movement of the entire volume a distance of approximately 80 ft, with the longest branch beyond that point conveying 1000 ft^3/min for an additional 70 ft. Assume further that the operating characteristics of the fan and the resistance of the coils, filters, and so on, allow a total supply duct resistance of 0.10-in. water-gage resistance pressure. The supply duct is not to be more than 12 in. deep.

solution The total length of the longest run is 80 + 70 = 150 ft:

$$\frac{100}{150} = 0.10 = 0.067 \text{ in. water gage}$$

Starting at this resistance at the bottom of Fig. 3-27, follow upward to the horizontal line representing 5000 ft^3/min. At this point, read the equivalent size of round duct required that is approximately 28 in. in diameter. Move diagonally upward to the right on the 28-in.-diameter line, and then across horizontally on this line in Fig. 3-27 to the vertical line representing the 12-in. side of a rectangular duct. At this point, read 60 in. as the width of the rectangular duct required on the intersecting curve line.

Thus, for the main duct run, the duct size will be 60 × 12 in. For the branch conveying 1000 ft^3/min, from the point where the 0.067-in. resistance line intersects the 1000-ft^3/min line, read 16 in. as the equivalent round duct required. Following through on Fig. 3-27 as for a larger duct, read 12 × 18 in. as the size of the branch duct.

Duct runs should take into account the number of bends and offsets. Obstructions of this kind are usually represented in terms of the equivalent length of straight duct necessary to produce the same resistance value. Where conditions require sharp or right-angle bends, vane elbows composed of a number of curved deflectors across the air-stream should be used.

Fans and Blowers

Various types of fans are used in air-conditioning applications and are classified as propeller, tube-axial, vane-axial, and centrifugal. The propeller and tube-axial fans consist of a propeller or disk-type wheel mounted inside a ring or plate and driven by a belt or direct drive motor.

A vane-axial fan consists of a disk-type wheel mounted within a cylinder. A set of air-guide vanes is located before or after the wheel and is

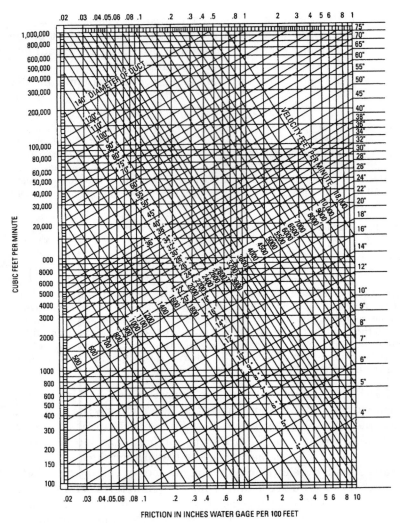

Figure 3-27   Graphic representation of air-duct areas.

belt-driven or direct drive. The centrifugal fan is a fan rotor or wheel within a scroll-type housing. This type of blower is better known as a squirrel-cage unit. Whenever possible, the fan wheel should be directly connected to the motor shaft. Where fan speeds are critical, a belt drive is employed, and various size pulleys are used.

The various devices used to supply air circulation in air-conditioning applications are known as fans, blowers, exhausts, or propellers. The different types of fans may be classified with respect to their construction as follows:

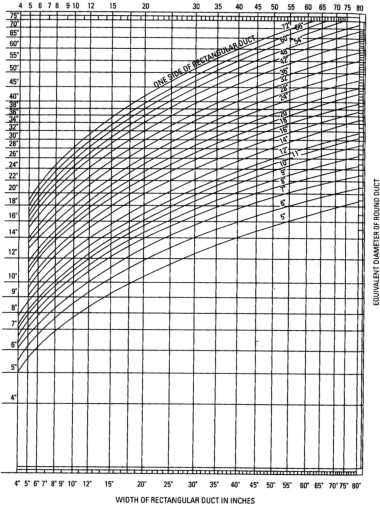

Figure 3-27 (*Continued*)

- Propeller
- Tube-axial
- Vane-axial
- Centrifugal

A *propeller fan* consists essentially of a propeller or disk-type wheel within a mounting ring or plate and includes the driving-mechanism supports for either belt or direct drive. A *tube-axial fan* consists of a

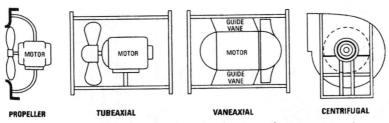

PROPELLER   TUBEAXIAL   VANEAXIAL   CENTRIFUGAL

Figure 3-28   Various fan classifications showing mounting arrangement.

propeller or disk-type wheel within a cylinder and includes the driving-mechanism supports for either belt drive or direct connection. A *vane-axial fan* consists of a disk-type wheel within a cylinder and a set of air guide vanes located before or after the wheel. It includes the driving-mechanism supports for either belt drive or direct connection. A *centrifugal fan* consists of a fan rotor or wheel within a scroll-type housing and includes the driving-mechanism supports for either belt drive or direct connection. Figure 3-28 shows the mounting arrangements.

Fan performance may be stated in various ways, with the air volume per unit time, total pressure, static pressure, speed, and power input being the most important. The terms, as defined by the National Association of Fan Manufacturers, are as follows:

- *Volume* handled by a fan is the number of cubic feet of air per minute expressed as fan-outlet conditions.

- *Total pressure* of a fan is the rise of pressure from fan inlet to fan outlet.

- *Velocity pressure* of a fan is the pressure corresponding to the average velocity determination from the volume of air-flow at the fan outlet area.

- *Static pressure* of a fan is the total pressure diminished by the fan-velocity pressure.

- *Power output* of a fan is expressed in horsepower and is based on fan volume and the fan total pressure.

- *Power input* of a fan is expressed in horsepower and is measured as horsepower delivered to the fan shaft.

- *Mechanical efficiency* of a fan is the ratio of power output to power input.

- *Static efficiency* of a fan is the mechanical efficiency multiplied by the ratio of static pressure to the total pressure.

- *Fan-outlet area* is the inside area of the fan outlet.

- *Fan-inlet area* is the inside area of the inlet collar.

TABLE 3-4 **Volume of Air Required**

Space to be ventilated	Air changes per hour	Minutes per change
Auditoriums	6	10
Bakeries	20	3
Bowling alleys	12	5
Club rooms	12	5
Churches	6	10
Dining rooms (restaurants)	12	5
Factories	10	6
Foundries	20	3
Garages	12	5
Kitchens (restaurants)	30	2
Laundries	20	3
Machine shops	10	6
Offices	10	6
Projection booths	60	1
Recreation rooms	10	6
Sheet-metal shops	10	6
Ship holds	6	10
Stores	10	6
Toilets	20	3
Tunnels	6	10

Air volume

The volume of air required is determined by the size of the space to be ventilated and the number of times per hour that the air in the space is to be changed. Table 3-4 shows the recommended rate of air change for various types of spaces.

In many cases, existing local regulations or codes will govern the ventilating requirements. Some of these codes are based on a specified amount of air per person and on the air required per square foot of floor area. Table 3-4 should serve as a guide to average conditions. Where local codes or regulations are involved, they should be taken into consideration. If the number of persons occupying the space is larger than would be normal for such a space, the air should be changed more often than shown.

Horsepower requirements

The horsepower required for any fan or blower varies directly as the cube of the speed, provided that the area of the discharge orifice remains unchanged. The horsepower requirements of a centrifugal fan generally decrease with a decrease in the area of the discharge orifice if the speed remains unchanged. The horsepower requirements of a propeller fan increase as the area of the discharge orifice decreases if the speed remains unchanged.

Figure 3-29 Drive and mounting arrangement for various types of propeller fans.

Drive methods

Whenever possible, the fan wheel should be directly connected to the motor shaft. This can usually be accomplished with small centrifugal fans and with propeller fans up to about 60 in. in diameter. The deflection and the critical speed of the shaft, however, should be investigated to determine whether or not it is safe.

When selecting a motor for fan operation, it is advisable to select a standard motor one size larger than the fan requirements. It should be kept in mind, however, that direct-connected fans do not require as great a safety factor as that of belt-driven units. It is desirable to employ a belt drive when the required fan speed or horsepower is in doubt, since a change in pulley size is relatively inexpensive if an error is made (see Fig. 3-29).

Directly connected small fans for various applications are usually driven by single-phase AC motors of the split-phase, capacitor, or shaded-pole type. The capacitor motor is more efficient electrically and is used in districts where there are current limitations. Such motors, however, are usually arranged to operate at one speed. With such a motor, if it is necessary to vary the air volume or pressure of the fan or blower, the throttling of air by a damper installation is usually made.

In large installations (such as when mechanical draft fans are required), various drive methods are used:

- A slip-ring motor to vary the speed.
- A constant-speed, directly connected motor, which, by means of moveable guide vanes in the fan inlet, serves to regulate the pressure and air volume.

Fan selection

Most often, the service determines the type of fan to use. When operation occurs with little or no resistance, and particularly when no duct system is required, the propeller fan is commonly used because of its simplicity and economy in operation. When a duct system is involved, a centrifugal or axial type of fan is usually employed. In general, centrifugal and axial fans are comparable with respect to sound effect, but the axial fans are somewhat lighter and require considerably less space. The following information is usually required for proper fan selection:

- Capacity requirement in cubic feet per minute
- Static pressure or system resistance
- Type of application or service
- Mounting arrangement of system
- Sound level or use of space to be served
- Nature of load and available drive

The various fan manufacturers generally supply tables or characteristic curves that ordinarily show a wide range of operating particulars for each fan size. The tabulated data usually include static pressure, outlet velocity, revolutions per minute, brake horsepower, tip or peripheral speed, and so on.

Fan applications

The numerous applications of fans in the field of air-conditioning and ventilation are well known, particularly to engineers and air-conditioning repair and maintenance personnel. The various fan applications are as follows:

- Exhaust fans
- Circulating fans
- Cooling-tower fans
- Kitchen fans
- Attic fans

Exhaust fans are found in all types of applications, according to the American Society of Heating and Ventilating Engineers. Wall fans are predominantly of the propeller type, since they operate against little or no resistance. They are listed in capacities from 1000 to 75,000 ft³/min. They are sometimes incorporated in factory-built penthouses and roof caps or provided with matching automatic louvers. Hood exhaust fans involving ductwork are predominantly centrifugal, especially in handling hot or corrosive fumes.

Spray-booth exhaust fans are frequently centrifugal, especially if built into self-contained booths. Tubeaxial fans lend themselves particularly well to this application where the case of cleaning and of suspension in a section of ductwork is advantageous. For such applications, built-in cleanout doors are desirable.

Circulating fans are invariably propeller or disk-type units and are made in a vast variety of blade shapes and arrangements. They are designed for appearance as well as utility. *Cooling-tower fans* are predominantly the propeller type. However, axial types are also used for packed towers, and occasionally a centrifugal fan is used to supply draft. *Kitchen fans* for domestic use are small propeller fans arranged for window or wall mounting and with various useful fixtures. They are listed in capacity ranges of from 300 to 800 ft^3/min.

Attic fans are used during the summer to draw large volumes of outside air through the house or building whenever the outside temperature is lower than that of the inside. It is in this manner that the relatively cool evening or night air is utilized to cool the interior in one or several rooms, depending on the location of the air-cooling unit. It should be clearly understood, however, that the attic fan is not strictly a piece of air-conditioning equipment since it only moves air and does not cool, clean, or dehumidify. Attic fans are used primarily because of their low cost and economy of operation, combined with their ability to produce comfort cooling by circulating air rather than conditioning it.

Fan operation

Fans may be centrally located in an attic or other suitable space (such as a hallway), and arranged to move air proportionately from several rooms. A local unit may be installed in a window to provide comfort cooling for one room only when desired. Attic fans are usually propeller types and should be selected for low velocities to prevent excessive noise. The fans should have sufficient capacity to provide at least 30 air changes per hour.

To decrease the noise associated with air-exchange equipment, the following rules should be observed:

- The equipment should be properly located to prevent noise from affecting the living area.

- The fans should be of the proper size and capacity to obtain reasonable operating speed.

- Equipment should be mounted on rubber or other resilient material to assist in preventing transmission of noise to the building.

If it is unavoidable to locate the attic air-exchange equipment above the bedrooms, it is essential that every precaution be taken to reduce the equipment noise to the lowest possible level. Since high-speed AC motors are usually quieter than low-speed ones, it is often preferable to use a high-speed motor connected to the fan by means of an endless V-belt, if the floor space available permits such an arrangement.

Attic-fan installation

Because of the low static pressures involved (usually less than 1/8 in. of water), disk or propeller fans are generally used instead of the blower or housed types. It is important that the fans have quiet operating characteristics and sufficient capacity to give at least 30 air changes per hour. For example, a house with 10,000 ft^3 content would require a fan with a capacity of 300,000 ft^3/h or 5000 ft^3/min to provide 30 air changes per hour.

The two general types of attic fans in common use are *boxed-in fans* and *centrifugal fans.* The boxed-in fan is installed within the attic in a box or suitable housing located directly over a central ceiling grille or in a bulkhead enclosing an attic stair. This type of fan may also be connected by means of a direct system to individual room grilles. Outside cool air entering through the windows in the downstairs room is discharged into the attic space and escapes to the outside through louvers, dormer windows, or screened openings under the eaves.

Although an air-exchange installation of this type is rather simple, the actual decision about where to install the fan and where to provide the grilles for the passage of air up through the house should be left to a ventilating engineer. The installation of a multiblade centrifugal fan is shown in Fig. 3-30. At the suction side, the fan is connected to exhaust ducts leading to grilles, which are placed in the ceiling of the two bedrooms. The air exchange is accomplished by admitting fresh air through

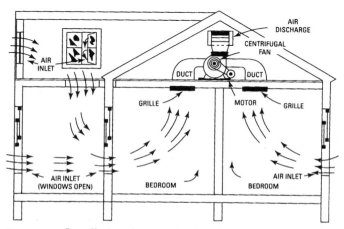

Figure 3-30   Installation of a centrifugal fan in a one-family dwelling.

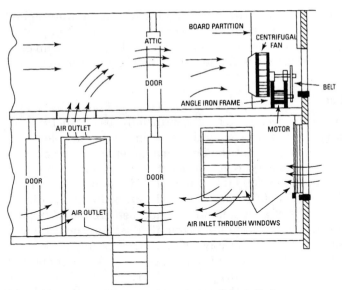

Figure 3-31   Belt-driven fan in a typical attic installation.

open windows and up through the suction side of the fan; the air is finally discharged through louvers as shown.

Another installation is shown in Fig. 3-31. This fan is a centrifugal curved-blade type, mounted on a light angle-iron frame, which supports the fan wheel, shaft, and bearings. The air inlet in this installation is placed close to a circular opening, which is cut in an airtight board partition that serves to divide the attic space into a suction and discharge chamber. The air is admitted through open windows and doors and is then drawn up the attic stairway through the fan into the discharge chamber.

Attic fan operation

The routine of operation to secure the best and most efficient results with an attic fan is important. A typical operating routine might require that, in the late afternoon when the outdoor temperature begins to fall, the windows on the first floor and the grilles in the ceiling or the attic floor be opened and the second-floor windows kept closed. This will place the principal cooling effect in the living rooms. Shortly before bedtime, the first-floor windows may be closed and those on the second floor opened to transfer the cooling effect to the bedrooms. A suitable time clock may be used to shut the motor off before arising time.

Refrigeration

Refrigeration is the process of removing heat from where it is not wanted. Heat is removed from food to preserve its quality and flavor. It is removed from room air to establish human comfort. There are innumerable applications in the industrial world where heat is removed from a certain place or material to accomplish a desired effect.

During refrigeration, unwanted heat is transferred mechanically to an area where it is not objectionable. A practical example of this is the window air conditioner that cools air in a room and exhausts hot air to the outdoors.

The liquid called the refrigerant is fundamental to the heat transfer accomplished by a refrigeration machine. Practically speaking, a commercial refrigerant is any liquid that will evaporate and boil at relatively low temperatures. During evaporation or boiling, the refrigerant absorbs the heat. The cooling effect felt when alcohol is poured over the back of your hand illustrates this principle.

In operation, a refrigeration unit allows the refrigerant to boil in tubes that are in contact, directly or indirectly, with the medium to be cooled. The controls and engineering design determine the temperatures reached by a specific machine.

Historical Development

Natural ice was shipped from the New England states throughout the western world from 1806 until the early 1900s. Although ice machines were patented in the early 1800s, they could not compete with the natural ice industry. Artificial ice was first commercially manufactured in the southern United States in the 1880s.

Figure 4-1 One of the first commercial home refrigerators. (*Courtesy of General Electric.*)

Domestic refrigerators were not commercially available until about 1920 (see Fig. 4-1). During the 1920s, the air-conditioning industry also got its start with a few commercial and home installations. The refrigeration industry has now expanded to touch most of our lives. There is refrigeration in our homes, and air conditioning in our place of work, and even in our automobiles. Refrigeration is used in many industries, from the manufacture of instant coffee to the latest hospital surgical techniques.

Structure of Matter

To be fully acquainted with the principles of refrigeration, it is necessary to know something about the structure of matter. Matter is anything that takes up space and has weight. Thus, matter includes everything but a perfect vacuum.

There are three familiar physical states of matter: solid, liquid, and gas or vapor. A solid occupies a definite amount of space. It has a definite shape. The solid does not change in size or shape under normal conditions.

A liquid takes up a definite amount of space, but does not have any definite shape. The shape of a liquid is the same as the shape of its container.

A gas does not occupy a definite amount of space and has no definite shape. A gas that fills a small container will expand to fill a large container.

Matter can be described in terms of our five senses. We use our senses of touch, taste, smell, sound, and sight to tell us what a substance is. Scientists have accurate methods of detecting matter.

Elements

Scientists have discovered 105 building blocks for all matter. These building blocks are referred to as elements. Elements are the most basic materials in the universe. Ninety-four elements, such as iron, copper, and nitrogen, have been found in nature. Scientists have made 11 others in laboratories. Every known substance, solid, liquid, or gas, is composed of elements. It is very rare for an element to exist in a pure state. Elements are nearly always found in combinations called *compounds*. Compounds contain more than one element. Even such a common substance as water is a compound, rather than an element (see Fig. 4-2).

Atom

An atom is the smallest particle of an element that retains all the properties of that atom, that is, all hydrogen atoms are alike. They are different from the atoms of all other elements. However, all atoms have certain things in common: They all have an inner part—the nucleus. This is composed of tiny particles called *protons* and neutrons. An atom also has an outer part. It consists of other tiny particles, called *electrons,* which orbit around the nucleus (see Figs. 4-3 and 4-4).

Neutrons have no electrical charge, but protons have a positive charge. Electrons are particles of energy and have a negative charge. Because of these charges, protons and electrons are particles of energy. That is, these charges form an electric field of force within the atom. Stated very simply, these charges are always pulling and pushing each other. This makes energy in the form of movement.

The atoms of each element have a definite number of electrons, and they have the same number of protons. A hydrogen atom has one electron and

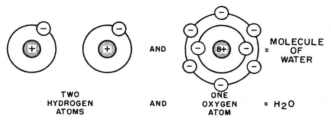

Figure 4-2   Two or more atoms linked are called a molecule. Here two hydrogen atoms and one oxygen atom form a molecule of the compound water (H_2O).

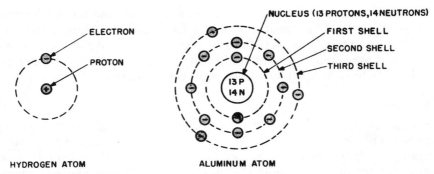

Figure 4-3 Atoms contain protons, neutrons, and electrons.

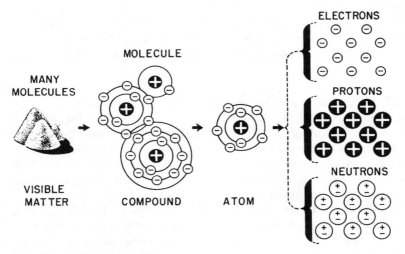

Figure 4-4 Molecular structure.

one proton. An aluminum atom has 13 of each. The opposite charges—negative electrons and positive protons—attract each other and tend to hold electrons in orbit. As long as this arrangement is not changed, an atom is electrically balanced. When chemical engineers know the properties of atoms and elements they can then engineer a substance with the properties needed for a specific job. Refrigerants are manufactured in the same way.

Properties of Matter

It is important for a refrigeration technician to understand the structure of matter. With this knowledge, the person can understand the factors that affect the structure of matter. These factors can be called the properties of matter. These properties are chemical, electrical, mechanical, or thermal (related to heat) in nature. Some of these properties are force, weight, mass, density, specific gravity, and pressure.

Force is described as a push or a pull on anything. Force is applied to a given area. Weight is the force of gravity pulling all matter toward the center of the earth. The unit of weight in the English system is the pound. The basic unit of mass in the metric system is the gram. Mass is the amount of matter present in a quantity of any substance. Mass is not dependent on location. A body has the same mass whether here on the earth, on the moon, or anywhere else. The weight does change at other locations. In the metric system, the kilogram (kg) (kilo means 1, 000) is the unit most often used for mass. In the English system, the slug is the unit of mass.

Density is the mass per unit of volume. Densities are comparative figures. That is, the density of water is used as a base and is set at 1.00. All other substances are either more or less dense than water.

The densities of gases are determined by a comparison of volumes. The volume of 1 lb of air is compared to the volume of 1 lb of another gas. Both gases are under standard conditions of temperature and pressure.

The specific gravity of a substance is its density compared to the density of water. Specific gravity has many uses. It can be used as an indicator of the amount of water in a refrigeration system. Testing methods are discussed in later chapters.

Pressure

Pressure is a force that acts on an area. Stated in a formula, it becomes:

$$P = \frac{F}{A}$$

where F = force
A = area
P = pressure

The unit of measurement of pressure in the English system is the pound per square foot or pounds per square inch (psi). The metric unit of pressure is the kilopascal (kPa). Pressure-measuring elements translate changes or differences in pressure into motion. The three types most commonly used are the diaphragm, the bellows, and the Bourbon spring tube.

Pressure-indicating devices

Pressure-indicating devices are most important in the refrigeration field. It is necessary to know the pressures in certain parts of a system to locate trouble spots.

The diaphragm is a flexible sheet of material held firmly around its perimeter so there can be no leakage from one side to the other

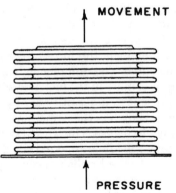

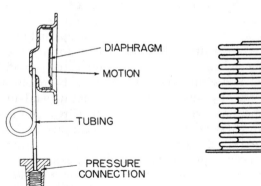

Figure 4-5 Pressure-sensing element, diaphragm type.

Figure 4-6 Pressure-sensing element, bellows type. (*Courtesy of Johnson.*)

(see Fig. 4-5). Force applied to one side of the diaphragm will cause it to move or flex. Diaphragms, in some cases, are made of flat sheet of material with a limited range of motion. Other diaphragms are at least one corrugation or fold. This allows more movement at the point where work is produced.

Some types of pressure controllers require more motion for per unit of force applied. To accomplish the desired result, the diaphragm is joined to the housing by a section with several convolutions or folds called *bellows*. Thus, the diaphragm moves in response to pressure changes. Each holds only a small amount (see Fig. 4-6). The bellow element may be assembled to extend or to compress as pressure is applied. The bellows itself act as a spring to return the diaphragm section to the original position when the pressure differential is reduced to zero. If a higher spring return rate is required, to match or define the measured pressure range, then an appropriate spring is added.

One of the most widely used types of pressure-measuring elements is the Bourdon spring tube, discussed in Chap. 1. It is readily adaptable to many types of instruments (see Fig. 4-7). The Bourdon tube is a flattened tube bent into a spiral or circular form closed at one end. When fluid pressure is applied within the tube, the tube tends to straighten or unwind. This produces motion, which may be applied to position an indicator or actuate a controller.

Pressure of liquid and gases

Pascal's law states that when a fluid is confined in a container that is completely filled, the pressure on the fluid is transmitted equally on all surfaces of the container. Similarly the pressure of a gas is the same on all areas of its container.

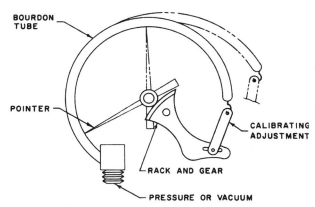

Figure 4-7 Pressure sensing element, Bourdon spring tube type. (*Courtesy of Johnson.*)

Atmospheric pressure

The layer of air that surrounds the earth is several miles deep. The weight of the air above exerts pressure in all directions. This pressure is called *atmospheric pressure*. Atmospheric pressure at sea level is 14.7 psi, which converts to 1.013×10^5 N/m^2.

The instrument used to measure atmospheric pressure is called a *barometer*. Two common barometers are the aneroid barometer and the mercury barometer. The aneroid barometer has a sealed chamber containing a partial vacuum. As the atmospheric pressure increases, the chamber is compressed causing the needle to move. As the atmospheric pressure decreases, the chamber expands, causing the needle to move in the other direction. A dial on the meter is calibrated to indicate the correct pressure.

The mercury barometer has a glass tube about 34-in. long. The tube holds a column of mercury. The height of this column reflects the atmospheric pressure. Standard atmospheric pressure at sea level is indicated by 29.92 in. of mercury, which converts to 759.96 mm.

Gage pressure

Gage pressure is the pressure above or below atmospheric pressure. This is the pressure measured with most gages. A gage that measures both pressure and vacuum is called a *compound gage*. Vacuum is the pressure that is below atmospheric pressure. A gage indicates zero pressure before you start to measure. It does not take the pressure of the atmosphere into account. In the customary system, gage pressure is measured in psi.

Absolute pressure

Absolute pressure is the sum of the gage pressure and atmospheric pressure. This is abbreviated as psia. A good example of this is the pressure in a car tire. This is usually 28 psi, which would be 42.7 psia.

For example:

$$\text{Gage pressure} = 28 \text{ psi}$$

$$\text{Atmospheric pressure} = 14.7 \text{ psi}$$

$$\text{Absolute pressure} = 42.7 \text{ psi}$$

The abbreviation for pounds per square inch gage is psig. The abbreviation for pounds per square inch absolute is psia. Absolute is found by adding 14.7 to the psig value. However, the atmospheric pressure does vary with altitude. In some cases, it is necessary to convert to the atmospheric pressure at the altitude where the pressure is being measured. This small difference can make a tremendous difference in correct readings of psia. To convert psi to kPa, the metric unit of pressure, multiply psi value by 6.9.

Compression ratio

Compression ratio is defined as the absolute head pressure divided by the absolute suction pressure.

$$\text{Compression ratio} = \frac{\text{absolute head pressure}}{\text{absolute suction pressure}}$$

Example 4-1   When the gage reading is zero or above,

$$\text{Absolute head pressure} = \text{gage reading} + 15 \text{ lb (14.7 actually)}$$

$$\text{Absolute suction pressure} = \text{gage reading} + 15 \text{ lb (14.7 actually)}$$

Example 4-2   When the low side reading is in vacuum range,

$$\text{Absolute head pressure} = \text{gage reading} + 15 \text{ lb (14.7 actually)}$$

$$\text{Absolute suction pressure} = \frac{30 - \text{gage reading in inches}}{2}$$

The calculation of compression ratio can be illustrated by the following.

Example 4-1 (continued)

$$\text{Head pressure} = 160 \text{ lb}$$

$$\text{Suction pressure} = 10 \text{ lb}$$

$$\text{Compression ratio} = \frac{\text{absolute head pressure}}{\text{absolute suction pressure}}$$

$$= \frac{160 + 15}{10 + 15} = \frac{175}{25} = 7:1$$

Example 4-2 (continued)

$$\text{Head pressure} = 160\,\text{lb}$$

$$\text{Suction pressure} = 10\,\text{in of vacuum}$$

$$\text{Absolute head pressure} = 160 + 15 = 175\,\text{lb}$$

$$\text{Absolute suction pressure} = \frac{30-10}{2} = \frac{20}{2} = 10$$

$$\text{Compression ratio} = \frac{\text{absolute head pressure}}{\text{absolute suction pressure}}$$

$$= \frac{175}{10} = 17.5:1$$

The preceding examples show the influence of back pressure on the compression ratio. A change in the head pressure does not produce such a dramatic effect. If the head pressure in both cases were 185 lb, the compression ratio in Example 4-1 would be 8:1 and in Example 4-2 it would be 20:1.

A high compression ratio will make a refrigeration system run hot. A system with a very high compression ratio may show a discharge temperature as much as 150°F (65.6°C) above normal. The rate of a chemical reaction approximately doubles with each 18°F (–7.8°C) rise in temperature. Thus, a system running an abnormally high head temperature will develop more problems than a properly adjusted system. The relationship between head pressure and back (suction) pressure, wherever possible, should be well within the accepted industry bounds of a 10:1 compression ratio.

It is interesting to compare, assuming a 175-lb heat pressure in both cases: Refrigerant 12 (R-12) versus Refrigerant 22 (R-22) operating at –35°F (–37°C) coil. At a –35°F (–37°C) coil, as described, the R-22 system would show a 10.9:1 compression ratio while the R-12 system would be at 17.4:1. The R-22 system is a borderline case. The R-12 system is not in the safe range and it would run very hot with all of the accompanying problems.

A number of other factors will produce serious high-temperature conditions. However, high compression ratio alone is enough to cause serious trouble. The thermometer shown in Fig. 4-8 reads temperature as a function of pressure. This device reads the pressure of R-22 and R-12. It also indicates the temperature in degrees Fahrenheit on the outside scale.

Temperature and Heat

The production of excess heat in a system will cause problems. Normally, matter expands when heated. This is the principle of thermal expansion. The linear dimensions increase, as does the volume. Removing heat from a substance causes it to contract in linear dimensions and in volume. This is the principle of the liquid in a glass thermometer.

Figure 4-8 Thermometer and pressure gage. (*Courtesy of Marsh.*)

Temperature is the measure of hotness or coldness on a definite scale. Every substance has temperature.

Molecules are always in motion. They move faster with a temperature increase and more slowly with a temperature decrease. In theory, molecules stop moving at the lowest temperature possible. This temperature is called *absolute zero*. It is approximately –460°F (–273°C).

The amount of heat in a substance is directly related to the amount of molecular motion. The absence of heat would occur only at absolute zero. Above that temperature, there is molecular motion. The amount of molecular motion corresponds to the amount of heat.

The addition of heat causes a temperature increase. The removal of heat causes a temperature decrease. This is true except when matter is going through a change of state.

Heat is often confused with temperature. Temperature is the measurement of heat intensity. It is not a direct measure of heat content. Heat content is not dependent on temperature. Heat content depends on the type of material, the volume of the material, and the amount of heat that has been put into or taken from the material. For example, one cup of coffee at 200°F (93.3°C) contains less heat than 1 gal of coffee at 200°F (93.3°C). The cup at 200°F (93.3°C) can also contain less heat than the gallon at a lower temperature of 180°F (82.2°C).

Specific heat

Every substance has a characteristic called *specific heat*. This is the measure of the temperature change in a substance when a given amount of heat is applied to it.

One Btu (British thermal unit) is the amount of heat required to raise the temperature of 1 lb of water by 1° at 39° F. With a few exceptions, such as ammonia gas and helium, all substances require less heat per pound than water to raise the temperature by 1°F.

Thus, the specific heat scale is based on water, which has a specific heat of 1.0. The specific heat of aluminum is 0.2. This means that 0.2 Btu will raise the temperature of 1 lb of aluminum by 1°F. One Btu will raise the temperature of 5 lb of aluminum by 1°F, or 1 lb of the same by 5°F.

Heat content

Every substance theoretically contains an amount of heat equal to the heat energy required to raise its temperature from absolute zero to its temperature at a given time. This is referred to as *heat content*, which consists of sensible heat and latent heat. Sensible heat can be felt because it changes the temperature of the substance. Latent heat, which is not felt, is seen as it changes the state of substance from solid to liquid or liquid to gas.

Sensible heat.   Heat that changes the temperature of a substance without changing its state when added or removed is called sensible heat. Its effect can be measured with a thermometer in degrees as the difference in temperatures of a substance (delta T, or ΔT).

If the weight and specific heat of a medium are known, the amount of heat added or removed in Btu can be computed by multiplying the sensible change (ΔT) by the weight of the medium and by its specific heat. Thus, the amount of heat required for raising the temperature of 1 gal of water (8.34 lb) from 140°F to 160°F is

$$\text{Sensible heat} = \Delta T \times \text{weight} \times \text{specific heat}$$

$$= (160 - 140) \times 8.34 \times 1 = 20 \times 8.34 = 166.8 \text{ Btu}$$

Latent heat.   The heat required to change the state of a substance without changing its temperature is called its latent heat, or hidden heat. Theoretically, any substance can be a gas, liquid, or solid, depending on its temperature and pressure. It takes heat to change a substance from a solid to a liquid or, from a liquid to a gas.

For example, it takes 144 Btu of latent heat to change 1 lb of ice at 32°F to 1 lb of water at 32°F. It takes 180 Btu of sensible heat to raise the temperature of 1 lb of water 180°F from 32°F to 212°F. It takes 970 Btu of latent heat to change 1 lb of water to steam at 212°F. When the opposite change is effected, equal amounts of heat are taken out or given up by the substance.

This exchange of heat, or the capability of a medium, such as water to take and give up heat, is the basis for most of the heating and air-conditioning industry. Most of the functions of the industry are concerned with adding or removing heat at a central point and distributing the heated or cooled medium throughout a structure to warm or cool the space.

Other sources of heat

Other heat in buildings comes principally from four sources: electrical energy, the sun, the outdoor air temperatures, and the building's occupants. Every kilowatt of electrical energy in use produces 3413 Btu/h, whether it is used in lights, the heating elements of kitchen ranges, toasters, or irons.

The sun is a source of heat. At noon, a square foot of surface directly facing the sun may receive 300 Btu/h, on a clear day. When outdoor air temperatures exceed the indoor space temperature. The outdoors become a source of heat. The amount of heat communicated depends on the size and number of windows, among other factors.

The occupants of a building are a source of heat, since body temperatures are higher than normal room temperatures. An individual seated and at rest will give off about 400 Btu/h in a 74°F (23.3°C) room. If the person becomes active, this amount of heat maybe increased by 2 or 3 times, depending upon the activity involved. Some of this heat is sensible heat, which the body gives off by convection and radiation. The remainder is latent heat, resulting from the evaporation of visible or invisible perspiration. The sensible heat increases the temperature of the room. The latent heat increases the humidity. Both add to the total heat in the room.

Refrigeration Systems

The refrigerator was not manufactured until the 1920s. Before that time, ice was the primary source of refrigeration. A block of ice was kept in the icebox. The icebox was similar to the modern refrigerator in construction. It was well insulated and had shelves to store perishables. The main difference was the method of cooling.

The iceman came about once a week to put a new 50- or 100-lb block of ice in the icebox. How much cooling effect does a 50-lb block of ice produce? The latent heat of melting for 1 lb of ice is 144 Btu. The latent heat of melting for a 50-lb block is 50 × 144, or 7200 Btu. The latent heat of melting for the 100-lb block was 14,400 Btu. The refrigeration was accomplished by convection in the icebox.

One of the first refrigerators is shown in Fig. 4-9. The unit on the top identified it as a refrigerator instead of an icebox. Some of these units, made in the 1920s, are still operating today.

Refrigeration from vaporization
(open system)

The perspiration on your body evaporates and cools your body. Water kept in a porous container is cooled on a hot day. The water seeps from the inside. There is a small amount of water on the outside surface. The surface water is vaporized—it evaporates.

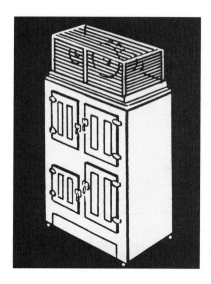

Figure 4-9 Early modification of
the icebox to make it a refrigerator
unit.

Much of the heat required for vaporization comes from the liquid in
the container. When heat is removed this way the liquid is cooled. The
heat is carried away with the vapor.

Basic refrigeration cycle

A substance changes state when the inherent amount of heat is varied.
Ice is water in a solid state and steam is a vapor state of water. A solid
is changed to a liquid and a liquid to a vapor by applying heat. Heat must
be added to vaporize or boil a substance. It must be taken away to
liquefy or solidify a substance. The amount of heat necessary will depend
on the substance and the pressure changes in the substance.

Consider, for example, an open pan of boiling water heated by a gas
flame. The boiling temperature of water at sea level is 212°F (100°C).
Increase the temperature of the flame and the water will boil away
more rapidly, although the temperature of the water will not change. To
heat or boil a substance, heat must be removed from another substance.
In this case, heat is removed from the gas flame. Increasing the temperature of the flame merely speeds the transfer of heat. It does not
increase the temperature of the water.

A change in pressure will affect the boiling point of a substance. As
the altitude increases above sea level, the atmospheric pressure and the
boiling temperature drop. For example, water will boil at 193°F (89.4°C)
at an altitude of 10,000 ft. At pressures below 100 psi, water has a boiling point of 338°F (170°C).

The relationship of pressure to refrigeration is shown in the following
example. A tank contains a substance that is vaporized at atmospheric

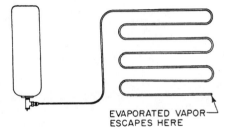

Figure 4-10 Basic step of refrigeration. (*Courtesy of Johnson.*)

EVAPORATED VAPOR
ESCAPES HERE

pressure. However, it condenses to a liquid when 100 lb of pressure is applied. The liquid is discharged from the tank through a hose and nozzle into a long coil of tubing to the atmosphere (see Fig. 4-10).

As the liquid enters the nozzle, its pressure is reduced to that of the atmosphere. This lowers its vaporization or boiling point. Part of the liquid vaporizes or boils using its own heat. The unevaporated liquid is immediately cooled as its heat is taken away. The remaining liquid takes heat from the metal coil or tank and vaporizes, cooling the coil. The coil takes heat from the space around it, cooling the space. This unit would continue to provide cooling or refrigeration for as long as the substance remains under pressure in the tank.

All of the other components of a refrigeration system are merely for reclaiming the refrigeration medium after it has done its job of cooling. The other parts of a refrigeration system, in order of assembly, are tank or liquid receiver, expansion valve, evaporator coil, compressor, and condenser.

Figure 4-11 illustrates a typical refrigeration system cycle. The refrigerant is in a tank or liquid receiver under high pressure and in a liquid state. When the refrigerant enters the expansion valve, the pressure is lowered, and the liquid begins to vaporize. Complete evaporation

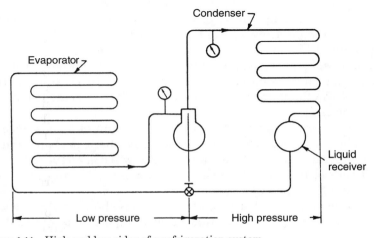

Figure 4-11 High and low sides of a refrigeration system.

takes place when the refrigerant moves into the evaporator coil. With evaporation, heat must be added to the refrigerant. In this case, the heat comes from the evaporator coil. As heat is removed from the coil, the coil is cooled. The refrigerant is now a vapor under low pressure. The evaporator section of the system is often called the low pressure, back pressure, or suction side. The warmer the coil, the more rapidly evaporation takes place and the higher the suction pressure becomes.

The compressor then takes the low-pressure vapor and builds up the pressure sufficiently to condense the refrigerant. This starts the high side of the system. To return the refrigerant to a liquid state (to condense it), heat picked up in the evaporator coil and the compressor must be removed. This is the function of the condenser used with an air- or water-cooled coil. Being cooler than the refrigerant, the air or water absorbs its heat. As it cools, the refrigerant condenses into a liquid and flows into the liquid receiver or tank. Since the pressure of the refrigerant has been increased, it will condense at a lower temperature.

In some systems, the liquid receiver may be part of another unit such as the evaporator or condenser.

Capacity

Refrigeration machines are rated in tons of refrigeration. This rating indicates the size and ability to produce cooling energy in a given period. One ton of refrigeration has cooling energy equal to that produced by 1 ton of ice melting in 24 hours. Since it takes 288,000 Btu of heat to melt 1 ton of ice, a 1-ton machine will absorb 288,000 Btu in a 24-hour period.

Refrigerants

Theoretically, any gas that can be alternately liquefied and vaporized within mechanical equipment can serve as a refrigerant. Thus, carbon dioxide serves as a refrigerant on many ships. However, the piping and machinery handling it must be very heavy-duty.

Practical considerations have led to the use of several refrigerants that can be safely handled at moderate pressures by equipment having reasonable mechanical strength and with lines of normal size and wall thickness. While no substance possesses all the properties of an ideal refrigerant, the hydrocarbon (Freon) refrigerants come quite close.

Refrigerant 12 is made of carbon (C), chlorine (Cl), and fluorine (F). Its formula is CCl_2F_2. It is made of a combination of elements. R-22 is made of carbon (C), hydrogen (H), chlorine (Cl), and fluorine (F). Its formula $CHClF_2$ is slightly different from that of R-12.

Each of these manufactured refrigerants has its own characteristics, such as odor and boiling pressure.

Refrigerants are the vital working fluids in refrigeration systems. They transfer heat from one place to another for cooling air or water in air-conditioning installations.

Many substances can be used as refrigerants, including water under certain conditions. The following are some common refrigerants:

- Ammonia: The oldest commonly used refrigerant, still used in some systems. It is very toxic.

- Sulfur dioxide: First to replace ammonia and to be used in small domestic machines. It is very toxic.

- Refrigerant 12: The first synthetic refrigerant to be used commonly. Used in a large number of reciprocating machines operating in the air-conditioning range. It is nontoxic.

- Refrigerant 22: Used in many of the same applications as R-12. Its lower boiling point and higher latent heat permit the use of smaller compressors and refrigerant lines. It is nontoxic.

- Refrigerant 40: Methyl chloride is used in the commercial refrigeration field, particularly in small installations. Today it is no longer used. It will explode when allowed to combine with air. It is nontoxic.

Refrigerant replacements and the atmosphere

Refrigerants such as ammonia are used for low-temperature systems. These include food and process cooling, ice rinks, and so forth. Propane has been used for some special applications. Now that chlorinated hydrocarbons have been determined to be harmful to the earth's ozone layer, R-11 (CCl_3F), R-12 (CCl_2F_2), and other similar compounds that were in common use along with the less harmful R-22 ($CHClF_2$) have had much attention in the press. Recent international protocols (standards) have set schedules for the elimination of damaging refrigerants from commercial use.

Replacements have been, and are being, developed. Part of the challenge is technical and part is economic. First, to find a fluid that has optimal characteristics and is safe is a challenge. Second, to encourage manufacture to produce and distribute the fluid at an affordable price and in sufficient quantities is another. R-123 ($CHCl_2CF_3$) has been developed as a near-equivalent replacement for R-11, with R-134a (CH_2FCF_3) replacing R-12. R-123 still comes under criticism for having some chlorine in it. R-134a can be bought at auto supplies stores for automobile air conditioners. Most new cars are required to have R-134a in their air-conditioning systems.

R-22 is used widely in residential and commercial air-conditioning scroll compressor systems. It too will be phased out someday (probably during the period 2020–2030). However, finding a suitable, widely accepted replacement has not come as quickly as first thought.

Refrigerants: New and Old

Refrigerants are used in the process of refrigeration. Refrigeration is a process whereby heat is removed from a substance or a space. A refrigerant is a substance that picks up latent heat when the substance evaporates from a liquid to a gas. This is done at a low temperature and pressure. A refrigerant expels latent heat when it condenses from a gas to a liquid at a high pressure and temperature. The refrigerant cools by absorbing heat in one place and discharging it in another area.

Desirable properties of a good refrigerant for commercial use are

- Low boiling point
- Safe nontoxic
- Easy to liquefy and moderate pressure and temperature
- High latent heat value
- Operation on a positive pressure
- Not affected by moisture
- Mixes well with oil
- Noncorrosive to metal

There are other qualities that all refrigerants have. These qualities are molecular weight, density, compression ration, heat value, and temperature of compression. These qualities will vary with the refrigerants. The compressor displacement and compressor type or design will also influence the choice of refrigerant.

Classification of Refrigerants

Refrigerants are classified according to their manner of absorption or extraction of heat from substances to be refrigerated. The classifications can be broken down into class 1, class 2, and class 3.

Class 1 refrigerants are used in the standard compression type of refrigeration systems. Class 2 refrigerants are used as immediate cooling agents between class 1 and the substance to be refrigerated. They do the same work for class 3. Class 3 refrigerants are used in the standard absorption type systems of refrigerating systems.

- Class 1: This class includes those refrigerants that cool by absorption or extraction of heat from the substances to be refrigerated by the absorption of their latent heats. Table 5-1 lists the characteristics of typical refrigerants.

- Class 2: The refrigerants in this class are those that cool substances by absorbing their sensible heats. They are air, calcium chloride brine, sodium chloride (salt) brine, alcohol, and similar nonfreezing solutions.

- Class 3: This group consists of solutions that contain absorbed vapors of liquefiable agents or refrigerating media. These solutions function through their ability to carry the liquefiable vapors. The vapors produce a cooling effect by the absorption of their latent heat. An example is aqua ammonia, which is a solution composed of distilled water and pure ammonia.

TABLE 5-1   Characteristics of Typical Refrigerants

Name	Boiling point (°F)	Heat of vaporization at boiling point Btu/lb. 1 atm
Sulfur dioxide	14.0	172.3
Methyl chloride	−10.6	177.8
Ethyl chloride	55.6	177.0
Ammonia	−28.0	554.7
Carbon dioxide	−110.5	116.0
Freezol (isobutane)	10.0	173.5
Freon 11	74.8	78.31
Freon 12	−21.7	71.04
Freon 13	−114.6	63.85
Freon 21	48.0	104.15
Freon 22	−41.4	100.45
Freon 113	117.6	63.12
Freon 114	38.4	58.53
Freon 115	−37.7	54.20
Freon 502	−50.1	76.46

Common refrigerants

Following are some of the more common refrigerants. Table 5-1 summarizes the characteristics to a selected few of the many refrigerants available for home, commercial and industrial use.

Sulfur dioxide. Sulfur dioxide (SO_2) is a colorless gas or liquid. It is toxic, with a very pungent odor. When sulfur is burned in air, sulfur dioxide is formed. When sulfur dioxide combines with water it produces sulfuric and sulfurous acids. These acids are very corrosive to metal. They have an adverse effect on most materials. Sulfur dioxide is not considered a safe refrigerant. Sulfur dioxide is not considered safe when used in large quantities. As a refrigerant, sulfur dioxide operates on a vacuum to give the temperatures required. Moisture in the air will be drawn into the system when a leak occurs. This means the metal parts will eventually corrode, causing the compressor to seize.

Sulfur dioxide (SO_2) boils at 14°F (–10°C) and has a heat of vaporization at boiling point (1 atm) of 172.3 Btu/lb. It has a latent heat value of 166 Btu/lb.

To produce the same amount of refrigeration, sulfur dioxide requires about one-third more vapor than Freon and methyl chloride. This means the condensing unit has to operate at a higher speed or the compressor cylinders must be larger. Since sulfur dioxide does not mix well with oil the suction line must be on a steady slant to the machine. Otherwise, the oil will trap out, constricting the suction line. This refrigerant is not feasible for use in some locations.

Methyl chloride. Methyl chloride (CHP) has a boiling point of –10.6°F (–23.3°C). It also has heat of vaporization at boiling point (at 1 atm) of 177.8 Btu/lb. It is a good refrigerant. However, because it will burn under some conditions some cities will not allow it to be used. It is easy to liquefy and has a comparatively high latent heat value. It does not corrode metal when in its dry state.

However, in the presence of moisture it damages the compressor. A sticky black sludge is formed when excess moisture combines with the chemical. Methyl chloride mixes well with oil. It will operate on a positive pressure as low as –10°F (–23°C). The amount of vapor needed to cause discomfort in a person is in proportion to the following numbers:

Carbon dioxide	100
Methyl chloride	70
Ammonia	2
Sulfur dioxide	1

That means methyl chloride is 35 times safer than ammonia and 70 times safer than sulfur dioxide.

Methyl chloride is hard to detect with the nose or eyes. It does not produce irritating effects. Therefore, some manufacturers add a 1 percent amount of *acrolein* as a colorless liquid with a pungent odor as a warning agent. It is produced by destructive distillation of fats.

Ammonia.   Ammonia (NH_3) is used most frequently in large industrial plants. Freezers for packing houses usually employ ammonia as a refrigerant. It is a gas with a very noticeable odor. Even a small leak can be detected with the nose. Its boiling point at normal atmospheric pressure is $-28°F$ ($-33°C$). Its freezing point is $-107.86°F$ ($-77.7°C$). It is very soluble in water. Large refrigeration capacity is possible with small machines. It has high latent heat [555 Btu at $18°F$ ($-7.7°C$)]. It can be used with steel fittings. Water-cooled units are commonly used to cool down the refrigerant. High pressures are used in the lines (125 to 200 psi). Anyone inside the refrigeration unit when it springs a leak is rapidly overcome by the fumes. Fresh air is necessary to reduce the toxic effects of ammonia fumes. Ammonia is combustible when combined with certain amounts of air (about one volume of ammonia to two volumes of air). It is even more combustible when combined with oxygen. It is very toxic. Heavy steel fittings are required since pressures of 125 to 200 psi are common. The units must be water cooled.

Carbon dioxide.   Carbon dioxide (CO_2) is a colorless gas at ordinary temperatures. It has a slight odor and an acid taste. Carbon dioxide is nonexplosive and nonflammable. It has a boiling point of $5°F$ ($-15°C$). A pressure of over 300 psi is required to keep it from evaporation. To liquefy the gas, a condenser temperature of $80°F$ ($26.6°C$) and a pressure of approximately 1000 psi are needed. Its critical temperature is $87.8°F$ ($31°C$). It is harmless to breathe except in extremely large concentrations. The lack of oxygen can cause suffocation under certain conditions of carbon dioxide concentration.

Carbon dioxide is used aboard ships and in industrial installations. It is not used in household applications. The main advantage of using carbon dioxide for a refrigerant is that a small compressor can be used. The compressor is very small since a high pressure is required for the refrigerant. Carbon dioxide is, however, very inefficient, compared to other refrigerants. Thus, it is not used in household units.

Calcium chloride.   Calcium chloride ($CaCl_2$) is used only in commercial refrigeration plants. Calcium chloride is used as a simple carrying medium for refrigeration.

Brine systems are used in large installations where there is danger of leakage. They are used also where the temperature fluctuates in the space to be refrigerated. Brine is cooled down by the direct expansion of the refrigerant. It is then pumped through the material or space to be cooled. Here, it absorbs sensible heat.

Most modern plants operate with the brine at low temperature. This permits the use of less brine, less piping or smaller diameter pipe, and smaller pumps. It also lowers pumping costs. Instead of cooling a large volume of brine to a given temperature, the same number of refrigeration units are used to cool a smaller volume of brine to a lower temperature. This results in greater economy. The use of extremely low-freezing brine, such as calcium chloride, is desirable in the case of the shell-type cooler. Salt brine with a minimum possible freezing point of –6°F (–21°C) may solidify under excess vacuum on the cold side of the refrigerating unit. This can cause considerable damage and loss of operating time. There are some cases, in which the cooler has been ruined.

Ethyl chloride.   Ethyl chloride (C_2H_5Cl) is not commonly used in domestic refrigeration units. It is similar to methyl chloride in many ways. It has a boiling point of 55.6°F (13.1°C) at atmospheric pressure. Critical temperature is 360.5°F (182.5°C) at a pressure of 784 lb absolute. It is a colorless liquid or gas with a pungent ethereal odor and a sweetish taste. It is neutral toward all metals. This means that iron, copper, and even tin and lead can be used in the construction of the refrigeration unit. It does, however, soften all rubber compounds and gasket material. Thus, it is best to use only lead for gaskets.

Freon Refrigerants

The Freon refrigerants have been one of the major factors responsible for the tremendous growth of the home refrigeration and air-conditioning industries. The safe properties of these products have permitted their use under conditions where flammable or more toxic refrigerants would be hazardous to use. There is a Freon refrigerant for every application—from home and industrial air-conditioning to special low-temperature requirements.

The unusual combination of properties found in the Freon compounds is the basis for the wide application and usefulness. Table 5-2 presents a summary of the specific properties of some of the fluorinated products. Figure 5-1 gives the absolute pressure and gage pressure of Freon refrigerants at various temperatures.

Molecular weights

Compounds containing fluorine in place of hydrogen have higher molecular weights and often have unusually low boiling points. For example, methane (CH_4) with a molecular weight of 16 has a boiling point of –258.5°F (–161.4°C) and is nonflammable. Freon 14 (CF_4) has

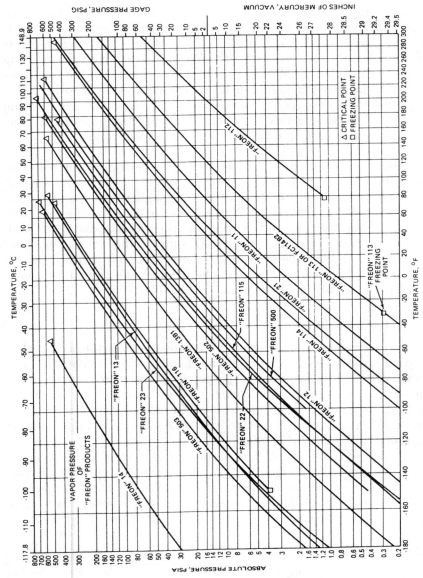

Figure 5-1 The absolute and gage pressures of Freon refrigerants.

TABLE 5-2 Physical Properties of Freon Products

		Freon 11	Freon 12	Freon 13	Freon 13B1	Freon 14
Chemical formula		CCl_3F	CCl_2F_2	$CClF_3$	$CBrF_3$	CF_4
Molecular weight		137.37	120.92	104.46	148.92	88.00
Boiling point at 1 atm	°C	23.82	−29.79	−81.4	−57.75	−127.96
	°F	74.87	−21.62	−114.6	−71.95	−198.32
Freezing point	°C	−111	−158	−181	−168	−184[2]
	°F	−168	−252	−294	−270	−299
Critical temperature	°C	198.0	112.0	28.9	67.0	−45.67
	°F	388.4	233.6	83.9	152.6	−50.2
Critical pressure	atm	43.5	40.6	38.2	39.1	36.96
	lb/sq in abs	639.5	596.9	561	575	543.2
Critical volume	cc/mol	247	217	181	200	141
	cu ft/lb	0.0289	0.0287	0.0277	0.0215	0.0256
Critical density	g/cc	0.554	0.588	0.578	0.745	0.626
	lb/cu ft	34.6	34.8	36.1	46.5	39.06
Density, liquid at 25°C (77°F)	g/cc	1.476	1.311	1.298@ −30°C (−22°F) 81.05	1.538	1.317@ −80°C (−112°F) 82.21
	lb/cu ft	92.14	81.84		96.01	
Density, sat'd vapor at boiling point	g/L	5.86	6.33	7.01	8.71	7.62
	lb/cu ft	0.367	0.395	0.438	0.544	0.476
Specific heat, liquid (Heat capacity) at 25°C (77°F)	cal/(g)(°C) or Btu/(lb)(°F)	0.208	0.232	0.247@ −30°C (−22°F)	0.208	0.294@ −80°C (−112°F)

TABLE 5-2 Physical Properties of Freon Products (Continued)

		Freon 11	Freon 12	Freon 13	Freon 13B1	Freon 14
Specific heat, vapor, at const pressure (1 atm) at 25°C (77°F)	cal/(g)(°C) or Btu/(lb)(°F)	0.142 @ 38°C (100°F)	0.145	0.158	0.112	0.169
Specific heat ratio at 25°C and 1 atm	C_p/C_v	1.137 @ 38°C (100°F)	1.137	1.145	1.144	1.159
Heat of vaporization at boiling point	cal/g Btu/lb	43.10 77.51	39.47 71.04	35.47 63.85	28.38 51.08	32.49 58.48
Thermal conductivity at 25°C (77°F) Btu/(hr)(ft)(°F) liquid vapor (1 atm)		0.0506 0.00451	0.0405 0.00557	0.0378 0.00501 } @ (−30°C) (−22°F.)	0.0234 0.00534	0.0361 0.00463 } @ −80°C (−112°F)
Viscosity at 25°C (77°F) liquid vapor (1 atm)	centipoise centipoise	0.415 0.0107	0.214 0.0123	0.170 0.0119 } @ (−30°C) (−22°F)	0.157 0.0154	0.23 0.0116 } @ (−80°C) (−112°F)
Surface tension at 25°C (77°F) dynes/cm		18	9	14@ −73°C −100°F	4	4@ −73°C (−100°F)
Refractive index of liquid at 25°C (77°F)		1.374	1.287	1.199@ −73°C (−100°F)	1.238	1.151@ −73°C (−100°F)
Relative dielectric strength at 1 atm and 25°C (77°F) (nitrogen = 1)		3.71	2.46	1.65	1.83	1.06

		Freon 21	Freon 22	Freon 23	Freon 112	Freon 113	Freon 114
Dielectric constant liquid		2.28 @ 29°C	2.13} @ 29°C (84°F)				
vapor (1 atm)		1.0036 @ 24°C	1.0032	1.0024 29°C @(84°F)		1.0012 @ 24.5°C (76°F)	
Solubility of "Freon" in water at 1 atm and 25°C (77°F)	wt %	0.11	0.028	0.009	0.03		0.0015
Solubility of water in "Freon" at 25°C (77°F)	wt %	0.011	0.009		0.0095 21°C (70°F)		
Toxicity		Group 5a	Group 6	Probably Group 6	Probably Group 6	Group 6	Probably Group 6
Chemical formula		$CHCl_2F$	$CHClF_2$	CHF_3	CCl_2F-CCl_2F	$CCl_2F-CClF_2$	$CClF_2-CClF_2$
Molecular weight		102.93	86.47	70.01	203.84	187.38	170.93
Boiling point at 1 atm	°C	8.92	-40.75	-82.03	92.8	47.57	3.77
	°F	48.06	-41.36	-115.66	199.0	117.63	38.78
Freezing point	°C	-135	-160	-155.2	26	-35	-94
	°F	-211	-256	-247.4	79	-31	-137
Critical temperature	°C	178.5	96.0	25.9	278	214.1	145.7
	°F	353.3	204.8	78.6	532	417.4	294.3
Critical pressure	atm	51.0	49.12	47.7	34^3	33.7	32.2
	lb/sq in abs	750	721.9	701.4	500	495	473.2
Critical volume	cc/mol	197	165	133	370^3	325	293
	cu ft/lb	0.0307	0.0305	0.0305	0.029	0.0278	0.0275

TABLE 5-2 Physical Properties of Freon Products (Continued)

Property		Freon 21	Freon 22	Freon 23	Freon 112	Freon 113	Freon 114
Critical density	g/cc	0.522	0.525	0.525	0.55	0.576	0.582
	lbs/cu ft	32.6	32.76	32.78	34	36.0	36.32
Density, liquid	g/cc	1.366	1.194	0.670	1.634} @ 30°C	1.565	1.456
at 25°C (77°F)	lbs/cu ft	85.28	74.53	41.82	102.1} (86°F)	97.69	90.91
Density, sat'd vapor	g/L	4.57	4.72	4.66	7.02	7.38	7.83
at Boiling Point	lbs/cu ft	0.285	0.295	0.291	0.438	0.461	0.489
Specific heat, liquid (heat capacity) at 25°C (77°F)	cal/(g)(°C) or Btu/(lb)(°F)	0.256	0.300	0.345 @ −30°C −22°F		0.218	0.243
Specific heat, vapor, at const pressure (1 atm) at 25°C (77°F)	cal/(g)(°C) or Btu/(lb)(°F)	0.140	0.157	0.176		0.161 @ 60°C (140°F)	0.170
Specific-heat ratio at 25°C and 1 atm	C_p/C_v	1.175	1.184	1.191 @ 0 pressure		1.080 @ 60°C (140°F)	1.084
Heat of vaporization	cal/g	57.86	55.81	57.23	37 (est)	35.07	32.51
at boiling point	Btu/b	104.15	100.45	103.02	67	63.12	58.53
Thermal conductivity[1] at 25°C (77°F) Btu/(hr) (ft) (°F) liquid		0.0592	0.0507	0.0569} @ −30°C		0.0434	0.0372
vapor (1 atm)		0.00506	0.00609	0.0060} (−22°F.)	0.040	0.0044 (0.5 atm)	0.0060

		(1)	(2)	(3)	(4)	(5)	(6)
Viscosity[1] at 25°C (77°F) liquid	centipoise	0.313	0.198	0.167 } @ -30°C	1.21	0.68	0.36
vapor (1 atm)	centipose	0.0114	0.0127	0.0118 } (-22°F.)		0.010 (0.1 atm)	0.0112
Surface tension at 25°C (77°F) dynes/cm		18	8	15 @ -73°C (-100°F)	23 @ 30°C (86°F)	17.3	12
Refractive index of liquid at 25°C (77°F)		1.354	1.256	1.215 @ -73°C (-100°F)	1.413	1.354	1.288
Relative-dielectric strength at 1 atm and 25°C (77°F) (nitrogen = 1)		1.85	1.27	1.04	5 (est)	3.9 (0.44 atm)	3.34
Dielectric constant liquid		5.34 @ 28°C	6.11 @ 24°C		2.54 @ 25°C (77°F)	2.41 @ 25°C (77°F)	2.26 @ 25°C
Vapor (1 atm)		1.0070 @ 30°C	1.0071 @ 25.4°C	1.0073 @ 25°C			1.0043 @26.8°C
Solubility of "Freon" in water at 1 atm and 25°C (77°F) wt %		0.95	0.30	0.10	0.012 (Sat'n Pres)	0.017 (Sat'n Pres)	0.013
Solubility of water in "Freon" at 25°C (77°F) wt %		0.13	0.13			0.011	0.009
Toxicity		much less than Group 4, somewhat more than Group 5[12]	Group 5a	probably Group 6	probably less than Group 4, more than Group 5	much less than Group 4, somewhat more than Group 5	Group 6

TABLE 5-2 Physical Properties of Freon Products (Continued)

FC 114B2	Freon 115	Freon 116	Freon 500	Freon 502	Freon 503
$CBrF_2-CBrF_2$	$CClF_2-CF_3$	CF_3-CF_3	a	b	c
259.85	154.47	138.01	99.31	111.64	87.28
47.26	-39.1	-78.2	-33.5	-45.42	-87.9
117.06	-38.4	-108.8	-28.3	-49.76	-126.2
-110.5	-106	-100.6	-159		
-166.8	-159	-149.1	-254		
214.5	80.0	19.7[4]	105.5	82.2	19.5
418.1	175.9	67.5	221.9	179.9	67.1
34.4	30.8	29.4[4]	43.67	40.2	43.0
506.1	453	432	641.9	591.0	632.2
329	259	225	200.0	199	155
0.0203	0.0269	0.0262	0.03226	0.02857	0.0284
0.790	0.596	0.612	0.4966	0.561	0.564
49.32	37.2	38.21	31.0	35.0	35.21
2.163 / 135.0	1.291 / 80.60	1.587 / 99.08 @ -73°C (-100°F)	1.156 / 72.16	1.217 / 75.95	1.233 / 76.95 @ -30°C (-22°F)
8.37	8.37	9.01	5.278	6.22	6.02
0.522	0.522	0.562	0.3295	0.388	0.374

0.166	0.285	0.232 @ $-73°C$ $(-100°F)$	0.258	0.293	0.287 @ $-30°C$ $(-22°F)$
	0.164	0.182 @ 0 pressure 1.085 (est) @ 0 pressure	0.175	0.164	0.16
	1.091	1.085 (est) @ 0 pressure	1.143	1.132	1.21 @ $-34°C$ $(-30°F)$
25 (est)	30.11	27.97	48.04	41.21	42.86
45 (est)	54.20	50.35	86.47	74.18	77.15
0.027	0.0302	0.045 } @ $-73°C$	0.0432	0.0373	0.0430 @ $-30°C$ $(-22°F)$
	0.00724	0.0098 } $(-100°F)$		0.00670	
0.72	0.193	0.30	0.192	0.180	0.144 @ $-30°C$ $(-22°F)$
	0.0125	0.0148	0.0120	0.0126	
18	5	16 @ $-73°C$ $(-100°F)$	8.4	5.9	6.1 @ $-30°C$ $(-22°F)$
1.367	1.214	1.206 @ $-73°C$ $(-100°F)$	1.273	1.234	1.209 @ $-30°C$ $(-22°F)$
4.02 (0.44 atm)	2.54	2.02		1.3	
2.34 @ 25°C $(77°F)$	1.0035 @ 27.4°C	1.0021 @ 23°C $(73°F)$		6.11 @ 25°C 1.0035 (0.5 atm)	
	0.006				
Group 5a	Group 6	probably Group 6	0.056 Group 5a	0.056 Group 5a	0.042 probably Group 6

FREON is Du pont's registered trademark for its fluorocarbon products

a. CCl_2F_2/CH_3CHF_2 (73.8/26.2% by wt.)
b. $CHClF_2/CClF_2CF_3$ (48.8/51.2% by wt.)
c. $CHF_3/CClF_3$ (40/60% by wt.)

171

a molecular weight of 88 and a boiling point of –198.4°F (–128°C) and is nonflammable. The effect is even more pronounced when chlorine is also present. Methylene chloride (CH_2Cl_2) has a molecular weight of 85 and boils at 105.2°F (40.7°C) while Freon 12 (CCl_2F_2) with a molecular weight of 121 boils at –21.6°F (–29.8°C). It can be seen that Freon compounds are high-density materials with low boiling points, low viscosity, and low surface tension. Freon includes products with boiling points covering a wide range of temperatures. See Table 5-3.

The high molecular weight of the Freon compounds also contributes to low vapor, specific heat values, and fairly low latent heats of vaporization. Tables of thermodynamic properties including enthalpy, entropy, pressure, density, and volume for the liquid and vapor are available from manufacturers.

TABLE 5-3   Fluorinated Products and Their Molecular Weight and Boiling Point

Freon Products

Product	Formula	Molecular weight	Boiling point °F	Boiling point °C
Freon 14	CF_4	88.0	–198.3	–128.0
Freon 503	$CHF_3/CClF_3$	87.3	–127.6	–88.7
Freon 23	CHF_3	70.0	–115.7	–82.0
Freon 13	$CClF_3$	104.5	–114.6	–81.4
Freon 116	$CF_3–CF_3$	138.0	–108.8	–78.2
Freon 13B1	$CBrF_3$	148.9	–72.0	–57.8
Freon 502	$CHClF_2/CClF_2–CF_3$	111.6	–49.8	–45.4
Freon 22	$CHClF_2$	86.5	–41.4	–40.8
Freon 115	$CClF_2–CF_3$	154.5	–37.7	–38.7
Freon 500	CCl_2F_2/CH_3CHF_2	99.3	–28.3	–33.5
Freon 12	CCl_2F_2	120.9	–21.6	–29.8
Freon 114	$CClF_2–CClF_2$	170.9	38.8	3.8
Freon 21	$CHCl_2F$	102.9	48.1	8.9
Freon 11	CCl_3F	137.4	74.9	23.8
Freon 113	$CCl_2F–CClF_2$	187.4	117.6	47.6
Freon 112	$CCl_2F–CCl_2F$	203.9	199.0	92.8

Other Fluorinated Compounds

FC 114B2	$CBrF_2–CBrF_2$	259.9	117.1	47.3
1,1-Difluoroethane*	$CH_3–CHF_2$	66.1	–13.0	–25.0
1,1,1-Chlorodifluoroethane[†]	$CH_3–CClF_2$	100.5	14.5	–9.7
Vinyl fluoride	$CH_2=CHF$	46.0	–97.5	–72.0
Vinylidene fluoride	$CH_2=CF_2$	64.0	–122.3	–85.7
Hexafluoroacetone	CF_3COCF_3	166.0	–18.4	–28.0
Hexafluoroisopropanol	$(CF_3)_2CHOH$	168.1	136.8	58.2

*Propellant or refrigerant 152a
[†]Propellant or refrigerant 142b

Freon compounds are poor conductors of electricity. In general, they have good dielectric properties.

Flammability

None of the Freon compounds are flammable or explosive. However, mixtures with flammable liquids or gases may be flammable and should be handled with caution. Partially halogenated compounds may also be flammable and must be individually examined.

Toxicity

Toxicity means intoxicating or poisonous. One of the most important qualities of the Freon fluorocarbon compounds is their low toxicity under normal conditions of handling and usage. However, the possibility of serious injury or death exists under unusual or uncontrolled exposures or in deliberate abuse by inhalation of concentrated vapors. The potential hazards of fluorocarbons are summarized in Table 5-4.

Skin effects

Liquid fluorocarbons with boiling points below 32°F (0°C) may freeze the skin, causing frostbite on contact. Suitable protective gloves and clothing give insulation protection. Eye protection should be used. In the event of frostbite, warm the affected area quickly to body temperature. Eyes should be flushed copiously with water. Hands may be held under armpits or immersed in warm water. Get medical attention immediately. Fluorocarbons with boiling points at or above ambient temperature tend to dissolve protective fat from the skin. This leads to skin dryness and irritation, particularly after prolonged or repeated contact. Such contact should be avoided by using rubber gloves or plastic gloves. Eye protection and face shields should be used if splashing is possible. If irritation occurs following contact, seek medical attention.

Oral toxicity

Fluorocarbons are low in oral toxicity as judged by single-dose administration or repeated dosing over long periods. However, direct contact of liquid fluorocarbons with lung tissue can result in chemical pneumonitis, pulmonary edema, and hemorrhage. Fluorocarbons 11 and 113, like many petroleum distillates, are fat solvents and can produce such an effect. If products containing these fluorocarbons were accidentally or purposely ingested, induction of vomiting would be contraindicated (medically wrong), In other words, *do NOT induce vomiting*.

TABLE 5-4 Potential Hazards of Fluorocarbons

Condition	Potential hazard	Safeguard
Vapors may decompose in flames or in contact with hot surfaces.	Inhalation of toxic decomposition products.	Good ventilation. Toxic decomposition products serve as warning agents.
Vapors are four to five times heavier than air. High concentrations may tend to accumulate in low places.	Inhalation of concentrated vapors can be fatal.	Avoid misuse. Forced-air ventilation at the level of vapor concentration. Individual breathing devices with air supply. Lifelines when entering tanks or other confined areas.
Deliberate inhalation to produce intoxication.	Can be fatal.	Do not administer epinephrine or other similar drugs.
Some fluorocarbon liquids tend to remove natural oils from the skin.	Irritation of dry, sensitive skin.	Gloves and protective clothing.
Lower boiling liquids may be splashed on skin.	Freezing.	Gloves and protective clothing.
Liquids may be splashed into eyes.	Lower boiling liquids may cause freezing. Higher boiling liquids may cause temporary irritation and if other chemicals are dissolved, may cause serious damage.	Wear eye protection. Get medical attention. Flush eyes for several minutes with running water.
Contact with highly reactive metals.	Violent explosion may occur.	Test the proposed system and take appropriate safety precautions.

Central nervous system (CNS) effects

Inhalation of concentrated fluorocarbon vapors can lead to CNS effects comparable to the effects of general anesthesia. The first symptom is a feeling of intoxication. This is followed by a loss of coordination and unconsciousness. Under severe conditions, death can result. If these symptoms are felt, the exposed individual should immediately go or be moved to fresh air. Medical attention should be sought promptly. *Individuals exposed to fluorocarbons should NOT be treated with adrenalin (epinephrine).*

Cardiac sensitization

Fluorocarbons can, in sufficient vapor concentration, produce cardiac sensitization. This is a sensitization of the heart to adrenaline brought about by exposure to high concentrations of organic vapors. Under severe exposure, cardiac arrhythmias may result from sensitization of the heart to the body's own levels of adrenaline. This is particularly so under conditions of emotional or physical stress, fright, panic, and so forth. Such cardiac arrhythmias may result in ventricular fibrillation and death. Exposed individuals should immediately go or be removed to fresh air. There, the hazard of cardiac effects will rapidly decrease. Prompt medical attention and observation should be provided following accidental exposures. A worker adversely affected by fluorocarbon vapors should *not* be treated with adrenalin (epinephrine) or similar heart stimulants since these would increase the risk of cardiac arrhythmias.

Thermal decomposition

Fluorocarbons decompose when exposed directly to high temperatures. Flames and electrical resistance heaters, for example, will chemically decompose fluorocarbon vapors. Products of this decomposition in air include halogens and halogen acids (hydrochloric, hydrofluoric, and hydrobromic), as well as other irritating compounds. Although much more toxic than the parent fluorocarbon, these decomposition products tend to irritate the nose, eyes, and upper respiratory system. This provides a warning of their presence. The practical hazard is relatively slight. It is difficult for a person to remain voluntarily in the presence of decomposition products at concentrations where physiological damage occurs.

When such irritating decomposition products are detected, the area should be evacuated and ventilated. The source of the problem should be corrected.

Applications of Freon Refrigerants

There is a Freon refrigerant for every application from home and industrial air-conditioning to special low-temperature requirements. Following are a few of the Freon refrigerants.

Freon 11 (CCl$_3$F) has a boiling point of 74.9°F (23.8°C) and is widely used in centrifugal compressors for industrial and commercial air-conditioning systems, It is also used for industrial process water and brine cooling. Its low viscosity and freezing point have also led to its use as a low-temperature brine.

Freon 12 (CCl$_2$F$_2$) has a boiling point of −21.6°F (−29.8°C) and is one of the most widely known and used Freon refrigerants. It is used principally in household and commercial refrigeration and air-conditioning. It is used for refrigerators, frozen food locker plants, water coolers, room and window air-conditioning units and similar equipment. It is generally used in reciprocating compressors ranging in size from fractional to 800 hp. It is also used in the smaller rotary-type compressors (see Fig. 5-2).

Freon 13 (CClF$_3$) has a boiling point of −114.6°F (−81.4°C) and is used in low-temperature specialty applications using reciprocating compressors and generally in cascade with Freon 12, Freon 22, or Freon 522.

Freon 22 (CHClF$_2$) has a boiling point of −41.4°F (−40.8°C) and is used in all types of household and commercial refrigeration and air-conditioning applications with reciprocating compressors. The outstanding thermodynamic properties of Freon 22 permit the use of

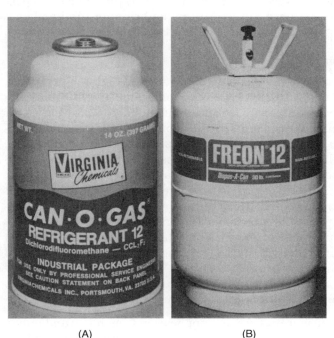

(A)   (B)

Figure 5-2 Freon can be purchased in a number of sizes. (*Courtesy of Virginia Chemical.*)

Figure 5-3 Freon 22 is marketed in containers of various sizes, such as 1-lb, 2-lb, and 15-lb cans. (*Courtesy of Virginia Chemical.*)

smaller equipment than is possible with similar refrigerants. This makes it especially attractive for uses where size is a problem (see Fig. 5-3).

Freon 113 (CCl$_2$F CClF$_2$) has a boiling point of 117.6°F (47.6°C). It is used in commercial and industrial air-conditioning and process water and brine cooling with centrifugal compression. It is especially useful in small tonnage applications.

Freon 114 (CClF$_2$ CClF$_2$) has a boiling point of 38.8°F (3.8°C). It is used in small refrigeration systems with rotary-type compressors. It is used in large industrial process cooling and air-conditioning systems using multistage centrifugal compressors.

Freon 500 (CCl$_2$F$_2$) is an azeotropic mixture. *Azeotropic* means that a mixture is liquid, maintains a constant boiling point, and produces a vapor of the same composition as the mixture with CH$_3$CHF$_2$. It is composed of 73.8 percent Freon 12 (CCl$_2$F$_2$) and 26.2 percent CH$_3$CHF$_2$. It boils at −28.3°F (−33.5°C). It is used in home and commercial air-conditioning in small and medium-size equipment and in some refrigeration applications.

Figure 5-4 Pressure gage for R-12, R-22, and R-502. (*Courtesy of Marsh.*)

Freon 502 is an azeotropic mixture also. It consists of 48.8 percent of Freon 22 and 51.2 percent of Freon 115, by weight. It boils at −49.72°F (−45.4°C). With Freon 502, refrigeration capacity is greater than with Freon 22. Note the pressure differences on the pressure gage in Fig. 5-4. Discharge temperatures are comparable to those found with Freon 12. It is finding new applications in low and medium-temperature cabinets for the display and storage of foodstuffs, in food freezing, and in heat pumps.

Freon 503 is an azeotropic mixture of CHF_3 and $CClF_3$. The weight ratio is 40 percent CHF_3 and 60 percent $CClF_3$. The boiling point of this mixture is −127.6°F (−88.7°C). It is used in low-temperature cascade systems.

Freon 13B1 $CBrF_3$ boils at −72°F (−57.8°C). It serves the temperature range between Freon 502 and Freon 13.

These are some of the refrigerants that are now under close scrutiny because of their chlorine content and their effect on the environment. Some have been banned and cannot be manufactured anywhere in the world. Others are being phased out gradually and replaced by a new combination of chemicals.

Reaction of Freon to Various Materials Found in Refrigeration Systems

Metals

Most of the commonly used construction metals, such as steel, cast iron, brass, copper, tin, lead, and aluminum, can be used satisfactorily with the Freon compounds under normal conditions of use. At high temperatures some of the metals may act as catalysts for the breakdown of the compound. The tendency of metals to promote thermal decomposition of the Freon compounds is in the following general order. Those metals that least promote thermal decomposition are listed first.

- Inconel
- Stainless steel
- Nickel   '
- 1340 steel
- Aluminum
- Copper
- Bronze
- Brass
- Silver

The above order is only approximate. Exceptions may be found for individual Freon compounds or for special conditions of use.

Magnesium alloys and aluminum containing more than 2 percent magnesium are not recommended for use in systems containing Freon compounds where water may be present. Zinc is not recommended for use with Freon 113. Experience with zinc and other Freon compounds has been limited and no unusual reactivity has been observed. However, it is more chemically reactive than other common construction metals. Thus, it would seem wise to avoid its use with the Freon compounds unless adequate testing is carried out.

Some metals may be questionable for use in applications requiring contact with Freon compounds for long periods of time or unusual conditions of exposure. These metals, however, can be cleaned safely with Freon solvents. Cleaning applications are usually for short exposures at moderate temperatures.

Most halocarbons may react violently with highly reactive materials, such as sodium, potassium, and barium in their free metallic form. Materials become more reactive when finely grinded or powdered. In this state, magnesium and aluminum may react with fluorocarbons, especially at higher temperatures. Highly reactive materials should not be brought into contact with fluorocarbons until a careful study is made and appropriate safety precautions are taken.

Plastics

A brief summary of the effect of Freon compounds on various plastic materials follows. However, compatibility should be tested for specific applications. Differences in polymer structure and molecular weight, plasticizers, temperature, and pressure may alter the resistance of the plastic toward the Freon compound.

Teflon TFE-fluorocarbon resin: No swelling observed when submerged in Freon liquids, but some diffusion found with Freon 12 and Freon 22.

Polychlorotrifluoroethylene: Slight swelling, but generally suitable for use with Freon compounds.

Polyvinyl alcohol: Not affected by the Freon compounds, but very sensitive to water. Used especially in tubing with an outer protective coating.

Vinyl: Resistance to the Freon compounds depends on vinyl type and plasticizer. Considerable variation is found. Samples should be tested before use.

Orlon acrylic fiber: Generally suitable for use with the Freon compounds.

Nylon: Generally suitable for use with Freon compounds, but may tend to become brittle at high temperatures in the presence of air or water. Tests at 250°F (121°C) with Freon 12 and Freon 22 showed the presence of water or alcohol to be undesirable. Adequate testing should be carried out.

Polyethylene: May be suitable for some applications at room temperatures. However, it should be thoroughly tested since greatly different results have been found with different samples.

Lucite acrylic resin (methacrylate polymers): Dissolved by Freon 22. However, it is generally suitable for use with Freon 12 and Freon 114 for short exposure. On long exposure, it tends to crack, craze, and become cloudy. Use with Freon 113 may be questionable. It probably should not be used with Freon 11.

Cast Lucite acrylic resin: It is much more resistant to the effect of solvents than extruded resin. It can probably be used with most of the Freon compounds.

Polystyrene: Considerable variation found in individual samples. However, it is generally not suited for use with Freon compounds. Some applications might be all right with Freon 114.

Phenolic resins: Usually not affected by the Freon compounds. However, composition of resins of this type may be quite different. Samples should be tested before use.

Epoxy resins: Resistant to most solvents and entirely suitable for use with the Freon compounds.

Cellulose acetate or nitrate: Suitable for use with Freon compounds.

Delrin-acetal resin: Suitable for use with Freon compounds under most conditions.

Elastomers: Considerable variation is found in the effect of the Freon compounds on elastomers. The effect depends on the particular compound and elastomer type. In nearly all cases a satisfactory combination can be found. In some instances the presence of other materials, such as oils, may give unexpected results. Thus, preliminary testing of the system involved is recommended.

Refrigerant Properties

Refrigerants can be characterized by a number of properties. These properties are pressure, temperature, volume, density, and enthalpy. Also, flammability, ability to mix with oil, moisture reaction, odor, toxicity, leakage tendency, and leakage detection are important properties that characterize refrigerants.

Freon refrigerants R-11, R-12, R-22, plus ammonia and water will be used to show their properties in relationship to the above mentioned categories. Freon R-11, R-12, and R-22 are common Freon refrigerants. The number assigned to ammonia is R-717, while water has the number R-718.

Pressure

The pressure of a refrigeration system is important. It determines how sturdy the equipment must be to hold the refrigerant. The refrigerant must be compressed and sent to various parts of the system under pressure. The main concern is keeping the pressure as low as possible. The ideal low-side pressure or evaporating, pressure should be as near atmospheric pressure (14.7 psi) as possible. This keeps down the price of the equipment. It also puts positive pressure on the system at all points. By having a small pressure, it is possible to prevent air and moisture from entering the system. In the case of vacuum or low pressure, it is possible for a leak to suck in air and moisture. Note the five refrigerants and their pressures in Table 5-5.

Freon R-11 is used in very large systems because it requires more refrigerant than others—even though it has the best pressure characteristics

TABLE 5-5   Operating Pressures

Refrigerant	Evaporating pressure (PSIG) at 5°F	Condensing pressure (PSIG) at 86°F
R-11	24.0 in. Hg	3.6
R-12	11.8	93.2
R-22	28.3	159.8
R-717	19.6	154.5
R-718	29.7	28.6

TABLE 5-6   Boiling Temperature

ASHRAE number	Type of refrigerant	Class of refrigerant	Boiling point °F (°C)
123	Single component	HCFC	82.2 (27.9)
11	Single component	CFC	74.9 (23.8)
245fa	Single component	HFC	59.5 (15.3)
236fa	Single component	HFC	29.5 (–1.4)
134a	Single component	HFC	–15.1 (–26.2)
12	Single component	CFC	–21.6 (–29.8)
401A	Zeotrope	HCFC	–27.7 (–32.2)
500	Azeotrope	CFC	–28.3 (–33.5)
409A	Zeotrope	HCFC	–29.6 (–34.2)
22	Single component	HCFC	–41.5 (–40.8)
407C	Zeotrope	HFC	–46.4 (–43.6)
502	Azeotrope	CFC	–49.8 (–45.4)
408A	Zeotrope	HCFC	–49.8 (–45.4)
404A	Zeotrope	HFC	–51.0 (–46.1)
507	Azetrope	HFC	–52.1 (–46.7)
402A	Zeotrope	HCFC	–54.8 (–48.2)
410A	Zeotrope	HFC	–62.9 (–52.7)
13	Single component	CFC	–114.6 (–81.4)
23	Single component	HFC	–115.7 (–82.1)
508B	Azeotrope	HFC	–125.3 (–87.4)
503	Azeotrope	CFC	–126.1 (–87.8)

of the group. Several factors must be considered before a suitable refrigerant is found. There is no ideal refrigerant for all applications.

Temperature

Temperature is important in selecting a refrigerant for a particular job. The boiling temperature is that point at which a liquid is vaporized upon the addition of heat. This, of course, depends upon the refrigerant and the absolute pressure at the surface of the liquid and vapor. Note that in Table 5-6, R-22 has the lowest boiling temperature. Water (R-718) has the highest boiling temperature. Atmospheric pressure is 14.7 psi.

Once again, there is no ideal atmospheric boiling temperature for a refrigerant. However, temperature-pressure relationships are important in choosing a refrigerant for a particular job.

Volume

Specific volume is defined as the definite weight of a material. Usually expressed in terms of cubic feet per pound, the volume is the reciprocal of density. The specific volume of a refrigerant is the number of cubic feet of gas that is formed when 1 lb of the refrigerant is vaporized. This is an important factor to be considered when choosing the size of refrigeration system components. Compare the specific volumes at 5°F (–15°C) of the five refrigerants we have chosen. Freon R-12 and R-22

TABLE 5-7   Specific Volumes at 5°F

Refrigerant	Liquid volume (ft³/lb)	Vapor volume (ft³/lb)
R-11	0.010	12.27
R-12	0.011	1.49
R-22	0.012	1.25
R-717	0.024	8.15
R-718 (water)	0.016	12 444.40

TABLE 5-8   Liquid Density at 86°F

Refrigerant	Liquid density (lb/ft³)
R-11	91.4
R-12	80.7
R-22	73.4
R-717	37.2
R-718	62.4

(the most often used refrigerants) have the lowest specific volumes as vapors. Refer to Table 5-7.

Density

Density is defined as the mass or weight per unit of volume. In the case of a refrigerant, it is the weight in terms of volume given in pounds per cubic foot. Note in Table 5-8 that the density of R-11 is the greatest. The density of R-717 (ammonia) is the least.

Enthalpy

Enthalpy is the total heat in a refrigerant. The sensible heat plus the latent heat makes up the total heat. Latent heat is the amount of heat required to change the refrigerant from a liquid to a gas. The latent heat of vaporization is a measure of the heat per pound that the refrigerant can absorb from an area to be cooled. It is, therefore, a measure of the cooling potential of the refrigerant circulated through a refrigeration system (see Table 5-9). Latent heat is expressed in Btu per pound.

TABLE 5-9   Enthalpy [Btu/lb. at 5°F (−15°C)]

Refrigerant	Liquid enthalpy	+	Latent heat of vaporization	=	Vapor enthalpy
R-11	8.88	+	84.00	=	92.88
R-12	9.32	+	60.47	=	78.79
R-22	11.97	+	93.59	=	105.56
R-717	48.30	+	565.00	=	613.30
R-718 (at 40°F)	8.05	+	1071.30	=	1079.35

Flammability

Of the five refrigerants mentioned so far, the only one that is flammable is ammonia. None of the Freon compounds is flammable or explosive. However, mixtures with flammable liquids or gases may be flammable and should be handled carefully. Partially halogenated compounds may also be flammable and must be individually examined. If the refrigerant is used around fire, its flammability should be carefully considered. Some city codes specify which refrigerants cannot be used within city limits.

Capability of mixing with oil

Some refrigerants mix well with oil. Others, such as ammonia and water, do not. The ability to mix with oil has advantages and disadvantages. If the refrigerant mixes easily, parts of the system can be lubricated easily by the refrigerant and its oil mixture. The refrigerant will bring the oil back to the compressor and moving parts for lubrication.

There is a disadvantage to the mixing of refrigerant and oil. If it is easily mixed, the refrigerant can mix with the oil during the off cycle and then carry off the oil once the unit begins to operate again. This means that the oil needed for lubrication is drawn off with the refrigerant. This can cause damage to the compressor and moving parts. With this condition, there is foaming in the compressor crankcase and loss of lubrication. In some cases, the compressor is burned out. Procedures for cleaning up a burned out motor will be given later.

Moisture and refrigerants

Moisture should be kept out of refrigeration systems. It can corrode parts of the system. Whenever low temperatures are produced, the water or moisture can freeze. If freezing of the metering device occurs, then refrigerant flow is restricted or cut off. The system will have a low efficiency or none at all. The degree of efficiency will depend upon the amount of icing or the part affected by the frozen moisture.

All refrigerants will absorb water to some degree. Those that absorb very little water permit free water to collect and freeze at low-temperature points. Those that absorb a high amount of moisture will form corrosive acids and corrode the system. Some systems will allow water to be absorbed and frozen. This causes corrosion.

Hydrolysis is the reaction of a material, such as Freon 12 or methyl chloride, with water. Acid materials are formed. The hydrolysis rate for the Freon compounds as a group is low compared with other halogenated compounds. Within the Freon group, however, there is considerable variation. Temperature, pressure, and the presence of other materials also greatly affect the rate. Typical hydrolysis rates for the Freon compounds and other halogenated compounds are given in Table 5-10.

TABLE 5-10   Hydrolysis Rate in Water (Gram per Litre of Water per Year)

| Compound | 1 atm pressure 86°F | | Saturation pressure 122°F with steel |
	Water alone	With steel	
CH_3Cl	*	*	110
CH_2Cl_2	*	*	55
Freon 113	<0.005	ca. 50[†]	40
Freon 11	<0.005	ca. 10[†]	28
Freon 12	<0.005	0.8	10
Freon 21	<0.01	5.2	9
Freon 114	<0.005	1.4	3
Freon 22	<0.01	0.1	*
Freon 502	<0.01[††]	<0.01[††]	

[*]Not measured
[†]Observed rates vary
[††]Estimated

With water alone at atmospheric pressure, the rate is too low to be determined by the analytical method used. When catalyzed by the presence of steel, the hydrolysis rates are detectable but still quite low. At saturation pressures and a higher temperature, the rates are further increased.

Under neutral or acidic conditions, the presence of hydrogen in the molecule has little effect on the hydrolytic stability. However, under alkaline conditions compounds containing hydrogen, such as Freon 22 and Freon 21, tend to be hydrolyzed more rapidly.

Odor

The five refrigerants are characterized by their distinct odor or the absence of it. Freon R-11, R-12, and R-22 have a slight odor. Ammonia (R-717) has a very acrid odor and can be detected even in small amounts. Water (R-718), of course, has no odor.

A slight odor is needed in a refrigerant so that its leakage can be detected. A strong odor may make it impossible to service equipment. Special gas masks may be needed. Some refrigerated materials may be ruined if the odor is too strong. About the only time that an odor is preferred in a refrigerant is when a toxic material is used for a refrigerant. A refrigerant that may be very inflammable should have an odor so that its leakage can be detected easily to prevent fire or explosions.

Toxicity

Toxicity is the characteristic of a material that makes it intoxicating or poisonous. Some refrigerants can be very toxic to humans, others may not be toxic at all. The halogen refrigerants (R-11, R-12, and R-22) are harmless in their normal condition or state. However, they form a highly toxic gas when an open flame is used around them.

TABLE 5-11   Molecular Weight of Selected Refrigerants

Refrigerant	Molecular weight
R-11	137.4
R-12	120.9
R-22	86.5
R-717 (ammonia)	17.0
R-718 (water)	18.0

Water, of course, is not toxic. However, ammonia can be toxic if present in sufficient quantities. Make sure the manufacturer's recommended procedures for handling are followed when working with refrigerants.

Tendency to leak

The size of the molecule makes a difference in the tendency of a refrigerant to leak. The greater the molecular weight, the larger the hole must be for the refrigerant to escape. A check of the molecular weight of a refrigerant will indicate the problem it may present to a sealed refrigeration system. Table 5-11 shows that R-11 has the least tendency to leak, whereas ammonia is more likely to leak.

Detecting Leaks

There are several tests used to check for leaks in a closed refrigeration system. Most of them are simple. Following are some useful procedures:

- Hold the joint or suspected leakage point under water and watch for bubbles.
- Coat the area suspected of leakage with a strong solution of soap. If a leak is present, soap bubbles will be produced.

Sulfur dioxide

To detect sulfur dioxide leaks, an ammonia swab may be used. The swab is made by soaking a sponge or cloth tied onto a stick or piece of wire in aqua ammonia. Household ammonia may also be used. A dense white smoke forms when the ammonia comes in contact with the sulfur dioxide. The usual soap bubble or oil test may be used when no ammonia is available.

If ammonia is used, check for leakage in the following ways:

- Burn a sulfur stick in the area of the leak. If there is a leak, a dense white smoke will be produced. The stronger the leak, the denser the white smoke.
- Hold a wet litmus paper close to the suspected leak area. If there is a leak, the ammonia will cause the litmus paper to change color.

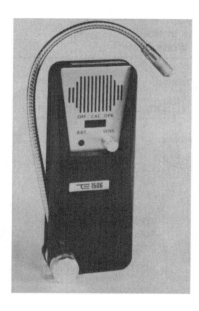

Figure 5-5   A handheld electronic leak detector. (*Courtesy of Thermal Engineering.*)

■ Refrigerants that are halogenated hydrocarbons (Freon compounds) can be checked for leakage with a halide leak test. This involves holding a torch or flame close to the leak area. If there is a refrigerant leak, the flame will turn green. In every instance, the room should be well ventilated when the torch test is made.

There is presently available an electronic detector for such refrigerant leaks. The detector gives off a series of rapid clicks if the refrigerant is present. The higher the concentration of the refrigerant, the more rapid the clicks (see Fig. 5-5).

Carbon dioxide

Leaks can be detected with a soap solution if there is internal pressure on the part to be tested. When carbon dioxide is present in the condenser water, the water will turn yellow with the addition of bromothymol blue.

Ammonia

Leaks are detected (in small amounts of ammonia) when a lit sulfur candle is used. The candle will give off a very thick, white smoke when it contacts the ammonia leak. The use of phenolphthalein paper is also considered a good test. The smallest trace of ammonia will *cause* the moistened paper strip to turn pink. A large ammonia will cause the phenolphthalein paper to turn a vivid scarlet.

Figure 5-6 A halide gas leak detector. (*Courtesy of Turner.*)

Methyl chloride

Leaks are detected by a leak-detecting halide torch (see Fig. 5-6). Some torches use alcohol for fuel and produce a colorless flange. When a methyl chloride leak is detected, the flame turns green. A brilliant blue flame is produced when large or stronger concentrations are present. In every instance, the room should be well ventilated when the torch test is made. The combustion of the refrigerant and the flame produces harmful chemicals. If a safe atmosphere is not present, the soap bubble test or oil test should be used to check for leaks.

As mentioned, methyl chloride is hard to detect with the nose or eyes. It does not produce irritating effects. Therefore, some manufacturers add a 1 percent amount of *acrolein* as a warning agent. Acrolein is a colorless liquid (C_3H_4O) with a pungent odor.

Ban on Production and Imports of Ozone-Depleting Refrigerants

In 1987, the Montreal Protocol, an international environmental agreement, established requirements that began the worldwide phaseout of ozone depleting CFCs which are the chlorofluorocarbons. These requirements were later modified. This lead to the phaseout, in 1996, of CFC production in all developed nations. In 1992, an amendment to the Montreal Protocol established a schedule for the phaseout of HCFCs. These are the refrigerants called hydrochlorofluorocarbons.

HCFCs are substantially less damaging to the ozone layer than CFCs. However, they still contain ozone-destroying chlorine. The Montreal Protocol, as amended, is carried out in the United States through Title VI of the Clean Air Act. This act is implemented by the EPA or Environmental Protection Agency.

An HCFC, known as R-22, has been the refrigerant of choice for residential heat pump and air-conditioning systems for more than four decades. Unfortunately for the environment, releases of R-22 that results from system leaks contribute to ozone depletion. In addition, the manufacture of R-22 results in a by-product that contributes significantly to global warming.

As the manufacture of R-22 is phased out over the coming years as part of the agreement to end production of HCFCs, manufacturers of residential air-conditioning systems are beginning to offer equipment that uses ozone-friendly refrigerants. Many homeowners may be misinformed about how much longer R-22 will be available to service their central AC systems and heat pumps. The future availability of R-22, and the new refrigerants that are replacing R-22 will be covered here. The EPA document assists consumers in deciding what to consider when purchasing a new AC system or heat pump, or when having an existing system repaired.

Phaseout Schedule for HCFCs, Including R-22

Under the terms of the Montreal Protocol, the United States agreed to meet certain obligations by specific dates. That will affect the residential heat pump and air-conditioning industry:

January 1, 2004: In accordance with the terms of the Protocol, the amount of all HCFCs that can be produced nationwide had to be reduced by 35 percent by 2004. In order to achieve this goal, the United States ceased production of HCFC-141b, the most ozone damaging of this class of chemicals, on January 1, 2003. This production ban should greatly reduce nationwide use of HCFCs as a group and make it likely that the 2004 deadline had a minimal effect on R-22 supplies.

January 1, 2010: After 2010, chemical manufacturers may still produce R-22. But this is to service existing equipment and not for use in new equipment. As a result, heating, ventilation, and air-conditioning (HVAC) system manufacturers will only be able to use preexisting supplies of R-22 in the production of new air conditioners and heat pumps. These existing supplies will include R-22 recovered from existing equipment and recycled by licensed reclaimers.

January 1, 2020: Use of existing refrigerant, including refrigerant that has been recovered, and recycled, will be allowed beyond 2020 to service existing systems. However, chemical manufacturers will no longer be able to produce R-22 to service existing air conditioners and heat pumps.

What does the R-22 phaseout mean for consumers? The following paragraphs are an attempt to answer this question.

Availability of R-22

The Clean Air Act does not allow any refrigerant to be vented into the atmosphere during installation, service, or retirement of equipment. Therefore, R-22 must be

- Recovered and recycled (for reuse in the same system) reclaimed (reprocessed to the same purity levels as new R-22)
- Destroyed

After 2020, the servicing of R-22-based systems will rely on recycled refrigerants. It is expected that reclamation and recycling will ensure that existing supplies of R-22 will last longer and be available to service a greater number of systems. As noted above, chemical manufacturers will be able to produce R-22 for use in new air-conditioning equipment until 2010, and they can continue production of R-22 until 2020 for use in servicing that equipment. Given this schedule, the transition away from R-22 to the use of ozone-friendly refrigerants should be smooth. For the next 20 years or more, R-22 should continue to be available for all systems that require R-22 for servicing.

Cost of R-22

While consumers should be aware that prices of R-22 may increase as supplies dwindle over the next 20 or 30 years, EPA believes that consumers are not likely to be subjected to major price increases within a short time period. Although there is no guarantee that service costs of R-22 will not increase, the lengthy phaseout period for R-22 means that market conditions should not be greatly affected by the volatility and resulting refrigerant price hikes that have characterized the phaseout of R-12, the refrigerant used in automotive air-conditioning systems and replaced by R-134a.

Alternatives to R-22

Alternatives for residential air-conditioning will be needed as R-22 is gradually phased out. Non-ozone-depleting alternative refrigerants are being introduced. Under the Clean Air Act, EPA reviews alternatives to ozone-depleting substances (ODSs) like R-22 in order to evaluate their effects on human health and the environment. The EPA has reviewed several of these alternatives to R-22 and has compiled a list of substitutes that the EPA has determined are acceptable. One of these substitutes is R-410A, a blend of hydrofluorocarbons (HFCs), substances that do not contribute to depletion of the ozone layer, but, like R-22, contribute to global warming. R-410A is manufactured and sold under various trade names, including Genetron AZ 20, SUVA 410A, and Puron. Additional refrigerants on the list of acceptable substitutes include R-134a and R-407C. These two refrigerants are not yet available for residential applications in the United States, but are commonly found in residential air-conditioning systems and heat pumps in Europe. EPA will continue to review new non-ozone-depleting refrigerants as they are developed.

Servicing existing units

Existing units using R-22 can continue to be serviced with R-22. There is no EPA requirement to change or convert R-22 units for use with a non-ozone-depleting substitute refrigerant. In addition, the new substitute refrigerants cannot be used without making some changes to system components. As a result, service technicians who repair leaks to the system will continue to charge R-22 into the system as part of that repair.

Installing new units

The transition away from ozone-depleting R-22 to systems that rely on replacement refrigerants like R-410A has required redesign of heat pump and air-conditioning systems. New systems incorporate compressors and other components specifically designed for use with specific replacement refrigerants. With these significant product and production process changes, testing and training must also change. Consumers should be aware that dealers of systems that use substitute refrigerants should be schooled in installation and service techniques required for use of that substitute refrigerant.

Servicing your system

Along with prohibiting the production of ozone-depleting refrigerants, the Clean Air Act also mandates the use of common sense in handling refrigerants. By containing and using refrigerants responsibly—that is, by recovering, recycling, and reclaiming, and by reducing leaks—their ozone depletion and global warming consequences are minimized. The Clean Air Act outlines specific refrigerant containment and management practices for HVAC manufacturers, distributors, dealers, and technicians. Properly installed home comfort systems rarely develop refrigerant leaks, and with proper servicing, a system using R-22, R-410A or another refrigerant will minimize its impact on the environment. While EPA does not mandate repairing or replacing small systems because of leaks, system leaks can not only harm the environment, but also result in increased maintenance costs.

One important thing a homeowner can do for the environment, regardless of the refrigerant used, is to select a reputable dealer that employs service technicians who are EPA-certified to handle refrigerants. Technicians often call this certification "Section 608 certification," referring to the part of the Clean Air Act that requires minimizing releases of ozone-depleting chemicals from HVAC equipment.

Purchasing new systems

Another important thing a homeowner can do for the environment is to purchase a highly energy-efficient system. Energy-efficient systems result in cost savings for the homeowner. Today's best air conditioners use much less energy to produce the same amount of cooling as air conditioners made in the mid-1970s. Even if your air conditioner is only 10 years old, you may save significantly on your cooling energy costs by replacing it with a newer, more efficient model. Products with EPA's Energy Star label can save homeowners 10 to 40 percent on their heating and cooling bills every year. These products are made by most major manufacturers and have the same features as standard products, but also incorporate energy saving technology. Both R-22 and R-410A systems may have the Energy Star label. Equipment that displays the Energy Star label must have a minimum seasonal energy efficiency ratio (SEER). The higher the SEER specification, the more efficient the equipment.

Energy efficiency, along with performance, reliability and cost, should be considered in making a decision. And don't forget that when purchasing a new system, you can also speed the transition away from ozone-depleting R-22 by choosing a system that uses ozone-friendly refrigerants.

Air-Conditioning and Working with Halon

Several regulations have been issued under Section 608 of the Clean Air Act to govern the recycling of refrigerants in stationary systems and to end the practice of venting refrigerants to the air. These regulations also govern the handling of halon fire extinguishing agents. A Web site and both the regulations themselves and fact sheets are available from the EPA Stratospheric Ozone Hotline at 1-800-296-1996.

NOTE: The handling and recycling of refrigerants used in motor vehicle air-conditioning systems are governed under Section 609 of the Clean Air Act.

General information

April 13, 2005: EPA finalized a rulemaking amending the definition of refrigerant to make certain that it only included substitutes that consisted of a class I or class II ozone-depleting substance. This rulemaking also amended the venting prohibition to make certain that it remains illegal to knowingly vent nonexempt substitutes that do not consist of a class I or class II ODS, such as R-134a and R-410A.

January 11, 2005: EPA published a final rule extending the leak repair required practices and the associated reporting and record keeping requirements to owners and/or operators of comfort cooling, commercial refrigeration, or industrial process refrigeration appliances containing more than 50 lb of a substitute refrigerant, if the substitute contains a class I or class II ozone-depleting substance. In addition, EPA has defined leak rate in terms of the percentage of the appliance's full charge that would be lost over a consecutive 12-month period, if the current rate of loss were to continue over that period. EPA now requires calculation of the leak rate every time that refrigerant is added to an appliance.

March 12, 2004: EPA finalized rulemaking sustaining the Clean Air Act prohibition against venting HFC and perfluorocarbon (PFC) refrigerants. This rulemaking found that the known venting of HFC and PFC refrigerants during the maintenance, service, repair, and disposal of air-conditioning and refrigeration equipment (i.e., appliances) remained illegal under Section 608 of the Clean Air Act. The ruling also restricted the sale of HFC refrigerants that consist of an ODS to EPA-certified technicians. However, HFC refrigerants and HFC refrigerant blends that did not consist of an ODS are not covered under "The Refrigerant Sales Restriction," a brochure that documents the environmental and financial reasons to replace CFC chillers with new, energy-efficient equipment. A partnership of governments, manufacturers, NGOs (nongovernmental organizations), and others have endorsed the brochure to eliminate uncertainty and underscore the wisdom of replacing CFC chillers.

Leak Repair

The leak repair requirements, promulgated under Section 608 of the Clean Air Act Amendments of 1990, require that when an owner or operator of an appliance that normally contains a refrigerant charge of more than 50 lb discovers that refrigerant is leaking at a rate that would exceed the applicable trigger rate during a 12-month period, the owner or operator must take corrective action.

TABLE 5-12   Trigger Leak Rates

Appliance type	Trigger leak rate
Commercial refrigeration	35%
Industrial process refrigeration	35%
Comfort cooling	15%
All other appliances	15%

Trigger rates

For all appliances that have a refrigerant charge of more than 50 lb, the following leak rates for a 12-month period are applicable (Table 5-12):

In general, owners or operators must either repair leaks within 30 days from the date the leak was discovered, or develop a dated retrofit/retirement plan within 30 days and complete actions under that plan within 1 year from the plan's date. However, for industrial process refrigeration equipment and some federally owned chillers, additional time may be available.

Industrial process refrigeration is defined as complex customized appliances used in the chemical, pharmaceutical, petrochemical, and manufacturing industries. These appliances are directly linked to the industrial process. This sector also includes industrial ice machines, appliances used directly in the generation of electricity, and in ice rinks. If at least 50 percent of an appliance's capacity is used in an industrial process refrigeration application, the appliance is considered industrial process refrigeration equipment and the trigger rate is 35 percent.

Industrial process refrigeration equipment and federally owned chillers must conduct initial and follow-up verification tests at the conclusion of any repair efforts. These tests are essential to ensure that the repairs have been successful. In cases where an industrial process shutdown is required, a repair period of 120 days is substituted for the normal 30-day repair period. Any appliance that requires additional time may be subject to record keeping/reporting requirements.

When additional time is necessary

Additional time is permitted for conducting leak repairs where the necessary repair parts are unavailable or if other applicable federal, state, or local regulations make a repair within 30 to 120 days impossible. If owners or operators choose to retrofit or retire appliances, a retrofit or retirement plan must be developed within 30 days of detecting a leak rate that exceeds the trigger rates. A copy of the plan must be kept on site. The original plan must be made available to EPA upon request. Activities under the plan must be completed within 12 months (from the date of the plan). If a request is made within 6 months from the expiration of the initial 30-day period, additional time beyond the 12-month period is available for owners or operators of industrial process refrigeration equipment and federally owned chillers in the following cases: EPA will permit additional time to the extent reasonably necessary where a delay is caused by the requirements of other applicable federal, state, or local regulations; or where a suitable replacement refrigerant, in accordance with the regulations promulgated under Section 612, is not available; and EPA will permit one additional 12-month period where an appliance is custom built and the supplier of the appliance or a critical component has quoted a delivery time of more than 30 weeks from when the order was placed, (assuming the order was placed in a timely manner). In some cases, EPA may provide additional time beyond this extra year where a request is made by the end of the ninth month of the extra year.

Relief from retrofit/retirement

The owners or operators of industrial process refrigeration equipment or federally owned chillers may be relieved from the retrofit or repair requirements if

- second efforts to repair the same leaks that were subject to the first repair efforts are successful, or

- within 180 days of the failed follow-up verification test, the owners or operators determine the leak rate is below 35 percent. In this case, the owners or operators must notify EPA as to how this determination will be made, and must submit the information within 30 days of the failed verification test.

System mothballing

For all appliances subject to the leak repair requirements, the timelines may be suspended if the appliance has undergone system mothballing. *System mothballing* means the intentional shutting down of a refrigeration appliance undertaken for an extended period of time where the refrigerant has been evacuated from the appliance or the affected isolated section of the appliance to at least atmospheric pressure. However, the timelines pick up again as soon as the system is brought back online.

EPA-Certified Refrigerant Reclaimers

The EPA listing of reclaimers is updated when additional refrigerant reclaimers are approved. Reclaimers appearing on this list are approved to reprocess used refrigerant to at least the purity specified in appendix A to 40 CFR part 82, subpart F (based on ARI Standard 700, "Specifications for Fluorocarbon and Other Refrigerants"). Reclamation of used refrigerant by an EPA-certified reclaimer is required in order to sell used refrigerant not originating from and intended for use with motor vehicle air conditioners.

The EPA encourages reclaimers to participate in a voluntary third-party reclaimer certification program operated by the Air-Conditioning and Refrigeration Institute (ARI). The volunteer program offered by the ARI involves quarterly testing of random samples of reclaimed refrigerant. Third-party certification can enhance the attractiveness of a reclaimer's program by providing an objective assessment of its purity.

Newer Refrigerants

Since the world has become aware of the damage the Freon refrigerants can do to the ozone layer, there has been a mad scramble to obtain new refrigerants that can replace all those now in use. There are some problems with adjusting the new and especially existing equipment to the properties of new refrigerant blends.

It is difficult to directly replace R-12 for instance. It has been the mainstay in refrigeration equipment for years. However, the automobile air-conditioning industry has been able to reformulate R-12 to produce an acceptable substitute, R-134a. There are others now available to substitute in the more sophisticated equipment with large amounts of refrigerants. Some of these will be covered here.

Freon Refrigerants

The Freon family of refrigerants has been one of the major factors responsible for the impressive growth of not only the home refrigeration and air-conditioning industry, but also the commercial refrigeration industry. The safe properties of these products have permitted their use under conditions where flammable or more toxic refrigerants would be hazardous.

Classifications

Following were commonly used Freon refrigerants:

Freon 11: Freon 11 has a boiling point of 74.8°F (24°C) and has wide usage as a refrigerant in indirect industrial and commercial air-conditioning systems employing single or multistage centrifugal compressors with capacities of 100 ton and above. Freon 11 is also employed as brine for low-temperature applications. It provides relatively low operating pressures with moderate displacement requirements.

Freon 12: The boiling point of Freon 12 is −21.7°F (−30°C). It is the most widely known and used Freon refrigerant. It is used principally in household and commercial refrigeration and air-conditioning units, for refrigerators, frozen-food cabinets, ice cream cabinets, food-locker plants, water coolers, room and window air-conditioning units, and similar equipment. It is generally used in reciprocating compressors, ranging in size from fractional to 800 hp. Rotary compressors are useful in small units. The use of centrifugal compressors with Freon 12 for large air-conditioning and process-cooling applications is increasing.

Freon 13: The boiling point of Freon 13 is −144.6°F (−98°C). It is used in low-temperature specialty applications employing reciprocating compressors and generally in cascade with Freon 12 or Freon 22.

Freon 21: Freon 21 has a boiling point of 48°F (8.8°C). It is used in fractional-horsepower household refrigerating systems and drinking-water coolers employing rotary vane-type compressors. Freon 21 is also used in comfort-cooling air-conditioning systems of the absorption type where dimethyl ether or tetraethylene glycol is used as the absorbent.

Freon 22: The boiling point of Freon 22 is −41.4°F (−40.7°C). It is used in all types of household and commercial refrigeration and air-conditioning applications with reciprocating compressors. The outstanding thermodynamic properties of Freon 22 permit the use of smaller equipment than is possible with similar refrigerants, making it especially suitable where size is a problem.

Freon 113: The boiling point of Freon 113 is 117.6°F (47.5°C). It is used in commercial and industrial air-conditioning and process water and brine cooling with centrifugal compression. It is especially useful in small-tonnage applications.

Freon 114: The boiling point of Freon 114 is 38.4°F (3.5°C). It is used as a refrigerant in fractional-horsepower household refrigerating systems and drinking-water coolers employing rotary vane-type compressors. It is also used in indirect industrial and commercial air-conditioning systems and in industrial process water and brine cooling to −70°F (−56.6°C) employing multistage centrifugal-type compressors in cascade of 100 tons refrigerating capacity and larger.

Freon 115: The boiling point of Freon 115 is −37.7°F (−38.72°C). It is especially stable, offering a particularly low discharge temperature in reciprocating compressors. Its capacity exceeds that of Freon 12 by as much as 50 percent in low-temperature systems. Its potential applications include household refrigerators and automobile air-conditioning.

Freon 502: Freon 502 is an azeotropic mixture composed of 48.8 percent Freon 22 and 51.2 percent Freon 115 by weight. It boils at −50.1°F (−45.61°C). Because it permits achieving the capacity of Freon 22 with discharge temperatures comparable to Freon 12, it is finding new reciprocating compressor applications in low-temperature display cabinets and in storing and freezing of food.

Properties of Freons

The Freon refrigerants are colorless and almost odorless, and their boiling points vary over a wide range of temperatures. Those Freon refrigerants that are produced are nontoxic, noncorrosive, nonirritating, and nonflammable under all conditions of usage. They are generally prepared by replacing chlorine or hydrogen with fluorine. Chemically, Freon refrigerants are inert and thermally stable up to temperatures far beyond conditions found in actual operation. However, Freon is harmful when allowed to escape into the atmosphere. It can deplete the ozone layer around the earth and cause more harmful ultraviolet rays to reach the surface of the earth.

Physical properties.   The pressures required in liquefying the refrigerant vapor affect the design of the system. The refrigerating effect and specific volume of the refrigerant vapor determine the compressor displacement. The heat of vaporization and specific volume of the liquid refrigerant affect the quantity of refrigerant to be circulated through the pressure-regulating valve or other system device.

Flammability.   Freon is nonflammable and noncombustible under conditions where appreciable quantities contact flame or hot metal surfaces. It requires an open flame at 1382°F (750°C) to decompose the vapor. Even at this temperature, only the vapor decomposes to form hydrogen chloride and hydrogen fluoride, which are irritating but are readily dissolved in water. Air mixtures are not capable of burning and contain no elements that will support combustion. For this reason, Freon is considered nonflammable.

Amount of liquid refrigerant circulated.   It should be noted that the Freon refrigerants have relatively low heat values, but this must not be considered a disadvantage. It simply means that a greater volume of liquid must be circulated per unit of time to produce the desired amount of refrigeration. It does not concern the amount of refrigerant in the system. Actually, it is a decided advantage (especially in the smaller- or low-tonnage systems) to have a refrigerant with low heat values. This is because the larger quantity of liquid refrigerant to be metered through the liquid-regulating device will permit the use of more accurate and more positive operating and regulating mechanisms of less sensitive and less critical adjustments. Table 5-13 lists the quantities of liquid refrigerant metered or circulated per minute under standard ton conditions.

Volume (piston) displacement.   For reason of compactness, cost of equipment, reduction of friction, and compressor speed, the volume of gas that must be compressed per unit of time for a given refrigerating effect, in general, should be as low as possible. Freon 12 has relatively a low-volume displacement, which makes it suitable for use in reciprocating compressors, ranging from the smallest size to those of up to

TABLE 5-13   Quantities of Refrigerant Circulated per Minute Under Standard Ton Conditions

Refrigerant	Pounds expanded per minute	Ft^3/lb liquid 86°F	Liquid expanded per minute, in^3	Specific gravity liquid 86°F (water-1)
Freon 22	2.887	0.01367	67.97	1.177
Freon 12	3.916	0.0124	83.9	1.297
Freon 114	4.64	0.01112	89.16	1.443
Freon 21	2.237	0.01183	45.73	1.360
Freon 11	2.961	0.01094	55.976	1.468
Freon 113	3.726	0.01031	66.48	1.555

800-ton capacity, including compressors for household and commercial refrigeration. Freon 12 also permits the construction of compact rotary compressors in the commercial sizes. Generally, low-volume displacement (high-pressure) refrigerants are used in reciprocating compressors; high-volume displacement (low-pressure) refrigerants are used in large-tonnage centrifugal compressors; intermediate-volume (intermediate-pressure) refrigerants are used in rotary compressors. There is no standard rule governing this usage.

Condensing pressure.   Condensing (high-side) pressure should be low to allow construction of lightweight equipment, which affects power consumption, compactness, and installation. High pressure increases the tendency toward leakage on the low side as well as the high side when pressure is built up during idle periods. In addition, pressure is very important from the standpoint of toxicity and fire hazard.

In general, a low-volume displacement accompanies a high-condensing pressure, and a compromise must usually be drawn between the two in selecting a refrigerant. Freon 12 presents a balance between volume displacement and condensing pressure. Extra-heavy construction is not required for this type of refrigerant, and so there is little or nothing to be gained from the standpoint of weight of equipment in using a lower-pressure refrigerant.

Evaporating pressure.   Evaporating (low-side) pressures above atmospheric pressure are desirable to avoid leakage of moisture-laden air into the refrigerating systems and permit easier detection of leaks. This is especially important with open-type units. Air in the system will increase the head pressures, resulting in inefficient operations, and may adversely affect the lubricant. Moisture in the system will cause corrosion and, in addition, may freeze out and stop operation of the equipment.

In general, the higher the evaporating pressure, the higher the condensing pressure under a given set of temperatures. Therefore, to keep head pressures at a minimum and still have positive low-side pressures, the refrigerant selected should have a boiling point at atmospheric pressure as close as possible to the lowest temperature to be produced under ordinary operating conditions. Freon 12, with a boiling point of $-21.7°F$ ($-29.83°C$), is close to ideal in this respect for most refrigeration applications. A still lower boiling point is of some advantage only when lower operating temperatures are required.

Refrigerant Characteristics

The freezing point of a refrigerant should be below any temperature that might be encountered in the system. The freezing point of all refrigerants, except water [$32°F$ ($0°C$)] and carbon dioxide [$-69.9°F$ ($-56.61°C$), triple point], are far below the temperatures that might be encountered in their use. Freon-12 has a freezing point of $-247°F$ ($-155°C$). See App. A for more details on refrigerants.

Critical temperature

The critical temperature of a refrigerant is the highest temperature at which it can be condensed to a liquid, regardless of a higher pressure. It should be above the highest condensing temperature that might be encountered. With air-cooled condensers, in general, this would be above $130°F$ ($54.44°C$). Loss of efficiency caused by superheating of the refrigerant vapor on compression and by throttling expansion of the liquid is greater when the critical temperature is low.

All common refrigerants have satisfactorily high critical temperatures, except carbon dioxide [$87.8°F$ ($31°C$)] and ethane [$89.8°F$ ($32.11°C$)].

These two refrigerants require condensers cooled to temperatures below their respective critical temperatures, thus generally requiring water.

Hydrofluorocarbons.   There are some hydrofluorocarbon refrigerants (such as R-134a) that are made to eliminate the problems with refrigerants in the atmosphere caused by leaks in systems. The R-134a is a non-ozone-depleting refrigerant used in vehicle air-conditioning systems. DuPont's brand name is Suva, and the product is produced in a plant located in Corpus Christi, Texas, as well in Chiba, Japan. According to DuPont's Web site, R-134a was globally adopted by all vehicle manufacturers in the early 1990s as a replacement for CFC-12. The transition to R-134a was completed by the mid-1990s for most major automobile manufacturers. Today, there are more than 300 million cars with air conditioners using the newer refrigerant.

Latent heat of evaporation

A refrigerant should have a high latent heat of evaporation per unit of weight so that the amount of refrigerant circulated to produce a given refrigeration effect may be small. Latent heat is important when considering its relationship to the volume of liquid required to be circulated. The net result is the refrigerating effect. Since other factors enter into the determination of these, they are discussed separately.

The refrigerant effect per pound of refrigerant under standard ton conditions determines the amount of refrigerant to be evaporated per minute. The refrigerating effect per pound is the difference in Btu content of the saturated vapor leaving the evaporator [5°F (–15°C)] and the liquid refrigerant just before passing through the regulating valve [86°F (30°C)]. While the Btu refrigerating effect per pound directly determines the number of pounds of refrigerant to be evaporated in a given length of time to produce the required results, it is much more important to consider the volume of the refrigerant vapor required rather than the weight of the liquid refrigerant. By considering the volume of refrigerant necessary to produce standard ton conditions, it is possible to make a comparison between Freon 12 and other refrigerants so as to provide for the reproportioning of the liquid orifice sizes in the regulating valves, sizes of liquid refrigerant lines, and so on.

A refrigerant must not be judged only by its refrigerating effect per pound, but the volume per pound of the liquid refrigerant must also be taken into account to arrive at the volume of refrigerant to be vaporized. Although Freon 12 has relatively low refrigerating effect, this is not a disadvantage, because it merely indicates that more liquid refrigerant must be circulated to produce the desired amount of refrigeration. Actually, it is a decided advantage to circulate large quantities of liquid refrigerant because the greater volumes required will permit the use of less sensitive operating and regulating mechanisms with less critical adjustment.

Refrigerants with high Btu refrigerating effects are not always desirable, especially for household and small commercial installations, because of the small amount of liquid refrigerant in the system and the difficulty encountered in accurately controlling its flow through the regulating valve. For household and small commercial systems, the adjustment of the regulating-valve orifice is most critical for refrigerants with high Btu values.

Specific heat

A low specific heat of the liquid is desirable in a refrigerant. If the ratio of the latent heat to the specific heat of a liquid is low, a relatively high proportion of the latent heat may be used in lowering the temperature of the liquid from the condenser temperature to the evaporator temperature. This results in a small net refrigerating effect per pound of refrigerant circulated and, assuming other factors remain the same, reduces the capacity and lowers the efficiency. When the ratio is low, it is advantageous to precool the liquid before evaporation by heat interchange with the cool gases leaving the evaporator.

In the common type of refrigerating systems, expansion of the high-pressure liquid to a lower-pressure, lower-temperature vapor and liquid takes place through a throttling device such as an expansion valve. In this process, energy available from the expansion is not recovered as useful work. Since it performs no external work, it reduces the net refrigerating effect.

Power consumption

In a perfect system operating between 5 and −86°F (−15 and −65.55°C) conditions, 5.74 Btu is the maximum refrigeration obtainable per Btu of energy used to operate the refrigerating system. This is the theoretical maximum *coefficient of performance* on cycles of maximum efficiency (e.g., the Carnet cycle). The minimum horsepower would be 0.821 hp/ton of refrigeration. The theoretical coefficient of performance would be the same for all refrigerants if they could be used on cycles of maximum efficiency.

However, because of engineering limitations, refrigerants are used on cycles with a theoretical maximum coefficient of performance of less than 5.74. The cycle most commonly used differs in its basic form from (1) the Carnet cycle, as already explained in employing expansion without loss or gain of heat from an outside source, and (2) in compressing adiabatically (compression without gaining or losing heat to an outside source) until the gas is superheated above the condensing medium temperature. These two factors, both of which increase the power requirement, vary in importance with different refrigerants. But, it so happens that when expansion loss is high, compression loss is generally low, and vice versa. All common refrigerants (except carbon dioxide and water) show about the same overall theoretical power requirement on a 5 to −86°F cycle. At least the theoretical differences are so small that other factors are more important in determining the actual differences in efficiency.

The amount of work required to produce a given refrigerating effect increases as the temperature level to which the heat is pumped from the cold body is increased. Therefore, on a 5 to −86°F cycle, when gas is superheated above 86°F temperature on compression, efficiency is decreased and the power requirement increased unless the refrigerating effect caused by superheating is salvaged through the proper use of a heat interchanger.

Volume of liquid circulated

Volumes of liquid required to circulate for a given refrigerant effect should be low. This is to avoid fluid-flow (pressure-drop) problems and to keep down the size of the required refrigerant change. In small-capacity machines, the volume of liquid circulated should not be so low as to present difficult problems in accurately controlling its flow through expansion valves or other types of liquid metering devices.

With a given net refrigerating effect per pound, a high density of liquid is preferable to a low volume. However, a high density tends to increase the volume circulated by lowering the net refrigerating effect.

Handling Refrigerants

One of the requirements of an ideal refrigerant is that it must be nontoxic. In reality, however, all gases (with the exception of pure air) are more or less toxic or asphyxiating. It is, therefore, important that wherever gases or highly volatile liquids are used, adequate ventilation be provided, because even nontoxic gases in air produce a suffocating effect.

Vaporized refrigerants (especially ammonia and sulfur dioxide) bring about irritation and congestion of the lungs and bronchial organs, accompanied by violent coughing, vomiting, and, when breathed in sufficient quantity, suffocation. It is of the utmost importance, therefore, that the serviceman subjected to a refrigerant gas find access to fresh air at frequent intervals to clear his lungs. When engaged in the repair of ammonia and sulfur dioxide machines, approved gas masks and goggles should be used. Carrene, Freon (R-12), and carbon dioxide fumes are not irritating and can be inhaled in considerable concentrations for short periods without serious consequences.

It should be remembered that liquid refrigerant would refrigerate or remove heat from anything it meets when released from a container. In the case of contact with refrigerant, the affected or injured area should be treated as if it has been frozen or frostbitten.

Storing and handling refrigerant cylinders

Refrigerant cylinders should be stored in a dry, sheltered, and well-ventilated area. The cylinders should be placed in a horizontal position, if possible, and held by blocks or saddles to prevent rolling. It is of utmost importance to handle refrigerant cylinders with care and to observe the following precautions:

- Never drop the cylinders, or permit them to strike each other violently.
- Never use a lifting magnet or a sling (rope or chain) when handling cylinders. A crane may be used when a safe cradle or platform is provided to hold the cylinders.
- Caps provided for valve protection should be kept on the cylinders at all times except when the cylinders are actually in use.
- Never overfill the cylinders. Whenever refrigerant is discharged from or into a cylinder, weigh the cylinder and record the weight of the refrigerant remaining in it.
- Never mix gases in a cylinder.
- Never use cylinders for rollers, supports, or for any purpose other than to carry gas.
- Never tamper with the safety devices in valves or on the cylinders.
- Open the cylinder valves slowly. Never use wrenches or tools except those provided or approved by the gas manufacturer.

- Make sure that the threads on regulators or other unions are the same as those on the cylinder valve outlets. Never force a connection that does not fit.

- Regulators and gages provided for use with one gas must not be used on cylinders containing a different gas.

- Never attempt to repair or alter the cylinders or valves.

- Never store the cylinders near highly flammable substances (such as oil, gasoline, or waste).

- Cylinders should not be exposed to continuous dampness, saltwater, or salt spray.

- Store full and empty cylinders apart to avoid confusion.

- Protect the cylinders from any object that will produce a cut or other abrasion on the surface of the metal.

Lubricants[*]

Lubricant properties can be evaluated. It can be determined if the product is right for the job. Three basic properties are

- Viscosity
- Lubricity
- Chemical stability

They must be satisfactory to protect the compressor. The correct viscosity is needed to fill the gaps between parts and flow correctly where it is supposed to go. Generally speaking, smaller equipment with smaller gaps between moving parts requires a lighter viscosity, and larger equipment with bigger parts needs heavier viscosity oils. Lubricity refers to the lubricant's ability to protect the metal surfaces from wear.

Good chemical stability means the lubricant will not react to form harmful chemicals such as acids, sludges, and so forth that may block tubing or there may be carbon deposits. The interaction of lubricant and refrigerant can cause potential problems as well.

Miscibility defines the temperature region where refrigerant and oil will mix or separate. If there is separation of the oil from the refrigerant in the compressor it is possible that the oil is not getting to metal parts that need it. If there is separation in the evaporator or other parts of the system it is possible that the oil does not return to the compressor and eventually there is not enough oil to protect it.

Solubility determines if the refrigerant will thin the oil too much. That would cause it to lose its ability to protect the compressor. The thinning effect also influences oil return.

[*]Courtesy of National Refrigerants

Once you mix a blend at a given composition, the pressure-temperature relationships follow the same general rules as for pure components. For example, the pressure goes up when the temperature goes up. For three blends containing different amounts of A and B, the pressure curve is similarly shaped, but in the resulting pressure will be higher for the blend which contains more of the A or higher pressure component.

Some refrigerant blends that are intended to match some other product. R-12 is a good example. It will rarely match the pressure at all points in the desired temperature range. What is more common is the blend will match in one region and the pressures will be different elsewhere.

In the example, the blend with concentration C1 matches the CFC at cold evaporator temperatures, but the pressures run higher at condenser conditions. The blend with composition C2 matches closer to room temperature. And, it may show the same pressure in a cylinder being stored, for example. The operation pressures at evaporator and condenser temperatures, however, will be somewhat different. Finally, the blend at C3 will generate the same pressures at hot condenser conditions, but the evaporator must run at lower pressures to get the same temperature (see Fig. 5-7).

It can be seen later that the choice of where the blend matches the pressure relationship can solve, or cause, certain retrofit-related problems.

Generally speaking, the R-12 retrofit blends have higher temperature glide. They do not match the pressure/temperature/capacity of R-12 across the wide temperature application range which R-12 once did. In other words, one blend does not fit all. Blends which match R-12 at

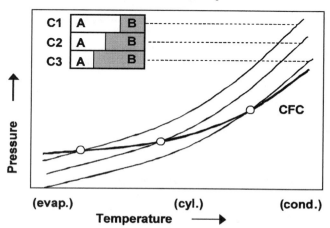

Figure 5-7 New refrigerants variable—composition. (*Courtesy of National Refrigerants.*)

colder evaporator temperatures may generate higher pressures and discharge temperatures when used in warmer applications or in high ambient temperatures. These are called *refrigeration blends.*

In refrigeration, it is often an easier and cheaper retrofit job if you can match evaporator pressures to R-12 and split the glide. That is because you can get similar box temperatures in similar run times. And, you would probably not need to change controls or the TXVs ,which are sensitive to pressure.

Blends, which match R-12 properties in hot conditions, such as in automotive AC condensers, may lose capacity or require lower suction pressures when applied at colder evaporator temperatures. These are called *automotive blends.*

For automotive air-conditioning many of the controls and safety switches are related to the high side pressure. If the blend generates higher discharge pressures you could short cycle more often and lose capacity in general. It is better to pick the high side to match R-12 and let the low side run a little lower pressure.

R-134a Refrigerant

The blended refrigerant R-134a is a long-term HFC alternative with similar properties to R-12. It has become the new industry standard refrigerant for automotive air-conditioning and refrigerator/freezer appliances. R-134a refrigerating performance will suffer at lower temperatures (below $-10°F$). Some traditional R-12 applications have used alternatives other than 134a for lower temperatures.

R-134a requires polyolester (POE) lubricants. Traditional mineral oils and alkyl benzenes do not mix with HFC refrigerants and their use with R-134a may cause operation problems or compressor failures. In addition, automotive AC systems may use poly alkaline glycols (PAGs), which are typically not seen in stationary equipment.

Both POEs and PAGs will absorb moisture, and hold onto it, to a much greater extent than traditional lubricants. The moisture will promote reactions in the lubricant as well as the usual problems associated with water—corrosion and acid formation. The best way to dry a wet HFC system is to rely on the filter drier. Deep vacuum will remove "free" water, but not the water that has absorbed into the lubricant.

R-134a applications

Appliances, refrigeration both commercial and self-contained equipment, centrifugal chillers, and automotive air-conditioning utilize R-134a. Retrofitting equipment with a substitute for R-12 is sometimes difficult there are a number of considerations to be examined before undertaking the task.

R-12 Systems—General Considerations

1. For centrifugal compressors it is recommended that the manufacturer's engineering staff become involved in the project—special parts or procedures may be required. This will ensure proper capacity and reliable operation after the retrofit.

2. Most older, direct expansion systems can be retrofit to R-401A, R-409A, R-414B, or R-416A (R-500 to R-401B or R-409A), so long as there are no components that will cause fractionation within the system to occur.

3. Filter driers should be changed at the time of conversion.

4. System should be properly labeled with refrigerant and lubricant type.

R-12 medium-high-temperature refrigeration (>0°F evaporation)

1. See Recommendation Table (this can be found on National Refrigerants Web site—click on Technical Manual) for blends that work better in high ambient heat conditions.

2. Review the properties of the new refrigerant you will use, and compare them to R-12. Prepare for any adjustments to system components based on pressure difference or temperature glide.

3. Filter driers should be changed at the time of conversion.

4. System should be properly labeled with refrigerant and lubricant type.

R-12 low-temperature refrigeration (<20° F evaporation)

1. See Recommendation Table for blends that have better low-temperature capacity.

2. Review the properties of the new refrigerant you will use, and compare them to R-12. Prepare for any adjustments to system components based on pressure difference or temperature glide.

3. Filter driers should be changed at the time of conversion.

4. System should be properly labeled with refrigerant and lubricant type.

Another blended refrigerant that can be used to substitute for R-12 is 401A . It is a blend of R-22, 152a, and 124. The pressure and system capacity match R-12 when the blend is running an average evaporator temperature of 10 to 20°F (−12.22 to −6.6°C).

Applications for this refrigerant are as a direct expansion refrigerate for R-12 in air-conditioning systems and in R-500 systems.

R-401B

This blend refrigerant is similar to R-401A except, it is higher in R-22 content. This blend has higher capacity at lower temperatures and matches R-12 at –20°F (–6.6°C). It also provides a closer match to R-500 at air-conditioning temperatures.

Applications for R-401B are in normally lower temperature R-12 refrigeration locations and in transport refrigeration and in R-500 as a direct expansion refrigerant in air-conditioning systems.

R-402A

A blend of R-22 and R-125 with hydrocarbon R-290 (propane) added to improve mineral oil circulation. This blend is formulated to match R-502 evaporator pressures, yet it has higher discharge pressure than 502. Although the propane helps with oil return, it is still recommended that some mineral oil be replaced with alkyl benzene.

Applications are in low-temperature (R-502) refrigeration locations. Retrofitting is used for R-502 substituting.

R-402B

Similar to R-402A, but with less R-125 and more R-22. This blend will generate higher discharge temperatures, which makes it work particularly well in ice machines. Applications are in ice machines where R-502 was used extensively.

Frequently Asked Questions

For what refrigerants are the R-60a, R-70a, and R-80a rated? See Fig. 5-8 for recovery units.

R-100 Recovery system

(A)

Refrigerant recovery cylinders

(B)

Figure 5-8   (*a*) Recovery units; (*b*) recovery unit cylinders. (*Courtesy of Ritchie Engineering.*)

They are tested by Underwriters Laboratories, Inc. (UL) to ARI 740-98 and approved for medium-pressure refrigerants R-12, R-134a, R-401C, R-406A, R-500; medium-high-pressure refrigerants R-401A, R-409A, R-401B, R-412A, R-411A, R-407D, R-22, R-411B, R-502, R-407C, R-402B, R-408A, R-509; and high-pressure refrigerants R-407A, R-404A, R-402A, R-507, R-407B, R-410A.

Q. Why should I purchase a recovery system?

A. With the Yellow Jacket name on a hose, you know you have got the genuine item for performance backed by more than 50 years. Now, you will also find the name on refrigerant recovery systems that are based on RRTI and RST, as is proven designs. RRTI was one of the original recovery companies and helped DuPont design its original unit. With the purchase of RST in 1998, Ritchie Engineering combined Yellow Jacket standards of manufacturing and testing with the RST track record of tough reliability.

Q. Is American Refrigeration Institute (ARI) the only testing agency?

A. No. ARI is only a certifying agency that hires another agency to perform the actual testing. UL is also EPA-approved as a testing and certifying agency. Yellow Jacket systems are UL-tested for performance. Some Yellow Jacket systems are tested to CSA, CUL, CE, and TUV safety standards which go beyond the ARI performance standards.

Q. Can I compare systems by comparing their ARI or UL ratings?

A. Yes, ARI and UL test standards should be the same. And remember that manufacturers can change the conditions under which they test their own machines to give the appearance of enhanced performance. Only ARI and UL test results provide consistent benchmarks and controls on which to make objective comparisons.

Q. How dependable are Yellow Jacket refrigerant recovery systems?

A. Yellow Jacket recovery systems get pushed to the limits: day in and day out in dirty conditions, on roof tops, and sometimes in freezing or high ambient temperatures. Yellow Jacket equipment has been tested at thousands of cycles, and are backed with the experience of units in the field since 1992.

Q. Is pumping liquid the fastest way to move refrigerant?

A. Yes, and the R-50a, R-60a, R-70a, R-80a, and R-100 monitor liquid flow at a rate safe for the compressor. In Yellow Jacket lab testing, over 80,000 lb of virgin liquid R-22 were continuously and successfully pumped. That is over 2500 standard 30-lb tanks, or the equivalent of refrigerant in over 25,000 typical central AC systems.

Q. What is the push/pull recovery method?

A. Many technicians use this recovery method, particularly on large jobs. By switching the R-50a, R-60a, R-70a, R-80a, or R-100 discharge valve to "recover" during push/pull recovery, the condenser is bypassed, increasing the "push" pressure and speeding the recovery.

Q. Why do the R-50a through 100 feature a built-in filter?

A. Every recovery machine requires an in-line filter to protect the machine against the particles and "gunk" that can be found in failed refrigeration systems. For your convenience, the R-50a, R-60a, R-70a, R-80a, and R-100 series incorporate a built-in 200-mesh filter that you can clean, and if necessary, replace. The filter traps 150-μm particles and protects against the dirtiest systems to maximize service life. In case of a burnout, an acid-core filter/drier is mandatory. The Yellow Jacket filter is built-in to prevent breaking off like some competitive units with external filters.

Q. What is auto purge and how does it work?

A. At the end of each cycle, several ounces of refrigerant can be left in the recovery machine to possibly contaminate the next job, or be illegally vented. Many competitive recovery machines require switching hoses, tuning the unit off and on, or other time-consuming procedures. The R-50a, R-60a, R-70a, R-80a, and R-100 can be quickly purged "on the fly" by simply closing the inlet valve, and switching the discharge valve to purge. In a few seconds, all residual refrigerant is purged and you are finished. Purging is completed without switching off the recovery unit.

Q. Can increased airflow benefit recovery cylinder pressure?

A. Yes. For reliable performance in the high ambient temperatures, Yellow Jacket units are engineered with a larger condenser and more aggressive fan blade with a greater pitch. This allows the unit to run cooler and keeps the refrigerant cooler in the recovery cylinder.

Q. Can I service a Yellow Jacket system in the field?

A. Although Yellow Jacket systems feature either a full 1- or an optional 2-year warranty, there are times when a unit will need a tune-up. The service manual with every unit includes a wide variety of information such as tips to speed recovery, troubleshooting guides, and parts listings. On the side of every unit, you will find hookup instructions, a quick start guide, and simple tips for troubleshooting. And if ever in doubt, just call 769-8370 and ask for customer service.

All service and repair parts are readily available through your nearest Yellow Jacket wholesaler.

Q. Are there any recovery systems certified for R-410A?

A. ARI 740-98 has been written but not yet enacted by the EPA. The Yellow Jacket R-60a, R-70a, and R-80a series have also been UL tested and certified for high-pressure gases such as R-410A that are covered under ARI 740-98.

Q. What features should I demand in a system to be used for R-410A?

A. Look for the following three features as a minimum:

- High-volume airflow through an oversized condenser to keep the unit running cooler and help eliminate cutouts in high ambient temperatures

- Single automatic internal high-pressure switch for simple operation

- Constant pressure regulator (CPR) valve rated to 600 psi for safety and it eliminates the need to monitor and regulate the unit during recovery

Q. What is a CPR valve?

A. The CPR valve is the feature that makes the Yellow Jacket R-70a and R-80a the first truly automatic recovery systems for every refrigerant. The single 600-psi-rated high-pressure switch covers all refrigerants and eliminates the need for a control panel with two selector switches for R-410A.

The CPR valve automatically reduces the pressure of the refrigerant being recovered. Regardless of which refrigerant, input is automatically regulated through a small orifice that allows refrigerant to flash into vapor for compression without slugging the compressor. Throttling is not required.

Under normal conditions, you could turn the machine to the "liquid" and "recover" settings. The machine will complete the job while you work elsewhere.

Q. With the Yellow Jacket R-70a design, do I have to manually reset a pressure switch between medium and high-pressure gases?

A. No. Some competitive machines require you to choose between medium and high-pressure gas settings before recovery. You will see the switch on their control panels. With the Yellow Jacket R-70a, the single internal automatic high-pressure switch makes the choice for you. That is why only R-70a is truly automatic.

Frequently Asked Questions about Pumps

Q. How can I select the right pump cfm?

A. See Fig. 5-9 for a sampling of pumps. The following guidelines are for domestic through commercial applications.

Q. Can I use a vacuum pump for recovery?

A. A vacuum pump removes water vapor and is *not* for refrigerant recovery. Connecting a vacuum pump to a pressurized line will damage the pump and vent refrigerant to the atmosphere, which is a crime.

System, tons	Pump, cfm
1–10	1.5
10–15	2.0
15–30	4.0
30–45	6.0
45–60	8.0
60–90	12.0
90–130	18.0
Over 130	24.0

Large capacity supervac™ pumps 12, 18, 24 cfm

Figure 5-9 Vacuum pumps. (*Courtesy of Ritchie Engineering.*)

Q. How much of a vacuum should I pull?

A. A properly evacuated system is at 2500 μm or less. This is 1/10 of 1 in. and impossible to detect without an electronic vacuum gage. For most refrigeration systems, American Society of Heating, Refrigerating, and Air-Conditioning Engineers (ASHRAE) recommends pulling vacuum to 1000 μm or less. Most system manufacturers recommend pulling to an even lower number of micrometers.

Q. Do I connect an electronic vacuum gage to the system or pump?

A. To monitor evacuation progress, connect it to the system with a vacuum/charge valve.

Q. Why does the gage micron (micrometer) reading rise after the system is isolated from the vacuum pump?

A. This indicates that water molecules are still detaching from the system's interior surfaces. The rate of rise indicates the level of system contamination and if evacuation should continue.

Q. Why does frost form on the system exterior during evacuation?

A. Because ice has formed inside. Use a heat gun to thaw all spots. This helps molecules move off system walls more quickly toward the pump.

Q. How can I speed evacuation?

A. ▪ Use clean vacuum pump oil. Milky oil is water saturated and limits pump efficiency.

▪ Remove valve cores from both high and low fittings with a vacuum/charge valve tool to reduce time through this orifice by at least 20 percent.

▪ Evacuate both high and low sides at the same time. Use short, 3/8-in. diameter and larger hoses.

SuperEvac systems can reduce evacuation time by over 50 to 60 percent. SuperEvac pumps are rated at 15 micrometers (or less) to pull a vacuum quickly. Large inlet allows you to connect a large diameter hose. With large oil capacity, SuperEvac pumps can remove more moisture from systems between oil changes.

Q. What hose construction is best for evacuation?

A. Stainless steel. There is no permeation and outgasing.

Frequently Asked Questions about Fluorescent Leak Scanners

Q. Does the ultra violet (UV) scanner light work better than an electronic leak detector? See Fig. 5-10.

A. No one detection system is better for all situations. But, with a UV lamp you can scan a system more quickly and moving air is never a problem. Solutions also leave a telltale mark at every leak site. Multiple leaks are found more quickly.

Q. How effective are new light emitting diode (LED) type UV lights? See Fig. 5-11.

A. LEDs are small, compact lights for use in close range. Most effective at 6-in. range. The model with two blue UV and three UV bulbs has a slightly greater range. Higher power Yellow Jacket lights are available.

Q. Can LED bulbs be replaced?

A. No. The average life is 110,000 h.

Q. Are RediBeam lamps as effective as the System II lamps?

A. The RediBeam lamp has slightly less power to provide lightweight portability. But with the patented reflector and filter technologies, the RediBeam 100-W bulb produces sufficient UV light for pinpointing leaks.

UV LEAK DETECTION TOOLS

53515 MACH IV FLEXIBLE UV LIGHT

The MACH IV flexible UV light has 4 TRUE UV LEDS delivering a brilliant fluorescent glow. Gets into tight areas easily.

53012 UV SWIVEL HEAD LIGHT

- 12V/100 WATT
- 180° Swivel Head Gets Into Tight Places
- Instant ON
- Heavy Duty Metal Construction
- Includes UV Enhancing Safety Glasses

53312 UV MINI LIGHT

- 12V/50 WATT
- High Intensity
- Compact & Lightweight
- Instant ON
- 16 ft. (5m) Cord
- Includes UV Enhancing Safety Glasses

RECHARGEABLE UV LIGHT

53411
- High Intensity 12V/50 WATT
- Cordless with Rechargeable Battery Cartridge and Charger
- Compact & Portable
- Includes UV Enhancing Safety Glasses

53412
- Same as 53411 (less Battery Charger and glasses)

53413 BATTERY CARTRIDGE

- 12V/50 WATT
- This powerful rechargeable battery cartridge holds a charge equal to 30 minutes of continuous use

BATTERY CHARGER

| 53414 | 110V/60 HZ |
| 53414-220 | 220V/50-60 HZ |

LIGHT PART#	LIGHT BULB PART#	LENS PART#
53012	53012-B	53012-L
53112	53012-B	53110-L
53312	53312-B	53312-L
53411	53312-B	53312-L
53515	53515-B	–

ACCESSORIES

53314 DYE REMOVER - 4 oz

53315 SERVICE LABELS - 25 per pack. Bright yellow label indicating that the system has been charged with UV Dye.

92398 UV ENHANCING SAFETY GLASSES This is a must for protection against ultraviolet light during leak detection.

53809 MINI DYE INJECTOR

"Cartridge Type" Dye Injector (10 Appl.) Concentrated Dye.

53223 "CARTRIDGE TYPE" UNIVERSAL DYE/OIL INJECTOR

Adding dye with the new 53223 is fast and easy, simply connect the injector to the low side of the A/C system and twist the handle to the next application line. The replaceable cartridge provides 25 applications of universal dye that is compatible with R12/22/502 and R134a systems. The injector hose comes complete with a R134a coupler and auto shut-off valve adapter for 1/4"FL systems.

"REFILLABLE" UNIVERSAL DYE/OIL INJECTOR

To inject the oil or dye, simply connect the injector to the low side of the A/C system and twist the handle until you reach the desired amount. The injector hose comes complete with a R134a coupler and auto shut-off valve adapter for 1/4"FL systems.

53123	**REFILLABLE UNIVERSAL DYE INJECTOR** with 2 oz Bottle of Concentrated A/C Dye (25 Appl.)
53123-A	**REFILLABLE UNIVERSAL DYE/OIL INJECTOR** (without dye)
53134	**REFILLABLE DYE INJECTOR** with (R134a 13mm Connection) and 2 oz Bottle of Concentrated A/C Dye (25 Appl.)
53322	**REFILLABLE DYE INJECTOR** with (1/4" Auto Shut-off Valve Connection) and 2 oz Bottle of Concentrated A/C Dye (25 Appl.)

ULTRAVIOLET DYES

Standard Universal A/C Dyes

Pack 1/4 oz -	1 Application
92699	Standard Universal Dye Six 1/4 oz Packs (6 pcs)
8 oz Bottle -	32 Applications
92708	R12/22/502 Standard Universal Dye
32 oz Bottle -	128 Applications
92732	R12/22/502 Standard Universal Dye

Concentrated Universal Dyes

Replaceable Cartridge -	25 Applications
53825	Universal Dye Cartridge
Replaceable Cartridge -	10 Applications
53810	Concentrated Dye Cartridge
2 oz Bottle -	25 Applications
53625	Universal Dye

Figure 5-10 Leak scanners (fluorescent type). (*Courtesy of Mastercool.*)

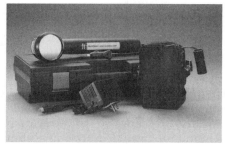

Universal UV system

Figure 5-11 LED ultraviolet lights. (*Courtesy of Ritchie Engineering.*)

Q. Does the solution mix completely in the system?
A. Solutions are combinations of compatible refrigeration oil and fluorescent material designed to mix completely with the oil type in the system.

Q. How are solutions different?
A. Solutions are available with mineral, alkylbenzene, PAG, or polyol ester base stock to match oil in the system.

Q. What is universal solution?
A. It is made from polyol ester and mixes well with newer oils. It also works in mineral oil systems, but can be harder to see.

Q. What is the lowest operating temperature?
A. It is −40°F (−40°C) for all solutions. However, in alkylbenzene systems it is −100° F (−73.3°C).

Q. Does solution stay in the system?
A. Yes. When future leaks develop, just scan for the sources. In over 6 years of testing, the fluorescent color retained contrast. When the oil is changed in the system, scanner solution must be added to the new oil.

Q. Is the solution safe?
A. Solutions were tested for 3 years before introduction and have been performance proven in the field since 1989. Results have shown the solutions are safe for technicians, the environment, and all equipment when used as directed. Solutions are pure and do not contain lead, chromium, or chlorofluorocarbon (CFC) products. Presently, solutions are approved and used by major manufacturers of compressors, refrigerant, and equipment.

Q. How do I determine oil type in system?
A. Many times the oil is known due to the type of refrigerant or equipment application. Systems should be marked with the kind of oil used. Always tag system when oil type is changed.

Q. In a system with a mix of mineral and alkylbenzene oil, which scanner solution should be used?

A. Base your choice of solution on whatever oil is present in the larger quantity. If you do not know which oil is in greater quantity, assume it to be alkylbenzene.

Q. How do I add solution to the system?

A. In addition to adding solution using injectors or mist infuser, you have other possibilities. If you do not want to add more gas to the system, connect the injector between the high and low side allowing system pressure to do the job. Or, remove some oil from the system, then add solution to the oil and pump back in.

Q. How is the solution different from visible colored dyes?

A. Unlike colored dyes, Yellow Jacket fluorescent solutions mix completely with refrigerant and oil and do not settle out. Lubrication, cooling capacity, and unit life are not affected; and there is no threat to valves or plugging of filters. The solutions will also work in a system containing dytel.

Q. How do I test the system?

A. Put solution into a running system to be mixed with oil and carried throughout the system. Nitrogen charging for test purposes will not work since nitrogen will not carry the oil. To confirm solution in the system, shine the lamp into the system's sight glass. Another way is to connect a hose and a sight glass between the high and low sides, and monitor flow with the lamp. The most common reason for inadequate fluorescence is insufficient solution in the system.

Q. Can you tell me more about bulbs?

A. 115-V systems are sold with self-ballasted bulbs in the 150-W range. Bulbs operate in the 365-nm long range UV area and produce the light necessary to activate the fluorescing material in the solution. A filter on the front of the lamp allows only "B" band rays to come through. "B" band rays are not harmful.

Q. What is the most effective way to perform an acid test?

A. Scanner solution affects the color of the oil slightly. Use a two-step acid-test kit which factors out the solution in the oil, giving a reliable result.

Q. Can fluorescent product be used in nonrefrigerant applications?

A. Yes, in many applications.

AccuProbe Leak Detector
with Heated Sensor Tip

This is the only tool you need for fast, easy, and certain leak detection. The heated sensor tip of the Yellow Jacket Accuprobe leak detector

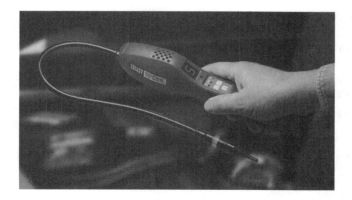

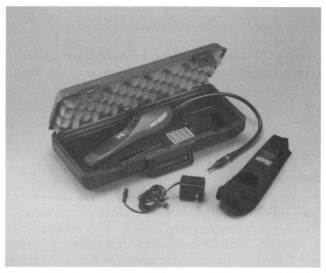

Figure 5-12 AccuProbe leak detector. (*Courtesy of Ritchie Engineering.*)

positively identifies the leak source for all refrigerants. That includes CFCs such as R-12 and R-502, hydrochlorofluorocarbons (HCFCs) as small as 0.03 oz/year, and hydrofluorocarbon (HFC) leaks of 0.06 oz/year, even R-404A, and R-410A (see Fig. 5-12).

Frequency of flashes in the tip and audible beeping increases the closer you get to the leak source. You zero-in and the exclusive smart alarm LED shows how big or small the leak on a scale of 1 to 9 is. Maximum value helps you determine if the leak needs immediate repair.

Service life of the replaceable sensor is more than 300 h with minimal cleaning and no adjustments. Replaceable filters help keep out moisture and dust that can trigger false sensing and alarms.

Three sensitivity levels include ultrahigh to detect leaks that could be missed with other detection systems. Additional features and benefits are given below:

- Detect all HFCs including R-134a, R-404A, R-410A, and R-407C and R-507; all HCFCs such as R-22 and all CFCs.
- Audible beeping can be muted.
- Extended flexible probe for easy access in hard-to-reach areas.
- Sensor not poisoned by large amount of refrigerant and does not need recalibration.
- Sensor failure report mode.
- Weighs only about 15 oz for handling comfort and ease.
- Carrying holster and hard, protective case included.

Bottle of non-ozone-depleting chemical included for use as a leak standard to verify proper functioning of sensor and electronic circuitry

Technology Comparison—Heated Sensor or Negative Corona?

Heated sensor leak detectors

When the heated sensing element is exposed to refrigerant, an electrochemical reaction changes the electrical resistance within the element causing an alarm. The sensor is refrigerant specific with superior sensitivity to all HFCs, HCFCs, and CFCs and minimal chance of false alarms. When exposed to large amounts of refrigerant which could poison other systems, the heated sensor clears quickly and does not need recalibration before reuse (see Fig. 5-13).

Negative corona leak detectors

In the sensor of an old-style corona detector, high voltage applied to a pointed electrode creates a corona. When refrigerant breaks the corona arc, the degree of breakage generates the level of the alarm. This technology has good sensitivity to R-12 and R-22, but only fair for R-134a, and poor for R-410A, R-404A, and R-407C. Sensitivity decreases with exposure to dirt, oils, and water. And false alarms can be triggered by dust, dirt specks, soap bubbles, humidity, smoke, small variations in the electrode emission, high levels of hydrocarbon vapors, and other non-refrigerant variables (see Fig. 5-14).

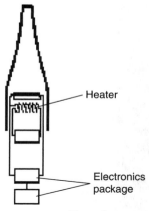

Figure 5-13 Heated sensor leak detector. (*Courtesy of Ritchie Engineering.*)

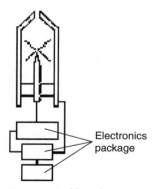

Figure 5-14 Negative corona leak detector. (*Courtesy of Ritchie Engineering.*)

Tips for Detecting System Leaks

1. Inspect entire air-conditioning system for signs of oil leakage, corrosion cracks, or other damage. Follow the system in a continuous path so no potential leaks are missed.

2. Make sure there is enough refrigerant in a system (about 15 percent of system capacity or 50 psi/min) to generate pressure to detect leaks.

3. Check all service access port fittings. Check seals in caps.

4. Move detector probe at 1 in./s within 1/4 in. of suspected leak area.

5. Refrigerant is heavier than air, so position probe below test point.

6. Minimize air movement in area to make it easier to pinpoint the leak.

7. Verify an apparent leak by blowing air into the suspected leak.

8. When checking for evaporator leaks, check for gas in condensate drain tube.

9. Use heated sensor type detector for difficult-to-detect R-134a, R-410A, R-407C, and R-404A.

New Combustible Gas Detector— with Ultrasensitive, Long-Life Sensor

Handheld precision equipment detects all hydrocarbon and other combustible gases including propane, methane, butane, industrial solvents, and more (see Fig. 5-15).

Figure 5-15 Combustible gas detector. (*Courtesy of Ritchie Engineering.*)

- Sensitivity, bar graph, and beeping to signal how much and how close.
- Unit is preset at normal sensitivity, but you can switch to high or low. After warm-up you will hear a slow beeping. Frequency increases when a leak is detected until an alarm sounds when moving into high gas concentration. The illuminated bar graph indicates leak size.
- If no leak is detected in an area you suspect, select high sensitivity. This will detect even low levels in the area to confirm your suspicions. Use low sensitivity as you move the tip over more defined areas, and you will be alerted when the tip encounters the concentration at the leak source.
- Ultrasensitive sensor detects less than 5 ppm methane and better than 2 ppm for propane. They perform equally well on a complete list of detectable gases including acetylene, butane, and isobutane.
- Automatic calibration and zeroing.
- Long-life sensor easily replaced after full service life.

Applications
- Gas lines/pipes
- Propane filling stations
- Gas heaters
- Combustion appliances
- Hydrocarbon refrigerant
- Heat exchangers

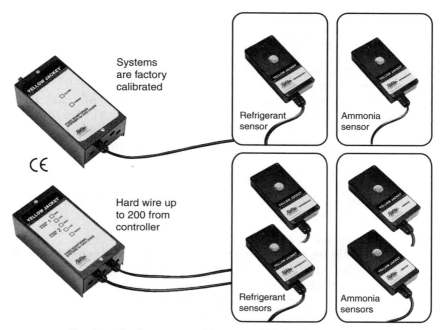

Figure 5-16 Fixed monitoring systems. (*Courtesy of Ritchie Engineering.*)

- Marine bilges
- Manholes
- Air quality
- Arson residue (accelerants)

Frequently Asked Questions about Fixed Monitoring Systems (See Fig. 5-16)

Q. Are calibrated leak testers available to confirm that the monitor is calibrated correctly?

A. The Yellow Jacket calibrated leak is a nonreactive mixture of R-134a or NH_3 and CO_2. The nonreturnable cylinders contain 10 L of test gas. The cylinders require a reusable control valve and flow indicator. Test gases can be ordered for 100 ppm or 1000 ppm mixtures.

Q. What refrigerants will the leak monitors detect?

A. Leak monitors will detect most CFC, HFC, and HCFCs such as R-11, R-12, R-13, R-22, R-113, R-123, R-134a, R-404A, R-407C, R-410A, R-500, R-502, and R-507. Yellow Jacket also has leak monitors available for ammonia and hydrocarbon-based refrigerants.

Q. Can the leak monitor be calibrated for specific applications?

A. Yes, the Yellow Jacket leak monitor can be calibrated for specific applications. Contact customer service for your specific need.

Q. If the unit goes into alarm, can it switch on the fan? Can it turn off the system at the same time?

A. The leak monitor has a pair of dry, normally open/normally closed contacts that can handle 10 A at 115 V. When the sensor indicates a gas presence higher than the set point, it opens the closed contacts and closes the open contacts which will turn equipment on or off.

Q. After a unit goes into alarm and the contacts close, what can it be connected to?

A. The open contacts can shut the system down, call a phone number, turn on a fan, or emergency light.

Q. How does the sensor work?

A. When the sintered metal oxide surface within the sensor absorbs gas molecules, electrical resistance is reduced in the surface allowing electrons to flow more easily. The system controller reads this increase in conductivity and signals an alarm. Metal oxide technology is proven for stability and performance.

Q. What is the detection sensitivity level of Yellow Jacket fixed monitors?

A. The dual sensitivity system has a low alarm level of about 100 ppm and a high level of about 1000 ppm for most CFC, HFC, and HCFC products. The high level for R-123 is an exception at about 300 ppm. Ammonia detection levels are about 100 ppm low and about 150 ppm high. The alarm level of all Yellow Jacket single-level systems is about 100 ppm.

Detection levels are preset at the factory to cover most situations. If necessary, however, you can order a custom level, or adjust the set point on site.

Q. What gas concentration must be detected?

A. Depends on the refrigerant. For a more thorough answer, terms established by U.S. agencies must first be understood:

- IDLH— immediately dangerous to life and health

- TWA—time-weighted average concentration value over an 8-h work day or a 40-h work week (OSHA or NIOSH levels)

- STEL—short-term exposure level measured over 15 min (NIOSH)

- Ceiling concentration—should not be exceeded in a working day (OSHA)

Obviously, the first consideration is IDLH. For most refrigerants, the IDLH is relatively high (e.g., R-12 is 15,000 ppm), and such a concentration would be unusual in a typical refrigerant leak situation. Leak detection, however, is still an immediate condition, so the STEL should be the next consideration, followed then by the 8-h TWA or ceiling concentration. R-22, for example, has a STEL of 1250 ppm and a TWA of 1000 ppm. (The TWA for most refrigerants is 1000 ppm.)

The draft UL standard for leak monitors requires gas detection at 50 percent of the IDLH. In other words, R-12 with a IDLH value of 15,000 ppm must be detected at 7500 ppm. As with most refrigerants, the TWA is 1000 ppm.

All of the foregoing suggests that for most CFC, HFC, and HCFC, detection at 1000 ppm provides a necessary safety margin for repair personnel. Ammonia with a significantly lower IDLH of 300 ppm and a TWA of 25 ppm requires detection at 150 ppm to comply with 50 percent IDLH requirements. R-123 has a TWA of 50 ppm and an IDLH of 1000 ppm, therefore detection at 100 ppm provides a good margin of safety. A monitor with a detection threshold of about 100 ppm for any gas provides an early warning so that repairs can be made quickly. This can save refrigerant, money, and the environment.

Q. How frequently should the system be calibrated?

A. Factory calibration should be adequate for 5 to 8 years. Routine calibration is unnecessary when used with intended refrigerants. Yellow Jacket sensors can not be poisoned, show negligible drift, and are stable in long term. You should, however, routinely check performance.

Q. Can there be a false alarm?

A. For monitoring mixtures, the semiconductor must be able to respond to molecularly similar gases. With such sensitivity, false alarms can be possible. Engineered features in Yellow Jacket monitors help minimize false alarms:

- *The two-level system waits about 30 s until it is "certain" that gas is present before signaling.*

- *At about 100 to 1000-ppm calibration level, false alarms are unlikely.*

To prevent an unnecessary alarm, turn off the unit or disable the siren during maintenance involving refrigerants or solvents. Temperature, humidity, or transient gases may occasionally cause an alarm. If in doubt, check with the manufacturer.

Q. What are alternative technologies for monitoring and detecting refrigeration gas leaks?

A. Infrared technology is sensitive down to 1 ppm. This level is not normally required for refrigeration gases and is also very expensive compared to semiconductor technology. As an infrared beam passes through an air sample, each substance in the air absorbs the beam differently. Variations in the beam indicate the presence of a particular substance. The technique is very gas specific and in a room of mixed refrigerants, more than one system would be required. To get over this problem, newer models work on a broad band principle so they can see a range of gases. As a result they do not generally operate below 50 ppm and can experience false alarms.

Electrochemical cells can be used for ammonia. These cells are very accurate, but are expensive, and are normally used to detect low levels (less than 500 ppm), and perform for about 2 years.

With air sampling transport systems, tubing extends from the area(s) to be monitored back to a central controller/ sensor.

Micropumps move air through the system eliminating a number of on-site sensors, but there may be problems. Air in the area of concern is sampled at intervals rather than monitored continuously. This can slow the response to changing conditions.

- *Dirt can be sucked into the tubes, blocking filters.*

- *Gases can be absorbed by the tube or leak out of the tube providing a concentration at the sensor lower than in the monitored area.*

- *Gases can leak into the tube in transit rather than the area monitored. The reading would be misleading.*

6

Solenoids and Relays

Magnetism is something that everyone knows about and has worked with, even when very young. Magnetism is the property of certain materials that permits them to produce or conduct magnetic lines of force.

Magnetic lines of force surround a current-carrying conductor or any atomic particle in motion. Even in the atom, magnetic fields result from the spinning motions of the electrons. Such magnetic lines of force can also interact with electric fields or other magnetic fields.

Every time you press a doorbell, an electromagnet causes it to ring or chime. Electromagnets are created by sending electricity through a coil. Permanent magnets are made from materials that will retain magnetism. That is, they keep their ability to attract iron (and other materials). They do not need electricity to function. Electromagnets lose their magnetism when the electric current is removed. Both permanent and temporary (electromagnet) magnets have their use in heating, air-conditioning, and refrigeration-control circuits.

Permanent Magnets

One type of permanent magnet is the natural magnet, the lodestone. It is also possible to make permanent magnets. For example, if you stroke a piece of high-carbon steel with a lodestone, the steel becomes magnetized (see Fig. 6-1).

Under normal conditions, the steel stroked with the lodestone keeps its magnetism permanently. When this happens, the molecules that form the steel bar line up in the same direction (see Fig. 6-2). The molecules that line themselves up in this way point toward the ends, or

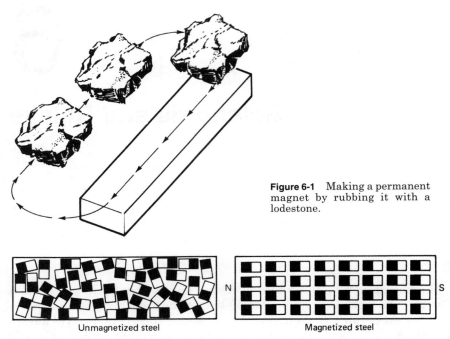

Figure 6-1 Making a permanent magnet by rubbing it with a lodestone.

N   S

Unmagnetized steel   Magnetized steel

Figure 6-2 A magnetic field aligns iron molecules to produce a permanent magnet.

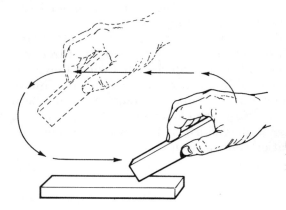

Figure 6-3 Permanent magnet used to make another permanent magnet.

poles, of the magnet. Each magnet has two poles, which are identified as north and south.

A permanent magnet can be used to create other magnets. Figure 6-3 shows a permanent magnet used to stroke a bar of high-carbon steel. This creates a second magnet.

Magnets made of high-carbon steel sometimes lose their magnetism. This happens if the magnetized steel bar receives a strong shock, for instance, if the bar is dropped or struck with a hammer. When this happens, the

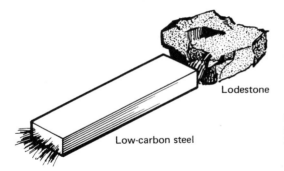

Figure 6-4 A temporary magnet made by placing the lodestone next to a piece of low-carbon steel.

molecules lose their alignment. This leads to a loss of magnetism. Magnetism can also be lost if a magnet is heated to high temperatures.

A magnet may be formed by rubbing a piece of steel with another magnet. This is called magnetic induction. This was the only way known of making magnets until electromagnetism was discovered.

Temporary Magnets

A bar of low-carbon steel attracts iron particles when it is in direct contact with another magnet. For example, you can place a lodestone against one pole of a bar of low-carbon steel as shown in Fig. 6-4. When this is done, the bar attracts iron particles. Removing the lodestone removes the magnetism. Then the iron particles fall away. The same happens if you place a permanent magnet against one end of a bar of low-carbon steel.

Low-carbon steel does not retain, or keep, magnetism. This material can serve as a temporary magnet. But it cannot be used for a permanent magnet.

Electromagnets

Magnetism also results when electric currents are passed through iron materials. Electromagnetism is important to electricity.

Magnetic Theory

It is important to know the basics about how magnetism behaves for understanding how certain pieces of equipment operate. Scientists still do not know exactly how magnets behave. There are no laws or fixed rules for magnetism. Instead, there are a number of theories. These theories explain the behavioral patterns that can be observed.

Magnetic permeability

The ability of a material to become magnetized is called *permeability*. The magnetic force that attracts iron materials is called *magnetic flux*. The greater the permeability of a material, the higher its magnetic flux.

TABLE 6-1   Selected Permeabilities

Material	Permeability
Bismuth	0.999833
Quartz	0.999985
Water	0.999991
Copper	0.999995
Liquid oxygen	1.00346
Oxygen (STP)*	1.0000018
Aluminum	1.0000214
Air (STP)*	1.0000004
Nickel	40
Cobalt	50
Iron	7,000
Permalloy	74,000

*STP, standard temperature and pressure.

Numbers are used to indicate the permeability of materials. The higher the number, the more easily a material is magnetized. The relative permeability of a number of materials is listed in Table 6-1. At lower levels, materials are considered to be nonmagnetic. This applies to aluminum and all materials listed above it in Table 6-1.

Materials with higher ratings, including nickel, cobalt, iron, and permalloy, have high permeability. Permalloy, a combination of materials, has extremely high permeability. When a bar of permalloy is held in the north-south direction, it becomes a magnet. When the bar is turned to face the east-west direction, it loses the magnetism.

Shapes of magnets

Magnets can be made in a variety of shapes. Some of these are shown in Fig. 6-5. One common shape for magnets is the rod or bar. Another shape is similar to the letter U. This is sometimes called a horseshoe magnet.

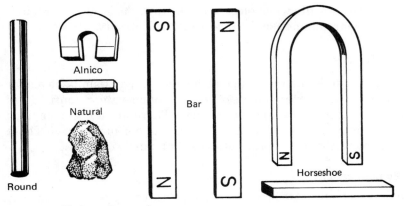

Figure 6-5 Various shapes of magnets.

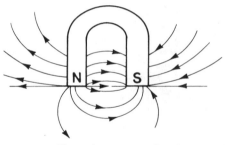

Figure 6-6 Horseshoe-magnet flux pattern.

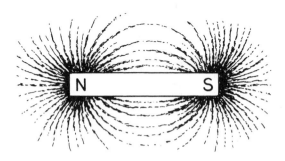

Figure 6-7 Bar magnet flux field.

Magnets made in a horseshoe shape have an advantage, their ends, or poles, are close together. Thus, horseshoe magnets have a strong field of magnetic attraction, or flux field (see Fig. 6-6). By comparison, a bar magnet of the same strength has a flux field with less attraction. This is demonstrated in Fig. 6-7. The horseshoe magnet comes in handy when we start studying electrical measurement instruments.

Poles of magnets

Dip a bar magnet into a pile of iron filings. When it is withdrawn, the filings will gather at the ends of the magnet (see Fig. 6-8). This demonstrates that magnetism is strongest at the ends. These are the poles of a magnet.

Each magnet has two poles. However, there is a difference between the poles. If you suspend a magnet from a string, it turns until one pole points north and the other south (see Fig. 6-9). (This will occur if you do not have any large deposits of steel around, such as steel storage cabinets or a building with a steel framework.) This is the principle behind the operation of a compass: one pole of a magnet is north-seeking. A bar magnet is suspended on a bearing to reduce friction. The north-seeking pole points north and the other pole is direct south. Therefore, the poles of a magnet are referred to as the north pole and the south pole.

If you point the north poles of two bar magnets at each other, they are not attracted. Actually, there is a reverse force pushing the magnets

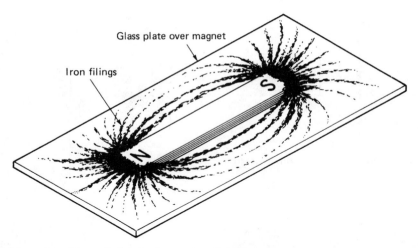

Figure 6-8   Magnetic flux is strongest at the poles.

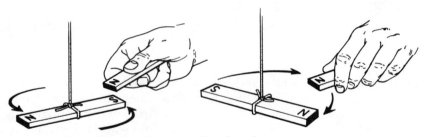

Figure 6-9   Magnetic poles align to north and south.

apart. This force repels them from each other. This is a rule of magnetism: like poles repel.

If you place the north pole of one magnet near the south pole of another, there is a strong attraction. This follows another rule of magnetism: unlike, or opposite, poles attract. This means that the north-seeking pole of a magnet is actually the south pole (see Fig. 6-10). The pattern of attraction around a bar magnet is shown in Fig. 6-11.

The relationship between electric current and magnetism was discovered by Hans Christian Oersted in 1820. This relationship is direct. When current flows through a conductor, there is a magnetic field around the conductor. The direction of the current flow determines the force field of an electromagnet. Currents flowing in the same direction set up connecting fields of force (see Fig. 6-12). Currents flowing in opposite directions set up repelling force fields (see Fig. 6-13).

The polarity of an electromagnet is determined by the direction of the current (see Fig. 6-14). This is important in an electrical meter.

The strength of an electromagnetic field is proportional to the current flowing through the conductor. More current means more magnetism.

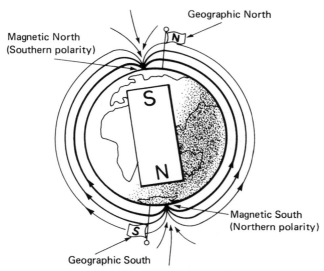

Figure 6-10   Polar attraction of magnets.

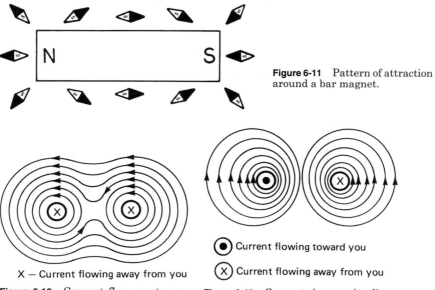

Figure 6-11   Pattern of attraction around a bar magnet.

X — Current flowing away from you

(•) Current flowing toward you

(X) Current flowing away from you

Figure 6-12   Current flow creates an electromagnetic field.

Figure 6-13   Currents in opposite directions create repelling fields.

Magnetism in a coil of wire

The magnetic field that surrounds a conductor depends on the form into which the wire is shaped (see Fig. 6-15). The magnetic field surrounding a single loop of wire is shown in Fig. 6-15a. Additional loops are shown in Fig. 6-15b. These loops form a spiral or helix. This shape is also

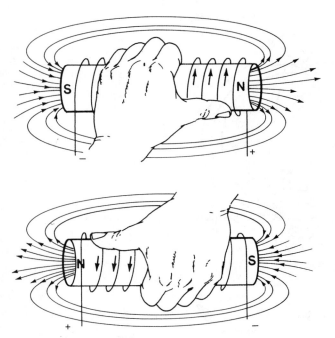

Figure 6-14 The left-hand rule shows how the polarity of an electromagnet is determined by the current flow. Grasp the coil with your fingers pointing in the direction of the current flow: the thumb of the left hand then points toward the north pole.

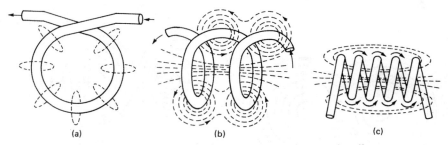

Figure 6-15 Magnetic fields form around various shapes of magnetic coils.

known as a helical coil. Figure 6-15*b* shows magnetic flux around loops of wire wound next to one another. The more loops in the wire, the stronger the magnetic field becomes. This is shown in Fig. 6-15*c*.

The strength of an electromagnet or coiled conductor depends on the amount of current flowing in the coil and the number of turns of wire. Keep in mind that the direction of current flow determines the magnetic polarity (see Figs. 6-13 and 6-14).

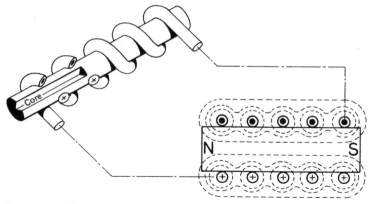

Figure 6-16   Electromagnets can be made by winding wire around a soft iron core.

Electromagnets

The usual form of electromagnet is a coil of wire wound around a soft iron core. Thus, the core will lose its magnetism when the source of energy is removed (see Fig. 6-16). The iron core provides an easy path for the magnetic field created by the coil. An electromagnet with an iron core forms a stronger magnet than the same coil of wire without the iron core.

The size of the iron core can help determine the strength of an electromagnet. For full strength, the core should be large enough to absorb all the magnetism from the coil. When the coil creates more magnetism than the core can absorb, it is called *saturation*. If the core is too small, there is too much magnetism to be absorbed. All the magnetism of the coil can be used when the core has a capacity slightly larger than the magnetic flux created.

Using electromagnetism

The generation and use of electricity are directly related to electromagnetism. Some of these uses are in the solenoid and relay. These two items are useful in heating, air-conditioning, and refrigeration equipment-control circuits. It is important to understand how they work, since they can be the source of much troubleshooting time.

The Solenoid

A magnetic field seeks the path of minimum reluctance, just as an electric current seeks the path of least resistance. Reluctance and resistance are related. Resistance refers to the opposition to current flow in a circuit. Reluctance refers to the opposition of the flow of a magnetic field. The lower the reluctance, the greater is the attraction of materials to a magnetic field.

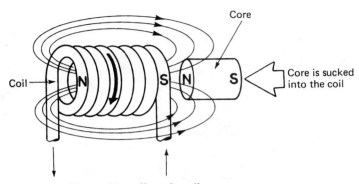

Core

Core is sucked
into the coil

Coil

Figure 6-17 The sucking effect of a coil.

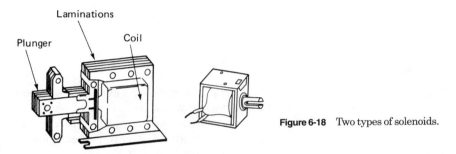

Laminations

Plunger

Coil

Figure 6-18 Two types of solenoids.

A solenoid is a device that uses these principles. A helical coil of wire produces a magnetic field. An iron core fits loosely within the coil of wire. When the current is off, the core rests outside the area of the coil. When the current is applied, the core is sucked into the coil. This is referred to as the *sucking effect* of a coil (see Fig. 6-17).

The sucking effect is often used in devices that require a small amount of physical movement. One type of solenoid is shown in Fig. 6-18. A common use for solenoids is in a device called solenoid valves (see Fig. 6-19). This is a device that opens and closes to permit the flow of liquids or gases. In most valves, for example, you turn a handle or knob to start or stop the flow of gas. Solenoid valves are used widely as safety devices. They are found in gas lines, air lines, and water lines. Many types are used in air-conditioning, refrigeration, and heating systems.

The solenoid in the valve shown in Fig. 6-19 is closed when there is no current through its coil. The valve stem is held in a closed position by a spring that applies light pressure. When the valve is closed, gas cannot flow. Current is applied when a person or thermostat turns on a heater. The current draws the movable core, or plunger, into the coil. This opens the gate to the flow of gas. When the current is turned off, the spring moves the valve system back into a closed position.

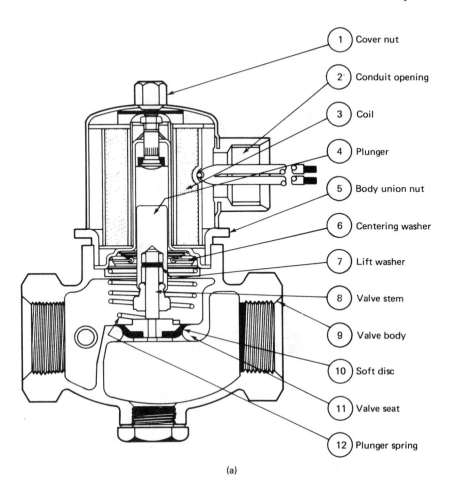

1. Cover nut
2. Conduit opening
3. Coil
4. Plunger
5. Body union nut
6. Centering washer
7. Lift washer
8. Valve stem
9. Valve body
10. Soft disc
11. Valve seat
12. Plunger spring

(a)

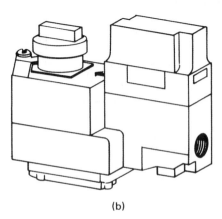

(b)

Figure 6-19 (*a*) Cutaway view of a solenoid valve. (*b*) Solenoid valve used for hot air furnace gas control. (*c*) Relay used to start a refrigerator or compressor.(*d*) Schematic for a refrigerator.

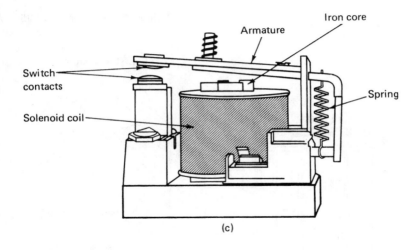

(c)

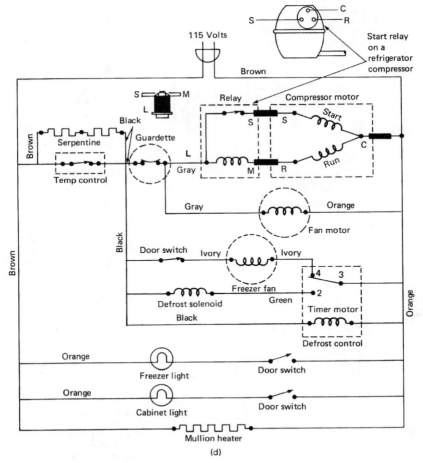

(d)

Figure 6-19 (*Continued*)

Switch contacts

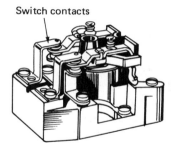

Figure 6-20 Heavy-duty power relay.

This same principle is used widely in solenoid-type relays. A solenoid relay is like an electrically operated switch. The coil in this type of relay pulls a core piece that has a number of electrical contacts. The contacts are designed so that they may either close or open electrical circuits. The relay contacts can themselves be designed to handle large amounts of current. But the coil of the control solenoid may operate on only a fraction of an ampere. The effect is that low-voltage, low-current electricity is used to control the flow of larger amounts of electricity.

Power relays

A heavy-duty power relay that operates from solenoid action is shown in Fig. 6-20. In this device, a spring pulls the solenoid core, or armature, away from the electrical contact when the current is off. When current flows, the armature is pulled toward the coil. An electrical contact connected to the armature is either closed or opened by this action. Note that the electromagnetism can be used either to open or close the relay. The action taken depends on the design of the relay. The relay contacts can be arranged for a variety of functions, such as SPST, SPDT, DPDT, or any other combination.

The advantage of a relay is that a substantial pulling power can be developed with a small coil current. The contacts can be made quite large and can handle or switch high values of electrical power. An extremely small amount of control power thus can be used to switch much higher voltages and currents in a safe manner (see Fig. 6-21).

Solenoid valves

The primary purpose of an electrically operated solenoid valve is to control automatically the flow of fluids, liquid or gas. There are two basic types of solenoid valves. The most common is the normally closed (NC) type, in which the valve opens when the coil is energized and closes when the coil is de-energized. The other type is the normally open (NO) valve, which opens when the coil is de-energized and closes when the coil is energized (see Fig. 6-22).

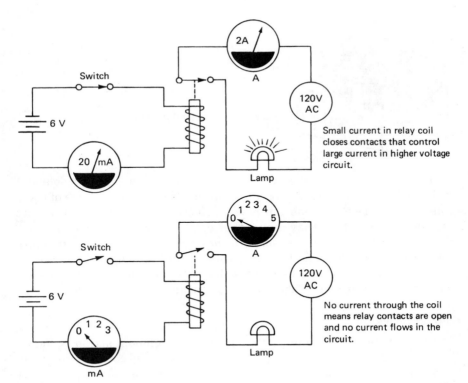

Small current in relay coil closes contacts that control large current in higher voltage circuit.

No current through the coil means relay contacts are open and no current flows in the circuit.

Figure 6-21 Relay in a circuit.

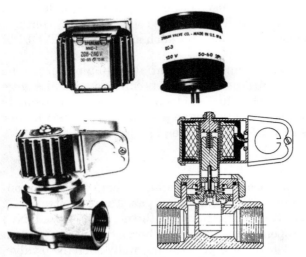

Figure 6-22 Solenoid coil and step-down transformer. (*Courtesy of Sporlan.*)

Principles of operation

Solenoid valve operation is based on the theory of the electromagnets. The solenoid valve coil sets up a magnetic field when electrical current is flowing through it. If a magnetic metal, such as iron or steel, is introduced into the magnetic field, the pull of the field will raise the metal and center it in the hollow core of the coil. By attaching a stem to the magnetic metal plunger, this principle is used to open the port of the valve. When the electrical circuit to the coil is broken, the magnetic field collapses and the stem and plunger either fall by gravity or are pushed down by the kick-off spring.

Some solenoid valves are designed with a hammer-blow effect. When the coil is energized, the plunger starts upward before the stem. The plunger then picks up the stem by making contact with a collar at the top. The momentum of the plunger assists in opening the valve against the unbalanced pressure across the port.

Applications

In many cases, valves are used for controlling the flow of refrigerants in liquid or suction lines or in hot-gas defrost circuits. They are equally suitable for many other less common forms of refrigerant control.

Liquid-line service.   The primary purpose of a solenoid valve in a refrigerant liquid line is to prevent flow into the evaporator during the off cycle. On multiple systems, a solenoid valve may be used in each liquid line leading to the individual evaporators.

The application of a liquid-line solenoid valve depends mainly on the method of wiring the valve with the compressor control circuit. It may be wired so the valve is energized only when the compressor is running. This type of application is shown in Fig. 6-23.

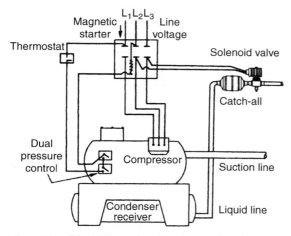

Figure 6-23   Liquid-line solenoid valve application.

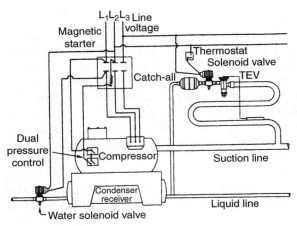

Figure 6-24   A pump-down control uses a thermostat to control the solenoid valve. (*Courtesy of Sporlan.*)

Another application, known as pump-down control, uses a thermostat to control the solenoid valve (see Fig. 6-24 for a wiring and valve location schematic). When the thermostat is satisfied, the valve closes and the compressor continues to run until a substantial portion of the refrigerant has been pumped from the evaporator. A low-pressure cutout control is used to stop the compressor at a predetermined evaporator pressure. When the thermostat again calls for refrigeration, the solenoid valve opens, causing the evaporator pressure to rise and the compressor to start. This arrangement can be used on either single or multiple evaporators.

Suction-line service.   There are several applications, particularly on suction lines, where pressure drops in the range of 2 to 4 lb/in.^{2} (psi) cannot be tolerated. Therefore, only valves that are capable of opening at very low pressure drops are suitable for this type of use. Some valves incorporate the floating disc principle and are ideally suited for such special applications. They are capable of opening full at pressure drops of 0.1 psi and below.

Larger-capacity valves are suitable for suction service when supplied with internal parts that are mechanically connected. With this arrangement, the piston is connected to the stem and plunger assembly, and when the coil is energized, the plunger assists in supporting the piston. As a result, the pressure drop through the valve is reduced to a bare minimum. Valves with the direct-connected assembly are designated usually with a prefix "D" to the type number.

When these valves are required for suction service, they are supplied with a spring support under the piston. The spring counterbalances the major portion of the piston's weight, and therefore the valve will open with far less pressure drop than normal. Valves with the counterbalance spring are identified by the prefix "S" added to the type number.

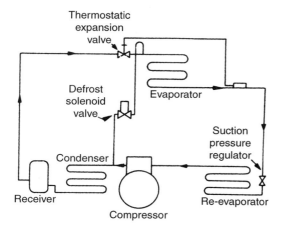

Figure 6-25 Solenoid valves used for hot-gas defrost. (*Courtesy of Sporlan.*)

High-temperature applications

In some high-temperature applications, a high-temperature coil construction is required. The temperature of the fluid or gas flowing through the solenoid valve will generally determine whether a high-temperature coil is necessary.

Hot-gas defrost service.   Several piping arrangements are used for hot-gas defrost systems, one of which is shown in Fig. 6-25. A portion of the compressor discharge gas is passed through the solenoid valve into the evaporator. The solenoid valve may be controlled either manually or automatically for this duty. Hot-gas defrost valve selection requires a slightly different approach from the simple pressure drop versus tonnage. Be sure to consider the evaporator temperature correction factors to make certain that the valve selected has adequate capacity.

Normally open solenoid valves have many uses. Perhaps the most popular is their adaptation to heat-reclaiming systems. The use of one normally closed valve and one normally open valve to shunt the discharge gas to either the outdoor condenser or the indoor heat-reclaiming coil provides positive opening and closing action.

This eliminates the problem found in some three-way valves, which have a tendency to leak hot gas into the heat-reclaiming coil when not required. When this leakage occurs during the cooling season, it imposes an extra load on the cooling system that wastes energy, rather than conserving it.

If leakage occurs during the heating season when all the discharge gas should be going to the reheat coil, a good portion of the liquid charge could become logged in the inactive condenser. For a simple schematic of a heat-reclaiming cycle, see Fig. 6-26. Many original equipment manufacturers (OEMs) have developed their own reheat cycle, which may be completely different from the one illustrated. In addition, some may

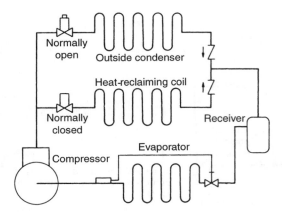

Figure 6-26 Heat-reclaiming cycle schematic. (*Courtesly of Sporlan.*)

incorporate head pressure control as well, so it is always advisable to consult the manufacturer's bulletin regarding its particular design.

Transformers for low-voltage controls.   The use of low-voltage control systems is becoming more widespread as the demand for residential air conditioning increases. This necessitates the use of a transformer for voltage reduction, normally to 24 V. The selection of a transformer is not accomplished by merely selecting one that has the proper voltage requirements. The volt-ampere (VA) rating is equally important. To determine the VA requirements for a specific solenoid valve, refer to the manufacturer's data. Insufficient transformer capacity will result in reduced operating power or a lowering of the maximum operating pressure differential (MOPD) of the valve. MOPD and safe working pressure (SWP) are usually noted on the valve nameplate (see Fig. 6-27).

If more than one solenoid valve and/or other accessories are operated from the same transformer, the transformer VA rating must be determined by adding the individual accessories' VA requirements.

Figure 6-27 Negative corona detector. (Ritchie). (*Courtesy of Sporlan.*)

7

Electric Motors

Motors

Most AC motors are of the induction type. They are, in general, simpler and cheaper to build than equivalent DC machines. They have no commutator, slip rings, or brushes, and there is no electrical connection to the rotors. Only the stator winding is connected to the AC source, and, then, as their name implies, induction produces the currents in the rotor. A common and particularly simple form of rotor for this type of motor is the squirrel-cage rotor (see Fig. 7-1). It is so named because of its resemblance to a treadmill-type squirrel cage. The induction motor is based on a rotating magnetic field. This is achieved by using multiple stator field windings (poles), each pair of which is excited by an AC voltage of the same amplitude and frequency as, but phase-displaced from, the voltage supplying the neighboring pair. Figure 7-2 shows how the magnetic field rotates in a four-pole induction motor, where the voltages to the two pairs of poles are 90° out of phase with each other. When the rotor is placed in the stator's rotating field, the induced currents set up their own fields, which react with the stator's field and push the rotor around.

Note that some rotors are skewed. The skew of a rotor refers to the amount of angle between the conductor slots and the end face of the rotor laminations. Normally, the conductors are in a nearly straight line, but for high torque applications the rotor is skewed, which increases the angle of the conductors. The term *full skew* refers to the maximum practical amount (see Fig. 7-3). Figure 7-4 shows how the rotor is located in reference to the stator and the end bells that hold it in place.

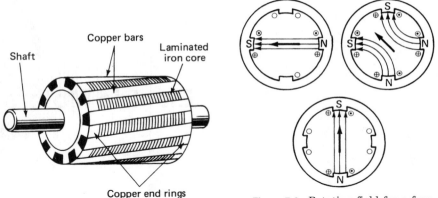

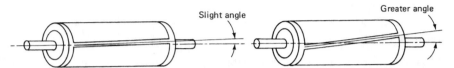

Figure 7-1 Squirrel cage rotor.

Figure 7-2 Rotating field for a four-pole stator.

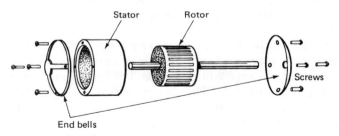

Figure 7-3 Skewed die-cast rotor. Note the angle of the conductor slots.

Figure 7-4 Exploded view of motor.

A number of types of AC motors are available. The types presented here are those most often encountered when working with heating, air-conditioning, and refrigeration equipment.

The shaded-pole induction motor is a single-phase motor. It uses a unique method to start the rotor turning. The effect of a moving magnetic field is produced by constructing the stator in a special way (see Fig. 7-5).

Portions of the pole piece surfaces are surrounded by a copper strap called a shading coil. The strap causes the field to move back and forth across the face of the pole piece. In Fig. 7-6, a numbered sequence and points on the magnetization curve are shown. As the alternating stator field starts increasing from zero Fig. 7-6(1), the lines of force expand across the face of the pole piece and cut through the strap. A voltage is

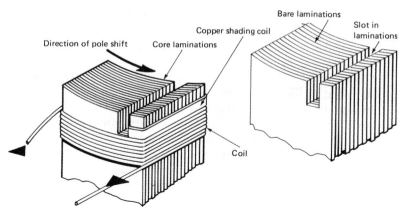

Figure 7-5   Shading the poles of a shaded-pole motor.

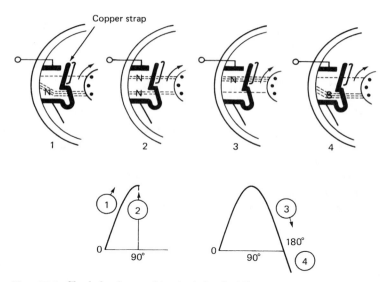

Figure 7-6   Shaded poles used in shaded-pole AC motors.

induced in the strap. The current that results generates a field that opposes the cutting action (and decreases the strength) of the main field. This action causes certain actions. As the field increases from zero to a maximum of 90°, a large portion of the magnetic lines of force is concentrated in the unshaded portion of the pole Fig. 7-6(1). At 90° the field reaches its maximum value. Since the lines of force have stopped expanding, no electromagnetic field (EMF) is induced in the strap, and no opposite magnetic field is generated. As a result, the main field is uniformly distributed across the poles as shown in Fig. 7-6(2).

From 90° to 180°, the main field starts decreasing or collapsing inward. The field generated in the strap opposes the collapsing field. The effect is to concentrate the lines of force in the shaded portion of the poles, as shown in Fig. 7-6(3).

Note that from 0° to 180° the main field has shifted across the pole face from the unshaded to the shaded portion. From 180° to 360°, the main field goes through the same change as it did from 0° to 180°. However, it is now in the opposite direction Fig. 7-6(4). The direction of the field does not affect the way the shaded pole works. The motion of the field is the same during the second half-hertz as it was during the first half-hertz.

The motion of the field back and forth between shaded and unshaded portions produces a weak torque. This torque is used to start the motor. Because of the weak starting torque, shaded-pole motors are built in only small sizes. They drive such devices as fans, timers, and blowers.

Reversibility

Shaded-pole motors can be reversed mechanically. Turn the stator housing and shaded poles end for end. These motors are available from 1/25 to 1/2 hp.

Uses

As previously mentioned, this type of motor is used as a fan motor in refrigerators and freezers and in some types of air-conditioning equipment where the demand is not too great. It can also be used as part of a timing device for defrost timers and other sequenced operations.

The fan and motor assembly are located behind the provisions compartment in the refrigerator, directly above the evaporator in the freezer compartment. The suction-type fan pulls air through the evaporator and blows it through the provisions compartment air dot and freezer compartment fan grille. Figure 7-7 shows a shaded-pole motor with a

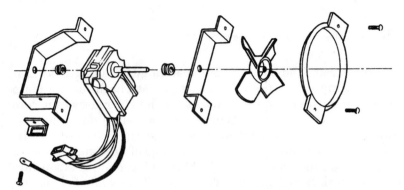

Figure 7-7 Fan, motor, and bracket assembly. (*Courtesy of Kelvinator.*)

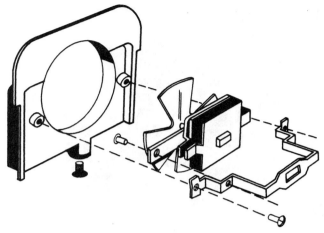

Figure 7-8 Fan and fan motor bracket assembly. (*Courtesy of Kelvinator.*)

molded plastic fan blade. For maximum air circulation, the location of the fan on the motor shaft is most important. Mounting the fan blade too far back or too far forward on the motor shaft, in relation to the evaporator cover, will result in improper air circulation. The freezer compartment fan must be positioned with the lead edge of the fan 1/4 in. in front of the evaporator cover. The fan assembly shown in Fig. 7-8 is used on top freezer, no-frost, fiberglass-insulated model refrigerators. The freezer fan and motor assembly is located in the divider partition directly under the freezer air duct.

Split-phase motor

The field of a single-phase motor, instead of rotating, merely pulsates. No rotation of the rotor takes place. A single-phase pulsating field may be visualized as two rotating fields revolving at the same speed, but in opposite directions. It follows, therefore, that the rotor will revolve in either direction at nearly synchronous speed, if it is given an initial impetus in either one direction or the other. The exact value of this initial rotational velocity varies widely with different machines. A velocity higher than 15 percent of the synchronous speed is usually sufficient to cause the rotor to accelerate to the rated or running speed. A single-phase motor can be made self-starting if means can be provided to give the effect of a rotating field.

To get the split-phase motor running, a run winding and a start winding are incorporated into the stator of the motor. Figure 7-9 shows the split-phase motor with the end cap removed so you can see the starting switch and governor mechanism.

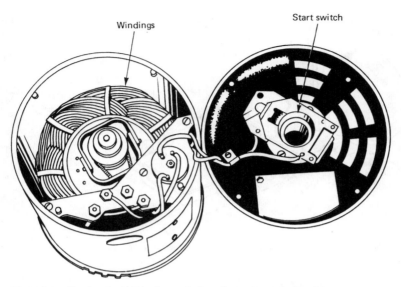

Windings

Start switch

Figure 7-9   Single-phase starting switch and governor mechanism.

This type of motor is difficult to use with air-conditioning and refrig-eration equipment inasmuch as it has very little starting torque and will not be able to start a compressor since it presents a load to the motor immediately upon starting. This type of motor, however, is very useful in heating equipment (see Fig. 7-10).

Figure 7-10   Single-phase split-phase furnace motor. (*Courtesy of Westinghouse.*)

Getting the motor started.   One of the most important parts of the single-phase electric motor is the start mechanism. A special type is needed for use with single-phase motors. A centrifugal switch is used to take a start winding out of the circuit once the motor has come up to within 75 percent of its run speed. The split-phase, capacitor-start, and other variations of these types all need the start mechanism to get them running.

The stator of a split-phase motor has two types of coils; one is called the run winding and the other the start winding. The run winding is made by winding the enamel-coated copper wire through the slots in the stator punchings.

The start winding is made in the same way except that the wire is smaller. Coils that form the start windings are positioned in pairs in the stator directly opposite each other and between the run windings. When you look at the end of the stator, you see alternating run windings and start windings (see Fig. 7-9).

The run windings are all connected together, so the electrical current must pass through one coil completely before it enters the next coil, and so on through all the run windings in the stator. The start windings are connected together in the same way, and the current must pass through each in turn (see Fig. 7-11).

The two wires from the run windings in the stator are connected to terminals on an insulated terminal block in one end bell where the power cord is attached to the same terminals. One wire from the start winding is tied to one of these terminals also. However, the other wire from the start winding is connected to the stationary switch mounted in the end bell. Another wire then connects this switch to the opposite terminal on the insulated block. The stationary switch does not revolve, but is placed so that the weights in the rotating portion of the switch, located on the

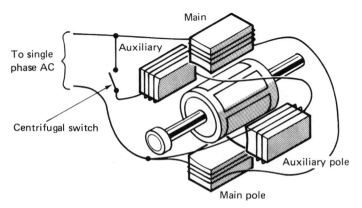

Figure 7-11   Single-phase induction motor.

rotor, will move outward when the motor is up to speed and open the switch to stop electrical current from passing through the start winding.

The motor then runs only on the main winding until such time as it is shut off. Then, as the rotor decreases in speed, the weights on the rotating switch again move inward to close the stationary switch and engage the start winding for the next time it is started.

Reversibility. The direction of rotation of the split-phase motor can be changed by reversing the start winding leads.

Uses. This type of motor is used for fans, furnace blowers, oil burners, office appliances, and unit heaters.

Repulsion-start induction-run motor

The repulsion-induction motor starts on one principle of operation and, when almost up to speed, changes over to another type of operation. Very high twisting forces are produced during starting by the repulsion between the magnetic pole in the armature and the same kind of pole in the adjacent stator field winding. The repulsing force is controlled and changed so that the armature rotational speed increases rapidly, and, if not stopped, would continue to increase beyond a practical operating speed. It is prevented by a speed-actuated mechanical switch that causes the armature to act as a rotor that is electrically the same as the rotor in single-phase induction motors. That is why the motor is called a repulsion-induction motor.

The stator of this motor is constructed very much like that of a split-phase or capacitor-start motor, but only run or field windings are mounted inside. End bells keep the armature and shaft in position and hold the shaft bearings.

The armature consists of many separate coils of wire connected to segments of the commutator. Mounted on the other end of the armature are governor weights that move push rods that pass through the armature core. These rods push against a short-circuiting ring mounted on the shaft on the commutator end of the armature. Brush holders and brushes are mounted in the commutator end bell, and the brushes, connected by a heavy wire, press against segments on opposite sides of the commutator (see Fig. 7-12).

When the motor is stopped, the action of the governor weights keeps the short-circuiting ring from touching the commutator. When the power is turned on and current flows through the stator field windings, a current is induced in the armature coils. The two brushes connected together form an electromagnetic coil that produces a north and south pole in the armature, positioned so that the north pole in the armature is next to a north pole in the stator field windings. Since like poles try to move apart, the repulsion produced in this case can be satisfied in only one way: the armature turns and moves the armature coil away from the field windings.

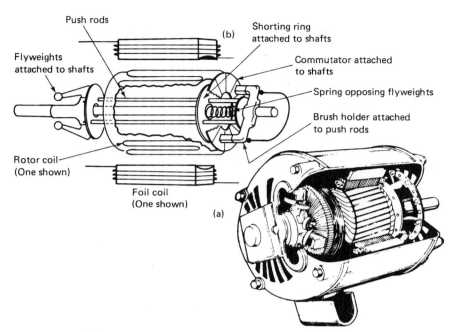

Figure 7-12   (*a*) Brush-lifting, repulsion-start, induction-run motor cutway; (*b*) brush-lifting, repulsion-start, induction motor armature details.

The armature turns faster and faster, accelerating until it reaches what is approximately 80 percent of the run speed. At this speed, the governor weights fly outward and allow the push rods to move. These push rods, which are parallel to the armature shaft, have been holding the short-circuiting ring away from the commutator. Now that the governor has reached its designed speed, the rods can move together electrically in the same manner that the cast aluminum disks did in the cage of the induction motor rotor. This means that the motor runs as an induction motor.

The repulsion-induction type of motor can start very heavy, hard-to-turn loads without drawing too much current. They are made from 1/2 to 20 hp. This type of motor is used for such applications as large air compressors, refrigeration equipment, and large hoists, and is particularly useful in locations where low line voltage is a problem.

This type of motor is no longer used in the refrigeration industry. Some older operating units may be found with this type of motor still in use.

Capacitor-start motor

The capacitor motor is slightly different from a split-phase motor. A capacitor is placed in the path of the electrical current in the start winding (see Fig. 7-13). Except for the capacitor, which is an electrical component that slows any rapid change in current, the two motors are same

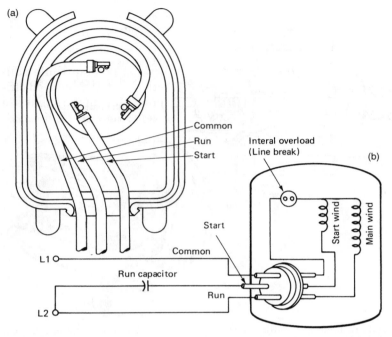

Figure 7-13 (*a*) Single-phase diagram for the AH air conditioner and heat-pump compressor, (*b*) terminal box showing the position of the terminals on the AH series of compressors. (*Courtesy of Tecumseh.*)

Figure 7-14 Capacitor start motor. (*Courtesy of Westinghouse.*)

electrically. A capacitor motor can usually be recognized by the capacitor can or housing that is mounted on the stator (see Fig. 7-14).

Adding the capacitor to the start winding increases the effect of the two-phase field described in connection with the split-phase motor. The capacitor means that the motor can produce a much greater twisting force when it is started. It also reduces the amount of electrical current required during starting to about 1.5 times the current required after the motor is up to speed. Split-phase motors require 3 or 4 times the current in starting that they do in running.

Reversibility.   An induction motor will not always reverse while running. It may continue to run in the same direction, but at a reduced efficiency. An inertia-type load is difficult to reverse. Most motors that are classified as reversible while running will reverse with a non-inertial-type load. They may not reverse if they are under no-load conditions or have a light lead or an inertial load.

One problem related to the reversing of a motor while it is still running is the damage done to the transmission system connected to the load. In some cases it is possible to damage a load. One way to avoid this is to make sure the right motor is connected to a load.

Reversing (while standing still) the capacitor-start motor can be done by reversing its start winding connections. This is usually the only time that a field technician will work on a motor. The available replacement motor may not be rotating in the direction desired, so the technician will have to locate the start winding terminals and reverse them in order to have the motor start in the desired direction.

Uses.   Capacitor motors are available in sizes from 1/6 to 20 hp. They are used for fairly hard starting loads that can be brought up to run speed in under 3 s. They may be used in industrial machine tools, pumps, air conditioners, air compressors, conveyors, and hoists.

Figure 7-15 shows a capacitor-start induction-run motor used in a compressor. This type uses a relay to place the capacitor in and out of the circuit. More about this type of relay will be discussed later. Figure 7-16 shows how the capacitor is located outside the compressor.

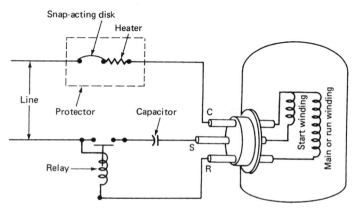

Figure 7-15   Capacitor-start induction-run motor used for a compressor. (*Courtesy of Tecumseh.*)

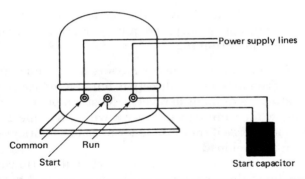

Start capacitor* sizes	
Compressor 1/8 hp	Capacitor size is 95 to 200 μF
Compressor 1/6 hp	Capacitor size is 95 to 200 μF
Compressor 1/4 hp	Capacitor size is 200 to 300 μF
Compressor 1/3 hp	Capacitor size is 250 to 350 μF
Compressor 1/2 hp	Capacitor size is 300 to 400 μF
Compressor 3/4 hp	Capacitor size is 300 to 400 μF

*Black case (Bakelite)

Figure 7-16 Location of start capacitor in a compressor circuit. (*Courtesy of Tecumseh.*)

Permanent split-capacitor motor

The permanent split-capacitor (PSC) motor is used in compressors for air-conditioning and refrigeration units. It has an advantage over the capacitor-start motor inasmuch as it does not need a centrifugal switch with its associated problems.

The PSC motor has a run capacitor in series with the start winding. Both run capacitor and start winding remain in the circuit during start and after the motor is up to speed. Motor torque is sufficient for capillary and other self-equalizing systems. No start capacitor or relay is necessary. The PSC motor is basically an air-conditioner compressor motor. It is also used in refrigerator compressors. It is very common through 3 hp. It is also available in the 4 and 5 hp sizes (see Fig. 7-17).

Theory of operation. The capacitor is inserted in series with the start winding (see Fig. 7-18). The phase shift produced by the capacitor is similar to that produced by the capacitor-start induction-run motor. The run capacitor is usually between 5 and 50 µF. Thus, it is smaller than the start capacitor used in the capacitor-start motor. This also means it will have less starting torque than the capacitor-start motor. However, the torque is enough to start the motor running even with a small load. The capacitor is in the circuit all the time. It is not removed by a relay or any other type of device. It is a permanent part of the circuit.

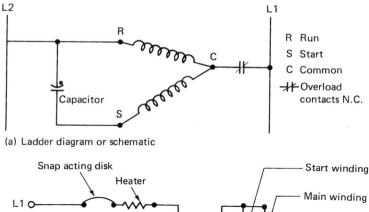

(a) Ladder diagram or schematic

R Run
S Start
C Common
⊣⊢ Overload
    contacts N.C.

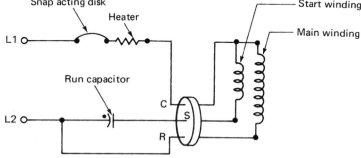

(b) Schematic drawing

Figure 7-17 Permanent split-capacitor motor schematic.

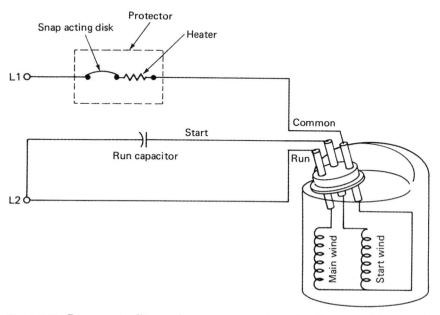

Figure 7-18 Permanent split-capacitor compressor schematic. (*Courtesy of Tecumseh.*)

After the compressor or fan is started and begins to run at speed, the motor produces a counter electromotive force (CEMF). The CEMF builds up to within a few volts of the applied voltage when the motor has reached full speed. As long as the difference between the applied voltage and the CEMF is small, very little current flows in the start winding. This is because the capacitor will allow more current to pass as the difference in applied voltage and CEMF gets larger, and less current to flow when the voltage difference is small. There is a small difference in voltage at full rpm, so the current in the start winding will be small. The 2 to 4 A does not constitute enough to cause damage to the compressor start winding (see Fig. 7-19).

Run capacitor sizes
1/8-hp compressor or motor uses a capacitor of 4 or 5 μF
1/6-hp compressor or motor uses a capacitor of 4 or 5 μF
1/2-hp compressor or motor uses a capacitor of 10 μF
1/2-to 2 hp compressor or motor uses a capacitor of 10 to 15 μF
3-hp compressor or motor uses 2 capacitors of 10 μF in parallel to equal 20 μF

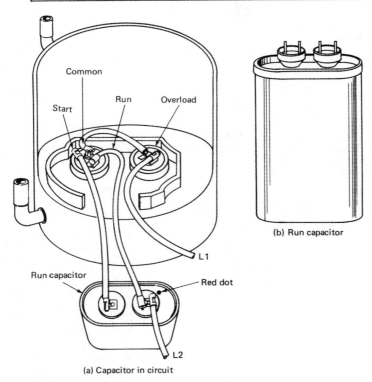

(b) Run capacitor

(a) Capacitor in circuit

Figure 7-19 PSC compressor hookup. (*Courtesy of Tecumseh.*)

The run capacitor left in the circuit aids in speed regulation. One thing to remember on this type of motor is to make sure the capacitor is placed in the circuit properly. The run capacitor has a red dot or some other marking on or near one of the two terminals. This is the outside foil of the capacitor. The run capacitor used here is in a grounded steel case that aids in the dissipation of heat (see Fig. 7-19b). The outside foil is near the capacitor case. If the insulation breaks down or shorts to the outside foil, it will make contact with the case. Excessive current will flow in the circuit when the short occurs. The main objective here is to keep the current as low as possible and have it do the least damage possible. That is why the red dot or mark is always placed where it is supposed to be. In Fig. 7-19, the red dot terminal should be connected to L2. Then, if it shorts, the foil touches the case, shorts to ground, causes excess current to flow, and trips the circuit breaker or blows the fuse.

However, if the red dot is connected incorrectly, the short will burn out the compressor start winding since it puts the winding directly across the power source by grounding one end of the winding. Make sure the capacitor's red dot is connected to the line side and not to the compressor start winding (see Fig. 7-19a).

Reversibility.   The PSC motor can be reversed if it has three wires leading from the case (see Fig. 7-20). To reverse simply connect either side of the capacitor to the line. However, with compressors it is best to leave them as is. They have been designed without the possibility of reversing in most cases. To reverse the four-wire type motor, transpose the black leads as shown in Fig. 7-21.

Uses.   Permanent split-capacitor motors may also be used for the fans that are mounted behind the condensers on air-conditioning units. They

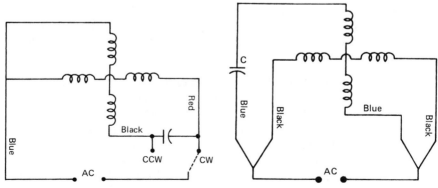

Figure 7-20 Permanent split-capacitor three lead schematic.

Figure 7-21 Permanent split-capacitor four-lead schematic.

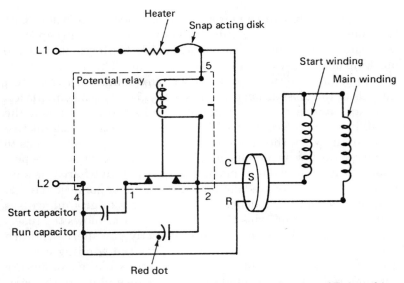

Figure 7-22   Capacitor start and run motor schematic. (*Courtesy of Tecumseh.*)

move the air past the condenser and thus remove the heat from the building being air conditioned. This type of motor can also be varied in speed by changing its windings. A number of color-coded windings are brought out so they may be connected to a switching arrangement for low, medium, and high speeds.

Capacitor-start capacitor-run motor

The capacitor start and run (CSR or CSCR) motor arrangement uses a start capacitor and a run capacitor in parallel with each other and in series with the motor start windings. This motor has high starting torque and runs efficiently. It is used in many refrigeration and air-conditioning applications up to 5 hp. A potential relay removes the start capacitor from the circuit after the motor is up to speed. Potential relays must be accurately matched to the compressor (see Fig. 7-22). Efficient operation depends on this.

Theory of operation.   The capacitor-start capacitor-run motor is used on air-conditioning and refrigeration units that need large starting torque. You will find the CSCR motor used on equipment that has an expansion valve and on air-conditioning systems when the permanent-split compressor has trouble starting. In some cases, the technician adds a capacitor in the field to make the PSC motor start more easily. This produces the same electrical characteristics as a CSCR type when the additional capacitor is hooked up to the existing arrangement. When the capacitor is added by the technician in the field, it is referred to as a hard-start kit.

Since the CSCR motor has the additional capacitor added during starting time, some way must be provided to move it once its purpose has been served. The potential relay is called upon to do the job. This type of relay will be discussed later in this chapter.

The start capacitor is available in sizes up to 600 μF. This is a large capacitor when compared to the run capacitor of up to 75 μF (see Fig. 7-22). The large capacitance value causes a larger phase shift between the start and run winding voltages. As the compressor motor starts to turn, the CEMF begins to build. The CEMF is present between the S and C terminals of the compressor. The potential relay coil is connected at terminals 2 and 5 (see Fig. 7-22). Note the symbol used for the coil of the potential relay. A PE (for potential relay) and a small resistor-like symbol show the coil part of the relay.

When the compressor reaches approximately 75 percent of its full rpm, the CEMF is strong enough to energize the potential relay coil. This pulls the contacts open. Contactors are shown as two parallel lines with a slant line through them with PR underneath. When the potential relay contacts open, the start capacitor is removed from the circuit. This leaves the run capacitor still in the circuit and in series with the start winding. The result is good starting torque and good running efficiency. The efficiency is increased inasmuch as the power factor is brought closer to unity or 1.00.

Reversibility.   This type of motor can be reversed by changing the leads from the start winding. It is, however, difficult to do if the compressor is sealed. In open-type motors, it is possible to reverse the direction of rotation by simply reversing the connections of the start windings.

Uses.   This type of motor is used on equipment with a need for good starting torque. This includes some types of refrigeration equipment and some hard to start air-conditioning systems.

Three-phase motor

The three-phase motor does not need centrifugal switches or capacitors. The three phases of this type of power generate their own rotating magnetic field when applied to a stator with three sets of windings (see Fig. 7-23). The stator windings are placed 120° apart. The rotor is a form-wound type or a cage type. The squirrel-cage rotor is standard for motors smaller than 1 hp (see Fig. 7-24).

Theory of operation.   For the purpose of identification of phases, note that the three phases are labeled A, B, and C in Fig. 7-25. Phase B is displaced in time from A by 1/3 Hz, and phase C is displaced from phase B by 1/3 Hz. In the stator the different phase windings are placed adjacent to each other so that a B winding is next to an A, a C is next to a B, and then an A is next to a C, and so on around the stator (see Fig. 7-25).

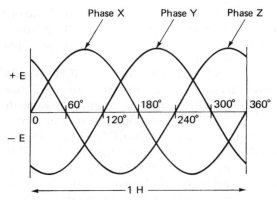

Figure 7-23   Three-phase AC waveform.

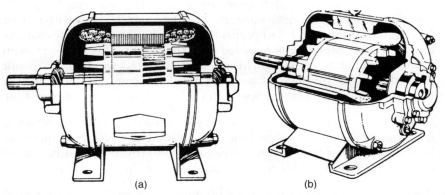

Figure 7-24   Cutaway view of a three-phase motor with (*a*) a half-etched squirrel cage rotor, and (*b*) a three-phase motor with a cast rotor.

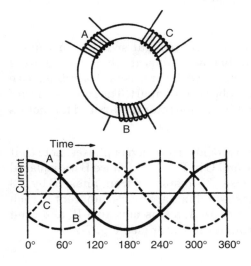

Figure 7-25   Three-phase current and coil placement in a motor.

The next step is to picture the magnetic fields produced by the three phases in just one group of A, B, and C windings for 1 Hz. Start with the A phase at its maximum positive peak current value. This A winding is a north pole at its maximum strength. As the cycle progresses, the magnetic pole at A will decrease to zero as the current changes direction. It will become a south pole. The strength of the field will increase until the current reaches its greatest negative value. This produces a maximum strength south pole, and then a decrease. It then passes through zero or neutral and becomes a maximum north pole at the end of the cycle.

The B phase winding does exactly the same thing, except that the rise and fall of the magnetic fields follow behind the A phase by 1/3 Hz, and the C phase winding magnetic field follows behind B by 1/3 Hz and the A phase by 2/3 Hz.

Assume that you can see only the maximum north poles produced by each phase. Your view of the motor is from the end of a complete stator connected to a three-phase power source. The north poles will move around the stator and appear to be revolving because of the current relationship of the A, B, and C phases.

The three-phase motor has a rotor that consists of steel disks pressed onto the motor shaft. The slots or grooves are filled with aluminum and connected on the ends to form a cage for the electrical current (see Fig. 7-26). The rotating magnetic field in the stator induces current into this electrical cage and thereby sets up north and south poles in the rotor. These north and south poles then follow their opposite members in the stator and the shaft rotates. Polyphase induction motors are often called squirrel-cage motors because of this rotor construction. This rotating field makes it possible for the motor to start without capacitors or switches. Thus, it is a simpler motor to maintain and operate.

Reversibility.   Three-phase motors can be reversed while running. It is very hard on the bearings and the driven machine, but it can be done by reversing any two of the three connections. This is usually done by a switch specially designed for the purpose.

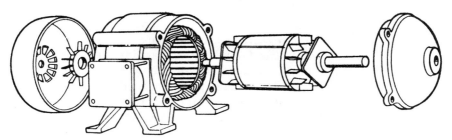

Figure 7-26   Three-phase motor with explosion-proof construction.

Open phase.   If a three-phase motor develops an open "leg" or one phase (two instead of three wires are coming into the motor terminal with power), it will slow down and hum noticeably. It will, however, continue to run in the same direction. If you try to start it with only two legs (or phases), it will not start but will rotate if started by hand (in fact, it will start in either direction). Once the other phase is connected, it will quickly come up to speed. The loss of one leg is usually due to a fuse in that leg blowing (i.e., if there are three individual fuses in the three-phase circuit, as is normally the case).

Uses.   Three-phase motors are used for machine tools, industrial pumps and fans, air compressors, and air-conditioning equipment. They are recommended wherever a polyphase power supply is available. They provide high starting and breakdown torque with smooth pull-up torque. They are efficient to operate and are designed for 208 to 220/460-V operation with horsepower ratings from 1/4 to the hundreds.

Capacitor Ratings

You have already been introduced to capacitors that are used for starting a motor and for improving the power factor and torque characteristics of the motor. Here we would like to take a look at the ratings and the proper care of capacitors for long-time operation.

   Never use a capacitor with a lower rating than specified on the original equipment. The voltage rating and the microfarad rating are important. A higher voltage rating than that specified is always usable. However, a voltage rating lower than that specified can cause damage. Make sure the capacitance marked on the capacitor in microfarads is as specified. Replace with a capacitor of the same size rated in μF, MF, UF, or MFD. All these abbreviations are used to indicate microfarads. See Table 7-1 for a listing of the start capacitors and their ratings for different voltages. Note that the capacitors are not exactly what they are rated for in microfarads. They have a tolerance and the limits are shown.

Start Capacitors and Bleeder Resistors

The development of high-power-factor, low-current, single-phase compressor motors that require start and run capacitors used with potential-type relays has created electrical peculiarities that did not exist in previous designs. In some situations, relay contacts may weld together, causing compressor motor failure. This phenomenon occurs due to the high voltage in the start capacitor discharging (arcing) across the potential relay contacts. To eliminate this, start capacitors are equipped with bleeder resistors across the capacitor terminals (see Fig. 7-27).

TABLE 7-1 Ratings and Test Limits for AC Electrolytic Capacitors

Capacity rating (microfarads)			110-V ratings		125-V ratings		220-V ratings	
Nominal	Limits	Average	Amps. at rated voltage, 60 Hz	Approx. max. watts	Amps. at rated voltage, 60 Hz	Approx. max. watts	Amps. at rated voltage, 60 Hz	Approx. max. watts
	25–30	27.5	1.04–1.24	10.9	1.18–1.41	14.1	2.07–2.49	43.8
	32–36	34	1.33–1.49	13.1	1.51–1.70	17	2.65–2.99	52.6
	38–42	40	1.56–1.74	15.3	1.79–1.98	19.8	3.15–3.48	61.2
	43–48	45.5	1.78–1.99	17.5	2.06–2.26	22.6	3.57–3.98	70
50	53–60	56.5	2.20–2.49	21.9	2.50–2.83	28.3	4.40–4.98	87.6
60	64–72	68	2.65–2.99	26.3	3.02–3.39	33.9	5.31–5.97	118.2
65	70–78	74	2.90–3.23	28.4	3.30–3.68	36.8	5.81–5.97	128.2
70	75–84	79.5	3.11–3.48	30.6	3.53–3.96	39.6	6.22–6.97	138
80	86–96	91	3.57–3.98	35	4.05–4.52	45.2	7.13–7.96	157.6
90	97–107	102	4.02–4.44	39.1	4.57–5.04	50.4	8.05–8.87	175.6
100	108–120	114	4.48–4.98	43.8	5.09–5.65	56.5	8.96–9.95	197
115	124–138	131	5.14–5.72	50.3	5.84–6.50	65		
135	145–162	154	6.01–6.72	62.8	6.83–7.63	85.8		
150	161–180	170	6.68–7.46	69.8	7.59–8.48	95.4		
175	189–210	200	7.84–8.74	81.4	8.91–9.90	111.4		
180	194–216	205	8.05–8.96	83.8	9.14–10.18	114.5		
200	216–240	228	8.96–9.95	93	10.18–11.31	127.2		
215	233–260	247	9.66–10.78	106.7	10.98–12.25	145.5		
225	243–270	257	10.08–11.20	110.9	11.45–12.72	151		
250	270–300	285	11.20–12.44	123.2	12.72–14.14	167.9		
300	324–360	342	13.44–14.93	147.8	15.27–16.96	201.4		
315	340–380	360	14.10–15.76	156				
350	378–420	399	15.68–17.42	172.5				
400	430–480	455	17.83–19.91	197.1				

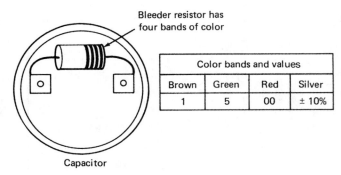

Color bands and values			
Brown	Green	Red	Silver
1	5	00	± 10%

Figure 7-27   Bleeder resistor across the capacitor terminals.

Bleeder resistor-equipped capacitors may not be available. Then a 2-W, 1500-Ω resistor can be soldered across the capacitor terminals. This does not interfere with the operation of the circuit, but does allow the capacitor to discharge across the resistor instead of across the relay switch points. This resistor is called a bleeder because it bleeds off the capacitor charge.

Run capacitors

The marked terminal of run capacitors should be connected to the R terminal of the compressor and thus to L2. Check the wiring diagram for the correct terminal. The run capacitor is in the circuit whenever the compressor is running. It is an oil-filled electrolytic capacitor that can take continuous use.

If the start capacitor is left in the circuit too long, in some cases the coil of the potential relay will open from vibration or use and cause the start capacitor to stay in the circuit longer than 10 to 15 s. This causes the electrolytic capacitor to explode or spew out its contents.

Motor Protectors

In most compressors there is a motor protector (see Fig. 7-28). The overload protector is inserted into the motor windings so that if they overheat the device will open the contacts of the switch. The bimetal element expands to cause the contacts to remove the power to the windings, These protectors are in addition to any circuit breakers that may be mounted outside the compressor.

Compressor Motor Relays

A hermetic compressor motor relay is an automatic switching device designed to disconnect the motor start winding after the motor has come up to running speed (see Fig. 7-29). The two types of motor relays

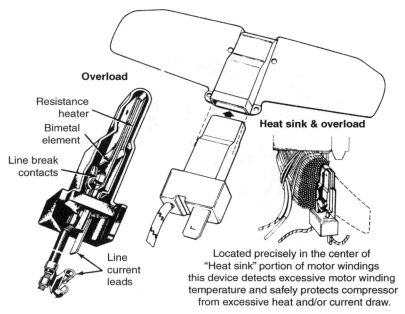

Overload

Resistance
heater

Bimetal
element

Line break
contacts

Line
current
leads

Heat sink & overload

Located precisely in the center of
"Heat sink" portion of motor windings
this device detects excessive motor winding
temperature and safely protects compressor
from excessive heat and/or current draw.

Figure 7-28 Motor internal line-break protector. (*Courtesy of Tecumseh.*)

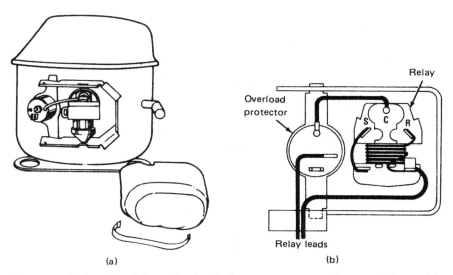

Relay

Overload
protector

S   C   R

Relay leads

(a)

(b)

Figure 7-29 (*a*) Location of the overload and relay on a compressor (*Courtesy of Tecumseh.*),
(*b*) starting relay and overload protector. (*Courtesy of Kelvinator.*)

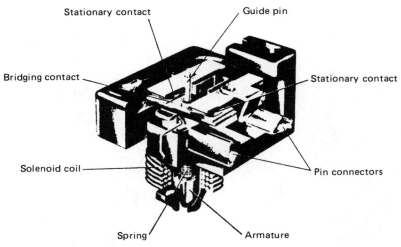

Figure 7-30   Current relay. (*Courtesy of Tecumseh.*)

used in refrigeration and air-conditioning compressors are the current and the potential type.

Current-type relay

The current-type relay is most often used with small refrigeration compressors up to 1 hp. Where power is applied to the compressor motor, the relay solenoid coil attracts the relay armature upward. This causes bridging contact and stationary contact to engage (see Fig. 7-30). This energizes the motor start winding. When the compressor motor comes up to running speed, the motor's main winding current is such that the relay solenoid coil de-energizes. This allows the relay contacts to drop open, which disconnects the motor start winding.

One thing to remember about this type of relay is its mounting. It should be mounted in a true vertical position so that the armature and bridging contact will drop free when the relay solenoid is de-energized.

Potential-type relay

This relay is generally used with large commercial and air-conditioning compressors (see Fig. 7-31). Motors may be capacitor-start, capacitor-run types up to 5 hp. Relay contacts are normally closed. The relay coil is wired across the start winding. It senses voltage change. Start winding voltages increase with motor speed. As the voltage increases to the specific pickup value, the armature pulls up, opening the relay contacts and de-energizing the start windings. After switching, there is still sufficient voltage induced in the start winding to keep the relay coil energized and the ray starting contacts open.

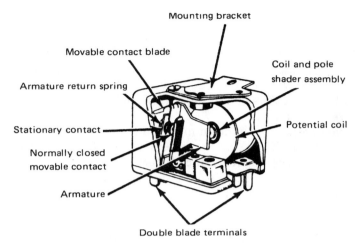

Figure 7-31 Potential-type relay. (*Courtesy of Tecumseh.*)

When the power is shut off to the motor, the voltage drops to zero. The coil is de-energized and the start contact is reset for the next start.

Many of these relays are extremely position sensitive. When changing a compressor relay, care should be taken to install the replacement in the same position as the original. Never select a replacement relay solely by horsepower or other generalized rating. Select the correct relay from the parts guide book furnished by the manufacturer. Visual inspection can distinguish the two relays. The current relay has heavy wire for the coil and the potential relay has fine wire for the coil.

Choosing Wire Size

There are two criteria for choosing wire size for installation of air-conditioning or refrigeration equipment. The size of the electrical conductor wire recommended for a given appliance circuit depends upon two things: limitation on voltage loss and minimum wire size.

Limiting voltage loss

Proper operation of an electrical device must be under the conditions for which it was designed. The wire size selected must be low in resistance per foot of length. This will ensure that the full load "line loss" of the total length of the circuit does not cause low voltage of the appliance terminals. Since the length of electrical feeders varies with each installation, wire sizing to avoid excessive voltage loss becomes the responsibility of the installing contractor. The National Electrical Code (NEC) or local code should be followed.

Minimum wire size

To avoid field wiring being damaged by tensile stress or overheating, national and local codes establish minimum wire sizes. The maximum amperage permitted for a given conductor limits internal heat generation so that temperature will not damage its insulation. This assumes proper fusing that will limit the maximum current flow so that the conductor will always be protected.

Wire size and voltage loss go hand in hand, so to speak. The larger the wire, the more current it can handle without voltage loss along the lines. Each conductor or wire has resistance. This resistance, measured in ohms per unit of wire length, increases as the cross-sectional area of the wire decreases. The size of the wire is indicated by gage number. The higher the gage number, the smaller the wire. American Wire Gage (AWG) is the standard used for wire size. Each gage number has a resistance value in ohms per foot of wire length. The resistance of aluminum wire is 64 percent greater than that for copper of the same gage number.

Wire selection

The wire size recommended for actual use should be the heavier of the two indicated by the procedures that will follow.

Local approval is usually necessary for any installation that has large current draws. The data presented here are based on the NEC. Much of the detail has been omitted in the interest of simplification. Thus, there may be areas of incompleteness not covered by a footnote or reference. In all cases it is recognized that final approval must come from the authority having local jurisdiction. The NEC sets forth minimum standards. It is an effort to establish some standard for safe operation of equipment.

Wire Size and Low Voltage

The voltage at which a motor or device should operate is stamped on the nameplate. This voltage indicates that the full capacity of the device is being utilized when that particular voltage is available. Motors operated at lower than rated voltage are unable to provide full horsepower without jeopardizing their service life. Electric heating units lose capacity even more rapidly at reduced voltages.

Low voltage can result in insufficient spark for oil burner ignition, reluctant starting of motors, and overheating of motors handling normal loads. Thus, it is not uncommon to protect electrical devices by selecting relays that will not close load circuits if the voltage is more than 15 percent below rating.

Air-Conditioning and Refrigeration Institute (ARI) certified cooling units are tested to ensure they will start and run at 10 percent above and 10 percent below their rating plate voltage. However, this does not imply that continuous operation at these voltages will not affect their capacity, performance, and anticipated service life. A large proportion of air-conditioning compressor burnouts can be traced to low voltage. Because the motor of a hermetic compressor is entirely enclosed within the refrigerant cycle, it is important that it not be abused either by overloading or undervoltage. Both of these can occur during peak load conditions. A national survey has shown that the most common cause of compressor low voltage is the use of undersized conductors between the utility lines and the condensing unit.

The size of the wire selected must be one that, under full load conditions, will deliver acceptable voltages to the appliance terminals. The NEC requires that conductors be sized to limit voltage drop between the outdoor-pole service tap and the appliance terminals to not in excess of 5 percent of rated voltage under full load conditions. This loss may be subdivided, with 3 percent permissible in service drops, feeders, meters, and overcurrent protectors at the distribution panel and the appliance (see Fig. 7-32).

In a 240-V service, the wire size selected for an individual appliance circuit should cause no more than 4.8-V drop under full load conditions. Even with this 5 percent limitation on voltage drop, the voltage at the equipment terminals is still very apt to be below the rating plate values (see Table 7-2).

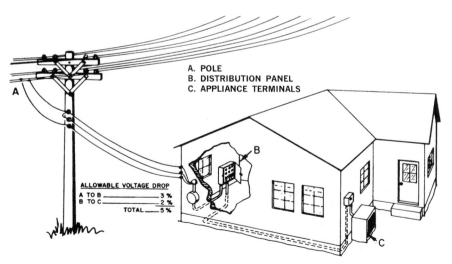

A. POLE
B. DISTRIBUTION PANEL
C. APPLIANCE TERMINALS

ALLOWABLE VOLTAGE DROP
A TO B _____ 3 %
B TO C _____ 2 %
TOTAL ____ 5 %

Figure 7-32   Voltage drops from post to air conditioner. (*Courtesy of Bryant.*)

TABLE 7-2   Permissible Maximum Voltage Drops

For a line voltage of	120	208	240	480
Feeders to distribution panel (3%)	3.6	6.24	7.2	14.4
Branch circuit to appliance (2%)	2.4	4.16	4.8	9.6
Total voltage drop fully loaded	6	10.4	12	24
Resultant* voltage at appliance	114	197.6	228	456

*Assumes full-rated voltage where feeders connect to utility lines. If utility voltage runs low, the overall voltage drop should be further reduced so as to make available at the appliance terminals a voltage as close as practical to that specified on the appliance rating plate.

Voltage drop calculations

Just as friction creates pressure loss in water flow through pipes, so does electrical resistance create voltage drop as current flows through a conductor. The drop increases with the length of the conductor (in feet), the current flow (in amperes), and the ohms of resistance per foot of wire. This relationship may be expressed as follows:

$$\text{Voltage drop} = \text{amperes} \times \text{ohms/foot} \times \text{length of conductor}$$

Figure 7-33 illustrates how voltage drop per 100 ft of copper conductor will increase with the amount of current drawn through the conductor. The wire size is indicated on the straight line. Match the amperes with the wire size. Then follow over to the left column to determine the voltage drop. For instance, Fig. 7-33 shows that there will be a 2.04-V drop per 100 ft of copper conductor for 20 A of current through a No. 10 wire.

The Effects of Voltage Variations on AC Motors

Motors will run at the voltage variations already mentioned. This does not imply such operation will comply with industry standards of capacity, temperature rise, or normally anticipated service life. Figure 7-34 shows general effects. Such effects are not guaranteed for specific motors.

The temperature rise and performance characteristics of motors sealed within hermetic compressor shells constitute a special case. These motors are cooled by return suction gas of varying quantity and temperature. Thus, Fig. 7-34 is not necessarily applicable to this specialized type of equipment.

The chart shows the approximate effect of voltage variations on motor characteristics. The reference base of voltage and frequency is understood to be that shown on the nameplate of the motor.

Some of the terms used in the chart are explained here.

$$\text{Normal slip} = \text{synchronous speed} - \text{the rating plate speed}$$

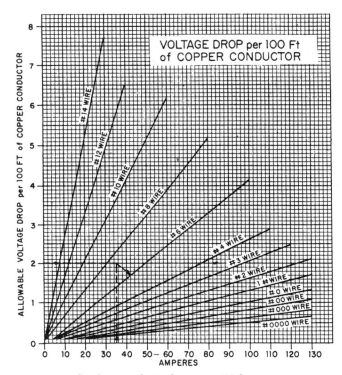

Figure 7-33 Conductor voltage drops per 100 ft.

Slip in the graph indicates the change in normal slip. Synchronous speeds for 60-Hz motors are

2-pole—3600 r/min or rpm

4-pole—1800 r/min or rpm

6-pole—1200 r/min or rpm

8-pole—900 r/min or rpm

Table 7-3 indicates the voltage drop that may be anticipated for various ampere flow rates through copper conductors of different gage size. Figure 7-33 provides the same data in graphic form.

These data are applicable to both single-phase and three-phase circuits. In each case, the wire length equals twice the distance from the power distribution panel to the appliance terminals, measured along the path of the conductors. This is twice the distance between B and C in Fig. 7-32, measured along the path of the conductors. For motorized appliances, particularly those that start under loaded conditions, the voltage at the appliance terminals should not drop more than 10 percent below rating plate values unless approved by the manufacturer.

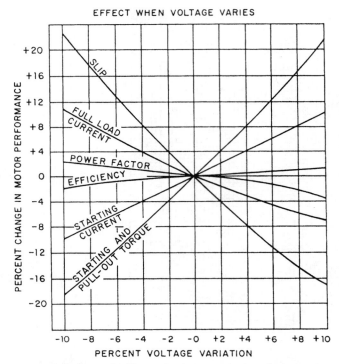

Figure 7-34 General effects of voltage variations in induction motor characteristics.

Thus, the voltage drop permissible in the load leads must anticipate any reduction below rated voltage that may be suffered under full load conditions at the point of power source connection (point A in Fig. 7-32).

Troublesome voltage losses may also occur elsewhere if electrical joints or splices are mechanically imperfect and create unanticipated resistance. Such connections may exist in the distribution panel, the meter socket, or even where outdoor power drops are clamped to the feeder lines on poles. Where there is a wide variation between no-load voltage and operating voltage, sources of voltage drop can be determined by taking voltmeter readings at various points in the circuit. These points might be ahead of the meter, after the circuit disconnect switch, at the appliance terminals, and at other locations.

Selecting Proper Wire Size

To provide adequate voltage at the appliance terminals, anticipate the minimum voltage that may exist at the distribution panel. Then determine the allowable voltage drop acceptable in the appliance circuit.

TABLE 7-3 Voltage Drop per 100 Ft of Copper Conductor of Wire Gage

Amperes*	No. 14	No. 12	No. 10	No. 8	No. 6	No. 4	No. 3	No. 2	No. 1	No. 0	No. 00	No. 000	No. 0000
5	1.29	0.81	0.51	0.32	0.21	0.13	0.11						
10	2.57	1.62	1.02	0.64	0.41	0.26	0.21	0.16	0.13	0.10			
15	3.86	2.43	1.53	0.96	0.62	0.39	0.31	0.24	0.19	0.15	0.12	0.10	0.10
20	5.14	3.24	2.04	1.28	0.82	0.52	0.41	0.32	0.26	0.20	0.16	0.13	0.13
25	6.43	4.05	2.55	1.60	1.03	0.65	0.51	0.41	0.32	0.26	0.20	0.16	0.15
30	7.71	4.86	3.06	1.92	1.23	0.78	0.62	0.49	0.39	0.31	0.24	0.19	0.18
35		5.67	3.57	2.24	1.44	0.91	0.72	0.57	0.45	0.36	0.28	0.22	0.20
40		6.48	4.08	2.56	1.64	1.04	0.82	0.65	0.52	0.41	0.32	0.26	0.23
45			4.59	2.88	1.85	1.17	0.92	0.73	0.58	0.46	0.36	0.29	0.26
50			5.10	3.20	2.05	1.30	1.03	0.81	0.65	0.51	0.41	0.32	0.31
60			6.12	3.84	2.46	1.56	1.23	0.97	0.77	0.61	0.49	0.38	0.36
70				4.48	2.87	1.82	1.44	1.13	0.90	0.71	0.57	0.45	0.41
80				5.12	3.28	2.08	1.64	1.30	1.03	0.82	0.65	0.51	0.46
90					3.69	2.34	1.85	1.46	1.16	0.92	0.73	0.58	0.51
100					4.10	2.59	2.05	1.62	1.29	1.02	0.81	0.64	0.56
110						2.85	2.26	1.78	1.42	1.12	0.89	0.70	0.61
120							2.46	1.94	1.55	1.22	0.97	0.77	0.66
130								2.10	1.68	1.33	1.05	0.83	0.71
140									1.81	1.43	1.13	0.90	0.77
150										1.53	1.22	0.96	
Ohms/100 ft													
Copper	0.257	0.162	0.1018	0.064	0.041	0.0259	0.0205	0.0162	0.0128	0.0102	0.0081	0.0064	0.0051
Aluminum	4.22	0.266	0.167	0.105	0.0674	0.0424	0.0336	0.0266	0.0129	0.0168	0.0133	0.0105	0.0084

*To determine voltage drop for aluminum conductors, enter the chart using 1.64 × actual amperes.

The conductor's lengths is twice the length of the branch leads, whether single or three phase.

Since resistance varies with temperature, it may be necessary to correct for wire temperature under load conditions if the ambient materially exceeds 80°F. If so, increase ampere values using the multiplier 1.0 + 0.002 × (ambient temperautre − 80°F).

Example: If current flow and environment result in conductors reaching 140°F under load conditions, the appliance ampere ratings should be increased by the multiplying factor 1.0 + 0.002 (140°F− 80°) = 1.0 + 0.12 = 1.12.

This should not exceed 2 percent of rated voltage. It should, for example, not exceed 4.1 V for 208-V service. Table 7-2 shows voltage drops for 120-, 208-, 240-, and 408-V service.

Determine the length of feed conductor. This is twice the length of the wire path from the source to the appliance. In Fig. 7-32, this is two times the distance from B to C measured along the path of the wire. If it is single-phase or three-phase, consider two conductors in establishing the total length of the circuit.

Determine the allowable drop per 100 ft of conductor.

Example 7-1 If for a 230-V installation, a 4.6-V drop is permissible, and the wire path is 115 ft from the distribution panel to the appliance (this makes 230 ft of conductor), then the allowable drop per 100 ft will be

$$\frac{4.6\,\text{V}\times100\,\text{ft}}{230\,\text{ft}} = 2.0\,\text{V}/100\,\text{ft}$$

Using either Table 7-3 or Figure 7-33, determine the gage wire required. When using the graph, select the gage number closest below and to the right.

Example 7-2 The full load value is 35 A. The allowable voltage drop is 2.0 V/100 ft. See Fig. 7-2.

Table solution (Table 7-3): Select No. 6 wire. This results in a drop of 1.44 V/100 ft.

Graph solution (Fig. 7-33): Intersection lies between No. 6 wire and No. 8 wire. Select the larger of the two; in this case it would be No. 6.

Unacceptable Motor Voltages

Occasionally, it becomes necessary to determine causes of unacceptable voltage conditions at motor terminals. Often this is necessary where excessive voltage drops are encountered as motors start. During this brief interval, the starting inrush current may approximate a motor's locked rotor amperage rating.

Table 7-4 shows the range of locked rotor amperes (LRA) per motor horsepower. LRA appears on the rating plates of hermetic compressors. Depending on the type of motor, its locked rotor amperage may be 2 to 6 times its rated full load current. Motor starting torque varies as the square of the voltage. Thus, only 81 percent of the anticipated torque is available if the voltage drops to 90 percent of the rating during the starting period.

The full load amperage value must be considered in choosing the proper wire size and making sure the motor has acceptable voltages. These are shown in Table 7-5.

TABLE 7-4 Range of Locked Rotor Amperes per Motor Horsepower

NEMA Code letter	115 1φ	208 1φ	208 3φ	230 1φ	230 3φ	460 1φ	460 3φ
A	0–27.4	0–15.1	0–9.1	0–13.7	0–7.9	0–6.9	0–4.0
B	27.5–30.9	15.2–17.0	9.2–9.8	13.8–15.5	8.0–9.0	7.0–7.7	4.1–4.5
C	31.0–34.8	17.1–19.4	9.9–11.2	15.6–17.4	9.1–10.1	7.8–8.7	4.6–5.0
D	34.9–39.2	19.5–21.6	11.3–12.5	17.5–19.6	10.2–11.3	8.8–9.8	5.1–5.7
E	39.3–43.5	21.7–24.0	12.6–13.9	19.7–21.7	11.4–12.5	9.9–10.9	5.8–6.3
F	43.6–48.7	24.1–26.9	14.0–15.5	21.8–24.4	12.6–14.1	11.0–12.2	6.4–7.0
G	48.8–54.8	27.0–30.3	15.6–17.5	24.5–27.4	14.2–15.8	12.3–13.7	7.1–7.9
H	54.9–61.7	30.4–33.7	17.6–19.5	27.5–30.6	15.9–17.7	13.8–15.3	8.0–8.8
J	61.8–69.6	33.8–38.4	19.6–22.2	30.7–34.8	17.8–20.1	15.4–17.4	8.9–10.1
K	69.7–78.4	38.5–43.3	22.3–25.0	34.9–39.2	20.2–22.6	17.5–19.6	10.2–11.3
L	78.5–87.1	43.4–48.0	25.1–27.7	39.3–43.2	22.7–25.2	19.7–21.8	11.4–12.6
M	87.2–97.4	48.1–53.8	27.8–31.1	43.3–48.7	25.3–28.7	21.9–24.4	12.7–14.1
N	97.5–109	53.9–60.0	31.2–34.6	48.7–54.5	28.3–31.5	24.5–27.3	14.2–15.8
P	110–122	60.1–67.2	34.7–38.8	54.6–61.0	31.6–35.2	27.4–30.5	15.9–17.6
R	123–139	67.3–76.8	38.9–44.4	61.1–69.6	35.3–40.2	30.6–34.8	17.7–20.1
S	140–157	76.9–86.5	44.5–50.0	69.7–78.4	40.3–45.3	34.9–39.2	20.2–22.6
T	158–174	86.6–96.0	50.1–55.5	78.5–87.0	45.4–50.2	39.3–43.5	22.7–25.1
U	175–195	96.1–108	55.6–56.4	87.1–97.5	50.3–56.3	44.5–48.8	25.2–28.2
V	196 and up	109 and up	56.5 and up	97.6 and up	56.4 and up	48.9 and up	28.3 and up

Note: Locked rotor amperes appear on rating plates of hermetic compressors.
The NEMA code letter appears on the motor rating plate.
Multiply above values by motor horsepower.

TABLE 7-5 Approximate Full-Load Amperage Values for AC Motors

Motor	Single phase*		Three-phase, squirrel cage induction		
HP	115 V	230 V	230 V	460 V	575 V
$^1/_6$	4.4	2.2			
$^1/_4$	5.8	2.9			
$^1/_3$	7.2	3.6			
$^1/_2$	9.8	4.9	2	1.0	0.8
$^3/_4$	13.8	6.9	2.8	1.4	1.1
1	16	8	3.6	1.8	1.4
$1^1/_2$	20	10	5.2	2.6	2.1
2	24	12	6.8	3.4	2.7
3	34	17	9.6	4.8	3.9
5	56	28	15.2	7.6	6.1
$7^1/_2$			22	11.0	9.0
10			28	14.0	11.0
15			42	21.0	17.0
20			54	27.0	22.0
25			68	34.0	27.0

*Does not include shaded pole.

Calculating Starting Current Values and Inrush Voltage Drops

Single-phase current

Wire size and inrush voltage drop can be calculated. The following formula can be used for single-phase current. For example, if a single-phase 230-V condensing unit, rated at 22-A full load and having a starting current of 91 A is located 125 ft from the distribution panel and so utilizes 250 ft of the No. 10 copper wire, the voltage drop expected during full load operation is calculated as follows:

Refer to the lower lines of Table 7-3. Note that the resistance of No. 10 copper wire is 0.1018 Ω/100 ft.

$$\text{Voltage drop} = \frac{22\,\text{A} \times 0.1018 \times 250\,\text{ft}}{100} = 5.6\,\text{V}$$

(Note that 5.6 V exceeds the 2 percent loss factor, which is 4.6 V.) If the full 3 percent loss (6.9 V) allowed ahead of the meter is present, then the voltage at the load terminal of the meter will be 223.1 V (230 − 6.9 = 223.1). Subtract the voltage drop calculated above and there will be only 217.5 V at the unit terminals during full load operation.

Thus, 223.1 − 5.6 = 217.5. With a total loss of 5.4 percent, (230 − 217.5 = 12.5, or 5.47 percent), it is common practice to move to the next largest wire size. Therefore, for this circuit, AWG No. 8 wire should be used instead of No. 10.

Insofar as motor starting and relay operation are concerned, the critical period is during the initial instant of start-up when the inrush current closely approximates the locked rotor value. For the equipment described in the above example, the voltage drop experienced at 91-A flow for No. 10 wire is again excessive, indicating the wisdom of using No. 8 wire.

For No. 8 wire:

$$\text{Inrush voltage drop} = 91 \text{ A} \times \frac{0.064 \text{ } \Omega}{100 \text{ ft}} \times 250 \text{ ft}$$

$$= 14.56 \text{ V}$$

For No. 10 wire:

$$\text{Inrush voltage drop} = 91 \text{ A} \times \frac{0.1018 \text{ } \Omega}{100 \text{ ft}} \times 250 \text{ ft} \qquad (7\text{-}1)$$

$$= 23.159 \text{ V} \quad \text{or} \quad 23.2 \text{ V}$$

For a 230-V circuit, the 23.2 V slightly exceeds a 10 percent drop between the meter and the appliance. To this must be added the voltage drop incurred in the lead-in wires from the outdoor power line. This total must then be deducted from the power line voltage on the poles, which may be less than 230 V during utility peak load periods. Although the inrush current may exist for only an instant, this may be long enough to cause a starting relay to open, thus cutting off current to the motor. Without current flow, the voltage at the unit immediately rises enough to reclose the relay, so there is another attempt to start the motor. While the unit may get underway after the second or third attempt, such "chattering relay" operation is not good for the relay, the capacitors, or the motor.

For electrical loads such as lighting, resistance heating, and cooking, inrush current may be considered the equivalent of normal current flow. In the case of rotating machinery, it is only during that initial period or rotation that the start-up current exceeds that of final operation. The same is true of relays during the instant of "pull-in."

Three-phase circuits

Calculating the inrush voltage drop for three-phase circuits is the same as calculating the drop for single-phase circuits. Again, the value for circuit length equals twice the length of an individual conductor. Since more conductors are involved, the normal current and the starting current per conductor are smaller for a motor of a given size. Thus, lighter wire may be used.

Example 7-3 Using the same wire length as in the single-phase example and the lower values of 13.7-A full load and 61-A starting inrush per conductor for the three-phase rating of the same size compressor, the use of No. 10 conductor results in

$$\text{Normal voltage drop} = 13.7 \, \text{A} \times \frac{0.1018 \, \Omega}{100 \, \text{ft}} \times 250 \, \text{ft}$$

$$= 3.5 \, \text{V}$$

$$\text{Inrush voltage drop} = 61 \, \text{A} \times \frac{0.1018 \, \Omega}{100 \, \text{ft}} \times 250 \, \text{ft} \qquad (7\text{-}2)$$

$$= 15.5 \, \text{V}$$

Inrush voltage drop

The actual inrush current through an appliance usually is somewhat less than the total of locked-rotor-current values. Locked rotor current is measured with rated voltage at the appliance terminals. Since voltage drop in the feed lines reduces the voltage available at the terminals, less than rated voltage can be anticipated across the electrical components. Consequently, inrush currents and voltage drops are somewhat less. This fact is illustrated in the following, which is based on the same installation as that in the previous single-phase examples. However, here the actual locked rotor current of 101 A is used. The formula can be found in Eq. (7-2).

Code Limitations on Amperes per Conductor

Varied mechanical conditions are encountered in field wiring. Thus the NEC places certain limitations on the smallness of conductors installed in the field. Such limitations apply regardless of conductor length. They ensure the following:

- That the wire itself has ample strength to withstand the stress of pulling it through long conduits and chases. With specific exceptions, no wire lighter than No. 14 copper is permitted for field wiring of line voltage power circuits.

- By stipulating the maximum amperage permissible for each wire gage, self-generated heat can be limited to avoid temperature damage to wire insulation. If wiring is installed in areas of high ambient temperature, the amperage rating may need to be reduced.

- By stipulating the maximum amperage of overload protectors for circuits, current flow is limited to safe values for the conductor used. Some equipment has momentary starting currents that trip-out overload protectors sized on the basis of full load current. Here heavier fusing is permissible—but only under specific circumstances. Current flow limitations for each gage protect wire insulation from damage due to overheating.

Heat Generated within Conductors

Heat generation due to current flow through the wire is important for the following two reasons.

1. Temperature rise increases the resistance of the wire and, therefore, the voltage drop in the circuit. Under most conditions of circuit usage, this added resistance generates additional heat in the wires. Finally, a temperature is reached where heat dissipation from the conductors equals the heat that they generate. It is desirable to keep this equilibrium temperature low. The number of Btu generated can be found by both of the following formulas.

2. Temperature also damages wire insulation. The degree of damage is dependent upon the insulation's ability to withstand temperature under varying degrees of exposure, age, moisture, corrosive environment, mechanical abuse, and thickness.

Estimating the probable operating temperature of a conductor and its insulation is difficult. The rate of heat dissipation from the wiring surfaces varies with the ambient temperature, the proximity of other heat-generating conductors, the heat conductivity of the insulation and jacket material, the availability of cooling air, and other factors. Freestanding individual conductors dissipate heat more effectively. However, the typical situation of two or three conductors, each carrying equal current and enclosed in a common jacket, cable, or conduit, anticipates limitations as set forth by the NEC.

Circuit Protection

Circuits supplying power to appliances must incorporate some means for automatically disconnecting the circuit from the power source should there be abnormal current flow due to accidental grounding, equipment overload, or short circuits. Such overload devices should operate promptly enough to limit the buildup of damaging temperatures in conductors or in the electrical components of an appliance. However, devices selected to protect circuit-feeding motors must be slow enough to permit the momentary inrush of heavy starting current. They must then disconnect the circuit if the motor does not start promptly, as can happen under low-voltage conditions.

Devices heavy enough to carry continuously the motor starting current do not provide the overload protection desired. Likewise, heavily fused branch circuits do not adequately protect the low-amperage components that cumulatively require the heavy fusing. For this reason some literature lists maximum allowable fuse sizes for equipment. While electrical components of factory-built appliances are individually safeguarded, the field combining of two or more units on one circuit may create a problem more complex than that normally encountered. Remember that the final authority is the local electrical inspector.

Standard rule

With a few exceptions, the ampere capacity of an overload protector cannot exceed the ampacity values listed by wire size by the NEC. (Check the NEC for these exceptions.) If the allowable ampacity of a conductor does not match the rating of a standard size fuse or nonadjustable trip-circuit breaker, the device with the next largest capacity should be used. Some of the standard sizes of fuses and nonadjustable trip-circuit breakers are 15, 20, 25, 30, 35, 40, 45, 50, 60, 70, 80, 90, 100, 110, 125, 150, 175, 200, 225, 250, and 300 A.

Fuses

One-time single-element fuses

If a current of more than rated load is continued sufficiently long, the fuse link becomes overheated. This causes the center portion to melt. The melted portion drops away. However, due to the short gap, the circuit is not immediately broken. An arc continues and burns the metal at each end until the arc is stopped because of the very high increase in resistance. The material surrounding the link tends to break the arc mechanically. The center portion melts first, because it is farthest from the terminals that have the highest heat conductivity (see Fig. 7-35).

Fuses will carry a 10 percent overload indefinitely under laboratory-controlled conditions. However, they will blow promptly if materially overloaded. They will stand 150 percent of the rated amperes for the following time periods:

- 1 min (fuse is 30 A or less)
- 2 min (fuse of 31 to 60 A)
- 4 min (fuse of 61 to 100 A)

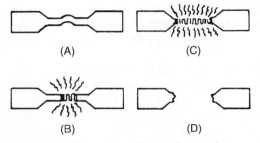

(A)   (C)

(B)   (D)

Figure 7-35 Illustration of how a fuse works.

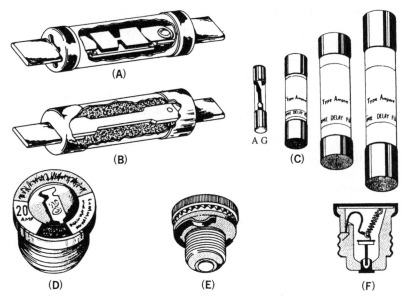

Figure 7-36   Types of fuses.

Time-delay two-element fuses

Two-element fuses use the burnout link described previously. They also use a low-temperature soldered connection that will open under overload. This soldered joint has mass, so it does not heat quickly enough to melt if a heavy load is imposed for only a short time. However, a small but continuous overload will soften the solder so that the electrical contact can be broken.

With this type of protection against light overloads, the fusible link can be made heavier, yet blow quickly to protect against heavy overloads. This results in fewer nuisance burnouts and equipment shutdowns. Two types of dual-element fuses are shown in Fig. 7-36.

Types of fuses

In addition to those fuses just described, there are three general categories based on shape and size.

- The AG (automotive glass) fuse consists of a glass cylinder with metallic end caps between which is connected a slender metal element that melts on current overload. This fuse has a length of 1-5/16 in. and a diameter of 1/4 in. It is available only for low amperages. While used in specific appliances, it is not used to protect permanently installed wiring.

- Cartridge fuses are like AG fuses. However, they are larger. The cylindrical tube is fiber, rather than glass. The metallic end pieces may be formed as lugs, blades, or cylinders to meet a variety of fuse box socket requirements. The internal metal fusible link may be enclosed in sand or powder to quench the burnout arc.

- Cartridge fuses are made in a variety of dimensions, based on amperage and voltage. Blade type terminals are common above 60 A. Fuses used to break 600-V arcs are longer than those for lower voltages. Fuses are available in many capacities other than the listed standard capacities, particularly in the two-element time-delay types. Often, they are so dimensioned as to not be interchangeable with fuses of other capacities.

- Plug fuses are limited in maximum capacity to 30 A. They are designed for use in circuits of not more than 150 V above ground. Two-element time-delay types are available to fit standard screw lamp sockets. They are also available with nonstandard threads made especially for various amperage ratings.

Thermostats

The thermostat (or temperature control) stops and starts the compressor in response to room temperature requirements. Each thermostat has a charged power element containing either a volatile liquid or an active vapor charge. The temperature sensitive part of this element (thermostat feeler bulb) is located in the return air stream. As the return air temperature rises, the pressure of the liquid or vapor inside the bulb increases. This closes the electrical contacts and starts the compressor. As the return air temperature drops, the reduced temperature of the feeler bulb causes the contacts to stop open and stops the compressor.

The advent of transistors and semiconductor chips or integrated circuits produced a more accurate method of monitoring and adjusting temperatures within a system. The microprocessor makes use of semiconductor and chip's abilities to compare temperatures. It can also program on and off cycles, as well as monitor the duration of each cycle. This leads to more accurate temperature control.

Figure 7-37 shows a processor-based thermostat. As you can see from the front of the control panel, you can adjust the program to do many things. It can also save energy, whether it is operating the furnace for heat or the air-conditioning unit for cooling. The units usually come with a battery so that the memory can retain whatever is programmed into it. The battery is also a backup for the clock so that the program is retained even if the line power is interrupted.

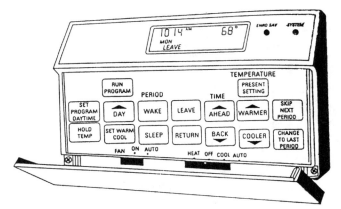

Figure 7-37 Microprocessor thermostat used for residential temperature and cooling control.

Thermostat as a control switch

The control switch (thermostat) may be located in the room to be cooled or heated, depending upon the particular switch selection point. The control switch (heat, off, cool, and auto) is of the sliding type and normally has four positions, marked HEAT, OFF, COOL, and AUTO. The thermostat is taken from its socket and programmed according to the manufacturer's directions. Then it is activated by plugging it into the wall socket and replacing a couple of screws to hold it in place.

To operate the unit as a ventilator, the switch on the left is marked "Fan" with an "On" and "Auto" choice to select the Fan operation. When a thermostat is installed for automatic cooling, the compressor and fans will cycle according to the dial requirements.

Figure 7-38 shows the electrical circuitry for a home heat-cool thermostat. Keep in mind that the thermostat should always be on the inside partition, never on an outside wall. Do not mount the instrument on a part of the wall that has steam or hot water pipes or warm air ducts behind it. The location should be such that direct sunshine or fireplace radiation cannot strike the thermostat. Be careful that the spot selected is not likely to have a floor lamp near it or a table lamp under it. Do not locate the thermostat where heat from kitchen appliances can affect it. Do not locate it on a wall that has a cold unused room on the other side.

After a thermostat has been mounted, it is wise to fill the stud space behind the instrument with insulating material. This is to prevent any circulation of cold air. Furthermore, the hole behind the thermostat for the wires should be sealed so that air cannot emerge from the stud space and blow across the thermostat element. It is quite common to find

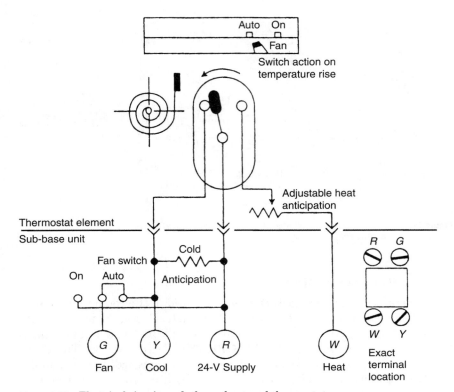

Figure 7-38 Electrical circuitry of a home heat-cool thermostat.

considerable air motion through this hole caused by a chimney effect in the stud space.

Servicemen

Servicemen who have good knowledge of refrigeration and air conditioning will be able to competently service air conditioners. Most air-conditioning units of present design contain compressors of the hermetic or sealed type. The only part that can be serviced in the field are the relay, control switch, fan, fan motor, start and run capacitors, air filters, and cabinet parts. The refrigerating system (consisting of the cooling unit, condensers, compressors, and connecting lines) generally cannot be serviced in the field. Most servicemen will find the newer electronic thermostats easy to program when following the manufacturer's instructions included with every thermostat and usually given to the homeowner at closing on the new house.

8

Condensers

Condensers are heat-transfer devices that are used to remove heat from hot refrigerant vapor. By using some method of cooling, the condenser changes the vapor to a liquid. There are three basic methods of cooling the condenser hot gases. The method used to cool the refrigerant and return it to the liquid state is used to categorize the two types of condensers:

- Air-cooled
- Water-cooled

Cooling towers can also be used to cool the refrigerant.

Most commercial and residential air-conditioning units are air-cooled. Water can be also used to cool the refrigerant. This is usually done where there is an adequate supply of fairly clean water. Industrial applications rely upon water to cool the condenser gases. The evaporative process is (cooling towers) also used to return the condenser gases to the liquid state.

Condensers

Air-cooled condensers

Figure 8-1 illustrates the refrigeration process within an air-cooled condenser. Figure 8-2 shows some of the various types of compressors and condensers mounted as a unit. These units may be located outside the cooled space. Such a location makes it possible to exhaust the heated air from the cooled space. Note that the condenser has a large-bladed fan that pushes air through the condenser fins. The fins are attached to coils of copper or aluminum tubing. Tubing houses the liquid and the gaseous vapors. When the blown air contacts the fins, it cools them. Heat from the compressed gas in the tubing is thus transferred to the cooler fin.

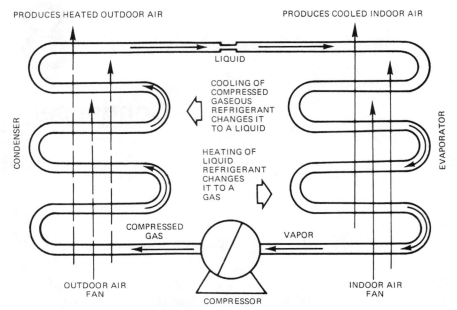

Figure 8-1   Refrigeration cycle.

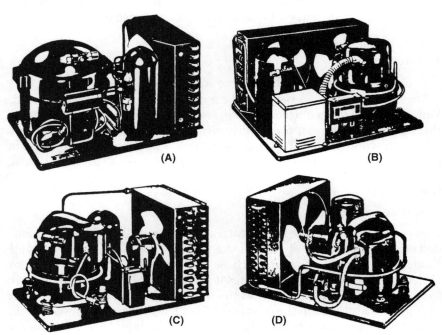

Figure 8-2   A condenser, fan, and compressor. Self-contained, in one unit. (*Courtesy of Tecumseh.*)

Heat given up by the refrigerant vapor to the condensing medium includes both the heat absorbed in the evaporator and the heat of compression. Thus, the condenser always has a load, that is, the sum of these two heats. This means the compressor must handle more heat than that generated by the evaporator. The quantity of heat (in Btu) given off by the condenser is rated in *heat per minute per ton* of evaporator capacity. These condensers are rated at various suction and condensing temperatures.

The larger the condenser area exposed to the moving air stream, the lower will be the temperature of the refrigerant when it leaves the condenser. The temperature of the air leaving the vicinity of the condenser will vary with the load inside the area being cooled. If the evaporator picks up the additional heat and transfers it to the condenser, then the condenser must transmit this heat to the air passing over the surface of the fins. The temperature rise in the condensing medium passing through the condenser is directly proportional to the condenser load. It is inversely proportional to the quantity and specific heat of the condensing medium.

To exhaust the heat without causing the area being cooled to heat up again, it is common practice to locate the condenser outside of the area being conditioned. For example, for an air-conditioned building, the condenser is located on the rooftop or on an outside slab at grade level (see Fig. 8-3).

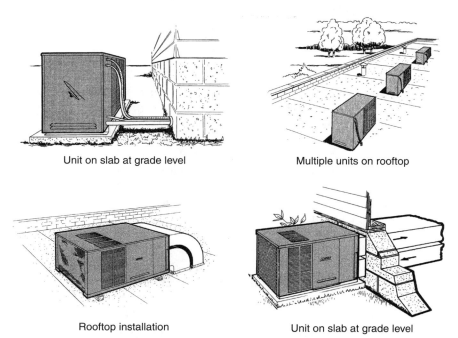

Unit on slab at grade level   Multiple units on rooftop

Rooftop installation   Unit on slab at grade level

Figure 8-3   Condensers mounted on rooftops and at grade level. (*Courtesy of Lennox.*)

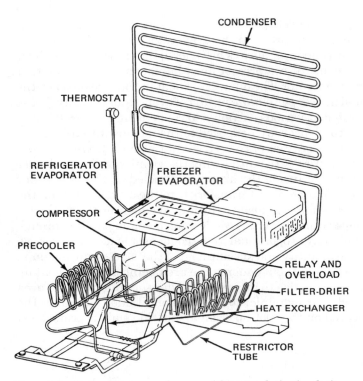

Figure 8-4 Flat, coil-type condenser, with natural air circulation.
Used in domestic refrigeration. (*Courtesy of Sears.*)

Some condensers are cooled by natural airflow. This is the case in domestic refrigerators. Such natural convection condensers use either plate surface or finned tubing (see Fig. 8-4).

Air-cooled condensers that use fans are classified as chassis mounted and remote. The chassis-mounted type is shown in Fig. 8-2. Here, the compressor, fan, and condenser are mounted as one unit. The remote type is shown in Fig. 8-3. Remote air-cooled condensers can be obtained in sizes that range from 1 to 100 ton. The chassis-mounted types are usually limited to 1 ton or less.

Water-cooled condensers

Water is used to cool condensers. One method is to cool condensers with water from the city water supply and then exhaust the water into the sewer after it has been used to cool the refrigerant. This method can be expensive and, in some instances, is not allowed by law. When there is a sewer problem, a limited sewer-treatment plant capacity, or drought, it is impractical to use this cooling method.

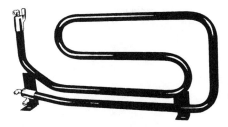

Figure 8-5 Coaxial, water-cooled condenser. Used with refrigeration and air-conditioning units where space is limited.

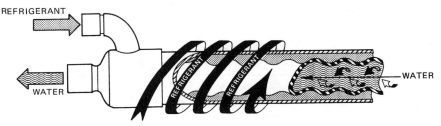

Figure 8-6 A typical counterflow path inside a coaxial water-cooled condenser. (*Courtesy of Packless.*)

The use of recirculation to cool the water for reuse is more practical. However, in recirculation, the power required to pump the water to the cooling location is part of the expense of operating the unit.

There are three types of water-cooled condensers:

- Double-tube
- Shell-and-coil
- Shell-and-tube

The double-tube type consists of two tubes, one inside the other (see Fig. 8-5). Water is piped through the inner tube. Refrigerant is piped through the tube that encloses the inner tube. The refrigerant flows in the opposite direction than the water (see Fig. 8-6).

This type of coaxial water-cooled condenser is designed for use with refrigeration and air-conditioning condensing units where space is limited. These condensers can be mounted vertically, horizontally, or at any angle.

They can be used with cooling towers also. They perform at peak heat of rejection with water pressure drop of not more than 5 psi, utilizing flow rates of 3 gallons per minute per ton.

The typical counterflow path shows the refrigerant going in a 105°F (41°C) and the water going in at 85°F (30°C) and leaving at 95°F (35°C) (see Fig. 8-7).

The counter-swirl design, shown in Fig. 8-6, gives a heat-transfer performance of superior quality.

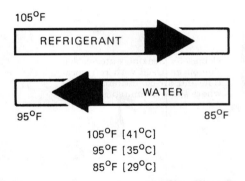

105°F [41°C]
95°F [35°C]
85°F [29°C]

Figure 8-7 Water and refrigerant temperatures in a counterflow, water-cooled condenser. (*Courtesy of Packless.*)

TABLE 8-1 Possible Metal Combinations Found in Water-Cooled Condensers

Shell metal	Tubing metal
Steel	Copper
Copper	Copper
Steel	Cupronickel
Copper	Cupronickel
Steel	Stainless steel
Stainless steel	Stainless steel

The tube construction provides for excellent mechanical stability. The water-flow path is turbulent. This provides a scrubbing action that maintains cleaner surfaces. The construction method shown also has very high system pressure resistance.

The water-cooled condenser shown in Fig. 8-5 can be obtained in a number of combinations. Some of these combinations are listed in Table 8-1. Copper tubing is suggested for use with fresh water and with cooling towers. The use of cupronickel is suggested when salt water is used for cooling purposes.

Convolutions to the water tube result in a spinning, swirling water flow that inhibits the accumulation of deposits on the inside of the tube. This contributes to the antifouling characteristics in this type of condenser. Figure 8-8 shows the various types of construction for the condenser.

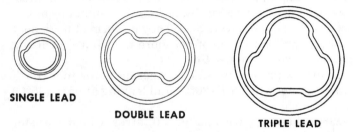

Figure 8-8 Different types of tubing fabrication, located inside the coaxial-type water-cooled condenser. (*Courtesy of Packless.*)

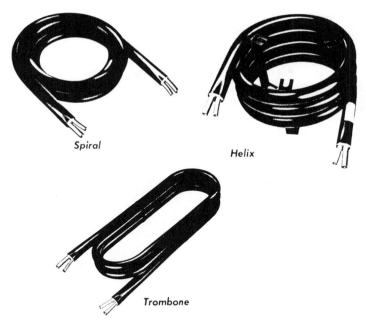

Figure 8-9   Three configurations of coaxial, water-cooled condensers. (*Courtesy of Packless.*)

This type of condenser may be added as a booster to standard air-cooled units. Figure 8-9 shows some of the variations in the configuration of this type of condenser:

- Spiral
- Helix
- Trombone

Note the inputs for water and refrigerant. Using a cooling tower to furnish the water that contacts the outside tube can further cool the condenser. Also, a water tower can be used to cool the water sent through the inside tube for cooling purposes. This type of condenser is usable where refrigeration or air-conditioning requirements are 1/3 ton to 3 ton.

Placing a bare tube or a finned tube inside a steel shell makes the shell-and-coil condenser (see Fig. 8-10). Water circulates through the coils. Refrigerant vapor is injected into the shell. The hot vapor contacts the cooler tubes and condenses. The condensed vapor drains from the coils and drops to the bottom of the tank or shell. From there it is recirculated through the refrigerated area by way of the evaporator. In most cases, placing chemicals into the water cleans the unit. The chemicals have a tendency to remove the deposits that build up on the tubing walls.

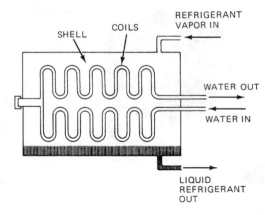

SHELL COILS REFRIGERANT VAPOR IN

WATER OUT

WATER IN

Figure 8-10 The shell-and-coil condenser.

LIQUID REFRIGERANT OUT

Chillers

A chiller is part of a condenser. Chillers are used to cool water or brine solutions. The cooled (chilled) water or brine is fed through pipes to evaporators. This cools the area in which the evaporators are located. This type of cooling, using chilled water or brine, can be used in large air-conditioning units. It can also be used for industrial processes where cooling is required for a particular operation.

Figure 8-11 illustrates such an operation. Note how the compressor sits atop the condenser. Chillers are the answer to requirements of 200 to 1600 ton of refrigeration. They are used for process cooling, comfort air-conditioning, and nuclear power plant cooling. In some cases, they are used to provide ice for ice-skating rinks. The arrows in Fig. 8-11 indicate the refrigerant flow and the water or brine flow through the large pipes. Figure 8-12 shows the machine in a cutaway view. The following explanation of the various cycles will provide a better understanding of the operation of this type of equipment.

Refrigeration cycle

The machine compressor continuously draws large quantities of refrigerant vapor from the cooler, at a rate determined by the size of the guide-vane opening. This compressor suction reduces the pressure within the cooler, allowing the liquid refrigerant to boil vigorously at a fairly low temperature [typically, 30 to 35°F (−1 to 2°C)].

Liquid refrigerant obtains the energy needed for the change to vapor by removing heat from the water in the cooler tubes. The cold water can then be used in the air-conditioning process.

After removing heat from the water, the refrigerant vapor enters the first stage of the compressor. There, it is compressed and flows into the

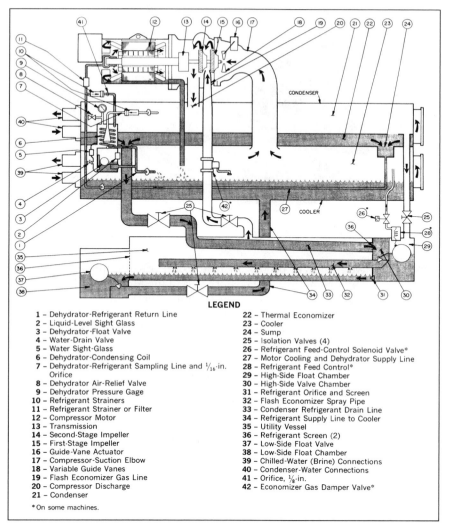

Figure 8-11 The chiller, compressor, condenser, and cooler are combined in one unit. (*Courtesy of Carrier.*)

second stage of the compressor. Here it is mixed with flash-economizer gas and further compressed.

Compression raises the refrigerant temperature above that of the water flowing through the condenser tubes. When the warm [typically 100 to 105°F (38 to 41°C)] refrigerant vapor contacts the condenser tubes, the relatively cool condensing water [typically, 85 to 95°F (29 to 35°C)] removes some of the heat and the vapor condenses into a liquid.

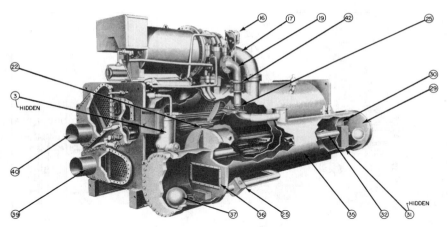

Figure 8-12   Cutaway view of a chiller.

Further heat removal occurs in the group of condenser tubes that form the thermal economizer. Here, the condensed liquid refrigerant is subcooled by contact with the coolest condenser tubes. These are the tubes that contain the entering water.

The subcooled liquid refrigerant drains into a high-side valve chamber. This chamber maintains the proper fluid level in the thermal economizer and meters the refrigerant liquid into a flash economizer chamber. Pressure in this chamber is intermediate between condenser and cooler pressures. At this lower pressure, some of the liquid refrigerant flashes to gas, cooling the remaining liquid. The flash gas, having absorbed heat, is returned directly to the compressor's second stage. Here, it is mixed with gas already compressed by the first stage impeller. Since the flash gas must pass through only half the compression cycle to reach condenser pressure, there is a savings in power.

The cooled liquid refrigerant in the economizer is metered through the low-side valve chamber into the cooler. Because pressure in the cooler is lower than economizer pressure, some of the liquid flashes and cools the remainder to evaporator (cooler) temperature. The cycle is now complete.

Motor-cooling cycle

Refrigerant liquid from a sump in the condenser (No. 24 in Fig. 8-11) is subcooled by passage through a line in the cooler (No. 27 in Fig. 8-11). The refrigerant then flows externally through a strainer and variable orifice (No. 11 in Fig. 8-11) and enters the compressor motor end. Here it sprays and cools the compressor rotor and stator. It then collects in the base of the motor casing. Here, it drains into the cooler. Differential pressure between the condenser and cooler maintains the refrigerant flow.

Dehydrator cycle

The dehydrator removes water and noncondensable gases. It indicates any water leakage into the refrigerant (see No. 6 in Fig. 8-11).

This system includes a refrigerant-condensing coil and chamber, water drain valve, purging valve, pressure gage, refrigerant float valve, and refrigerant piping.

A dehydrator-sampling line continuously picks up refrigerant vapor and contaminants, if any, from the condenser. Vapor is condensed into a liquid by the dehydrator-condensing coil. Water, if present, separates and floats on the refrigerant liquid. The water level can be observed through a sight glass.

Water may be withdrawn manually at the water drain valve. Air and other noncondensable gases collect in the upper portion of the dehydrator-condensing chamber. The dehydrator gage indicates the presence of air or other gases through a rise in pressure. These gases may be manually vented through the purging valve.

A float valve maintains the refrigerant liquid level and pressure difference necessary for the refrigerant-condensing action. Purified refrigerant is returned to the cooler from the dehydrator float chamber.

Lubrication cycle

The oil pump and oil reservoir are contained within the unishell. Oil is pumped through an oil filter-cooler that removes heat and foreign particles. A portion of the oil is then fed to the compressor motor-end bearings and seal. The remaining oil lubricates the compressor transmission, compressor thrust and journal bearings, and seal. Oil is then returned to the reservoir to complete the cycle.

Controls

The cooling capacity of the machine is automatically adjusted to match the cooling load by changes in the position of the compressor inlet-guide vanes (see Fig. 8-13).

A temperature-sensing device in the circuit of the chilled water leaving the machine cooler continuously transmits signals to a solid-state module in the machine-control center. The module, in turn, transmits the amplified and modulated temperature signals to an automatic guide vane actuator.

A drop in the temperature of the chilled water leaving the circuit causes the guide vanes to move toward the closed position. This reduces

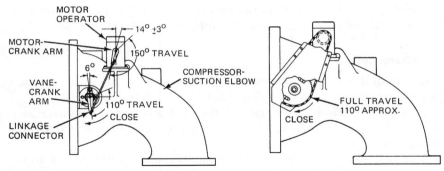

Figure 8-13 Vane motor-crank angles. These are shown as No. 16 and No. 17 in Fig. 8-11.

the rate of refrigerant evaporation and vapor flow into the compressor. Machine capacity decreases. A rise in chilled water temperature opens the vanes. More refrigerant vapor moves through the compressor and the capacity increases.

The modulation of the temperature signals in the control center allows precise control of guide vane response, regardless of the system load.

Solid-state capacity control

In addition to amplifying and modulating the signals from chilled water sensor to vane actuator, the solid-state module in the control center provides a means for preventing the compressor from exceeding full load amperes. It also provides a means for limiting motor current down to 40 percent of full load amperes to reduce electrical demand rates.

A throttle adjustment screw eliminates guide vane hunting. A manual capacity control knob allows the operator to open, close, or hold the guide-vane position when desired.

Cooling Towers

Cooling towers are used to conserve or recover water. In one design, the hot water from the condenser is pumped to the tower. There, it is sprayed into the tower basin. The temperature of the water decreases, as it gives up heat to the air circulating through the tower. Some of the towers are rather large, since they work with condensers yielding 1600 ton of cooling capacity (see Fig. 8-14).

Most of the cooling that takes place in the tower results from the evaporation of part of the water as it falls through the tower.

The lower the wet bulb temperature of the incoming air, the more efficient the air is in decreasing the temperature of the water being fed into the tower. The following factors influence the efficiency of the cooling tower:

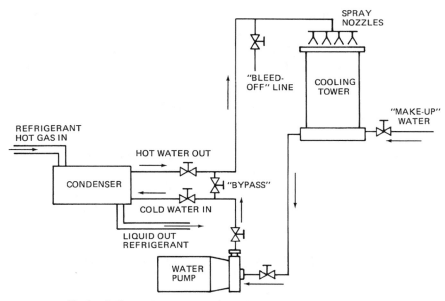

Figure 8-14   Recirculating water system using a tower.

- Mean difference between vapor pressure of the air and pressure in the tower water
- Length of exposure time and amount of water surface exposed to air
- Velocity of air through the tower
- Direction of air flow relative to the exposed water surface (parallel, transverse, or counter)

Theoretically, the lowest temperature to which the water can be cooled is the temperature of the air (wet bulb) entering the tower. However, in practical terms, it is impossible to reach the temperature of the air In most instances, the temperature of the water leaving the tower will be no lower than 7 to 10°F (–14 to –12°C) above the air temperature.

The range of the tower is the temperature of the water going into the tower and the temperature of the water coming out of the tower. This range should be matched to the operation of the condenser for maximum efficiency.

Cooling systems terms

The following terms apply to cooling-tower systems.

Cooling range is the number of degrees in Fahrenheit through which the water is cooled in the tower. It is the difference between the temperature of the hot water entering the tower and the temperature of the cold water leaving the tower.

Approach is the difference in degree Fahrenheit between the temperature of the cold water leaving the cooling tower and the wet-bulb temperature of the surrounding air.

Heat load is the amount of heat "thrown away" by the cooling tower in Btu per hour (or per minute). It is equal to the pounds of water circulated multiplied by the cooling range.

Cooling-tower pump head is the pressure required to lift the returning hot water from a point level with the base of the tower to the top of the tower and force it through the distribution system.

Drift is the small amount of water lost in the form of fine droplets retained by the circulating air. It is independent of, and in addition to, evaporation loss.

Bleed-off is the continuous or intermittent wasting of a small fraction of circulating water to prevent the build-up and concentration of scale-forming chemicals in the water.

Make-up is the water required to replace the water that is lost by evaporation, drift, and bleed-off.

Design of cooling towers

Classified by the air-circulation method used, there are two types of cooling towers. They are either natural-draft or mechanical-draft towers. Figure 8-15 shows the operation of the natural-draft cooling tower. Figure 8-16 shows the operation of the mechanical-draft cooling tower. The forced-draft cooling tower shown in Fig. 8-17 is just one example of the mechanical-draft designs available today.

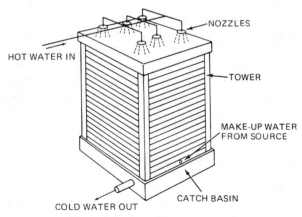

Figure 8-15   Natural-draft cooling tower.

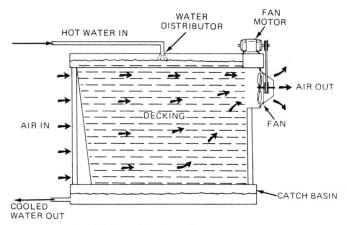

Figure 8-16   Small induced-draft cooling tower.

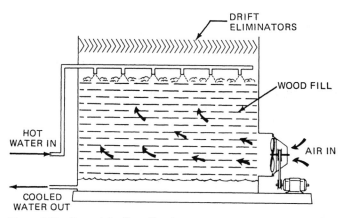

Figure 8-17   Forced-draft cooling tower.

Cooling tower ratings are given in tons. This is based on heat-transfer capacity of 250 Btu/(min·ton). The normal wind velocity taken into consideration for tower design is 3 mi/h. The wet bulb temperature is usually 80°F (27°C) for design purposes. The usual flow of water over the tower is 4 gal/min for each ton of cooling desired. Several charts are available with current design technology. Manufacturers supply the specifications for their towers. However, there are some important points to remember when use of a tower is being considered:

1. In tons of cooling, the tower should be rated at the same capacity as the condenser.

2. The wet-bulb temperature must be known.

3. The temperature of the water leaving the tower should be known. This would be the temperature of the water entering the condenser.

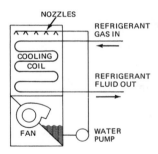

Figure 8-18 Evaporative cooler has no fill deck. The water-cooling process fluid directly. (*Courtesy of Marley.*)

Towers present some maintenance problems. These stem primarily from the water used in the cooling system. Chemicals are employed to control the growth of bacteria and other substances. Scale in the pipes and on parts of the tower also must be controlled. Chemicals are used for each of these controls. This problem will be discussed in the next chapter.

Evaporative Condensers

The evaporative condenser is a condenser and a cooling tower combined. Figure 8-18 illustrates how the nozzles spray water over the cooling coil to cool the fluid or gas in the pipes. This is a very good water conservation tower. In the future, this system will probably become more popular. The closed-circuit cooler should see increased use because of dwindling water supplies and more expensive treatment problems. The function of this cooler is to process the fluid in the pipes. This is a sealed contamination-free system. Instead of allowing the water to drop onto slats or other deflectors, this unit sprays the water directly onto the cooling coil.

New Developments

All-metal towers with housing, fans, fill, piping, and structural members made of galvanized or stainless steel are now being built. Some local building codes are becoming more restrictive with respect to fire safety. Low maintenance is another factor in the use of all-metal towers.

Engineers are beginning to specify towers less subject to deterioration due to environmental conditions. Thus, all-steel or all-metal towers are called for. Already, galvanized steel towers have made inroads into the air-conditioning and refrigeration market. Stainless-steel towers are being specified in New York City, northern New Jersey, and Los Angeles. This is primarily due to a polluted atmosphere, which can lead to early deterioration of nonmetallic towers and, in some cases, metals.

Figure 8-19 shows a no-fans design for a cooling tower. Large quantities of air are drawn into the tower by cooling water as it is injected through spray nozzles at one end of a venturi plenum. No fans are needed.

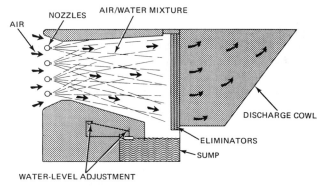

Figure 8-19 Cooling tower with natural draft properties. There are no moving parts in the cooling tower. (*Courtesy of Marley.*)

Effective mixing of air and water in the plenum permits evaporative heat transfer to take place without the fill required in conventional towers.

The cooled water falls into the sump and is pumped through a cooling-water circuit to return for another cycle. The name applied to this design is Baltimore Aircoil. In 1981, towers rated at 10 to 640 ton with 30 to 1920 gal/min were standard. Using prestrainers in the high-pressure flow has minimized the nozzle-clogging problem. There are no moving parts in the tower. This results in very low maintenance costs.

Air-cooled condensers are reaching 1000 ton in capacity. Air coolers and air condensers are quite attractive for use in refineries and natural-gas-compressor stations. They are also used for cooling in industry, as well as for commercial air-conditioning purposes. Figure 8-19 shows how the air-cooled condensers are used in a circuit system that is completely closed. These are very popular where there is little or no water supply.

Temperature Conversion

A cooling tower is a device for cooling a stream of water. Evaporating a portion of the circulating stream does this. Such cooled water may be used for many purposes, but the main concern here is its utilization as a heat sink for a refrigeration-system condenser. A number of types of cooling towers are used for industrial and commercial purposes. They are usually regarded as a necessity for large buildings or manufacturing processes. Some of these types have already been mentioned, but the following will bring you more details on the workings of cooling towers and their differences.

Cooling water concerns that must be addressed for the health of those who operate and maintain the systems. There is the potential for harboring and for the growth of pathogens in the water basin or related surface. This may occur mainly during the summer and also during idle periods. When the temperature falls in the 70 to 120°F (21 to 49°C) there are periods when the unit will not be operational and will sit idle. Dust from the air will settle in the water and create an organic medium for the culture of bacteria and pathogens. Algae will grow in the water—some need sunlight, others grow without. Some bacteria feed on iron. The potential for pathogenic culture is there, and cooling-tower design should include some kind of filtration and/or chemical sterilization of the water.

Types of Towers

The atmospheric type of tower does not use a mechalnical device, such as a fan, to create air flow through the tower. There are two main types of atmospheric towers—large and small. The large hyperbolic towers are equipped with "fill" since their primary applications are with electric power plants. The steam driven alternator has very high temperature steam to reduce to water or liquid state.

Atmospheric towers are relatively inexpensive. They are usually applied in very small sizes. They tend to be energy intensive because of the high spray pressures required. The atmospheric towers are far more affected by adverse wind conditions than are other types. Their use on systems requiring accurate, dependable cold water temperatures is not reocmmended (see Fig. 8-20).

Mechanical-draft towers, such as in Fig. 8-21, are categorized as either forced-draft towers or induced draft. In the forced-draft type the fan is located in the ambient air stream entering the tower. The air is also brought through or induced to enter the tower by a fan as shown in Fig. 8-22. In the induced-draft draws air through the tower by an induced draft.

Forced-draft towers have high air-entrance velocities and low edit velocities. They are extremely susceptible to recirculation and are therefore considered to have less performance stability than induced-draft towers. There is concern in northern climates as the forced-draft fans

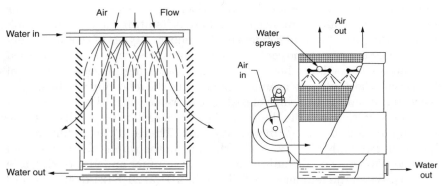

Figure 8-20   Atmospheric tower.   **Figure 8-21**   Forced-draft counterflow tower.

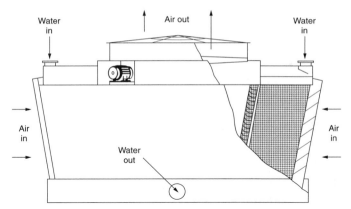

Figure 8-22   Induced-draft cross flow tower.

located in the cold entering ambient air stream can become subject to severe icing. The resultant imbalance comes when the moving air, laden with either natural or recirculated moisture, becomes ice.

Usually forced draft towers are equipped with centrifugal blower-type fans. These fans require approximately twice the operating horsepower of propeller-type fans. They have the advantage of being able to operate against the high static pressures generated with ductwork. So equipped, they can be installed either indoors or within a specifically designed enclosure that provides sufficient separation between the air intake and discharge locations to minimize recirculation (see Fig. 8-23).

Crossflow towers

Crossflow towers, as shown in Fig. 8-24, have a fill configuration through which the air flows horizontally. That means it is across the downward

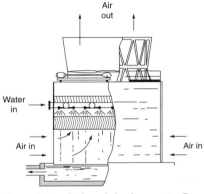

Figure 8-23   Induced-draft counterflow tower.

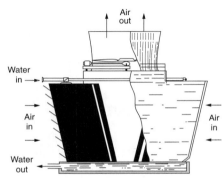

Figure 8-24   Double-flow crossflow tower.

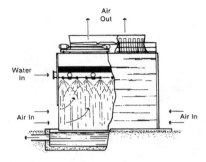

Figure 8-25   Spray-filled counter-flow tower.

fall of the water. The water being cooled is delivered to hot-water inlet basins. The basins are located above the fill areas. The water is distributed to the fill by gravity through metering orifices in the basins' floor. This removes the need for a pressure-spray distribution system. And, it places the resultant gravity system in full view for maintenance.

A cooling tower is a specialized heat exchanger (see Fig. 8-25). The two fluids, air and water, are brought into direct contact with each other. This is to affect the transfer of heat. In the spray-filled tower such as Fig. 8-25 this is accomplished by spraying a flowing mass of water into a rain-like pattern. Then an upward-moving mass flow of cool air is induced by the action of the fan.

Fluid cooler

The fluid cooler is one of the most efficient systems for industrial and HVAC applications (see Fig. 8-26). By keeping the cooling process fluid

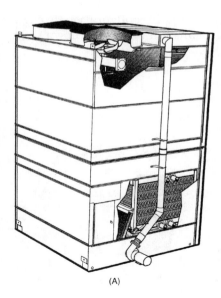

(A)

Figure 8-26a   MH fluid cooler, rear view. (*Courtesy of Marley.*)

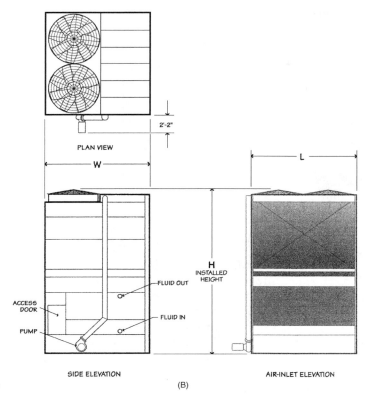

Figure 8-26b MH fluid cooler, various views. (*Courtesy of Marley.*)

Figure 8-26c MH fluid cooler, front view. (*Courtesy of Marley.*)

and in a clean, closed loop it combines the function of a cooling tower and heat exchanger into one system. It is possible to provide superior operational and maintenance benefits.

The fluid-cooler coil is suitable for cooling water, oils, and other fluids. It is compatible with most oils and other fluids when the carbon-steel

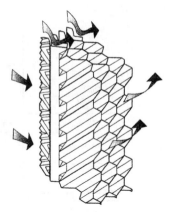

Figure 8-27 MH fluid-cooler coil. (*Courtesy of Marley.*)

Figure 8-28 MH fluid-cooler fill media. (*Courtesy of Marley.*)

coil in a closed, pressurized system. Each coil is constructed of continuous steel tubing, formed into a serpentine shape and welded into an assembly (see Fig. 8-27). The complete asssembly is then hot-dipped in liquid tin to galvanize it after fabrication. The galvanized-steel coil has proven itself through the years. Paints and electrostatically applied coatings can't seem to approach galvanization for inceasing coil longevity. The coils can also be made of stainless steel.

Operation of the fluid cooler.   The fluid cooler uses a mechanically induced draft, crossflow technology. And, the fill media is located above the coil. The process fluid is pumped internally through the coil. Recirculating water is cooled as it passes over the fill media, as shown in Fig. 8-28. The process fluid is thermally equalized and redistributed over the outside of the coil. A small portion of recirculating water is evaporatd by the air drawn that is passing through the coil and fill media. This cools the process fluid. The coil section rejects heat through evaporative cooling. This process uses the fresh air stream and precooled recirculating spray water. Recirculated water falls from the coil into a collection basin. From the base it is then pumped back up to be distributed over the fill media.

For industrial and HVAC applications this is an ideal type of system. The process fluid is kept in a clean, closed loop. It combines the function of a cooling tower and heat exchanger into one system. This improves efficiency and has many maintenance benefits. The unit shown here has a capacity ranging from 100 to 650 ton in a compact enclosure. It is suitable for cooling a wide range of fluids from water and glycols, to quench oils and plating solutions.

Compressors

Refrigeration compressors can be classified according to the following:

- Number of cylinders
- Method of compression
- Type of drive
- Location of the driving force or motor

The method of compression may be *reciprocating centrifugal* or *rotary*. The location of the power source also classifies compressors. Independent compressors are belt driven. Semihermetic compressors have direct drive, with the motor and compressor in separate housings. The hermetic compressor has direct drive, with the motor and compressor in the same housing.

Reciprocating units have a piston in a cylinder. The piston acts as a pump to increase the pressure of the refrigerant from the low side to the high side of the system. A reciprocating compressor can have twelve or more cylinders (see Fig. 9-1).

The most commonly used reciprocating compressor is made for refrigerants R-22 and R-134a. These are for heating, ventilating, and air-conditioning and process cooling. The most practical refrigerants used today are R-134a and R-22. However, R-134a is gaining in acceptance, in view of the CFC regulations worldwide. As a matter of fact, some countries only accept R-134a today. Other environmentally acceptable refrigerants are R-404A and R-507. They are for low and medium temperature applications. R-470C is for medium temperatures and air-conditioning applications. Recently, R-410A has gained acceptance as an environmentally

Figure 9-1 Reciprocating compressor. (*Courtesy of Trane.*)

acceptable substitute for R-22, but only for residential and small equipment. R-410A is not a drop-in refrigerant for R-22.

There are three types of reciprocating compressors: open drive, hermetic, and semihermetic.

Two methods of capacity control are generally applied to the reciprocating refrigeration compressor used on commercial air-conditioning systems. Both methods involve mechanical means of unloading cylinders by holding open the suction valve.

The most common method of capacity control uses an internal multiple-step valve. This applies compressor oil pressure or high-side refrigerant pressure to a bellows or piston that actuates the unloader.

The second method of capacity control uses an external solenoid valve for each cylinder. The solenoid valve allows compressor oil or high-side refrigerant to pass to the unloader.

A centrifugal compressor is basically a fan or blower that builds refrigerant pressure by forcing the gas through a funnel-shaped opening at high speed (see Fig. 9-2).

Compressor capacity is controlled when the vanes are opened and closed. These vanes regulate the amount of refrigerant gas allowed to enter the fan or turbine (see Fig. 9-3). When the vanes restrict the flow of refrigerant, the turbine cannot do its full amount of work on the refrigerant. Thus, its capacity is limited. Most centrifugal machines can be limited to 10 to 25 percent of full capacity by this method. Some will operate at almost zero capacity. However, another method, though less common, is to control the speed of the motor that is turning the turbine.

Condensers

A condenser must take the superheated vapor from the compressor, cool it to its condensing temperature, and then condense it. This action is opposite to that of an evaporator. Generally, two types of condensers are used: air cooled and water cooled. Also see Chap. 7 for more on condensers.

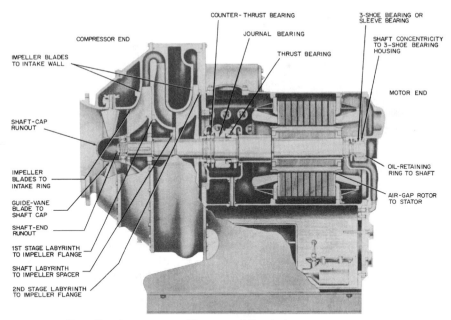

Figure 9-2 Centrifugal compressor. (*Courtesy of Carrier.*)

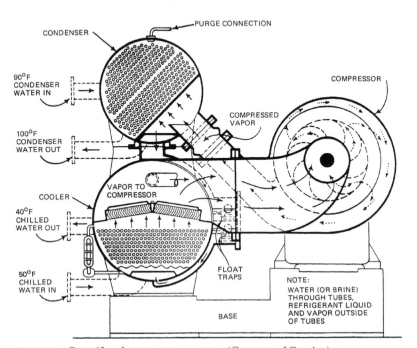

Figure 9-3 Centrifugal compressor system. (*Courtesy of Carrier.*)

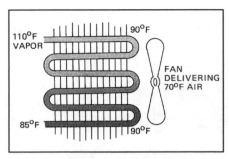

Figure 9-4 Schematic of air-cooled condenser. (*Courtesy of Johnson.*)

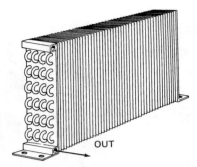

Figure 9-5 Air-cooled condenser. (*Courtesy of Johnson.*)

Air-cooled condensers

Air-cooled condensers are usually of the fin and tube type, with the refrigerant inside the tubes and air flowing in direct contact over the outside. Usually, a fan forces the air over the coil. This increases its cooling capabilities (see Figs. 9-4 and 9-5).

Water-cooled condensers

In the water-cooled condenser, the refrigerant is cooled with water within pipes (see Fig. 9-6). The tubing containing water is placed inside a pipe or housing containing the warm refrigerant. The heat is then transferred from the refrigerant through the tubing to the water. Water-cooled condensers are more efficient than air-cooled condensers. However, they must be supplied with large quantities of water. This water must be either discharged or reclaimed by cooling it to a temperature that makes it reusable.

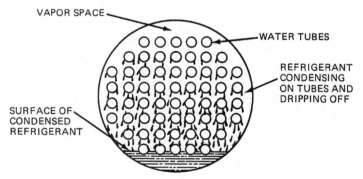

Figure 9-6 Cross-section of a shell and tube condenser. (*Courtesy of Johnson.*)

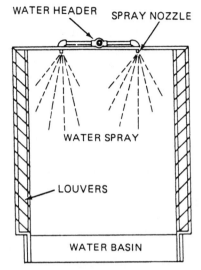

Figure 9-7 Spray-type cooling tower. (*Courtesy of Johnson.*)

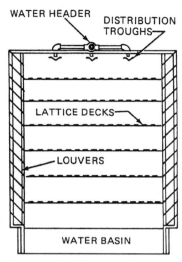

Figure 9-8 Deck-type cooling tower. (*Courtesy of Johnson.*)

A cooling tower usually accomplishes reclaiming (see Figs. 9-7 and 9-8). The tower chills the water by spraying it into a closed chamber. Air is forced over the spray. Cooling towers may be equipped with fans to force the air over the sprayed water.

Another device used to cool refrigerant is an evaporative condenser (see Fig. 9-9). Here, the gas-filled condenser is placed in an enclosure. Water is sprayed on it and air is forced over it to cool the condenser by evaporation.

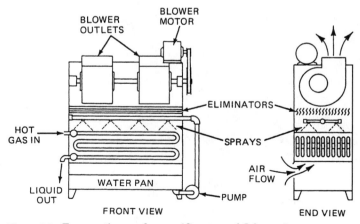

Figure 9-9 Evaporative condenser. (*Courtesy of Johnson.*)

Hermetic Compressors

A hermetic compressor is a direct compressor—a direct-connected motor compressor assembly enclosed within a steel housing. It is designed to pump low-pressure refrigerant gas to a higher pressure.

A hermetic container is one that is tightly sealed so no gas or liquid can enter or escape. Welding seals the container.

Tecumseh hermetic compressors have a low-pressure shell, or housing. This means that the interior of the compressor housing is subject only to suction pressure. It is not subject to the discharge created by the piston stroke. This point is emphasized to stress the hazard of introducing high-pressure gas into the compressor shell at pressures above 150 psig.

The major internal parts of a hermetic compressor are shown in Fig. 9-10. The suction is drawing into the compressor shell then to and through the electric motor that provides power to the crankshaft. The crankshaft revolves in its bearings, driving the piston or pistons in the cylinder or cylinders. The crankshaft is designed to carry oil from the oil pump in the bottom of the compressor to all bearing surfaces. Refrigerant gas surrounds the compressor crankcase and the motor as it is drawn through the compressor shell and into the cylinder or

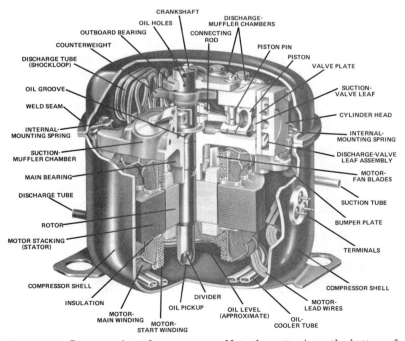

Figure 9-10 Cutaway view of a compressor. Note the motor is on the bottom of the compressor and the piston is on the top. (*Courtesy of Tecumseh.*)

cylinders, through the suction muffler and suction valves. The gas is compressed by the moving piston and is released through the discharge valves, discharge muffler, and compressor discharge tube.

Compressor types

Hermetic compressors have different functions. Some are used for home refrigeration. Some are used to produce air-conditioning. Others are used in home or commercial freezers. Hermetic compressors are also used for food display cases.

The serial number plate on the compressor tells several things about the compressor (see Fig. 9-11). Also notice, that several manufacturers made the motor for the compressor:

- A.O. Smith
- Aichi
- Delco
- Emerson
- General Electric
- Ranco
- Wagner
- Westinghouse

Makers of electric motors for compressors usually mark them for the compressor manufacturers. Newer models are rated in hertz (Hz) and very old models may be marked in cycles per second or cps instead of hertz. The following is a brief outline of some of the points to be remembered in servicing.

Pancake models are designated with a "P" as the first letter of serial number. They are made with 1/20- to 1/3-hp motors. All of them use an oil charge of 22 oz and R-134a or the latest replacement as a refrigerant. They have a temperature range of 20 to 55°F (–6 to 13°C). The smaller horsepower models are used where –30 to 10°F (–34 to –12°C) is required.

The T and AT compressor models have 1/6-, 1/5-, 1/4-, and 1/3-hp motors. All of these models use R-134a refrigerant or its equivalent replacement. The smaller sizes use a 38-oz oil charge, while the larger horsepower models use 32 oz. They have temperature ranges of –30 to 10°F (–34 to –12°C) and 20 to 55°F (–6.6 to 13°C).

The AE compressors are used for household refrigerators, freezers, dehumidifiers, vending machines, and water coolers (see Fig. 9-12). They are made in 1/20, 1/12, 1/8, 1/6, 1/5, and 1/4 hp units. The oil charge may be 10, 16, 20, or 23 oz. This AE compressor model line uses R-134a, or an acceptable substitute, and, in some cases, R-22 as a refrigerant. The older models still in use may have R-12 refrigerant. When servicing older equipment it is best to remember some of the older charges.

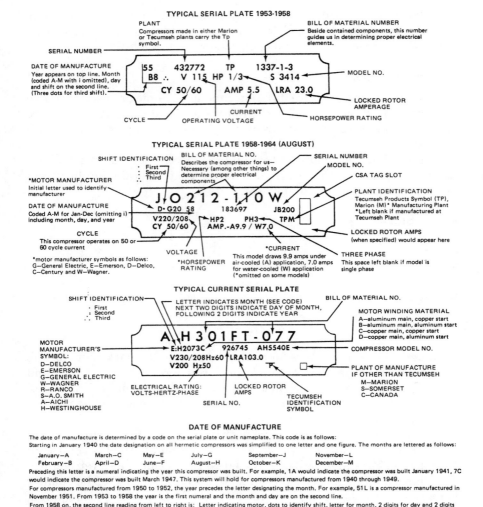

Figure 9-11a Serial plate information on Tecumseh's hermetic compressors.

Figure 9-13 shows the overload relay in its proper location with the cover removed to indicate the proper positioning. Figure 9-14 shows all parts assembled under the cover. The cover is secured to the fence with a bale strap.

This type of compressor may have a resistance-start induction-run motor (see Fig. 9-15). It may have a capacitor-start induction-run motor (see Fig. 9-16).

MODEL NUMBERS
Nomenclature Explained

EXAMPLE:

AE	A	4	4	40	Y	XA	XC
COMPRESSOR FAMILY	RELEASE VARIANT (GENERATION)	APPLICATION	NUMBER OF DIGITS IN RATED BTU CAPACITY	FIRST TWO DIGITS IN RATED BTU CAPACITY	REFRIGERANT	VOLTAGE	CONDENSING UNITS

AE
AG
AH
AJ
AK
AV
AW
AZ
HG
RG
RK
TP
TH

A = 1st
B = 2nd
C = 3rd
ETC.

In this example (4) total digits, with the first two (40), or 4,000 BTU capacity

PRIMARY REFRIGERANTS	
A = R12	
B = R410A	
C = R407C	
E = R22	
F = R417a	
J = R502	
Y = R134a	
Z = R404A/ R507	

See Unit Information on Page 2.

PRIMARY APPLICATION PARAMETERS		
EVAP. TEMPERATURE	RATING PT.	MOTOR STARTING TORQUE
1. Low	-10 F	Normal
2. Low	-10 F	High
3. High	+45 F	Normal
4. High	+45 F	High
5. Air Cond.*	+45 F	Normal
6. Medium	+20 F	Normal
7. Medium	+20 F	High
8. Air Cond.	+49 F	Normal
9. Commercial	+20 F	High
0. Commercial	+20 F	Normal
A. Medium/Low	+20 F	Normal

* Application "5" compressors, when applied to units, become application "4" in the units model number.

VOLTAGE CODES
XA = 115-60-1; 100-50-1
XB = 230-60-1; 200-50-1
XC = 220-240-50-1
XD = 208-230-60-1; 200-50-1
XF = 208-230-60-3; 200-240-50-3
XG = 460-60-3; 380-420-50-3
XH = 575-60-3; 480-520-50-3
XN = 208-230-60-1; 200-220-50-1
XP = 220-60-1; 200-50-1
XT = 200-230-60-3; 200-230-50-3
XU = 100-60-1; 100-50-1
XV = 265-60-1
AB = 115-60-1; 90-50-1
VA = 265-60-1; 240-220-50-1
Note: For explanation of voltages not listed, contact Tecumseh Compressor Company.

Figure 9-11b Model numbers. (*Courtesy of Tecumseh.*)

AEA4440YXA XC

1. E and G = Evaporative Condensate Units.
X = A holding character, reserved for future use.
Condensing Unit Features, see chart below.

The Letters I, O, and Q are eliminated.		Fan Cooled	Water Cooled	Air Water Cooled	Receiver Tank	BX Cable	Interconnect Compressor	See B/M	Accumulator
A	Standard Unit	●							
B	Std. Unit w/Receiver Tank	●			●	●			
C	Std. Unit w/ Tank & BX Cable	●			●	●			
D	Std. Unit w/ BX Cable	●				●			
E,F,K	Physical Design Variant (Conduit)	●						●	
G,H,J,L,P	Physical Design Variant (Standard)	●						●	●
M	Advanced Commercial Design	●			●	●			
N	Advanced Commercial Design	●				●			
S	Customer Special						●	●	
T	Interconnect Compressor								●
U	Water Cooled - Adv. Commercial Design		●		●	●			
V	Electrical Special (Conduit Design)	●				●		●	
W	Water-Cooled Unit		●		●	●	●		
X	Interconnect Unit	●			●	●			
Y	Air Water Cooled Unit			●		●			
Z	Electrical Special (Standard Unit)	●						●	
Evaporative Condensate Units									
EC	Large Evaporative Condensate Units Black Plastic Base	●							
ED	Large Evaporative Condensate Units Black Plastic Base	●							
EE	Large Evaporative Condensate Units Black Plastic Base	●							
GA	Small Evaporative Condensate Units Gray Plastic Base	●							
GB	Small Evaporative Condensate Units Gray Plastic Base	●		●					
GK	Small Evaporative Condensate Units Gray Plastic Base	●							
GL	Small Evaporative Condensate Units Gray Plastic Base	●							

Figure 9-11c Nomenclature explained. (*Courtesy of Tecumseh.*)

TECUMSEH
Compressor Group

Compressor Application Categories

Application	LT	MT	Commercial Temp	HT	Air Conditioning	
Rating Point Temp Range	-10°F -40°F to +10°F	+20°F -10°F to +30°F	+20°F -10°F to +45°F	+45°F +20°F to +55°F	Cooling +45°F +32°F to 55°F	Heating -10°F -15°F to 57°F
	Domestic Refrigerators	Draft-Beer Coolers	Draft-Beer Coolers	Draft-Beer Coolers	Air Conditioning	Heat Pumps
	Domestic Freezers	Ice Machines	Commercial Refrigerators	Ice Machines		
	Ice-Cream Cabinets	Commercial Refrigerators	Dehumidifiers	Commercial Refrigerators		
	Soft Ice Cream	Soft Ice-Cream Machines	Walk-In Coolers	Dehumidifiers		
	Mashines	Beverage Vendors w/Ice	Reach-In Refrigerators	Bulk Milk Coolers		
	Sluch Machines	Reach-In Refrigerators	Beverage Vendors w/Ice	Beverage Vendor - Ice		
	Beverage Vendor w/Ice	Walk-In Coolers	Dough Retarders	Reach-In Refrigerators		
	Reach-In Freezers	Dough Retarders	Ice-Storage Cabinets	Walk-In Coolers		
	Walk-In Freezers	Ice-Storage Cabinets	Beverage Dispensers	Dough Retarders		
	Bakery Freezers	Beverage Dispensers	Bottle Coolers	Florist Refrigerators		
	Ice Storage Cabinets	Bottle Coolers	Air Drier	Beverage Dispensers		
	Beverage Glass	Egg Coolers	Open Display Cases	Bottle Coolers		
	Chillers	Air Drier	Fur-Storage Refrigeration	Egg Coolers		
		Open Display Cases	Mortuary Refrigeration	Air Drier		
		Fur-Storage Refrigeration	Meat Cases	Open Display Cases		
		Meat Cases	Salad Cabinets	Fur-Storage Refrigeration		
		Salad Cabinets	Milk and Dairy Cases	Mortuary Refrigeration		
		Mile and Dairy Cases	Liquid Chillers	Drinking Water Coolers		
		Liquid Chillers		Salad Cabinets		
				Milk and Dairy Cases		
				Liquid Chillers		

Notes:

1. On high temperature systems, subtract 20°F from the desired temperature of the product to obtain approximate evaporating temperature of the compressor application, on medium temperature systems subtract 10 to 12°F, and on low temperature systems subtract 6 to 8°F.
2. For air conditioning, sutract 15°F from the desired room temperature, then apply point 1.
3. If in doubt, use the higher temperature pump.

Figure 9-11d Compressor application categories. (*Courtesy of Tecumseh.*)

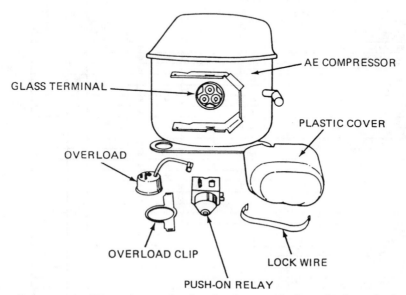

Figure 9-12 An AE compressor showing the glass terminal, overload, overload clip, push-on relay, plastic cover, and lock wire. (*Courtesy of Tecumseh.*)

Tecumseh AE Compressors

5,800 - 7,500 BTUH

- For room air-conditioning cooling
- External line-break overload
- Low sounds and vibration levels
- Adaptable to limited cabinet space
- Lightweight

Figure 9-12   (*Continued*)

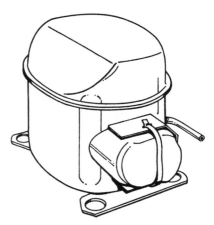

Figure 9-13 Overload and really in assembled positions. (*Courtesy of Tecumseh.*)

Figure 9-14 Completely assembled compressor. (*Courtesy of Tecumseh.*)

Model AK compressors are rated in Btu per hour. They have a 7000 to 12,000 Btu rating range. All model AK compressors are used for air-conditioning units. The refrigerant used is R-22. A 17-oz charge of oil is used on all models.

The AB compressors also are used for air-conditioning units. However, they are larger, starting with the 19,000 Btu/h rating and extending up to 24,000 Btu, or a 2-ton limit. Keep in mind that 12,000 Btu equals 1 ton. Refrigerant 22 is used with a 36-oz charge of oil.

The AU and AR compressors are made in 1/2, 3/4, 1, and 1-1/4 hp sizes. They are used primarily for air-conditioning units. Most of the models use R-22, except for a few models that use R-134a or its equivalent substitute. A 30-oz charge of oil is standard, except in one of the 1/2-hp models. Because of such exceptions, you must refer to the manufacturer's specifications chart to obtain the information relative to a specific model number within a series.

Resistance Start - Induction Run (RSIR)

● **Motor Type Can Be Determined By Looking At Model Nameplate**

  Information Located On The AS400 Bill Of Material (Shift F5)

● **Normal-Starting Ability (Approved For Self Equalizing Systems)**

● **Model Numbers Begin With: 0, 1, 3, 6**

● **Accessories Include Current Relay**

● **Connections To Relay: Three**

● **Relay Removes Start Winding From Circuit During Operation**

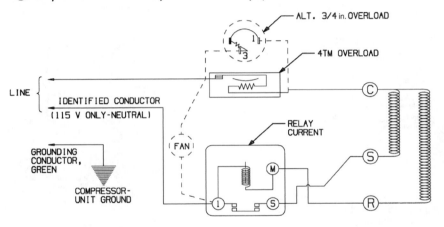

115 OR 230 V SCHEMATIC WIRING DIAGRAM - RSIR

Figure 9-15a Resistance-start induction-run motor for a compressor. (*Courtesy of Tecumseh.*)

The ISM (internal spring mount) series of compressors ranges in size from 1/8 to 1 hp. Their temperature range is from −30 to 10°F (−34 to −12°C) and from 20 to 55°F (−6 to 13°C). The oil charge is either 40 or 45 oz, depending upon the particular model.

The AH compressors are designed for residential and commercial air-conditioning and heat pump applications (see Fig. 9-17). They can be obtained with either three- or four-point mountings (see Fig. 9-18). The internal line-break motor protector is used. It is located precisely in the center of the heat sink position of the motor windings. Thus, it detects excessive motor winding temperature and safely protects the compressor from excessive heat and/or current flow (see Fig. 9-19).

PTC Start - Induction Run (PTCS/IR)

- Normal-Starting Ability (Approved For Self-Equalizing Systems)

- Model-Numbers Begin With 1 (AZ, TH, AE, TW, TP)

- Accessories Include PTC Relay

- Relay Removes Start Winding From Circuit During Operation

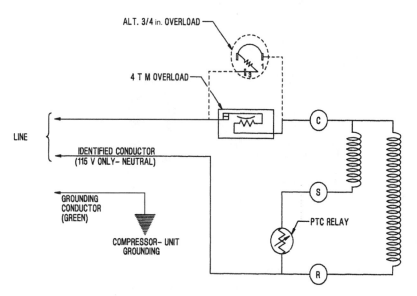

Figure 9-15b A 115-V or 230-V schematic diagram—PTCS-IR. (*Courtesy of Tecumseh.*)

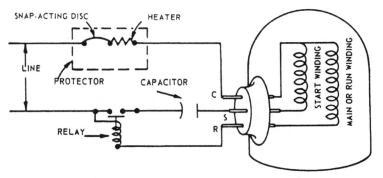

Figure 9-16a Capacitor-start induction-run motor for a compressor. (*Courtesy of Tecumseh.*)

PTC Start · Capacitor Run (PTCS/CR)

- Normal-Starting Ability (Aprroved For Self-Equalizing Systems)

- Model Numbers Begin With 1 (AZ, TH, TP)

- Accessories Include PTC Relay And Run Capacitor

- Relay Removes Start Winding From Circuit During Operation

- Run Capacitor Remains In The Circuit During Operation

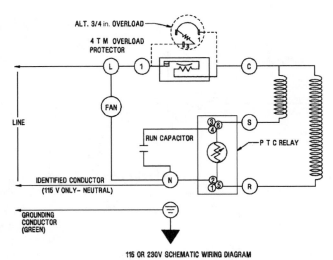

115 OR 230V SCHEMATIC WIRING DIAGRAM

Figure 9-16b PTC capacitor run (PTCS/CR). (*Courtesy of Tecumseh.*)

The snap on terminal cover assembly is shown in Fig. 9-20. It is designed for assembly without tools. The molded fiberglass terminal-cover may be secured or held in place by a bale strap.

This AH compressor series has a run capacitor in the circuit, as shown in Fig. 9-21. This compressor is designed for single-phase operation. Figure 9-22 shows the terminal box with the position of the terminals and the ways in which they are connected for *run, start,* and *common.*

The AH compressors are rated in Btu per hour. They range from 3500 to 40,000 Btu/h. These models use 45 oz of oil for the charge. They are used as air-conditioning units and for almost any other temperature range applications. They use R-134a or suitable substitute or R-22 for refrigerant.

The B and model compressors are available in 1/3, 1/2, 3/4, 1, 1-1/2, 1-3/4, and 2 hp units. All of them use a 45-oz charge of oil. They have a wide variety of temperature and air-conditioning applications. These

PTC Relay

- PTC = Positive Temperature Coefficent

- Resistance of PTC Material Increases with Increasing Temperature

- PTC Relay Sometimes used in Place of Current Relay

- Senses Current in the Motor-Start Winding

- Normally has Very Low Resistance

- When Power is Applied, Relay Resistance Increases, Removing Start Winding from Curcuit

- Relay Will Reset After a Cooldown Period of 3 to 10 S

- Usually Limited to use on Domestic Refrigerators and Freezers

Exploded View

Cover
Male Quick Connect Terminals
Pin Connector
PTC Pill Spring
PTC Pill
Base

Figure 9-16c PTC relay. (*Courtesy of Tecumseh.*)

models may have a B or a C preceding the model serial number. This is to indicate the series of compressors.

The AJ series of air-conditioning compressors ranges in size from 1100 to 19,500 Btu (see Fig. 9-23). An oil charge of 26 or 30 oz is standard, depending upon the model. They are mounted on three or four points (see Fig. 9-24). A snap-on terminal cover allows quick access to the connections under the cover (see Fig. 9-25). This particular model has an antislug feature that is standard on all AJM 12 and larger models (see Fig. 9-26). (An anti-slug feature keeps the liquid refrigerant moving.)

This type of compressor relies upon the permanent split-capacitor motor. In this instance, the need for both start and run capacitor is not presented. The start relay and the start capacitor are eliminated in this arrangement (see Fig. 9-27). With the PSC motor, the run capacitor acts as both a start and run capacitor. It is never disconnected. Both motor windings are always engaged while the compressor is starting and running.

PSC motors provide good running performance and adequate starting torque for low line voltage conditions. They reduce potential motor trouble since the electrical circuit is simplified (see Fig. 9-28).

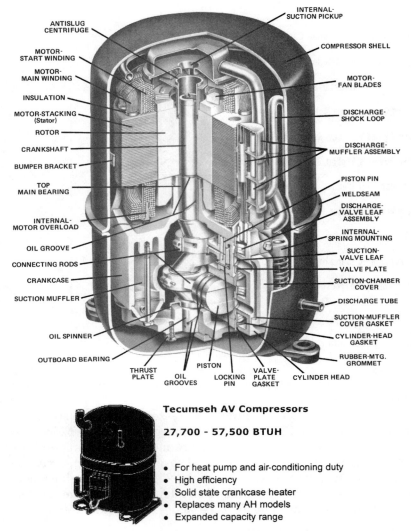

Figure 9-17 Construction details of the Tecumseh AH air-conditioning and heat pump compressors. (*Courtesy of Tecumseh.*)

The figure shows a run capacitor designed for continuous duty. It increases the motor efficiency while improving power and reducing current drain from the line. Do not operate the compressor without the designated run capacitor. Otherwise, an overload results in the loss of start and run performance. Adequate motor overload protection is not available either. A run capacitor in the circuit causes the motor to have some rather unique characteristics. Such motors have better pullout characteristics when a sudden load is applied.

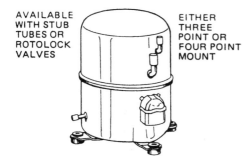

AVAILABLE
WITH STUB
TUBES OR
ROTOLOCK
VALVES

EITHER
THREE
POINT OR
FOUR POINT
MOUNT

Figure 9-18   External view of the
Tecumseh AH air-conditioning
and heat pump compressor with
its grommets and spacers.
(*Courtesy of Tecumseh.*)

MOUNTING GROMMETS AND SPACERS

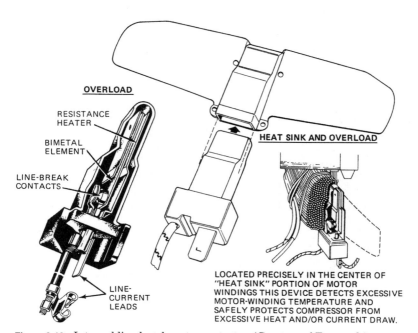

OVERLOAD

RESISTANCE
HEATER

BIMETAL
ELEMENT

LINE-BREAK
CONTACTS

HEAT SINK AND OVERLOAD

LINE-
CURRENT
LEADS

LOCATED PRECISELY IN THE CENTER OF
"HEAT SINK" PORTION OF MOTOR
WINDINGS THIS DEVICE DETECTS EXCESSIVE
MOTOR-WINDING TEMPERATURE AND
SAFELY PROTECTS COMPRESSOR FROM
EXCESSIVE HEAT AND/OR CURRENT DRAW.

Figure 9-19   Internal line-break motor protector. (*Courtesy of Tecumseh.*)

Figure 9-29 shows how this particular series of compressors is wired
for using the capacitor in the run and start circuit. Note the overload is
an external line breaker. This motor overload device is firmly attached
to the compressor housing. It quickly senses any unusual temperature
rise or excess current draw. The bimetal disk reacts to either excess

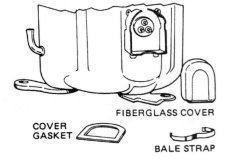

COVER
GASKET

FIBERGLASS COVER

BALE STRAP

Figure 9-20 Snap-on terminal cover assembly. (*Courtesy of Tecumseh.*)

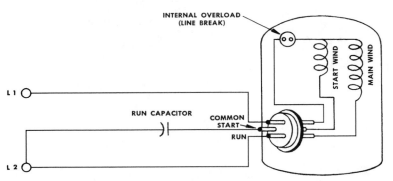

Figure 9-21 Single-phase diagram for the AH air conditioner and heat pump compressor. (*Courtesy of Tecumseh.*)

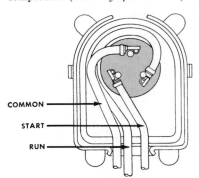

Figure 9-22 Terminal box showing the position of the terminals on the AH series of compressors. (*Courtesy of Tecumseh.*)

temperature or excess current draw. It flexes downward, thereby disconnecting the compressor from the power source (see Fig. 9-30).

The CL compressor series is designed for residential and commercial air-conditioning and heat pumps. These compressors are made in 2-1/2, 3, 3-1/2, 4, and 5 hp sizes. They can be operated on three-phase or single-phase power (see Fig. 9-31). Since this is one of the larger compressors, it has two cylinders and pistons. It needs a good protection system for

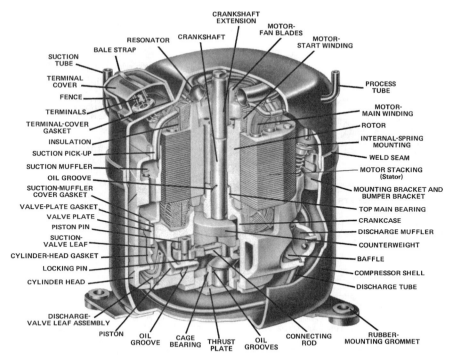

CRANKSHAFT
EXTENSION

MOTOR-
FAN BLADES

CRANKSHAFT

MOTOR-
START WINDING

RESONATOR

BALE STRAP

SUCTION
TUBE

TERMINAL
COVER

FENCE

TERMINALS

TERMINAL-COVER
GASKET

INSULATION

SUCTION PICK-UP

SUCTION MUFFLER

OIL GROOVE

SUCTION-MUFFLER
COVER GASKET

VALVE-PLATE GASKET

VALVE PLATE

PISTON PIN

SUCTION-
VALVE LEAF

CYLINDER-HEAD GASKET

LOCKING PIN

CYLINDER HEAD

DISCHARGE-
VALVE LEAF ASSEMBLY

PISTON

OIL
GROOVE

CAGE
BEARING

THRUST
PLATE

OIL
GROOVES

CONNECTING
ROD

RUBBER-
MOUNTING GROMMET

PROCESS
TUBE

MOTOR-
MAIN WINDING

ROTOR

INTERNAL-SPRING
MOUNTING

WELD SEAM

MOTOR STACKING
(Stator)

MOUNTING BRACKET AND
BUMPER BRACKET

TOP MAIN BEARING

CRANKCASE

DISCHARGE MUFFLER

COUNTERWEIGHT

BAFFLE

COMPRESSOR SHELL

DISCHARGE TUBE

Figure 9-23 Cutaway view of the AJ series of air-conditioning compressors. (*Courtesy of Tecumseh.*)

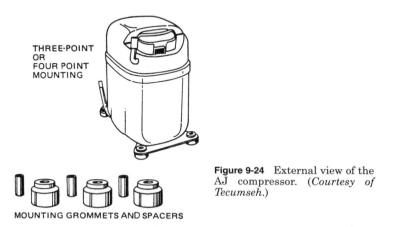

THREE-POINT
OR
FOUR POINT
MOUNTING

MOUNTING GROMMETS AND SPACERS

Figure 9-24 External view of the AJ compressor. (*Courtesy of Tecumseh.*)

the motor. This one has an internal thermostat to interrupt the control circuit to the motor contactor. The contactor then disconnects the compressor from the power source. Figure 9-32 shows the location of the internal thermostat.

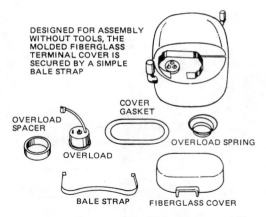

DESIGNED FOR ASSEMBLY
WITHOUT TOOLS, THE
MOLDED FIBERGLASS
TERMINAL COVER IS
SECURED BY A SIMPLE
BALE STRAP

COVER
GASKET

OVERLOAD
SPACER

OVERLOAD SPRING

OVERLOAD

BALE STRAP   FIBERGLASS COVER

Figure 9-25 Snap-on terminal cover assembly.
(Courtesy of Tecumseh.)

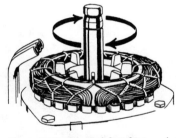

Figure 9-26 An antislug feature is standard on all AJ1M12 models and on larger models of the AJ series. *(Courtesy of Tecumseh.)*

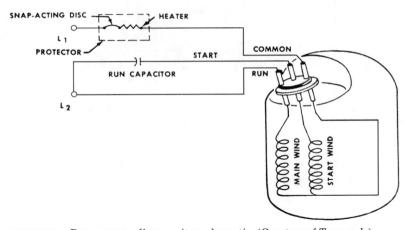

Figure 9-27 Permanent split-capacitor schematic. *(Courtesy of Tecumseh.)*

There is a supplementary overload in the compressor terminal box so it can be reached for service (see Fig. 9-33). A locked rotor or another condition producing excessive current draw causes the bimetal disk to flex upward. This opens the pilot circuit to the motor contactor.

The contactor then disconnects the compressor from the power source. Single-phase power requires one supplementary overload (see Fig. 9-34). Three-phase power requires two supplementary overloads (see Fig. 9-35). This CL line of compressors uses R-22 and R-12 (R-134a) or its suitable substitute refrigerants. They also use an oil charge of either 45 or 55 oz. In some cases, when the units are interconnected, they use 65 oz.

Run Capacitors

- Defined by Microfarad and Voltage Ratings

- When Substituting, Microfarad Rating Must be the Same, Voltage

  Rating can be Higher

- Aides in Starting and Improving the Efficiency of the Motor

- Continuous Duty - Always in Motor Circuit

Figure 9-28 Run capacitors. (*Courtesy of Tecumseh.*)

The H, J, and PJ compressors vary from 3/4 to 3 hp. They have a wide temperature range. All of these models use a 55-oz oil charge. R-22 and R-12 refrigerants, or R-12 substitutes like R-134a that meet environmental requirements, are used.

The F and PF compressors are 2, 3, 4, and 5 hp units. They use either 115 or 165 oz of oil. They, too, are used for a number of temperature ranges.

Newer Models Designations and Coding

The newer Tecumseh models are classified according to backpressure. For instance, the commercial backpressure (CBP) models start with No. 0 for no starting capacitor or No. 9 when the starting capacitor is required. These CBP models have an evaporator temperature range of −10 to 45°F (−23 to 7°C). Compressor capacity is measured at 20°F (−6.6°C) evaporator temperature.

High backpressure (HBP) model numbers start with No. 3 for no starting capacitor or No. 4 when starting capacitor is required. The evaporator temperature range for these models is 20 to 55°F (−6.6 to 12.7°C). Compressor capacity is measured at 45°F (7°C) evaporator temperature.

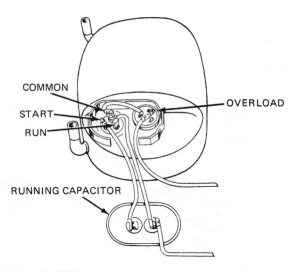

Permanent Split Capacitor (PSC)

● Normal-Starting Ability (Aprroved For Self-Equalizing Systems)

● Model Numbers Begin With: 5, 8

● Accessories Include Run Capacitor

● Run Capacitor Remains In The Circuit During Operation

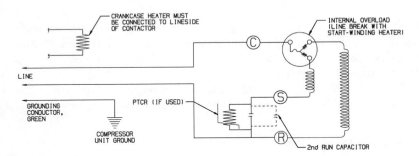

230 V-SCHEMATIC WIRING DIAGRAM
PSC - INTERNAL OVERLOAD (LINE BREAK) - START-WINDING HEATER - CRANKCASE HEATER

Figure 9-29   A PSC motor hookup. (*Courtesy of Tecumseh.*)

Air-conditioning models of a single cylinder reciprocating type are designated with AE, AK, and AJ. Model numbers starting with No. 5 are standard and those with No. 8 are high-efficiency models. The evaporator temperature range of these models is 32 to 55°F (0 to 12.7°C). Compressor capacity is measured at 45°F (7°C) for standard models or 49°F (9.4°C) evaporator temperature for high-efficiency models.

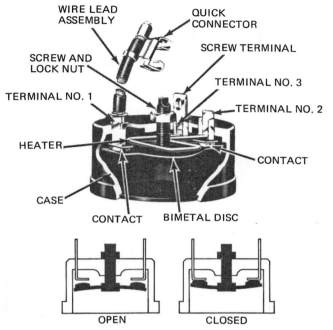

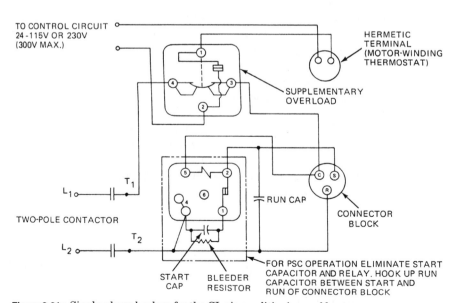

Figure 9-30 External line-break overload. (*Courtesy of Tecumseh.*)

Figure 9-31a Single-phase hookup for the CL air-conditioning and heat pump compressors. (*Courtesy of Tecumseh.*)

Three Phase

● Increased Starting Ability (Approved For NonSelf-Equalizing Systems)

● Available In Two And Three Cylinder Reciprocating Designs (AW, AV, AG)

● No Accessories Required For Starting

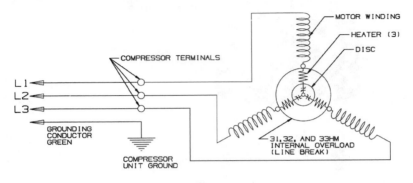

SCHEMATIC WIRING DIAGRAM - THREE PHASE
INTERNAL OVERLOAD (LINE BREAK).

Figure 9-31*b* Three-phase hookup. (*Courtesy of Tecumseh.*)

Air-conditioning and heat pump (AC/HP) compressors come in two and three cylinder reciprocating AC models (AW, AV, AG) plus rotary AC models.

The AC/HP model number starts with No. 5. The evaporator temperature range is −15 to 55°F (−26 to 12.7°C). Compressor capacity evaporator temperature is measured at 45°F (7.2°C) (AC) and −10°F (−23.3°C) for heat pump models.

Low backpressure models start with No. 1 for no starting capacitor or No. 2 with starting capacitor required models. The evaporator temperature range is −40 to 10°F (−40 to −12.2°C). Compressor capacity is measured at −10°F (−23.3°C) evaporator temperature.

Medium backpressure (MBP) model numbers start with No. 6 for no starting capacitor and No. 7 with starting capacitor required. The evaporator temperature range is −10 to 30°F (−23.3 to −1.1°C). Compressor capacity is measured at 20°F (−6.6°C) evaporator temperature.

Hermetic Compressor Motor Types

There are four general types of single-phase motors. Each has distinctly different characteristics. Compressor motors are designed for specific requirements regarding starting torque and running efficiency. These are two of the reasons why different types of motors are required to meet various demands.

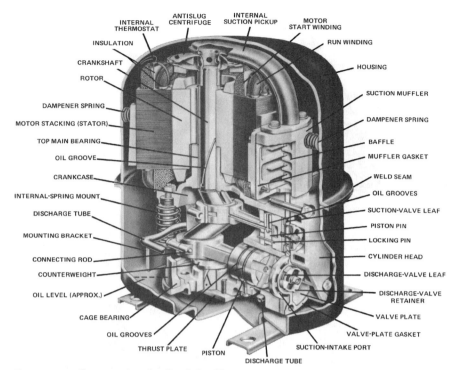

INTERNAL
THERMOSTAT

ANTISLUG
CENTRIFUGE

INTERNAL
SUCTION PICKUP

MOTOR
START WINDING

INSULATION

RUN WINDING

CRANKSHAFT

HOUSING

ROTOR

SUCTION MUFFLER

DAMPENER SPRING

DAMPENER SPRING

MOTOR STACKING (STATOR)

BAFFLE

TOP MAIN BEARING

MUFFLER GASKET

OIL GROOVE

WELD SEAM

CRANKCASE

OIL GROOVES

INTERNAL-SPRING MOUNT

SUCTION-VALVE LEAF

DISCHARGE TUBE

PISTON PIN

MOUNTING BRACKET

LOCKING PIN

CONNECTING ROD

CYLINDER HEAD

COUNTERWEIGHT

DISCHARGE-VALVE LEAF

OIL LEVEL (APPROX.)

DISCHARGE-VALVE
RETAINER

CAGE BEARING

VALVE PLATE

OIL GROOVES

VALVE-PLATE GASKET

THRUST PLATE

PISTON

SUCTION-INTAKE PORT

DISCHARGE TUBE

Figure 9-32a Construction details of the CL compressor. (*Courtesy of Tecumseh.*)

Tecumseh AG Compressors

46,000 - 71,280 BTUH

- For heat pump or central air-conditioning duty
- Internal line-break overload
- Internal pressure-relief valve
- Low sounds and vibration levels
- Replaces CL models and other brands

Figure 9-32b Later model AG Compressors. (*Courtesy of Tecumseh.*)

Resistance-start induction-run

The resistance-start induction-run (RSIR) motor is used on many small hermetic compressors through 1/3 hp. The motor has low starting torque. It must be applied to completely self-equalizing capillary tube systems such as household refrigerators, freezers, small water coolers, and dehumidifiers.

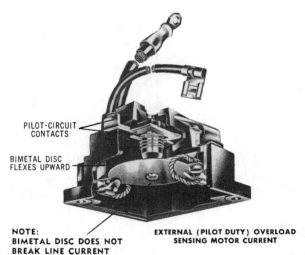

PILOT-CIRCUIT
CONTACTS

BIMETAL DISC
FLEXES UPWARD

NOTE:
BIMETAL DISC DOES NOT
BREAK LINE CURRENT

EXTERNAL (PILOT DUTY) OVERLOAD
SENSING MOTOR CURRENT

Figure 9-33 Cutaway view of the supplementary overload. *(Courtesy of Tecumseh.)*

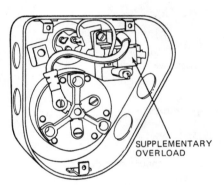

SUPPLEMENTARY
OVERLOAD

Figure 9-34 Location of the supplementary overload on the CL compressor series. *(Courtesy of Tecumseh.)*

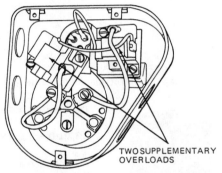

TWO SUPPLEMENTARY
OVERLOADS

Figure 9-35 Location of two supplementary overloads in the terminal box makes it applicable for three-phase power connections. *(Courtesy of Tecumseh.)*

This motor has a high-resistance-start winding that is not designed to remain in the circuit after the motor has come up to speed. A current relay is necessary to disconnect the start winding as the motor comes up to design speed (see Fig. 9-36).

Capacitor-start induction-run

The capacitor-start induction-run (CSIR) motor is similar to the RSIR. However, a start capacitor is included in series with the start winding to produce a higher starting torque. This motor is commonly used on commercial refrigeration systems with a rating through 3/4 hp (see Fig. 9-37).

PTC Start Device

⦿ **PTC = Positive Temperature Coefficent**

⦿ **Resistance of PTC Material Increases With Increasing Temperature**

⦿ **PTC Device By-Passes Run Capacitor Initially, Which Increases Starting Ability**

⦿ **PTC Resistance Increases, Putting The Run Capacitor Back Into Operation**

⦿ **Device Resets When The Compresor Is Shut Off**

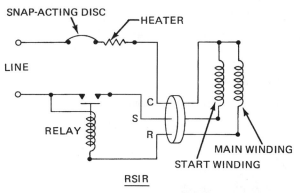

Figure 9-36 Resistance-start induction-run motor schematic. (*Courtesy of Tecumseh.*)

Capacitor-start and -run

The capacitor-start and -run (CSR) motor arrangement uses a start capacitor and a run capacitor in parallel with each other and in series with the motor start winding. This motor has high starting torque and

Capacitor Start - Induction Run (CSIR)

- Increased Starting Ability (Approved For NonSelf-Equalizing Systems)

- Model Numbers Can Begin With: 2, 4, 7, 9

- Accessories Include Current Relay And Start Capacitor

- Connections To Relay: Five

- Relay Removes Start Winding And Start Capacitor From Circuit

  During Operation

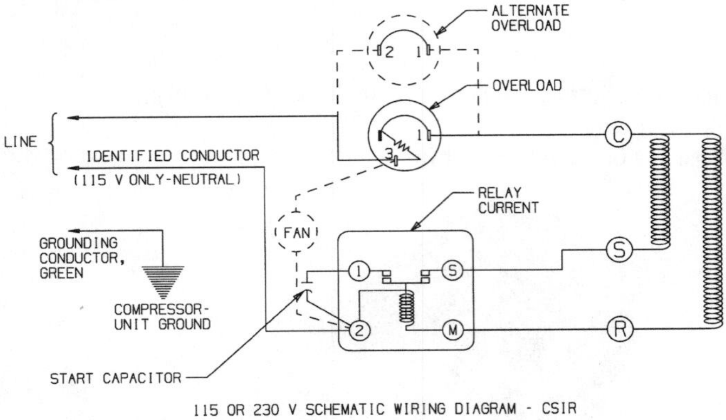

ALTERNATE OVERLOAD

OVERLOAD

LINE

IDENTIFIED CONDUCTOR
(115 V ONLY-NEUTRAL)

RELAY CURRENT

GROUNDING CONDUCTOR, GREEN

FAN

COMPRESSOR-UNIT GROUND

START CAPACITOR

115 OR 230 V SCHEMATIC WIRING DIAGRAM - CSIR

Figure 9-37a Capacitor-start induction-run motor schematic. (*Courtesy of Tecumseh.*)

Start Capacitors

- Defined by Microfarad and Voltage Ratings

- When Substituing, Microfarad Rating Must be the Same, Voltage Rating

  Can be Higher

- Used to Increase Starting Ability of the Motor

- Intermittent Duty - Must be Removed from Circuit Once CompressorStarts

- If Used with Run Capacitor, Start Capacitor Should Have Resistor Attached

Figure 9-37b Start capacitors. PTC start device. (*Courtesy of Tecumseh.*)

Capacitor Start - Run (CSR)

● Increased Starting Ability (Approved For NonSelf-Equalizing Systems)

● Model Numbers Can Begin With: 2, 4, 7, 9

● Accessories Include Potential Relay, Start, And Run Capacitors

● Connections To Relay: Five

● Relay Removes Start Winding And Start Capacitor From Circuit

   During Operation

● Run Capacitor Remains In The Circuit For PSC Operation (See Below)

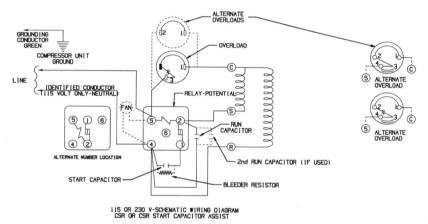

Figure 9-38 Capacitor start and run motor schematic. (*Courtesy of Tecumseh.*)

runs efficiently. It is used on many refrigeration and air-conditioning applications through 5 hp. A potential relay removes the start capacitor from the circuit after the motor is up to speed. Potential relays must be accurately matched to the compressor (see Fig. 9-38). Efficient operation depends on this.

Permanent split capacitor

The permanent split capacitor (PSC) has a run capacitor in series with the start winding. Both run capacitor and start winding remain in the circuit during start and after the motor is up to speed. Motor torque is sufficient for capillary and other self-equalizing systems. No start capacitor or relay is necessary. The PSC motor is basically an air-conditioning compressor motor. It is very common through 3 hp. It is also available in 4 and 5 hp sizes (see Fig. 9-39).

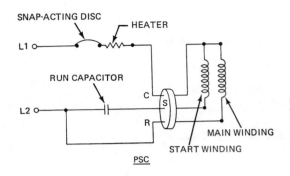

Figure 9-39   Permanent split-capacitor motor schematic. (*Courtesy of Tecumseh.*)

Compressor Motor Relays

A hermetic compressor motor relay is an automatic switching device designed to disconnect the motor start winding after the motor has attained a running speed.

There are two types of motor relays used in the refrigeration and air-conditioning compressors:

- Current-type relay
- Potential-type relay

Current-type relay

The current-type relay is generally used with small refrigeration compressors up to 3/4 hp. When power is applied to the compressor motor, the relay solenoid coil attracts the relay armature upward, causing bridging contact and stationary contact to engage (see Fig. 9-40). This energizes the motor start winding. When the compressor motor attains running speed, the motor main winding current is such that the relay solenoid coil de-energizes and allows the relay contacts to drop open. This disconnects the motor start winding.

The relay must be mounted in true vertical position so that the armature and bridging contact will drop free when the relay solenoid is de-energized.

Potential-type relay

This relay is generally used with large commercial and air-conditioning compressors. The motors may be capacitor-start capacitor-run types up to 5 hp. Relay contacts are normally closed. The relay coil is wired across the start winding. It senses voltage change. Start winding voltage increases with motor speed. As the voltage increases to the specific pick-up value, the armature pulls up, opening the relay contacts and

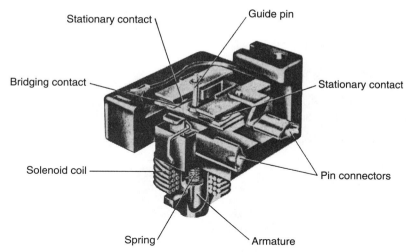

Stationary contact

Guide pin

Bridging contact

Stationary contact

Solenoid coil

Pin connectors

Spring

Armature

Figure 9-40 Current-type relay. This is generally used with small refrigeration compressors up to 3/4 hp. (*Courtesy of Tecumseh.*)

de-energizing the start winding. After switching, there is still sufficient voltage induced in the start winding to keep the relay coil energized and the relay starting contacts open. When power is shut off to the motor, the voltage drops to zero. The coil is de-energized and the start contacts reset (see Fig. 9-41).

Many of these relays are extremely position sensitive. When changing a compressor relay, care should be taken to install the replacement in the same position as the original. Never select a replacement relay solely by horsepower or other generalized rating. Select the correct relay from the parts guidebook furnished by the manufacturer.

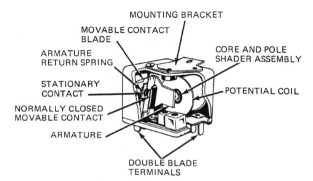

MOUNTING BRACKET

MOVABLE CONTACT BLADE

ARMATURE RETURN SPRING

CORE AND POLE SHADER ASSEMBLY

STATIONARY CONTACT

POTENTIAL COIL

NORMALLY CLOSED MOVABLE CONTACT

ARMATURE

DOUBLE BLADE TERMINALS

Figure 9-41a Potential-type relay. Usually found on large commercial and air-conditioning compressors up to 5 hp. (*Courtesy of Tecumseh.*)

Potential Relay

- ● Senses Voltage in the Motor Start Winding

- ● Relay Contacts are Normally Closed

- ● When Power is Applied, Relay Contacts Eventually Open, Removing Start Winding from Circuit

- ● Relay Contacts Close When Power to the Compressor is Shut Off

Figure 9-41*b*   Potential relay. (*Courtesy of Tecumseh.*)

Compressor Terminals

For the compressor motor to run properly it must have the power correctly connected to its terminals outside of the hermetic shell. There are several different types of terminals used on the various models of Tecumseh compressors.

Tecumseh terminals are always thought of in the order of common, start, and run. Read the terminals in the same way you would read the sentences on a book's page. Start at the top left-hand corner and read across the first line from left to right. Then, read the second line from left to right. In some cases three lines must be *read* to complete the identification process. Figure 9-42 shows the different arrangements of terminals. All Tecumseh compressors, except one model, follow one of these patterns. The exception is the old twin-cylinder, internal-mount compressor built at Marion. This was a 90° piston model designated with an "H" at the beginning of the model number (i.e., HA 100). The terminals were reversed on the H models and read run, start, and common (see Fig. 9-43). These compressors were replaced by the J-model series in 1955. All J models follow the usual pattern for common, start, and run.

Built-up terminals

Some built-up terminals have screw- and nut-type terminals for the attaching of wires (see Fig. 9-44). Others may have different arrangements. The pancake compressors built in 1953 and after have glass terminals that look something like those shown in Fig. 9-45. The terminal arrangement for S and C single-cylinder ISM (internal spring mount)

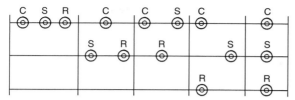

Figure 9-42 Identification of compressor terminals. (*Courtesy of Tecumseh.*)

Figure 9-43 Built-up terminals. These are on the obsolete twin-cylinder internal mount H models. (*Courtesy of Tecumseh.*)

Figure 9-44 Built-up terminals. These are on all external-mount B and C twin-cylinder models and on F, PF, and CF four-cylinder external-mount models. (*Courtesy of Tecumseh.*)

Figure 9-45 Built-up terminals on pancake compressors manufactured before 1952. (*Courtesy of Tecumseh.*)

Figure 9-46 Built-up terminals on S and C single-cylinder ISM models. (*Courtesy of Tecumseh.*)

Figure 9-47 Built-up terminals on twin-cylinder internal-mount J and PJ models. (*Courtesy of Tecumseh.*)

models resembles like that shown in Fig. 9-46. Models J and PJ with twin-cylinder internal mount have a different terminal arrangement. It looks like that shown in Fig. 9-47.

Glass quick-connect terminals

Figure 9-48 shows the quick-connect terminals used on S and C single-cylinder ISM models. The AK and CL models also use this type of

INTERNAL
THERMOSTAT
TERMINALS

Figure 9-48 Glass quick-connect terminals. (*Courtesy of Tecumseh.*)

MANY "CL" MODELS

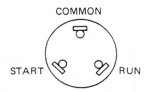

Figure 9-49 Glass quick-connect terminals. (*Courtesy of Tecumseh.*)

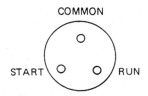

Figure 9-50 Glass terminals. For pancake-type compressors. (*Courtesy of
Tecumseh.*)

arrangement. Many of the CL models have the internal thermostat terminals located close by.

Quick-connect glass terminals are also used on AU and AR air-
conditioning models. The AE air-conditioning models also use glass
quick connects. Models AB, AJ, and AH also use glass quick connects,
but notice how their arrangement of common, start, and run varies
from that shown in Fig. 9-48. Figure 9-49 shows how the AU, AR, AE,
AB, AJ, and AH models terminate.

Glass terminals are also used on pancake-type compressors with P,
R, AP, and AR designations (see Fig. 9-50). The T and AT models, as well
as the AE refrigeration models, also use the glass terminals, but without the quick connect.

Keep in mind that you should never solder any wire or wire termination to a compressor terminal. Heat applied to a terminal is liable to
crack the glass terminal base or loosen the built-up terminals. This
will, in turn, cause a refrigerant leak at the compressor.

Motor Mounts

To dampen vibration, hold the compressor while in shipment, and cush-
ion horizontal thrust when the compressor starts or stops, some type of
mounting is necessary. Several different arrangements are used.
However, each of them uses a base plate. Also, some space is allowed
between the rubber grommet and the washer on the nut. The rubber
grommet absorbs most of the vibration (see Fig. 9-51).

In Fig. 9-52, you can see the use of a spring to prevent damage to the
rubber grommet. This is used for the heavier compressors. There are
usually three, but sometimes four of these rubber motor mounts on each
compressor model. One of the greatest uses of this type of mount is to
make sure that vibrations are not transferred to other parts of the
refrigeration system or passed on to the pipes. There, they would weaken
the soldered joints.

Crankcase Heaters

Most compressors have crankcase heaters. This is because most air-
conditioning and commercial systems are started up with a large part

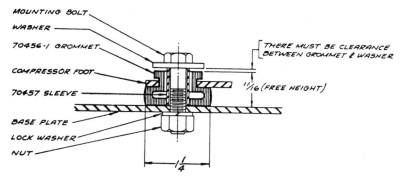

Figure 9-51 Mounting grommet assembly. (*Courtesy of Tecumseh.*)

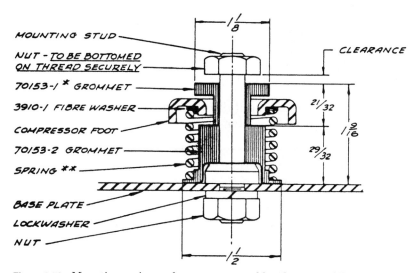

Figure 9-52 Mounting spring and grommet assembly. (*Courtesy of Tecumseh.*)

of the system refrigerant charge in the compressor. This is especially when the unit has been idle for some time or when the compressor is being started for the first time. On start-up, the refrigerant boils off, taking the oil charge with it. This means the compressor is forced to run for a long as 3 or 4 min until the oil charge circulates through the system and returns to the crankcase. Obviously, this shortens the service life of the compressor.

The solution is to charge the system so that little or no refrigerant collects in the crankcase and to operate the crankcase heater at least 12 h before start-up or after a prolonged down time.

Two types of crankcase heaters are in common use on compressors. The wrap around type is usually referred to as the "belly band." The other type is the run capacitance off-cycle heat method.

The wraparound heater should be strapped to the housing below the oil level and in close contact with the housing. A good heater will maintain the oil at least 10°F (5°C) above the temperature of any other system component. When the compressor is stopped it will maintain it at or above a minimum temperature of 80° F (27°C).

The run capacitance, off-cycle heat method, single-phase compressors are stopped by opening only one leg (L_1). Thus, the other leg to the power supply (L_2) of the run capacitor remains "hot." A trickle current through the start windings results, thereby warming the motor windings. Thus, the oil is warmed on the "off-cycle."

Make sure you pull the switch that disconnects the whole unit from the power source before working on such a system. Capacitance crankcase heat systems can be recognized by one or more of the following:

- Contactor or thermostat breaks only one leg to the compressor and condenser fan.
- Equipment carries a notice that power is on at the compressor when it is not running and that the main breaker should be opened before servicing.
- Run capacitor is sometimes split (it has three terminals) so that only part of the capacitance is used for off-cycle heating.

> CAUTION: Make sure you use an exact replacement when changing such dual-purpose run capacitors. The capacitor must be fused and carry a bleed resistor across the terminals.

The basic wiring diagram for a PSC compressor with a run capacitance off-cycle heat is shown in Fig. 9-53.

Electrical Systems for Compressor Motors

Most of the problems associated with hermetic compressors are electrical. Most of the malfunctions are in the current relay, potential relay, circuit breaker, or loose connections. In most cases, internal parts of the compressor housing can be checked with an ohmmeter.

Normal starting torque motors with a current-type relay

Normal starting torque motors (RSIR) with a current-type relay mounted on the compressor terminals require several tests that must be performed in the listed sequence. Figure 9-54 shows a two-terminal external overload device in series with the start and run windings.

The fan motor runs from point 1 on the current relay to point 3 on the overload device. L_2 has the relay coil inserted in series with the run

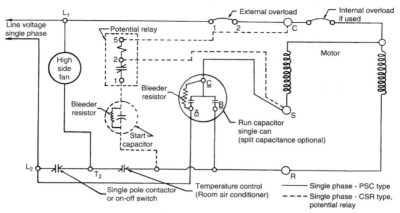

Figure 9-53 Single-phase CSR- or PSC-type compressor motor hookup with internal or external line-break overloads. (*Courtesy of Tecumseh.*)

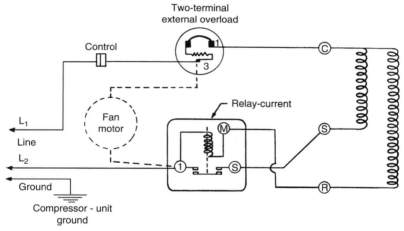

Figure 9-54a Normal starting motors (RSIR) with current relay mounted on the compressor terminals. (*Courtesy of Tecumseh.*)

winding. When the winding draws current, the solenoid is energized. This is done by the initial surge of current through the run winding. When the relay energizes with sufficient current, it closes the contacts (points 1 and S) and places the start winding in the circuit. The start winding stays in the circuit until the relay de-energizes. When the motor comes up to about 75 percent of its run speed, the relay de-energizes since the current through the run winding drops off. This change in current makes it a very sensitive circuit. The sensing relay must be in good operating condition. Otherwise, it will not energize or de-energize at the proper times.

Current Relay

- ● Senses Current in the Motor-Main Winding

- ● Relay Contacts are Normally Open

- ● When Power is Applied, Relay Contacts Close, Energizing the

  Start Winding

- ● Relay Contacts Open when the Motor Reaches Running Speed

Figure 9-54b Current relays. (*Courtesy of Tecumseh.*)

The start contacts on the current-type relay are normally open (see Fig. 9-54). Check the electrical system on this type of compressor system by using a voltmeter for obtaining line voltage reading. Then, use an ohmmeter to check continuity. That means the power must be off. Make sure the circuit breaker is off at the main power supply for this unit. If a fan is used [as shown in the dotted lines of Fig. 9-54(a)], make sure on lead is disconnected from the line. Next, check continuity across the following:

1. Check continuity across L_1 and point 3 of the overload. There is no continuity. Close control contacts by hand. If there is still no continuity, replace the control.

2. Check continuity across No. 3 and No. 1 on the overload. If there is no continuity, the protector may be tripped. Wait 10 min and check again. If there is still no continuity, the protector is defective. Replace it.

3. Pull the relay off the compressor terminals. Be sure to keep it in an upright position.

4. Check continuity across relay terminal 1 (or L) and S. If there is continuity, relay contacts are closed, when they should be open. Replace the relay.

5. Check continuity across No. 1 and M. If there is no continuity, replace the relay. The solenoid is open.

6. Check continuity across compressor terminals C and R. If there is no continuity, there is an open run winding. Replace the compressor.

7. Check continuity across compressor terminals C and S. If there is no continuity there is an open start winding. Replace the compressor.

8. Check continuity across compressor terminal C and the shell of the compressor. There is no continuity. This means the motor is grounded. Replace the compressor.

9. Check the winding resistance values against those published by the manufacturer of the particular model.

If all the tests prove satisfactory, and there is no capillary restriction, plus, the unit continues to fail to operate properly, change the relay. The new relay will eliminate any electrical problems, such as improper pickup or dropout that cannot be determined by the tests listed above. If a good relay fails to correct the difficulty, the compressor is inoperative due to internal defects. It must be replaced.

High starting torque motors with a current-type relay

High starting torque motors (CSIR) with a current-type relay mounted on the compressor terminals can be easily checked for proper operation. Remember from the previous type that the current-type relay normally has its contacts open.

Use a voltmeter first to check the power source. Use an ohmmeter to check continuity. Make sure the power is off and the fan motor circuit is open. The electrical system on this type of hermetic system can be checked as follows (see Fig. 9-55).

1. Check continuity across L_1 and overload's terminal 3. If there is no continuity, close the control contacts. If there is still no continuity, replace the control.

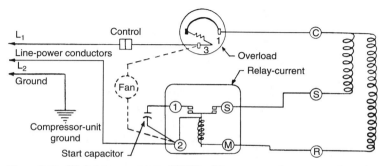

Figure 9-55 High torque motors (CSIR) with current relay mounted on the compressor terminals. (*Courtesy of Tecumseh.*)

2. Check continuity across No. 3 and No. 1 on the overload. If there is no continuity, the protector may be tripped. Wait for 10 min and check again. If there is still no continuity, the protector is defective. Replace it.

3. Pull relay off compressor terminals. *Keep it upright!*

4. Check continuity across relay terminals 1 and S. If there is continuity, the relay contacts are closed when they should be open. Replace the relay.

5. Check continuity across relay terminals 2 and M. If there is no continuity, replace the relay.

6. Check continuity across compressor terminals C and R. If there is no continuity, there is an open run winding. Replace the compressor.

7. Check continuity across compressor C and S. If there is no continuity, there is an open start winding. Replace the compressor.

8. Check continuity across C and shell of the compressor. If there is continuity, there is a grounded motor. Replace the compressor.

9. Check the winding resistance against the values given in the manufacturer's resistance tables.

10. Check continuity across relay terminals 1 and 2. Place the meter on the R × 1 scale. If there is continuity, there is a shorted capacitor. Replace the start capacitor. Place the meter on the R × 100,000 scale. If there is no needle deflection, there is an open capacitor. Replace the start capacitor.

If all the tests prove satisfactory, there is no capillary restriction, and the unit still fails to operate properly, change the relay. The new relay will eliminate electrical problems such as improper pickup and dropout. These cannot be determined with the tests listed above. If a good relay fails to correct the difficulty, the compressor is inoperative due to internal defects. It must be replaced.

**High starting torque motors with a
two-terminal external overload and
a remote-mounted potential relay**

High starting torque motors (CSIR) with a two-terminal external overload and a remote-mounted potential relay represent another type that must be checked. These are used in compressors for light air-conditioning units and also for commercial and residential refrigeration units.

In this type of motor the starting contacts on the potential-type relay are normally closed. The electrical system on this type of hermetic system can be seen in Fig. 9-56. Use a voltmeter to check the power

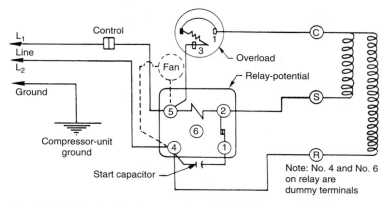

Figure 9-56 High starting torque motors (CSIR) with two-terminal external overload and potential relay mounted remote. (*Courtesy of Tecumseh.*)

source. Then use an ohmmeter, with the power turned off, to check continuity. Make sure leads 2S and 4R are disconnected. Open the fan circuit, if there is one.

Now, using the ohmmeter, check continuity across the following:

1. Check continuity across L and 3. If there is no continuity, close the control contacts. If there is still no continuity, replace the control.

2. Check continuity across No. 3 and No. 1 on the overload. If there is no continuity, the protector may be tripped. Wait 10 min and try again. If there is still no continuity, the protector is defective. Replace the protector.

3. Check continuity across No. 3 on the overload and No. 5 on the relay. If there is no continuity, check the leads between No. 3 on the protector and No. 5 on the relay.

4. Check continuity across No. 1 on the overload and C on the compressor. If there is no continuity, check the leads between No. 1 on the overload and C on the compressor.

5. Check continuity across C and S on the compressor. If there is no continuity, an open start winding is indicated. Replace the compressor.

6. Check continuity across C and R on the compressor. If there is no continuity, there is an open in the run winding of the compressor. Replace the compressor.

7. Check continuity across No. 5 on the relay and No. 2 on the relay. If there is no continuity, the solenoid's coil is open. The relay is defective. Replace the relay.

8. Check continuity across No. 2 and No. 1 on the relay. If there is no continuity, the contacts are open when they should be closed. Replace the relay.

9. Check continuity across No. 1 on the relay and No. 4 on the relay with the meter on the $R \times 1$ scale. If there is continuity, the capacitor is shorted. Replace the start capacitor. No needle deflection on the meter when it is on the $R \times 100,000$ scale means the capacitor is open. Replace the capacitor.

10. Check between C and the shell of the compressor. If there is continuity, there is a short. The motor is grounded. Replace the compressor.

11. Check the motor winding resistances against the manufacturer's specification sheet.

12. Check continuity between leads 2S and 4R and reconnect the unit. If all the tests prove satisfactory and the unit still does not operate properly, change the relay. The new relay will eliminate any electrical problems, such as improper pickup and dropout, which cannot be determined with the checks just performed. If a good relay fails to correct the difficulty, the compressor is inoperative due to internal defects. It must be replaced.

High starting torque motors with three-terminal overloads and remote-mounted relays

High starting torque motors (CSR) with three-terminal overloads and remote-mounted potential relays are another type of motor used in the hermetic compressor systems (see Fig. 9-57).

Starting contacts on the potential type of relay are normally closed. The electrical system power supply of the compressor can be checked. Use a voltmeter to check the power source.

Use the ohmmeter to check continuity across the following locations:

First, disconnect the leads so that no external wiring connects terminals 5-C, S-2 on the relay, and R-2 on the overload. Using the ohmmeter, check the following:

1. Check continuity across the control contacts—L_1 and C—on the compressor. The control contacts must be closed. If they are open, replace the compressor.

2. Check continuity across No. 5 and No. 2 on the relay. No continuity indicates an open potential coil. Replace the relay.

3. Check continuity across No. 2 and No. 1 on the relay. No continuity indicates an open contact situation. Replace the relay.

4. Check continuity across terminals C and S on the compressor. No continuity indicates an open start winding. Replace the compressor.

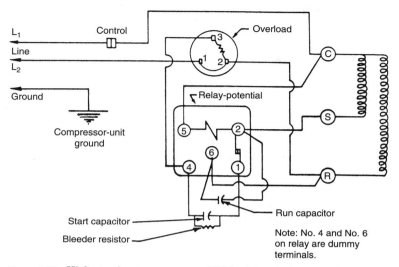

Figure 9-57 High starting torque motors (CSR) with a three-terminal external overload and potential relay mounted remote. (*Courtesy of Tecumseh.*)

5. Check continuity across terminals C and R on the compressor. No continuity indicates an open run winding. Replace the compressor.

6. Check continuity across No. 6 and No. 2 on the relay with the meter on the R × 1 scale. Continuity shows a shorted capacitor. Replace the run capacitor. Set the meter on the R × 100,000 scale. If there is no needle deflection, the capacitor is open. Replace the run capacitor.

7. Check continuity across No. 1 on the relay and No. 3 on the overload. Check as in step 6 above.

8. Check continuity across No. 1 and No. 3 on the overload. No continuity indicates the overload is open and should be replaced. However, it should have been given at least 10 min to replace itself properly.

9. Check continuity across C terminal on the compressor and the other ohmmeter lead to the shell of the compressor. Continuity indicates the motor has become grounded to the shell. Replace the compressor.

10. Check the resistance of the motor windings against the values given in the manufacturer's resistance tables.

11. Check continuity of the leads removed above and reconnect terminals 5 to C, S to 2 on the relay, and R and 2 on the overload.

If the tests prove satisfactory and the unit still does not operate properly, replace the relay. The new relay will eliminate any electrical problems, such as improper pickup and dropout, which cannot be determined with the checks just performed. If a good relay fails to correct the difficulty, the compressor is inoperative due to internal defects. It must be replaced.

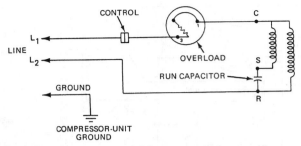

Figure 9-58 PSC motors with two-terminal external overload. (*Courtesy of Tecumseh.*)

PSC motor with a two-terminal external overload and run capacitor

Another type of motor used on compressors is the PSC (see Fig. 9-58). It has a two-terminal external overload and a run capacitor. It does not have a start capacitor or relay.

Use a voltmeter to check the source voltage. Then, using an ohmmeter, perform the following checks. Disconnect the run capacitor from terminals S and R before starting the tests.

1. L_1 and No. 3 on the overload show no continuity. Close the control contacts. If there is still no continuity, replace the control.

2. C and S terminals on the compressor show no continuity. This means the start winding is open. Replace the compressor.

3. C and R terminals on the compressor show no continuity. This means the run winding is open. A replacement compressor is needed to correct the problem.

4. C and 1 on the overload show no continuity. A defective lead from C to 1 is the probable cause.

5. No. 1 and No. 3 on the overload indicate no continuity. The protector may be tripped. Wait 10 min before checking again. If there is still no continuity, the protector is defective. Replace the overload protector.

6. C and the shell of the compressor show continuity. The motor is shorted to the shell or ground. Replace the compressor.

7. Check the motor windings against the manufacturer's tables.

8. Check across the run capacitor with the meter on the $R \times 1$ scale. If it shows continuity, the capacitor is shorted and must be replaced. Set the meter on $R \times 100,000$ scale. No needle deflection indicates that the capacitor is open and needs to be replaced.

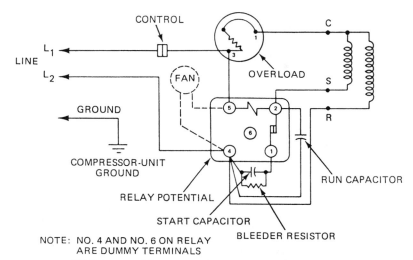

CONTROL

LINE L_1

L_2

FAN

OVERLOAD

C

S

R

GROUND

COMPRESSOR-UNIT
GROUND

RELAY POTENTIAL

START CAPACITOR

RUN CAPACITOR

BLEEDER RESISTOR

NOTE: NO. 4 AND NO. 6 ON RELAY
ARE DUMMY TERMINALS

Figure 9-59   PSC motors with two-terminal external overload with start components field installed. (*Courtesy of Tecumseh.*)

9. Reconnect the capacitor to the circuit at terminals S and R. The marked terminal should go to R.

If the above PSC tests reveal no difficulties, but the compressor does not operate properly, add the proper relay and start capacitor to provide additional starting torque. Figure 9-59 gives the proper wiring for a field installed relay and capacitor. If the unit still fails to operate, the compressor is inoperative due to internal defects. It must be replaced.

PSC motor with an internal overload (line-breaker)

Those PSC motors with an internal overload (line-breaker) are a little different from those just checked. Thus, the testing sequence varies somewhat. This compressor has an internal line break overload and a run capacitor. It does not have a start capacitor or relay (see Fig. 9-60).

1. Use a voltmeter to check the power source. Check the voltage at compressor terminals C and R. If there is no voltage, the control circuit is open.

2. Unplug the unit and check continuity across, the thermostat and/or contactor. Check the contactor holding coil.

3. If the line voltage is present between terminals C and R and the compressor does not operate, unplug the unit and disconnect the run capacitor from S and R.

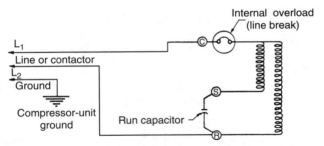

Figure 9-60 PSC motors with internal overload or line breaker. (*Courtesy of Tecumseh.*)

NOTE: The compressor shell must be at 130°F (54°C) or less for the following checks. This temperature can be read by a method termed a Tempstik. However, using the hand provides a less reliable guide. If it can remain in contact with the compressor shell without discomfort at a temperature of 130°F (54°C) or less, the motor is not overheated.

4. Using the ohmmeter, check the following:

a. Check continuity between R and S. If there is continuity, it can be assumed that both windings are intact. If there is no continuity, it can be assumed that one or both of the windings are open and the compressor should be replaced.

b. Check continuity between R and C. If there is no continuity, the internal overload is tripped. Wait for it to cool off and close. It sometimes takes more than an hour.

c. There is continuity between R and S, but no continuity between R and C (or S and C). If the motor is cool enough (below 130°F (54°C) to have closed the overload, then it can be assumed that the overload is defective. The compressor should be replaced.

d. Check continuity between the S terminal and the compressor shell and between the R terminal and the compressor shell. If there is continuity in either or both cases, the motor is grounded. The compressor should be replaced.

e. Check the motor winding resistance against the values given in the manufacturer's charts.

f. Check across the run capacitor with the meter on the R × 1 scale. If there is continuity, the capacitor is shorted and should be replaced.

g. Check across the run capacitor with the meter on the R × 100,000 scale. If there is no needle deflection on the meter, the capacitor is open and should be replaced.

5. Reconnect the run capacitor into the circuit at S and R.

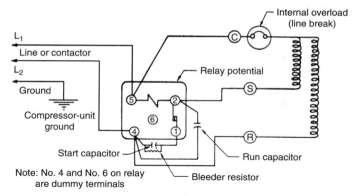

Figure 9-61 A CSR or PSC motor with start components and internal overload (line-breaker). (*Courtesy of Tecumseh.*)

CSR or PSC motor with the start components and an internal overload or line-breaker

The next combination is the CSR or PSC motor with the start components and an internal overload or line breaker. The run capacitor, start capacitor, and potential relay are the major components outside the compressor (see Fig. 9-61).

1. Using the voltmeter, check the power source. Check voltage at the compressor terminals C and R. If there is no voltage, the control circuit is open.

2. Unplug the unit and check continuity across the thermostat and/or contactor. Check the contactor holding coil.

3. If the line voltage is present between terminals C and R and the compressor does not operate, unplug the unit and disconnect the connections to the compressor terminals.

NOTE: The compressor shell must be at 130°F (54°C) or less for the following checks. A Tempstik can read this temperature. A less reliable guide is that at this temperature the hand can remain in contact with the compressor shell without discomfort.

---

4. Using the ohmmeter, check the following:

    a. Check continuity between R and S. If there is continuity, it can be assumed that both windings are intact. If there is no continuity, it can be assumed that one or both of the windings are open. The compressor should be replaced.

b. Check continuity between R and C. If there is no continuity, the internal overload is tripped. Wait for it to cool off and close. It sometimes takes more than an hour.

c. There is continuity between R and S, but no continuity between R and C (or S and C). If the motor is cool enough 130°F (54°C) to have closed the overload, then it can be assumed that the overload is defective. The compressor must be replaced.

d. Check continuity between the S terminal and the compressor shell and between R and the compressor shell. If there is continuity in either or both cases, the motor is grounded and the compressor should be replaced.

e. Check the motor winding resistance against the values given in the manufacturer's tables for the specific model being tested.

f. Check continuity across the run capacitor with the meter on the R × 1 scale. If there is continuity, the capacitor is shorted and should be replaced.

g. Check continuity across the run capacitor with the meter on the R × 100,000 scale. If there is no needle deflection, the capacitor is open and should be replaced.

5. Check continuity across 5 and 2 on the relay. No continuity indicates an open potential coil. Replace the relay. The electrolytic capacitor used for the start usually has its contents on the outside of the compressor housing. If the coil did not energize properly, it leaves the start capacitor in the circuit too long (only a few seconds). This means the capacitor will get too hot. When this happens, the capacitor will spew its contents outside the container.

6. Check continuity across No. 2 and No. 1 on the relay. No continuity shows an open contacts condition. Replace the relay.

7. Check continuity across No. 4 and No. 1 on the relay with the meter on the R × 1 scale. Continuity indicates a shorted capacitor. Replace with the meter on the R × 100,000 scale, if there is no needle deflection, the start capacitor is open. Replace the capacitor.

If all of the tests prove satisfactory and the unit still fails to operate properly, change the relay. If a new relay does not solve the problem, then it is fairly safe to assume that the compressor is defective and should be replaced.

Compressors with internal thermostat, run capacitor, and supplementary overload

Some compressors have an internal thermostat, a run capacitor, and a supplementary overload. However, they do not have a start capacitor or relay. The schematic for such a compressor is shown in Fig. 9-62.

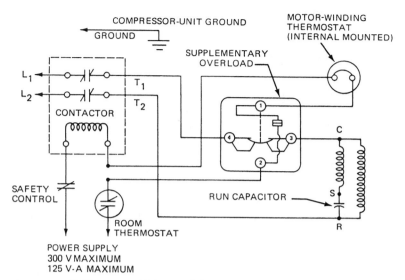

Figure 9-62 PSC motors with internal thermostat and supplementary external overload. (*Courtesy of Tecumseh.*)

The supplementary overload has normally closed contacts connected in series with the normally closed contacts of the internal thermostat in the motor. Operation of either of these devices will open the control circuit to drop out the contactor. Make sure the control thermostat and the system safety controls are closed. Using a voltmeter, check the power source at L_1, L_2, and the control circuit power supply. If the contactor is not energized, the contactor holding coil is defective or the control circuit is open in either the supplementary overload or the motor thermostat. Unplug the unit and disconnect the run capacitor from terminals S and R.

Using the ohmmeter, check the continuity across the following:

1. Check continuity across No. 3 and No. 4 on the overload. No continuity means the supplementary overload is defective. Replace it.

2. Check continuity across No. 1 and No. 2 on the overload. No continuity can mean the overload may be tripped. Wait 10 min. Test again. If there is still no continuity, the overload is defective. Replace overload.

3. Check continuity across the internal (motor winding) thermostat terminals at the compressor. See Fig. 9-48 for the location of the internal thermostat terminals. If there is no continuity, the internal thermostat may be tripped. Wait for it to cool down and close. It sometimes takes an hour. If the compressor is cool to the touch [below

130°F (54°C)] and there is still no continuity, internal thermostat circuitry is open and the compressor must be replaced.

4. Check continuity across terminals C and S on the compressor. No continuity indicates an open start winding. Replace the compressor.

5. Check continuity across terminals C and R. No continuity indicates an open run winding. Replace the compressor.

6. Check continuity across terminal C and the shell of the compressor. Continuity shows a grounded compressor. Replace the compressor.

7. Check the motor winding resistance with the chart given by the manufacturer.

8. Check continuity across the run capacitor with a meter on the $R \times 1$ scale. If there is continuity, the capacitor is shorted. Replace it. Place the meter on the $R \times 100,000$ scale. If there is no needle deflection, the capacitor is open. Replace the capacitor.

9. Reconnect the capacitor to the circuit at terminals S and R.

CSR or PSC motor with start components, internal thermostat, and supplementary external overload

Another arrangement for single-phase compressors is the CSR or PSC motor with start components, internal thermostat, and supplementary external overload (see Fig. 9-63).

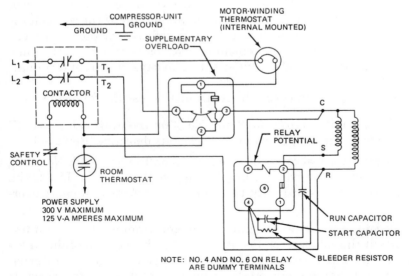

Figure 9-63 A CSR or PSC motor with start components and internal thermostat and supplementary external overload. (*Courtesy of Tecumseh.*)

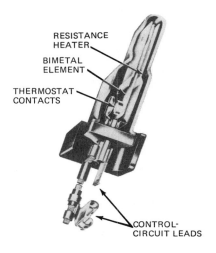

RESISTANCE HEATER

BIMETAL ELEMENT

THERMOSTAT CONTACTS

CONTROL-CIRCUIT LEADS

Figure 9-64 Internal thermostat embedded in the motor winding. (*Courtesy of Tecumseh.*)

This type of compressor is equipped with an internal thermostat, run capacitor, start capacitor, potential relay, and supplemental overload.

The supplemental overload has normally closed contacts connected in series with the normally closed contacts of the internal thermostat located in the motor. Operation of either of these devices will open the control circuit to drop out the contactor. Figure 9-64 shows the details of the internal thermostat.

Make sure the control thermostat and system safety controls are closed. Using the voltmeter, check the power source at L_1 and L_2. Also check the control circuit power supply with the voltmeter.

If the contactor is not energized, the contactor holding coil is defective or the control circuit is open in either the supplemental overload or the motor thermostat. Unplug the unit and disconnect the connections to the compressor terminals.

Use the ohmmeter and check for continuity across the following:

1. With the control circuit power supply off, check the continuity of the contactor holding coil.

2. Check continuity across No. 4 and No. 3 of the supplemental overload. No continuity means the overload is defective. Replace the overload.

3. Check continuity across No. 1 and No. 2 of the overload. No continuity means the overload may be tripped. Wait at least 10 minutes and test again. If there is still no continuity, the overload is defective. Replace the defective overload.

4. Check continuity across the internal thermostat terminals at the compressor. See Fig. 9-48 for the location of the terminals of the

internal thermostat. If there is no continuity, the internal thermostat may be tripped. Wait for it to cool off and close. It sometimes takes more than one hour to cool. If the compressor is cool to the touch (below 130°F (54°C), and there is still no continuity, the internal thermostat circuitry is open and the compressor must be replaced.

5. Check for continuity across terminals R and S on the compressor. If there is no continuity, one or both of the windings are open. Replace the compressor.

6. Check for continuity across terminal S and the compressor shell. Check for continuity across terminal R and the compressor shell. If there is continuity in either or both cases, the motor is grounded and the compressor should be replaced.

7. Check the motor winding resistances with the chart furnished by the compressor manufacturer.

8. Check for continuity across terminals 5 and 2 on the relay. No continuity indicates an open potential coil. Replace the relay.

9. Check for continuity across terminals 2 and 1 on the relay. No continuity indicates open contacts. Replace the relay.

10. Check for continuity across terminals 4 and 1 on the relay with the meter on the R × 1 scale. If continuity is read, it indicates a shorted capacitor. Replace the capacitor. Repeat with the meter on the R × 100,000 scale. No needle deflection indicates the start capacitor is open. Replace the start capacitor.

11. Discharge the run capacitor by placing a screwdriver across the terminals. Remove the leads from the run capacitor. With the meter set on the R × 1 scale, continuity across the capacitor terminals indicates a shorted capacitor. Replace the capacitor. Repeat the same test with the meter set on the R × 100,000 scale. No needle deflection indicates an open capacitor. Replace the run capacitor. If all the tests prove satisfactory and the unit still fails to operate properly, change the relay. The new relay will eliminate electrical problems such as improper pickup or dropout, which cannot be determined by the above tests. If a good relay fails to correct the difficulty, the compressor is inoperative due to internal defects. It must be replaced.

One other arrangement for a compressor using single-phase current is a CSR or PSC motor with start-components, internal thermostat, supplemental external overload, and start winding overload (see Fig. 9-65).

The diagnosis is identical to that described for the previous type of motor circuit. However, there is an additional start-winding overload in

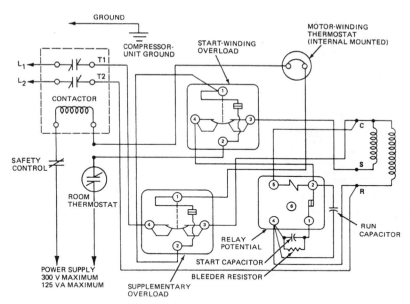

Figure 9-65   A CSR or PSC motor with start components and internal thermostat plus supplemental external overload and start winding overload. Note No. 4 and No. 6 on the relay are dummy terminals. (*Courtesy of Tecumseh.*)

the control circuit in series with the internal thermostat and supplemental overload. Check the start-winding overload in the same way the supplemental overload is checked.

Compressor Connections and Tubes

Tecumseh, like other compressor manufacturers, made compressors for many manufacturers of refrigerators, air-conditioning systems, and coolers. Because of this, the same compressor model may be found in the field in many suction and discharge variations. Each variation depends upon the specific application for which the compressor was designed.

Suction connections can usually be identified as the stub tube with the largest diameter in the housing. If two stubs have the same outside diameter, then the one with the heavier wall will be the suction connection. If both of the largest stub tubes have the same outside diameter and wall thickness, then either can be used as the suction connection. However, the one farthest from the terminals is preferred.

The stub tube not chosen for the suction connection may be used for processing the system. Compressor connections can usually be easily identified. However, occasionally some question arises concerning oil cooler tubes and process tubes.

Figure 9-66 Location of the oil cooling tubes inside the compressor shell. (*Courtesy of Tecumseh.*)

Oil cooler tubes are found only in low-temperature refrigeration models. These tubes connect to a coil or hairpin bend within the compressor oil sump (see Fig. 9-66). This coil or hairpin bend is not open inside the compressor. Its only function is to cool the compressor sump oil. The oil cooler tubes are generally connected to an individually separated tubing circuit in the air-cooled condenser.

Process tubes

Process tubes are installed in compressor housings at the factory as an aid in factory dehydration and charging. These can be used in place of the suction tube if they are of the same diameter and wall thickness as the suction tube.

Standard discharge tubing arrangements for Tecumseh hermetic compressors are shown in Fig. 9-67. Discharge tubes are generally in the same position within any model family. Suction and process tube positions may vary.

Other manufacturers of compressors

Besides Tecumseh, there are other manufacturers of compressors for the air-conditioning and refrigeration trade. One is Americold Compressor Corporation. Two of the models made by Americold are the M series and the A series (see Fig. 9-68). Both use the same overload relay and current relay connections (see Fig. 9-69). All of these models use R-12, or suitable substitute, as the refrigerant. They are made in sizes ranging from 1/10 through 1/4 hp. They weigh 21 to 25 lb. Figure 9-70 shows the location of the suction and discharge stubs as well as the process tube.

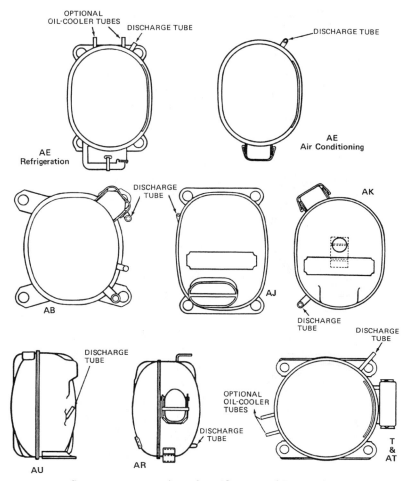

Figure 9-67 Compressor connection tubes. (*Courtesy of Tecumseh.*)

Rotary Compressors

The rotary compressor is made in two different configurations—the *stationary blade* rotary compressor and the *rotating blade* rotary compressor. The stationary blade rotary compressor is the type that has just been described. Both of these compressors have problems regarding lubrication. This problem has been partly solved.

Stationary blade rotary compressors

The only moving parts in a stationary blade rotary compressor are a steel ring, an eccentric or cam, and a sliding barrier (see Fig. 9-71).

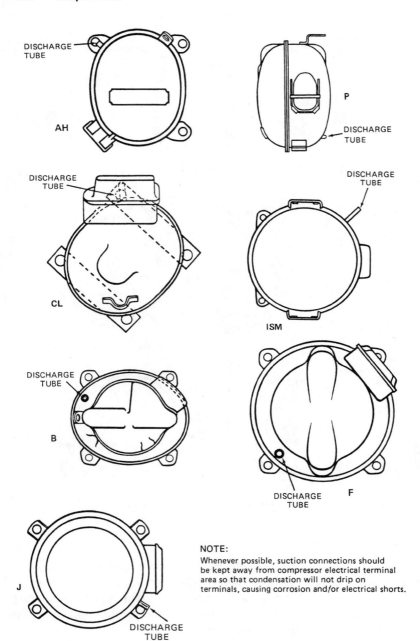

Figure 9-67 *(Continued)*

NOTE:
Whenever possible, suction connections should be kept away from compressor electrical terminal area so that condensation will not drip on terminals, causing corrosion and/or electrical shorts.

Tecumseh AJ Compressors

12,500 – 19,000 BTUH

- RAC and small central air conditioning only
- External line-break overload
- Low sounds and vibration levels
- Adaptable to limited cabinet space

Tecumseh AW Compressors

13,600 – 32,000 BTUH

- For heat pump and air-conditioning duty
- Internal line-break overload
- Internal pressure-relief valve

17,000 – 31,500 BTUH (60 H)   17,300 – 27,000 BTUH (50 H)

- For heat pump and air-conditioning duty
- Internal line-break overload
- Internal pressure-relief valve

Figure 9-67 *(Continued)*

Figure 9-72 shows how the rotation of the off-center cam compresses the gas refrigerant in the cylinder of the rotary compressor. The cam is rotated by an electric motor. As the cam spins it carries the ring with it. The ring rolls on its outer rim around the wall of the cylinder.

To be brought into the chamber, the gas must have a pathway. Note that in Fig. 9-73 the vapor comes in from the freezer and goes out to the condenser through holes that have been drilled in the compressor frame. Note that an offset rotating ring compresses the gas. Figure 9-74 shows how the refrigerant vapor in the compressor is brought from the freezer. Then, the exit port is opening. When the compressor starts to draw in the vapor from the freezer the barrier is held against the ring by a spring.

This barrier separates the intake and exhaust ports. As the ring rolls around the cylinder it compresses the gas and passes it on to the condenser (see Fig. 9-75). The finish of the compression portion of the stroke or operation is shown in Fig. 9-76. The ring rotates around the cylinder wall. The spring tension of the barrier's spring and the pressure of the cam being driven by the electric motor hold it in place. This type of compressor is not used as much as the reciprocating hermetic type of compressor.

Tecumseh AK Compressors

6,900 – 15,000 BTUH

- For room air-conditioning cooling and special heat-pump applications as determined by the original equipment manufacturer
- External line-break overload
- Low sound and vibration levels
- Adaptable to limited cabinet space

Tecumseh RK Compressors

8,100 – 17,700 BTUH (Rotary)

- Rotary design means low noise and vibration levels
- Air-conditioning and heat-pump application
- High efficiency
- Compact Design

Tecumseh SF Compressors

72,000 – 150,000 BTUH

- "Quadro-flex model"
- High-pressure housing
- No crankcase heater required
- Direct suction flow into cylinder adds to high efficiency
- Factory-installed suction screen

Figure 9-67   (*Continued*)

Rotating blade rotary compressors

The rotating blade rotary compressor has its roller centered on a shaft that is eccentric to the center of the cylinder. Two spring-loaded roller blades are mounted 180° apart. They sweep the sides of the cylinder. The roller is mounted so that it touches the cylinder at a point between the intake and the discharge ports. The roller rotates. In rotating, it pulls the vapor into the cylinder through the intake port. Here, the vapor is trapped in the space between the cylinder wall, the blade, and the point of contact between the roller and the cylinder. As the next blade passes the contact point, the vapor is compressed. The space or the vapor becomes smaller and smaller as the blade rotates.

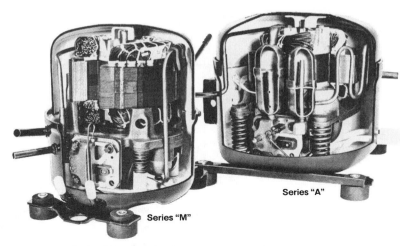

Figure 9-68 Series M and series A compressors made by Americold.

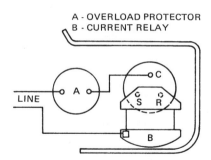

A - OVERLOAD PROTECTOR
B - CURRENT RELAY

Figure 9-69 Location of the terminals for the compressors and electrical connections on the Americold compressors. (*Courtesy of Americold.*)

Once the vapor has reached the pressure determined by the compressor manufacturer, it exits through the discharge port to the condenser.

On this type of rotating blade rotary compressor the seals on the blades present a particular problem. There also are lubrication problems. However, a number of rotary compressors are still in operation in home refrigerators.

Some manufacturers make rotary blade compressors for commercial applications. They are used primarily with ammonia. Thus, there is no copper or copper alloy tubing or parts. Most of the ammonia tubing and working metal is stainless steel.

Screw Compressors

Screw compressors operate more or less like pumps, and have continuous flow refrigerant compared to reciprocals. Reciprocal have pulsations. This

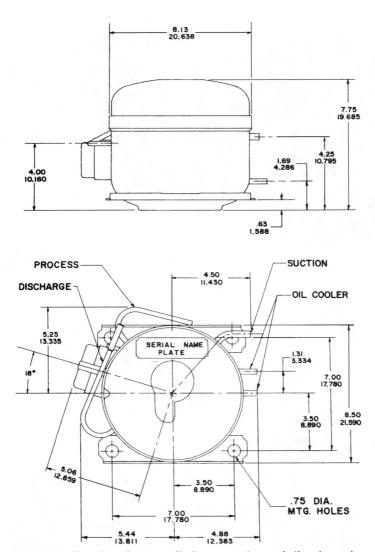

Figure 9-70 Location of process, discharge, suction, and oil cooler stubs on Americold compressors. (*Courtesy of Americold.*)

results in smooth compression with little vibration. Reciprocals, on the other hand, make pulsating sounds and vibrate. They can be very noisy.

Screw compressors have almost linear capacity-control mechanisms. That results in excellent part-load performance. Due to its smooth operation, low vibration screw compressors tend to have longer life than reciprocals.

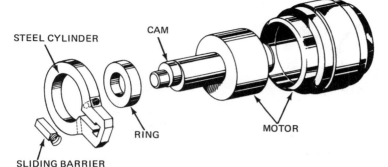

Figure 9-71   Parts of a rotary compressor. (*Courtesy of General Motors.*)

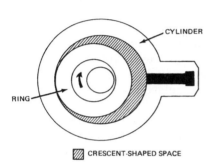

Figure 9-72   Operation of a rotary compressor. (*Courtesy of General Motors.*)

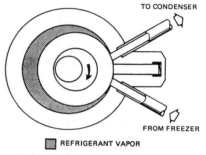

Figure 9-73   Beginning of the compression phase of a rotary compressor. (*Courtesy of General Motors.*)

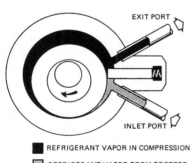

Figure 9-74   Beginning of the intake phase in a rotary compressor. (*Courtesy of General Motors.*)

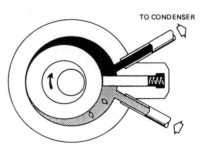

Figure 9-75   Compression and intake phases half completed in a rotary compressor. (*Courtesy of General Motors.*)

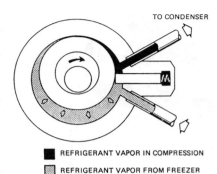

TO CONDENSER

Figure 9-76 Finish of the compression phase of the rotary compressor. (*Courtesy of General Motors.*)

■ REFRIGERANT VAPOR IN COMPRESSION

▨ REFRIGERANT VAPOR FROM FREEZER

Centrifugals are constant speed machines. These machines surge under certain operating conditions. This results in poor performance and high power consumption at part load. Screw compressors have proven themselves in tough refrigeration applications including on-board ships. Today, screw compressors practically dominate refrigerated ships, transporting fruits, vegetables, meats, and frozen foods across the ocean with good reliability. These compressors have replaced the traditional shipboard centrifugals.

Screw compressors were developed in Germany in the 1800s. They were patented in 1883 in Italy, but not in the United States until 1905. This type of compressor is a positive-displacement compressor. That means it uses a rotor driving another rotor (twin) or gate rotors (single) to provide the compression cycle. Both methods use injected fluids to cool the compressed gas, seal the rotor or rotors, and lubricate the bearings.

Single screw

A single screw compressor is shown in Fig. 9-77. The compression process starts with the rotors meshed at the inlet port of the compressor. The rotors turn. The lobes separate at the inlet port, increasing the volume between the lobes. This increased volume causes a reduction in pressure. Thus, drawing in the refrigerant gas. The intake cycle is completed when the lobe has turned far enough to be sealed off from the inlet port. As the lobe continues to turn, the volume trapped in the lobe between the meshing point of the rotors, the discharge housing, and the stator and rotors, is continuously decreased. When the rotor turns far enough, the lobe opens to the discharge port, allowing the gas to leave the compressor (see Fig. 9-78).

Twin screw

The twin screw is the most common type of screw compressor used today. It uses a double set of rotors (male and female) to compress the

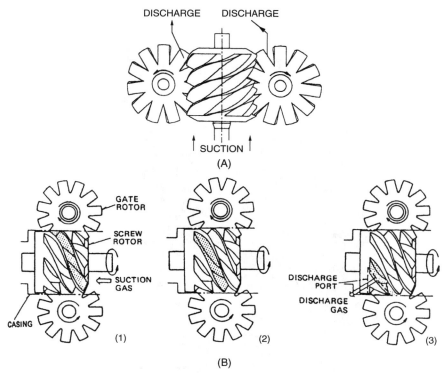

Figure 9-77   (*a*) Single screw compressor. (*b*) Monoscrew compression cycle: (1) suction, (2) compression, (3) discharge or exhaust. (*Courtesy of Single Screw Compressor, Inc.*)

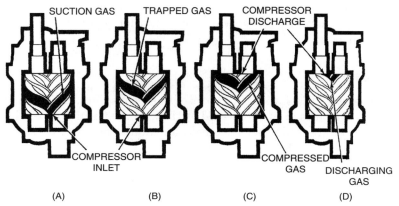

Figure 9-78   Twin-screw compression cycle: (*a*) intake of gas, (*b*) gas trapped in compressor housing and rotor cavities, (*c*) compression cycle, (*d*) compressed gas is discharged through the discharge port. (*Courtesy of Sullair Refrigeration.*)

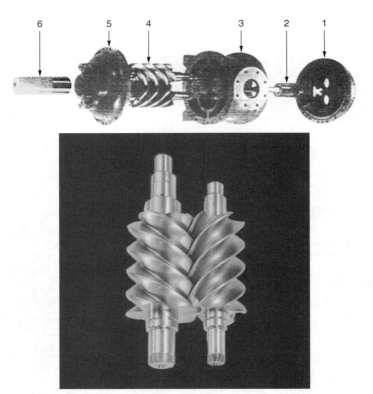

Figure 9-79 Twin-screw compressor parts: (1) discharge housing, (2) slide valve, (3) stator, (4) male and female rotors, (5) inlet housing, (6) hydraulic capacity control cylinder. (*Courtesy of Sullair Refrigeration.*)

refrigerant gas. The male rotor usually has four lobes. The female rotor consists of six lobes. Normally, this is referred to as a 4 + 6 arrangement. However, some compressors, especially air conditioners are using other variations, such as 5 + 7.

Making the Rotors

Not until the mid-1960s were the rotors cut using a symmetrical or circular profile. This was in turn replaced by the asymmetrical profile. This is a line-generated profile that improved the adiabatic efficiency of the screw compressor (see Fig. 9-79).

Scroll Compressors

The scroll compressor (Fig. 9-80) is being used by the industry in response to the need to increase the efficiency of air-conditioning equipment. This

Components:

OrbitingScroll Set

StationaryScroll Set

Crankshaft

Motor

Housing

Figure 9-80 A Copeland scroll compressor. (*Courtesy of Lennox.*)

is done in order to meet the U.S. Department of Energy Standards of 1992. The standards apply to all air conditioners. All equipment must have a Seasonal Energy Efficiency Ratio (SEER) of 10 or better. The higher the number, the more efficient the unit is. The scroll compressor seems to be the answer to more efficient compressor operation.

Scroll compression process

Figure 9-81 shows how the spiral-shaped members fit together. A better view is shown in Fig. 9-82. The two members fit together forming crescent-shaped gas pockets. One member remains stationary, while the second member is allowed to orbit relative to the stationary member.

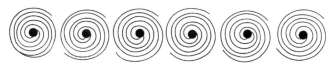

Figure 9-81   Scroll compression process. (*Courtesy of Lennox.*)

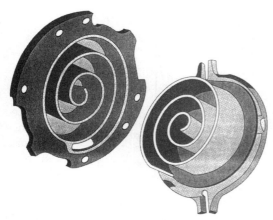

Figure 9-82 Two halves of the scroll compressor. (*Courtesy of Lennox.*)

This movement draws gas into the outer pocket created by the two members, sealing off the open passage. As the spiral motion continues, the gas is forced toward the center of the scroll form. As the pocket continuously becomes smaller in volume it creates increasingly higher gas pressures. At the center of the pocket, the high-pressure gas is discharged from the port of the fixed scroll member. During the cycle, several pockets of gas are compressed simultaneously. This provides a smooth, nearly continuous compression cycle.

This results in a 10 to 15 percent more efficient operation than with the piston compressors. A smooth, continuous compression process means very low flow losses. No valves are required. This eliminates all valve losses. Suction and discharge locations are separate. This substantially reduces heat transfer between suction and discharge gas. There is no reexpansion volume. This increases the compressor's heat pump capacity in low-ambient operation. Increased heat pump capacity in low ambient temperatures reduces the need for supplemental heat when temperatures drop.

During summer, this means less cycling at moderate temperatures. It also allows better dehumidification to keep the comfort level high. When temperatures rise, the scroll compressor provides increased capacity for more cooling.

During the winter, the scroll compressor heat pumps deliver more warm air to the conditioned space than conventional models.

Operation

The scroll compressor has no valves and low gas pulses. No valves and low gas pulses allow for smooth and quiet operation. It has fewer moving parts (only two gas compression parts as compared to 15 components in piston-type compressors) and no compressor start components are required. There is no accumulator or crankcase heater required. And a high-pressure cutout is not needed.

The radial compliance design features a superior liquid handling capacity. This allows small amounts of liquid and dirt to pass through without damaging the compressor. At the same time, this eliminates high stress on the motor and provides high reliability. Axial compliance allows the scroll tips to remain in continuous contact, ensuring minimal leakage. Performance actually gets better over the time because there are no seals to wear and causes gas leakage.

Scroll compressor models

Examples of air-conditioner units with the scroll compressor are the Lennox HP-20 and HS-22.

HP-20 model. The HP-20 is also designed for efficient use in heat pump installations (see Fig. 9-83). It has a large coil surface area to deliver more comfort per watt of electricity. A copper tube coil with aluminum fins provides effective heat transfer for efficient heating and cooling (see Fig. 9-84). The scroll compressor technology has been around for a long time. However, this was one of the first to make use of it in heat pumps.

HS-22 model. The HS-22 model also uses a Copeland-compliant scroll compressor (see Fig. 9-85). The insulated cabinet allows it to operate without disturbing the neighbors in closely arranged housing developments.

Scroll compressor

Strengths:

- Efficient gas compression
- Low sound and vibration levels
- Fewer parts, smaller size, lighter weight, more per pallet
- No internal suspension system

Weaknesses:

- Orbiting and stationary scrolls must match perfectly
- Used in air conditioning/commercial compressors under development
- Need to reduce vibration to unit using generous amounts of tubing

Figure 9-83 HP-20 Model with a heavy-duty scroll compressor: (1) cabinet, (2) coil area, (3) copper tubing, (4) fan, (5) scroll compressor. (*Courtesy of Lennox.*)

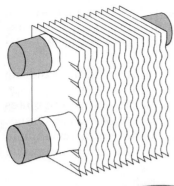

Figure 9-84 Copper tubing in the condenser with aluminum fins. (*Courtesy of Lennox.*)

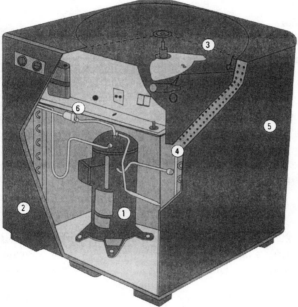

Figure 9-85 HS-22 Model: (1) scroll compressor, (2) cabinet, (3) fan, (4) copper tubing in condenser, (5) coil area in compressor, (6) filter-drier. (*Courtesy of Lennox.*)

The cabinet, with vertical air discharge, creates a unit that has sound ratings as low as 7.2 bells.

The condensing unit has a SEER rating as high as 13.5. The scroll compressor is highly efficient with a large double row condenser. This highly efficient coil increases the efficient use of energy even more.

10

Evaporators

Evaporators remove heat from the space being cooled. As the air is cooled, it condenses the water vapor. This must be drained. If the water condensing on the evaporator coil freezes when the temperature is below 32°F (0°C)—the refrigerator or freezer must work harder. Frozen water or ice acts as an insulator. It reduces the efficiency of the evaporator. When evaporators are operated below 32°F, they must be periodically defrosted. Defrosting eliminates frost buildup on the coils or the evaporator plates.

There are several types of evaporators. Three of the most common are

- *Coiled evaporators* are used in warehouses for refrigerating large areas.

- *Fin evaporators* are used in air-conditioning systems that are part of the furnace in a house (see Fig. 10-1). Finned evaporator have fans that blow air over thin metal surfaces.

- *Plate evaporators* use flat surfaces for their cooling surface (see Fig. 10-2). They are commonly used in freezers. If the object to be cooled or frozen is placed directly in contact with the evaporator plate, the cold is transferred more efficiently. Figure 10-3 shows a typical home refrigerator-cooling system. Note the evaporator.

Coiled Evaporator

Evaporator coils on air-conditioning units fall into two categories:

Finned-tube coil is placed in the air stream of the unit. Refrigerant vaporizes in it. The refrigerant in the tubes and the air flowing around the fins attached to the tubes draw heat from the air. This is commonly referred to as a direct expansion cooling system (see Fig. 10-4).

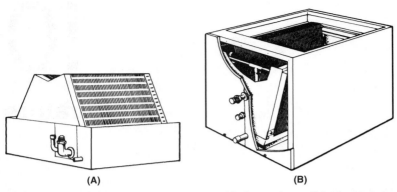

(A)   **(B)**

Figure 10-1   Evaporators. (*a*) Evaporator used in home air-conditioning systems where the unit is placed in the bonnet of the hot-air furnace. (*b*) Slanted evaporator used in home air-conditioning.

Figure 10-2   Plate evaporator.

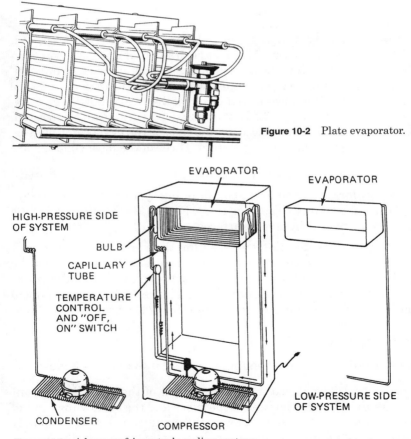

Figure 10-3   A home refrigerator's cooling system.

Figure 10-4 Finned coil evaporator. (*Courtesy of Johnson Controls.*)

Shell-and-tube chiller units are used to chill water for air-cooling purposes. Usually, the refrigerant is in tubes mounted inside a tank or
shell containing the water or liquid to be cooled. The refrigerant in the
tubes draws the heat through the tube wall and from the liquid as it
flows around the tubes in the shell. This system can be reversed.
Thus, the water would be in the tubes and the refrigerant would be
in the tank. As the gas passes through the tank over the tubes, it
would draw the heat from the water in the tubes (see Fig. 10-5).

Figure l0-5 shows how K-12 is used in a standard vapor-compression
refrigeration cycle. System water for air-conditioning and other uses is
cooled as it flows through the evaporator tubes. Heat is transferred
from the water to the low-temperature, low-pressure refrigerant. The
heat removed from the water causes the refrigerant to evaporate. The
refrigerant vapor is drawn into the first stage of the compressor at a rate
controlled by the size of the guide-vane opening. The first stage of the
compressor raises the temperature and pressure of the vapor. This
vapor, plus vapor from the flash economizer, flows into the second stage
of the compressor. There, the saturation temperature of the refrigerant
is raised above that of the condenser water.

This vapor mixture is discharged directly into the condenser. There,
relatively cool condenser water removes heat from the vapor, causing it
to condense again to liquid. The heated water leaves the system, returning to a cooling tower or other heat-rejection device.

A thermal economizer in the bottom section of the condenser brings
warm condensed refrigerant into contact with the inlet water tubes.

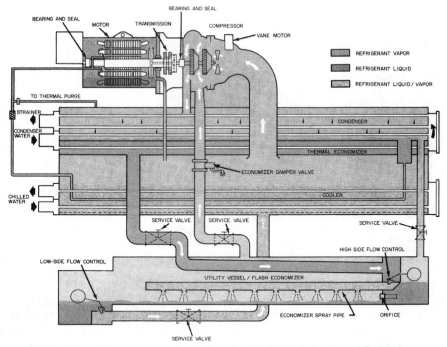

Figure 10-5   Complete operation of a shell-and-tube chiller. (*Courtesy of Carrier.*)

These are the coldest water tubes. They may hold water with a temperature as low as 55°F (13°C). This subcools the refrigerant so that when it moves on in the cycle, it has greater cooling potential. This improves cycle efficiency and reduces power per ton requirements. The liquefied refrigerant leaves the condenser through a plate-type control. It flows into the flash economizer or utility vessel. Here, the normal flashing of part of the refrigerant into vapor cools the remaining refrigerant. This flash vapor is diverted directly to the second stage of the compressor. Thus, it does not need to be pumped through the full compression cycle. The net effect of the flash economizer is energy savings and lower operating costs. A second plate-type control meters the flow of liquid refrigerant from the utility vessel back to the cooler, where the cycle begins again (see Fig. 10-6).

Application of Controls for Hot-Gas Defrost of Ammonia Evaporators

To defrost ammonia evaporators, it is sometimes necessary to check the plumbing arrangement and the valves used to accomplish the task. To enable hot-gas defrost systems to operate successfully, several factors

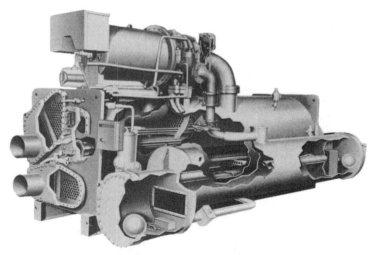

Figure 10-6   Cutaway view of the chiller portion of the shell-and-tube chiller shown in Fig. 10-5.

must be considered. There must be an adequate supply of hot gas. The gas should be at a minimum of 100 psig. The defrost cycle should be accurately timed. Condensate removal or storage must be provided. An automatic suction accumulator or heat reservoir should be used to protect compressors from liquid refrigerant slugs if surge drums or other evaporators are not adequate to handle the excess gas and condensates (see Fig. 10-7).

Controls must be used to direct and regulate the pressure and flow of ammonia and hot gas during refrigeration and defrost cycles.

Direct-expansion systems

Figure 10-7 shows a high temperature system [above 32°F (0°C)] with no drip-pan defrost. During the normal cooling cycle, controlled by a thermostat, the room temperature may rise above the high setting of the thermostat. This indicates a need for refrigeration. The liquid solenoid (valve A), pilot solenoid (valve B), and the dual-pressure regulator (valve D) open, allowing refrigerant to flow. When solenoid (valve D) is energized. The low-pressure adjusting bonnet controls the regulator. The regulator maintains the predetermined suction pressure in the evaporator.

When the room temperature reaches the low setting on the thermostat, there is no longer need for refrigeration. At this time, solenoid valve A and solenoid valve D close and remain closed until further refrigeration is required.

The hot-gas solenoid (valve C) remains closed during the normal refrigeration cycle. When the three-position selector switch is turned to "Defrost," liquid solenoid valve A and valve D, with a built-in pilot solenoid,

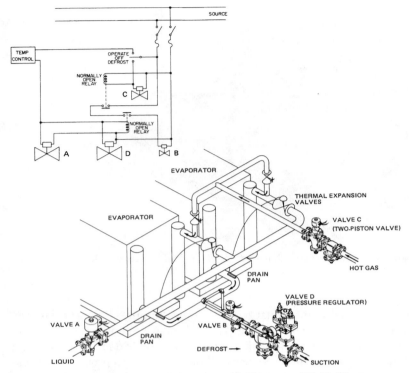

Figure 10-7 High-temperature defrost system. (*Courtesy of Hubbell.*)

close. This allows valve D to operate as a defrost pressure regulator on the high setting. The hot gas solenoid (valve C) opens to allow hot gas to enter the evaporator. When the defrost is complete, the system is switched back to the normal cooling cycle.

The system may be made completely automatic by replacing the manual switch with an electric time clock. Table 10-1 shows the valve sizes needed for this system.

Valves used in direct-expansion systems. The pilot solenoid (valve B) is a 1/8 in. ported solenoid valve that is direct operated and suitable as a liquid, suction, hot gas, or pilot valve at pressures to 300 lb.

Solenoid valve A is a one-piston, pilot-operated valve suitable for suction, liquid, or gas lines at pressures of 300 lb. It is available with a 9/16 in. or 3/4-in. port.

Solenoid valve C is a rugged, pilot-operated, two-piston valve with spring return for positive closing under the most adverse conditions. It is used for compressor unloading, and for liquid and hot-gas applications.

TABLE 10-1   Valve Sizing for High-Temperature System

Tons refrigerant	Liquid solenoid (in.)	Hot-gas solenoid (in.)	Pilot solenoid (in.)	Dual-pressure regulator (in.)
3	$^1/_2$	$^1/_2$	$^1/_4$	$^3/_4$
5	$^1/_2$	$^1/_2$	$^1/_4$	$^3/_4$
7	$^1/_2$	$^1/_2$	$^1/_4$	1
10	$^1/_2$	$^1/_2$	$^1/_4$	$1^1/_4$
12	$^1/_2$	$^1/_2$	$^1/_4$	$1^1/_4$
15	$^1/_2$	$^1/_2$	$^1/_4$	$1^1/_2$
20	$^1/_2$	$^1/_2$	$^1/_4$	$1^1/_2$
25	$^1/_2$	$^1/_2$	$^1/_4$	2
30	$^1/_2$	$^1/_2$	$^1/_4$	2
35	$^1/_2$	$^1/_2$	$^1/_4$	2
40	$^3/_4$	$^3/_4$	$^1/_4$	2
45	$^3/_4$	$^3/_4$	$^1/_4$	$2^1/_2$
50	$^3/_4$	$^3/_4$	$^1/_4$	$2^1/_2$

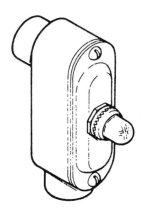

Figure 10-8   Pilot light assembly. (*Courtesy of Hubbell.*)

The dual-pressure regulator (valve D) is designed to operate at two predetermined pressures without resetting or adjustment. By merely opening and closing a pilot solenoid, it is capable of maintaining either the low- or high-pressure setting.

Figure 10-8 shows a pilot light assembly. It is placed on valves when it is essential to know their condition for troubleshooting procedures.

A low-temperature defrost system with water being used to defrost the drain pan is shown in Fig. 10-9.

Cooling cycle

During the normal cooling cycle controlled by a thermostat, as room temperature rises above the high setting on the thermostat there is a need for refrigeration. Liquid solenoid (valve A) and the built-in pilot

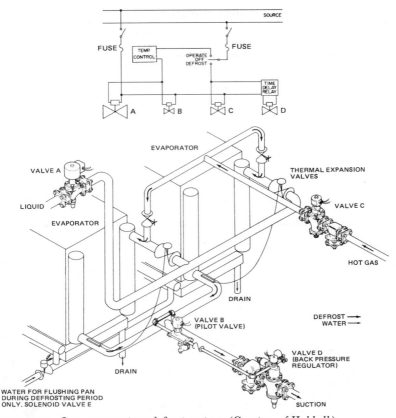

Figure 10-9   Low-temperature defrost system. (*Courtesy of Hubbell.*)

(valve D) open, allowing refrigerant to flow. The opening of the built-in pilot allows the presure to bypass the sensing chamber of valve D. This forces it to remain wide open with resultant minimum pressure drop through the valve.

When the room temperature drops to the low setting on the thermostat, there is no longer need for referation. Solenoid valve A and pilot vlave D close. They remain closed until refrigeration is again required. Hot-gas valve C and defrost water soldenoid valve E remain closed during the cooling cycle.

Defrost cycle

When the three-position selector switch is turned to defrost, solenoid valve A and pilot-solenoid valve D close as hot-gas valve C and evaporator-pilot valve B open. This allows hot gas to enter the evaporator. Valve D now acts as a back pressure regulator, maintaining a predetermined pressure above the freezing point. After a regulated delay, preferably

TABLE 10-2   Valve Sizing for Low-Temperature System

Tons refrigerant	Liquid solenoid (in.)	Hot-gas solenoid (in.)	Back-pressure regulator (in.)	Pilot solenoid (in.)	Defrost water (in.)
3	$^1/_2$	$^1/_2$	1	$^1/_4$	$^1/_2$
5	$^1/_2$	$^1/_2$	$1^1/_4$	$^1/_4$	$^3/_4$
7	$^1/_2$	$^1/_2$	$1^1/_4$	$^1/_4$	$^3/_4$
10	$^1/_2$	$^1/_2$	$1^1/_2$	$^1/_4$	1
12	$^1/_2$	$^1/_2$	$1^1/_2$	$^1/_4$	1
15	$^1/_2$	$^1/_2$	2	$^1/_4$	$1^1/_4$
20	$^1/_2$	$^1/_2$	2	$^1/_4$	$1^1/_4$
25	$^1/_2$	$^1/_2$	$2^1/_2$	$^1/_4$	$1^1/_2$
30	$^1/_2$	$^1/_2$	$2^1/_2$	$^1/_4$	$1^1/_2$
35	$^1/_2$	$^1/_2$	3	$^1/_4$	2
40	$^3/_4$	$^3/_4$	3	$^1/_4$	2
45	$^3/_4$	$^3/_4$	3	$^1/_4$	2
50	$^3/_4$	$^3/_4$	3	$^1/_4$	2

toward the end of the defrost cycle, the time delay relay allows the water solenoid to open. This causes water to spray over the evaporator, melting ice that may be lodged between coils and flushing the drain pan.

When the evaporator is defrosted, the system is returned to the cooling cycle by turning the three-position selector switch. The hot-gas solenoid (valve C) and built-in pilot (valve E) close as the liquid solenoid (valve A) opens.

This system can be made completely automatic by replacing the manual selector with an electric time clock. Table 10-2 shows some of the valve sizings for the low-temperature system.

Direct Expansion with Top Hot-Gas Feed

In the evaporator shown in Fig. 10-10, when the defrost cycle is initiated, the hot gas is introduced through the hot-gas solenoid valve to the manifold. It then passes through the balancing glove valve and the pan coil to a check valve that prevents liquid crossover. From the check valve, hot gas is directed to the top of the evaporator. Here, it forces the refrigerant and accululated oil from the relief regulator (valve A). This regulator has been de-energized to convert it to a relief regulator set at about 70 psig. It meters defrost condensate to the suction line and acculmulator.

Direct Expansion with Bottom Hot-Gas Feed

Compare the systems shown in Figs. 10-10 and 10-11. In the system shown in Fig. 10-11, the defrost hot gas is introduced into the bottom of the evaporator through the drain pan. The system operates similarly to that shown in Fig. 10-10. However, most of the liquid refrigerant is retained in the evaporators as defrost proceeds from the bottom to the top.

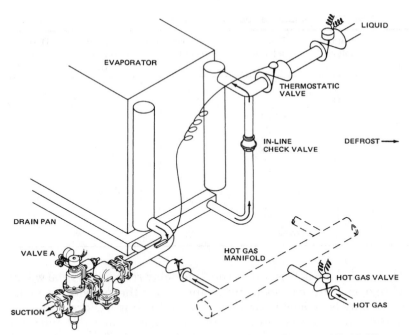

Figure 10-10 Direct-expansion evaporator—top feed. (*Courtesy of Hubbell.*)

Flooded Liquid Systems

Figure 10-12 shows a flood-gas and liquid leg shutoff (top hot-gas feed) system. Here, the gas-powered valve is used on both ends of the evaporator. It is a gas-powered check valve. At defrost, the normally closed type-A pilot solenoid is energized. Hot-gas pressure closes the gas-powered check valves. Hot gas flows through the solenoid, globe valves, pan coil, and in-line check valve into the top of the evaporator. Here, it purges the evaporator of fluids. The evaporator is discharged at the metered rate through valve B. that has been de-energized and acts as a regulator during defrost.

At the end of the defrost cycle, excess pressure will bleed from the relief line at a safe rate through the energized valve B. The gas-powered valves will not open the evaporator to the surge drum until the gas pressure is nearly down to the system pressure.

Flooded-gas leg shutoff (bottom hot-gas feed)

The system shown in Fig. 10-13 is similar to that shown in Fig. 10-12. However, the liquid leg of the evaporator dumps directly into the surge

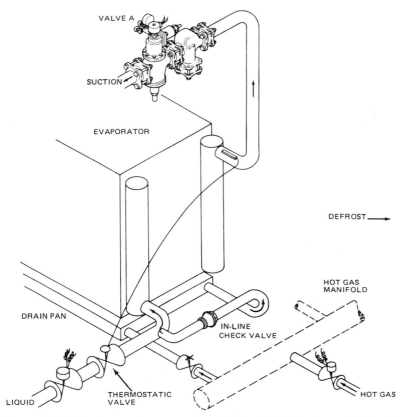

Figure 10-11 Direct-expansion evaporator—bottom feed. (*Courtesy of Hubbell.*)

drum without a relief valve. In this system, valve C is a defrost regulator. It is placed in the suction line, where it is normally open. During defrost, valve C is de-energized, converting to a defrost regulator. In such a system, it is recommended that a large-capacity surge drum or valve A be used as a bypass valve. This will bleed defrost pressure gradually around valve C into the suction line. Note how the in-line check valve is used to prevent cross flow.

Flooded-ceiling evaporator—liquid leg shutoff (bottom hot-gas feed)

Figure 10-14 illustrates a flooded-ceiling evaporator. Upon initiation of the defrost sequence, the hot gas solenoid (Number 1) is opened. Gas flows to gas-powered check valve, isolating the bottom of the surge tank from the evaporator. The hot gas flows through the pan coil and the in-line

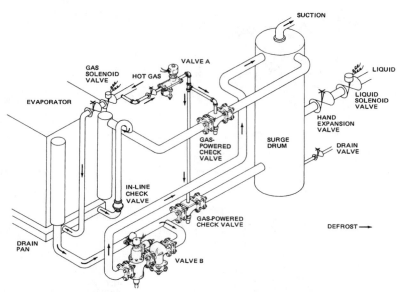

Figure 10-12   Gas and liquid leg shutoff—top feed. (*Courtesy of Hubbell.*)

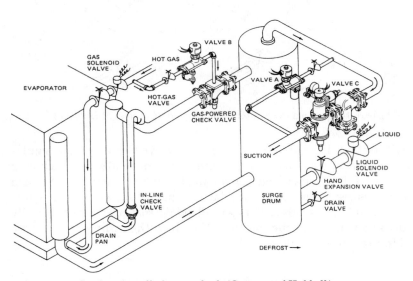

Figure 10-13   Gas leg shutoff—bottom feed. (*Courtesy of Hubbell.*)

check valve into the evaporator. Excess gas pressure is dumped into the surge tank. It will bleed through valve A. During defrost, this valve has been de-energized to perform as a relief regulator set at approximately 70 psig.

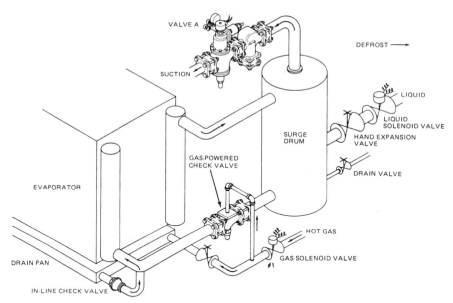

Figure 10-14   Ceiling evaporator, liquid leg shutoff—bottom feed. (*Courtesy of Hubbell.*)

Flooded-ceiling evaporator—liquid leg shutoff (top hot-gas feed)

Figure 10-15 shows a multiple flooded-evaporator system using input and output headers to connect the various evaporators and the surge drum. Note that, upon defrost, the fluid and condensate, are purged from the evaporator and surge drum into the remote accumulator through the regulator, which is a reseating safety valve. This is usually set at about 70 psig. The accumulator must be sized to accept the refrigerant, plus hot-gas condensate.

Flooded-ceiling blower (top hot-gas feed)

Figure 10-16 shows a modification of the system shown in Fig. 10-15. In the system shown in Fig. 10-16, top-fed hot defrost gas forces the evaporator fluid directly to the bottom of the large surge drum. The defrost regulator (valve A), which is normally open, is de-energized during the defrost to act as a relief regulator.

To minimize heating of the ammonia that accumulates in the surge drum during defrost, a thermostat bulb should be used to sense the temperature rise in the bottom header. This thermostat can be used to terminate the defrost cycle. Once again, the gas-powered check valve isolates the evaporator from the surge drum until the gas pressure is shut off.

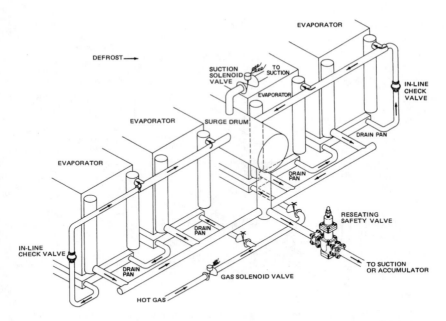

Figure 10-15   Ceiling evaporator, liquid leg shutoff—top feed. (*Courtesy of Hubbell.*)

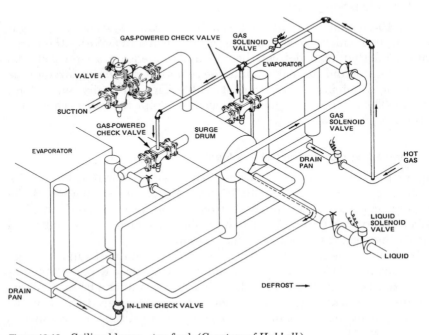

Figure 10-16   Ceiling blower—top feed. (*Courtesy of Hubbell.*)

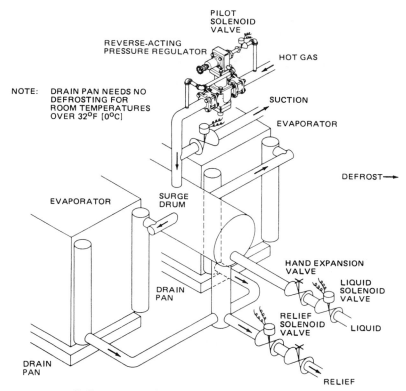

PILOT
SOLENOID
VALVE

REVERSE-ACTING
PRESSURE REGULATOR

HOT GAS

NOTE: DRAIN PAN NEEDS NO
DEFROSTING FOR
ROOM TEMPERATURES
OVER 32°F [0°C]

SUCTION

EVAPORATOR

DEFROST⟶

EVAPORATOR

SURGE
DRUM

HAND EXPANSION
VALVE

LIQUID
SOLENOID
VALVE

DRAIN
PAN

RELIEF
SOLENOID
VALVE

LIQUID

DRAIN
PAN

RELIEF

Figure 10-17   Ceiling blower—feed through surge drum. (*Courtesy of Hubbell.*)

Flooded-ceiling blower (hot-gas feed through surge drum)

Figure 10-17 shows a simple defrost. It is a setup for the refrigeration system shown in Fig. 10-16. However, both the evaporator and surge drum are emptied during the defrost, necessitating the use of an ample suction accumulator to protect the compressor. In this system the pilot-solenoid valve in conjunction with the reverse-acting pressure regulator limits the system pressure. This permits the use of a simple solenoid valve and globe valve for rate control in the relief line.

Flooded floor-type blower (gas and liquid leg shutoff)

Figure 10-18 illustrates a flooded floor unit suitable for operation down to −70°F (−57°C). The gas-pressure-powered valve used in this circuit has a solenoid pilot operator. This provides positive action with gas or liquid loads at high or low temperatures and pressures.

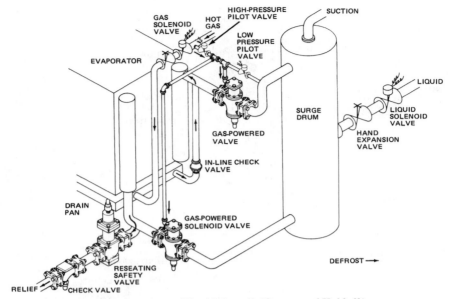

Figure 10-18   Floor blower—gas and liquid shutoff. (*Courtesy of Hubbell.*)

To defrost a group of evaporators without affecting the temperatures of the common surge drum, the gas-powered valve is used at each end of the evaporator. A reseating safety valve is a relief regulator. It controls the defrost pressure to the relief-line accumulator. A check valve prevents backflow into the relief line. The in-line check valve prevents crossover between adjacent evaporators.

At high temperatures [above −25°F (−31°C)], use of the gas-powered check valve in place of the gas-powered solenoid valve is recommended.

Flooded floor–type blower (gas leg shutoff)

The system shown in Fig. 10-19 is similar to that shown in Fig. 10-18. However, a single gas-type, pressure-powered valve is used. Overpressure at the surge drum is relieved by valve B, a defrost-relief regulator. This is normally wide open. It becomes a regulating valve when its solenoid is de-energized during defrost.

Defrost gas flows through the hot-gas solenoid when energized. It then flows through the glove valve and the in-line check valve to force the evaporator fluid into the surge drum.

An optional hot-gas thermostat bulb may be used to sense heating of the bottom of the evaporator. Thus, it can act as a backup for the timed defrost cycle.

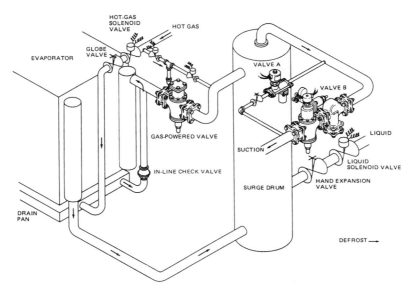

Figure 10-19 Floor blower—gas leg shutoff. (*Courtesy of Hubbell.*)

Liquid-Recirculating Systems

Liquid refrigerant recirculating systems are frequently fed by upward liquid flow through their evaporators. These systems are called bottom-fed. This is accomplished by either mechanical or gas-displacement recirculators during the refrigerant cycle (see Fig. 10-20).

In some systems, more than a single evaporator is fed from the same recirculator, as shown in Fig. 10-20. Then, a proper distribution of liquid between evaporators must be maintained to achieve efficient operation of each evaporator. This balance is usually accomplished by the insertion of adjustable glove valves or orifices into the liquid-feeder line. Similarly, adjustment of the glove valves or insertion of orifices is also often used properly to distribute hot gas during the defrost cycle.

Equalizing orifices or glove valves are not used if the hot gas used for defrosting is fed to the bottom of the evaporators as shown in Fig. 10-20. In such cases, most of the hot gas could flow through the circuits nearest the hot-gas supply line. The same would also happen in circuits where both vertical and horizontal headers are used, as in Fig. 10-21. The more remote circuits could remain full of cold liquid. Consequently, they would not defrost.

Supplying hot gas to the top of the evaporator forces liquid refrigerant down through the evaporator and out through a reseating safety valve relief regulator into the suction-line return to the accumulator (see Fig. 10-21). Reseating safety-valve relief regulators are usually set to relieve at 60 to 80 psig to provide rapid defrost.

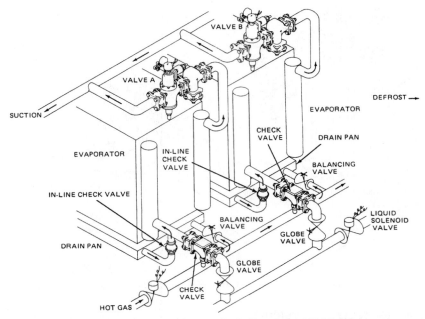

Figure 10-20 Flooded recirculator—bottom feed. (*Courtesy of Hubbell.*)

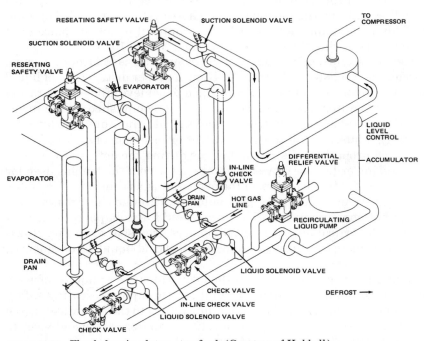

Figure 10-21 Flooded recirculator—top feed. (*Courtesy of Hubbell.*)

The use of check valves is important in flooded liquid-recirculating systems fed by mechanical gas-displacement liquid recirculators. The check valves are used where the pressure of the hot gas used for defrost is higher than the system pressure. The reseating safety-check valve must be used to stop this gas at high pressure from flowing back into the liquid supply-line.

Flooded recirculator (bottom hot-gas feed)

The multiple system shown in Fig. 10-20 shows a check valve mounted in each of the liquid-refrigerant branch lines. A single solenoid valve is used in the main refrigerant line. The defrost gas is bottom fed.

Flooded recirculator (top-gas feed)

The system illustrated in Fig. 10-21 shows a check valve mounted directly at the outlet of each of the liquid-solenoid valves. The defrost gas is top fed. This system permits selective defrosting of each evaporator. A single accumulator is used to protect the compressor during defrost, as well as to accumulate both liquid refrigerant and defrost condensate. This protection is accomplished by using a differential pressure–regulator valve in an evaporator bypass circuit.

The differential pressure–regulator valve will open sufficiently to relieve excess pressure across the compressor inlet. The pressure will discharge as excess pressure differential occurs. When the pressure differential is less than the regulator-valve setting, the regulator will be tightly closed.

Low-temperature ceiling blower

The low-temperature liquid recirculating system illustrated in Fig. 10-22 uses several controls. During the cooling cycle, Number 1 pilot valve is opened and Number 2 pilot valve is closed, holding the gas-powered solenoid valve wide open. This allows flow of liquid through the energized liquid-solenoid valve from the recirculator and then through the circuit of the unit. The in-line check valve installed between the drain-pan coil header and suction line prevents drainage of liquid into the drain-pan coil.

For defrost, the liquid solenoid valve is closed. The Number 1 pilot solenoid is de-energized. The Number 2 solenoid is opened, closing the gas-powered solenoid valve tightly. The hot-gas solenoid is energized. This allows distribution of the hot gas through the drain-pan coils, the in-line check valve, the top of the suction header, and the coil. The gas comes out the bottom of the liquid header.

Check valve A prevents the flow of the high-pressure gas in the liquid line. Therefore, the gas is relieved through the safety-valve relief regulator (B). This is set to maintain pressure in the evaporator to promote rapid, efficient defrost.

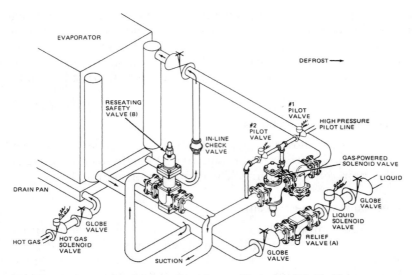

Figure 10-22 Low-temperature ceiling blower. (*Courtesy of Hubbell.*)

Year-Round-Automatic Constant Liquid Pressure Control System

The constant liquid control system is a means of increasing the efficiency of a refrigeration system that utilizes air-cooled, atmospheric, or evaporative condensers (see Fig. 10-23). This is accomplished by automatically maintaining a constant liquid pressure throughout the year to assure efficient operation. Constant liquid pressure on thermal expansion valves, float controls, and other expansion devices results in efficient low-side operation. Hot gas defrosting, liquid recirculation, or other refrigerant-control systems require constant liquid pressure for successful operation. Liquid pressure is reduced by cold weather and extremely low wet-bulb temperatures with low refrigeration loads.

To compensate for a decrease in liquid measure, it is necessary automatically to throttle the discharge to a predetermined point and regulate the flow of discharge pressure to the liquid line coming from the condenser and going to the receiver. Thus, predetermined pressure is applied to the top of the liquid in the receiver. The constant liquid-pressure control does this. In addition, when the compressor "start and stop" is controlled by pressure-stats, the pressure-operated hot-gas flow-control valve is a tight closing stop valve during stop periods. This permits efficient "start and stop" operation of the compressor by pressure control of the low side.

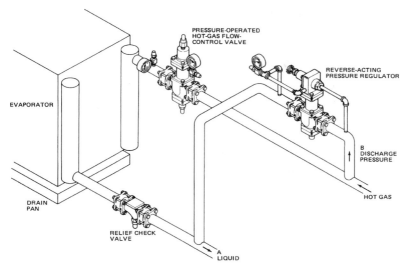

Figure 10-23   Year-round automatic control system. (*Courtesy of Hubbell.*)

The three valves in the system shown in Fig. 10-23 are

- The reverse acting pressure regulator
- The pressure-operated hot-gas flow-control valve
- The relief check valve

The function of the control system is to maintain a constant liquid pressure (A). The reverse acting pressure-regulator valve accomplishes this, which is a modulating-type valve. It maintains a constant predetermined pressure on the downstream side of the regulator. To maintain a constant pressure (A) it is necessary to maintain a discharge pressure (B) approximately 5 psi above (A). This is accomplished by the hot-gas control valve, which will maintain a constant pressure (B) on the upstream or inlet side of the regulator. Due to the design of the regulator, a constant supply of gas will be available at a predetermined pressure to supply the pressure regulator to maintain pressure (A). Excess hot gas is not required to maintain a fill flow into the condenser.

The relief check valve prevents pressure (A) from causing backflow into the condenser. When the compressor shuts down, the hot-gas flow-control valve closes tightly and shuts off the discharge line. This prevents gas from flowing into the condenser.

The check valve actually prevents the backflow of liquid into the condenser. Thus, liquid cannot back up into the condenser in extremely cold weather. Sufficient low-side pressure will be maintained to start the compressor when refrigeration is required.

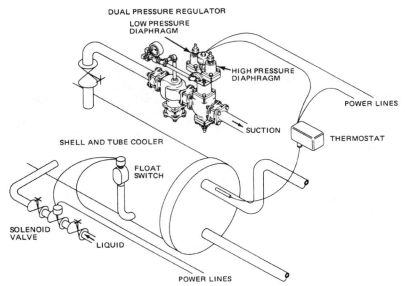

Figure 10-24   Dual-pressure regulator application. (*Courtesy of Hubbell.*)

Dual-Pressure Regulator

A dual-pressure regulator is shown in Fig. 10-24. It is used on a shell-and-tube cooler. The dual-pressure regulator is particularly adaptable for the control of shell-and-tube brine or water coolers, which at intervals may be subjected to increased loads. Such an arrangement is shown in Fig. 10-24.

The high-pressure diaphragm is set at a suction pressure suitable for the normal load. The low-pressure diaphragm is set for a refrigerant temperature low enough to take care of any intermittent additional loads on the cooler. In this case, a thermostat affects the transfer between low and high pressure. The remote bulb of the thermostat is located in the water- or brine-line leaving the cooler. A temperature increase at this bulb indicating an increase in load will cause the thermostat to open the electric pilot and transfer control of the cooler to the low-temperature diaphragm. Upon removal of the excess load, the thermostat will cause the electric pilot to close the low-pressure port. The cooler is then automatically transferred to the normal pressure for which the high-pressure diaphragm is set. The diaphragms may be set at any two evaporator pressures at which it is desirable to operate. Any electric switching device responsive to load change may be used to change from one evaporator pressure to the other.

Valves and Controls for Hot-Gas Defrost of Ammonia-Type Evaporators

The following valves and controls are used in the hot-gas defrost systems of ammonia-type evaporators:

Hot-gas or pilot-solenoid valve. The valve is a 1/8 in. ported solenoid valve. It is a direct-operated valve suitable as a liquid, suction, hot-gas, or pilot valve at pressures to 300 lb.

Suction-, liquid-, or gas-solenoid valve. The suction-solenoid valve is a one-piston, pilot-operated valve suitable for suction-, liquid-, or gas-lines at pressures to 300 lb. It is available with a 9/16 in. or 3/4 in. port.

Pilot-operated solenoid valve. The valve is a one-piston, pilot-operated solenoid valve used as a positive stop valve for applications above −30°F (−34°C) on gas or liquid.

Pilot-operated two-piston valve. The solenoid valve is a rugged, pilot-operated, two-piston valve with spring return for positive closing under the most adverse conditions. It is used for compressor unloader, suction, liquid, and hot-gas applications.

Gas-powered solenoid valve. The gas-powered solenoid valve is a power-piston type of valve that uses high pressure to force the valve open through the control of pilot valves. Because of the high power available to open these valves, heavy springs may be used to close the valves positively at temperatures down to −90°F (−68°C).

Dual-pressure regulator valve. The dual-pressure regulator valve is designed to operate at two predetermined pressures without resetting or adjustment. By merely opening and closing a pilot solenoid, either the low- or high-pressure setting is maintained.

Reseating safety valve. The reseating safety valve is generally used as a relief regulator to maintain a predetermined system pressure. The pressure maintained by the valve is adjustable manually.

Back-pressure regulator arranged for full capacity. The back-pressure regulator is normally used where pressure control of the evaporator is not required—as in a direct expansion system. A pilot solenoid is energized, allowing pressure to bypass the sensing chamber of the regulator holding the valve wide open. De-energizing the pilot valve allows the valve to revert to its function as a back-pressure regulator maintaining a preset pressure upstream of the valve. The valve performs both as a suction solenoid and as a relief regulator.

Differential relief valve. The differential relief valve is a modulating regulator for liquid or gas use. It will maintain a constant preset pressure differential between the upstream and downstream side of a regulator.

Reverse-acting pressure regulator. The reverse-acting pressure regulator is used to maintain a constant predetermined pressure downstream of the valve. When complete shutoff of the regulator is required, a pilot valve is installed in the upstream feeder line. When the solenoid valve is closed, the regulator closes tightly. When the solenoid valve is open, the regulator is free to operate as the pressure demands. With the solenoid installed as described above, this becomes a combination reverse-acting regulator and stop valve.

Gas-powered check valve. The gas-powered check valve is held in a normally open position by a strong spring. Gas pressure applied at the top of the valve closes the valve positively against the high system pressures. A manual opening stem is standard.

Check valve. The check valve is a spring-loaded positive check valve with manual opening stem. It is used to prevent backup of relatively high pressure into lower-pressure lines.

In-line check valve. The in-line check valve is used in multiple-branch liquid lines fed by a single solenoid valve. This check valve prevents circulation between evaporators during refrigeration. The in-line check valve is also used between drain pans and evaporators to prevent frosting of the drain pan during refrigeration.

These valves and controls are necessary. They cause defrosting operations to take place in large evaporators used for commercial jobs. Some manufacturing operations also call for large-capacity refrigeration equipment.

Back-Pressure Regulator Applications of Controls

In a refrigeration system designed to maintain a predetermined temperature at full load, any decrease in load would tend to lower below full-load temperature the temperature of the medium being cooled.

To maintain constant temperatures in applications having varying loads, means must be provided to change refrigerant temperature to meet varying-load requirements.

Refrigerant temperature is a function of evaporator pressure. Thus, the most direct means of changing refrigerant temperature to meet varying-load requirements is to vary the system pressure. This variation of system pressure is accomplished by adjusting the setting of a back-pressure regulator. A number of back-pressure valve controls are available. Some of them are mentioned here.

Refrigerant-powered compensating-type pilot valve

The upper portion of the valve head is similar to a standard pressure-regulating head. On the lower portion of the head another diaphragm is connected to the main diaphragm by a push rod. As the thermal bulb warms, the liquid in it expands, pushing up on the rod and opening the regulator. Because this is accomplished by an outside power source, the

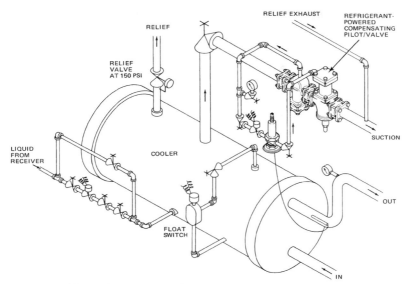

Figure 10-25   Thermal-compensating back-pressure regulator. (*Courtesy of Hubbell.*)

pressure drop through the head is reduced considerably. The valve head will function in connection with the regulator on a 1/2- to 3/4-lb overall pressure drop. The point at which the modulation or compensation takes place may be adjusted by turning the adjusting stem. By turning the stem in, the product temperature is increased. By turning the stem out, the product temperature is decreased. The back-pressure valve will remain wide open, taking advantage of the line-suction pressure until the product being cooled approaches the temperature at which modulation is to begin. The valve head will hold the temperature of the product to within ±1/2°F (0.28°C) of the desired temperature. In the case of failure of the thermal element, the valve head can be used as a straight back-pressure valve by readjusting it to the predetermined suction pressure at which the system is desired to operate (see Fig. 10-25).

Air-compensating back-pressure regulator

A standard regulator is reset by manually turning the adjusting stem, which increases the spring pressure on top of the diaphragm. In an air-compensated regulator, a change of pressure on top of the diaphragm is accomplished by introducing air pressure into the airtight bonnet over the diaphragm. As this air pressure is increased, the setting of the regulator will be increased. This will produce like changes of evaporator pressure and refrigerant temperature. The variations in air pressure are produced by the temperature changes of the thermostatic remote bulb placed in the stream of the medium being cooled as it leaves the evaporator.

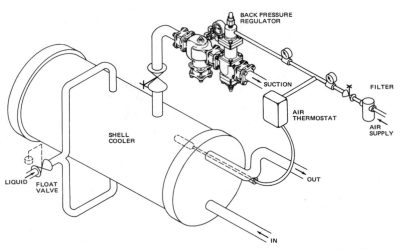

Figure 10-26   Air-compensating back-pressure regulator. (*Courtesy of Hubbell.*)

Temperature changes in the medium being cooled over the remote bulb of the thermostat will cause the thermostat to produce air pressures in the regulator bonnet within a range of 0 to 15 lb. This will cause the regulator to change the evaporator suction pressure in a like amount. A more definite understanding of this operation is obtained by assuming certain working conditions for the purpose of illustration. In cases where a larger range of modulation is required, a three-to-one air relay may be installed. This will permit a 45-lb range of modulation (see Fig. 10-26).

Electric-compensating back-pressure regulator

A standard regulator is reset manually by turning the adjusting stem, usually found at the top of the regulator. In an electrically compensated regulator, turning the stem to obtain different refrigerant pressures and temperatures in the evaporator is accomplished by a small electric motor. This motor rotates the adjusting stem in accordance with temperature variations in a thermostatic bulb placed in the medium being cooled as it leaves the evaporator. The adjusting stem, spring, and controlling diaphragm have been separated from their positions at the top of the regulator. They have been placed in a small remote unit mounted on a common base with the motor and gear drive. This compensating unit may be located in any convenient place within 20 ft of the main regulator. The unit is connected to it by two small pipelines. These convey the pressure changes set up by the control diaphragm.

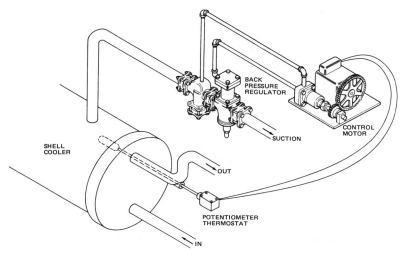

Figure 10-27 Electric-compensating back-pressure regulator. (*Courtesy of Hubbell.*)

The total arc of rotation of the motor and the large gear on the motor acting through the smaller pinion on the adjusting stem of the diaphragm unit will rotate the stem about two turns. This is sufficient to cause the regulator to vary the evaporator pressure through a total range of about 13 lb (see Fig. 10-27).

Valve Troubleshooting

Most of the problems in an evaporator system occur in the valves that make the defrost system operate properly. Every valve has its own particular problems. A differential pressure relief regulator valve is shown in Fig. 10-28.

A listing of its component parts should help you see the areas where trouble may occur. Table 10-3 lists possible causes and remedies.

The valve difficulties and remedies listed in Table 10-3 are for one particular type of valve. Manufacturers issue troubleshooting tables such as the one shown. These should be consulted when troubleshooting the valves of the evaporator system.

Noise in hot-gas lines

Noise in hot-gas lines between interconnected compressors and evaporator condensers may be eliminated by the installation of mufflers. This noise may be particularly noticeable in large installations and is usually caused by the pulsations in gas flow caused by the reciprocating action of the compressors and velocity of the gas through the hot-gas line from

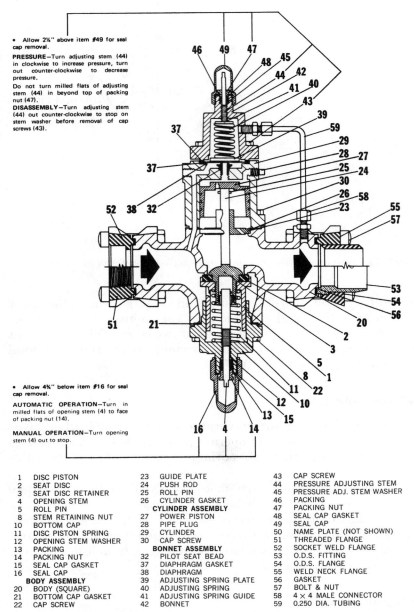

- Allow 2¼" above item #49 for seal cap removal.

PRESSURE—Turn adjusting stem (44) in clockwise to increase pressure, turn out counter-clockwise to decrease pressure.

Do not turn milled flats of adjusting stem (44) in beyond top of packing nut (47).

DISASSEMBLY—Turn adjusting stem (44) out counter-clockwise to stop on stem washer before removal of cap screws (43).

- Allow 4¾" below item #16 for seal cap removal.

AUTOMATIC OPERATION—Turn in milled flats of opening stem (4) to face of packing nut (14).

MANUAL OPERATION—Turn opening stem (4) out to stop.

1	DISC PISTON	23	GUIDE PLATE	43	CAP SCREW
2	SEAT DISC	24	PUSH ROD	44	PRESSURE ADJUSTING STEM
3	SEAT DISC RETAINER	25	ROLL PIN	45	PRESSURE ADJ. STEM WASHER
4	OPENING STEM	26	CYLINDER GASKET	46	PACKING
5	ROLL PIN		**CYLINDER ASSEMBLY**	47	PACKING NUT
8	STEM RETAINING NUT	27	POWER PISTON	48	SEAL CAP GASKET
10	BOTTOM CAP	28	PIPE PLUG	49	SEAL CAP
11	DISC PISTON SPRING	29	CYLINDER	50	NAME PLATE (NOT SHOWN)
12	OPENING STEM WASHER	30	CAP SCREW	51	THREADED FLANGE
13	PACKING		**BONNET ASSEMBLY**	52	SOCKET WELD FLANGE
14	PACKING NUT	32	PILOT SEAT BEAD	53	O.D.S. FITTING
15	SEAL CAP GASKET	37	DIAPHRAGM GASKET	54	O.D.S. FLANGE
16	SEAL CAP	38	DIAPHRAGM	55	WELD NECK FLANGE
	BODY ASSEMBLY	39	ADJUSTING SPRING PLATE	56	GASKET
20	BODY (SQUARE)	40	ADJUSTING SPRING	57	BOLT & NUT
21	BOTTOM CAP GASKET	41	ADJUSTING SPRING GUIDE	58	4 × 4 MALE CONNECTOR
22	CAP SCREW	42	BONNET	59	0.250 DIA. TUBING

Figure 10-28 Differential pressure relief regulators automatically maintain a preset differential between the upstream (inlet) and the downstream (outlet) side of the control valve. (*Courtesy of Hubbell.*)

TABLE 10-3   Troubleshooting a Differential Pressure Relief Regulator

Symptom	Probable cause	*Remedy
Erratic Operation. No adjustment.	Damaged pilot seat bead and/or diaphragms.	Replace.
Regulator remains open.	Dirt-binding power or disc pistons.	Clean, repair, and/or
	Dirt lodged in seat disc or pilot seat bead area.	replace damaged items.
	Tubing sensing downstream pressure blocked.	Remove obstruction.
	Manual opening stem holding disc piston open.	Turn opening stem to automatic position.
Short cycling, hunting, or chattering.	Regulator too large for load conditions.	Install properly sized metered orifice control.
	Power piston bleed hole enlarged.	Replace or contact factory for sizing.
	O.D. of power piston worn, creating excessive clearance.	Replace piston.
Excessive pressure drop.	Regulator too small for load.	Replace with correctly sized regulator.
	Passage to sensing chamber blocked.	Remove obstructions.
	Strainer blocked.	Clean strainer—replace screen if damaged.
No adjustment over 90 psig.	Range spring rated at 2 to 90 psig.	Order range kit rated at 75 to 300 psig.
No adjustment under 2 psig.	Range spring rated at 2 to 90 psig.	Order range kit rated at 25 in. vacuum to 50 psig.

*If repair requires metal removal—replace part.

the compressor. The proper location of a muffler is in a horizontal or down portion of the hot-gas line, immediately after leaving the compressor. It should never be installed in a riser. The problem of decreased system capacity due to excessive vertical lifts in the liquid line is usually solved by the installation of subcoolers. Check valves are used in the suction line of low-temperature fixtures when they are multiplied with high-temperature fixtures. Their use is most important when the condensing unit is regulated by low-pressure control.

Controlling Refrigerant: Valves, Tubing, and Filters

A number of devices are used in refrigeration and air-conditioning systems to control the flow of refrigerant. Proper selection, installation, and maintenance of these devices holds the key to efficient performance under varying conditions.

Metering Devices

Metering devices divide the high side from the low side of the refrigeration system. Acting as a pressure control, metering devices allow the correct amount of refrigerant to pass into the evaporator.

Hand-expansion valve

Of the several types of metering devices. the hand-expansion valve is the simplest (see Fig. 11-1). Used only on manually controlled installations, the hand-expansion valve is merely a needle valve with a fine adjustment stem. When the machine is shut down, the hand-expansion valve must be closed to isolate the liquid line.

Automatic-expansion valve

The automatic-expansion valve controls liquid flow by responding to the suction pressure of the unit acting on its diaphragm or bellows (see Fig. 11-2). When the valve opens, liquid refrigerant passes into the evaporator. The resulting increase in pressure in the evaporator closes the valve. Meanwhile, the compressor is pulling the gas away from the

AUTOMATIC
EXPANSION VALVE

Figure 11-1 Hand-expansion valve. (*Courtesy of Mueller Brass.*)

Figure 11-2 Automatic-expansion valve. (*Courtesy of Mueller Brass.*)

Figure 11-3 A thermostatic-expansion valve. (*Courtesy of Mueller Brass.*)

coils, reducing the pressure. This pressure reduction allows the expansion valve to open again. In operation, the valve never quite closes. The needle floats just off the seat and opens wide when the unit calls for refrigeration. When the machine is shut down, the pressure building up in the coils closes the expansion valve until the unit starts up.

Thermostatic-expansion valve

The thermostatic-expansion valve, used primarily in commercial refrigeration and air-conditioning, is a refinement of the automatic-expansion valve (see Fig. 11-3). A bellows or diaphragm responds to pressure from a remote bulb charged with a substance similar to the refrigerant in the system. The bulb is attached to the suction line near the evaporator outlet. It is connected to the expansion valve by a capillary tube.

In operation, the thermostatic-expansion valve keeps the frost line of the unit at the desired location by reacting to the superheat of the suction gas. Superheat cannot be present until all liquid refrigerant in the evaporator has been vaporized. Thus, it is possible to obtain a range of evaporator temperatures by adjusting the superheat control of the thermostatic-expansion valve.

The prime importance of this type of metering device is its ability to prevent the flood-back of slugs or liquid through the suction line to the compressor. If this liquid returns to the compressor, it could damage it. The compressor is designed to pump vapors, not liquids.

Capillary tubing

Small-bore capillary tubing is used as a metering device. It is used on everything from the household refrigerator to the heat pump. Essentially, it is a carefully measured length of very small diameter tubing.

It creates a predetermined pressure drop in the system. The capillary has no moving parts.

Because a capillary tube cannot stop the flow of refrigerant when the condensing unit stops, such a refrigeration unit will always equalize high-side and low-side pressures on the off cycle. For this reason, it is important that the refrigerant charge be of such a quantity that it can be held on the low side of the system without damage to the compressor. In a charge of several pounds, this "critical charge" of refrigerant may have to be carefully weighed.

An accumulator, or enlarged chamber, is frequently provided on a capillary tube system to prevent slugs of liquid refrigerant from being carried into the suction line.

Float valve

A float valve, either high-side or low-side, can serve as a metering device. The high-side float, located in the liquid line, allows the liquid to flow into the low side when a sufficient amount of refrigerant has been condensed to move the float ball. No liquid remains in the receiver. A charge of refrigerant just sufficient to fill the coils is put into the system on installation. This type of float, formerly used extensively, is now limited to use in certain types of industrial and commercial systems.

The low-side float valve keeps the liquid level constant in the evaporator. It is used in flooded-type evaporators where the medium being cooled flows through tubes in a bath of refrigerant. The low-side float is more critical in operation than the high-side float and must be manufactured more precisely. A malfunction will cause the evaporator to fill during shutdown. This condition will result in serious pounding and probable compressor trouble on start-up.

Needle valves, either diaphragm or packed type, may be used as hand-expansion valves. As such, they are usually installed in a bypass line around an automatic- or thermostatic-expansion valve. They are placed in operation when the normal control is out of order or is removed for repairs.

Fittings and Hardware

Modern refrigerants can escape through the most minute openings. Since porosity in a fitting could create such an opening, it is mandatory that porosity be eliminated from fittings and accessories that are to be used with refrigerants (see Fig. 11-4). One way to eliminate porosity in fittings is to either forge or draw them from brass rod. This creates a final grain structure that prevents the seepage of refrigerant due to porosity. The threads on fittings must be machined with some degree of accuracy to prevent leaks. Solder-type fittings should be made of wrought copper, brass rod, or brass forgings (see Fig. 11-5). This eliminates

Figure 11-4   45°-flare fitting. (*Courtesy of Mueller Brass.*)

Figure 11-5   Wrought copper, solder-to-solder fitting. (*Courtesy of Mueller Brass.*)

the possibility of leaks due to porosity of the metal. The tube is not weakened by the cutting of threads, as is the case with iron pipe. A soldered joint allows the use of a much lighter wall tube with complete safety and with significant cost savings.

One advantage of copper pipe over iron is the elimination of scale and corrosion. In service, a light coating of copper oxide forms on the outside of the copper tube. This coating prevents chemical attack. There is no "rusting out" of copper tube.

Copper tubing

For flare-fitting applications, seamless soft copper tube is recommended (see Fig. 11-6). This tube is furnished with sealed ends. It is supplied in 50-ft lengths in sizes from 1/8- through 3/4-in. outer diameter (OD) for flaring and through OD for soldering.

The chief demand for this tube is in sizes from 1/4 through 5/8-in. OD. Sizes smaller than 1/4 in. are seldom used in commercial refrigeration. To uncoil the tube without kinks, hold one free end against the floor or on a bench and uncoil along the floor or bench. The tube may be cut to length with a hacksaw or tube cutter. In either case, deburr the end before flaring. Bending is readily accomplished with either external- or internal-bending springs or lever-type bending tools.

Figure 11-6   Refrigeration service tube in a 50-ft coil. (*Courtesy of Mueller Brass.*)

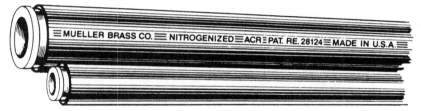

Figure 11-7   Nitrogenized ACR-copper tube. (*Courtesy of Mueller Brass.*)

ACR (air-conditioning and refrigeration) tube is frequently used. It is cleaned, degreased, dried, and end-sealed at the factory. This assures the user that he or she is installing a clean, trouble free tube. Some tubing is available with an inert gas (nitrogen) (see Fig. 11-7). The *nitrogenized* ACR tube is purged, charged with clean, dry nitrogen, and then sealed with reusable plugs. After cutting the tube, the remaining length can easily be replugged. The remaining nitrogen limits excess oxides during succeeding brazing operations. It comes in 20-ft lengths. Type-L hard tube has from 3/8- through 3-1/8-in. OD. Type-K tube is also available.

Where tubing will be exposed inside food compartments, tinned copper is recommended. Type-L, hard-temper copper tube is recommended for field installations using solder-type fittings. Type M is sufficiently strong for any pressures of the commonly used refrigerants. However, it is used chiefly in manufactured assemblies where external damage to the tube is not as likely as in field installations. For maximum protection against possible external damage to refrigerant lines, a few cities require the use of Type-K copper tube.

Line

Correct line sizes are essential to obtaining maximum efficiency from refrigeration equipment. In supermarkets, for example, the long lines running under the floor from the display cases to the machine room at the rear of the store must be fully engineered. Otherwise, problems of oil return, slugging, or erratic refrigeration are quite likely. Table 11-1 lists refrigerant line sizes. When available, the manufacturer's recommendations must be followed regarding step-sizing, risers, traps, and the like. Available information on Refrigerant 502 claims performance at temperatures below 5°F (−15°C) when compared with Refrigerant 22. The newer refrigerants like Refrigerant 502a are said to be equivalent in performance to the older designated refrigerant.

Solder

Each solder is designed for a certain job. For instance, 50-50 solder, which consists of 50 percent tin and 50 percent lead, will not function

TABLE 11-1 Sizes of Refrigerant Lines (*Courtesy of Mueller Brass.*)

Btu per hour	Refrigerant 12 Liquid line	R12 Suction 5°F (−15°C)	R12 Suction 40°F (4.4°C)	Refrigerant 22 Liquid line	R22 Suction 5°F (−15°C)	R22 Suction 40°F (4.4°C)	Refrigerant 40 Liquid line	R40 Suction 5°F (−15°C)	R40 Suction 40°F (4.4°C)	Refrigerant 502 Liquid line	R502 Suction 5°F (−15°C)	R502 Suction 40°F (4.4°C)
3,000	1/4	1/2	1/2	1/4	1/2	1/2	1/4	1/2	1/2	1/4	1/2	1/2
6,000	3/8	5/8	5/8	3/8	5/8	5/8	1/4	1/2	1/2	3/8	5/8	5/8
9,000	3/8	7/8	5/8	3/8	7/8	5/8	3/8	5/8	5/8	3/8	7/8	5/8
12,000	3/8	1 1/8	7/8	3/8	7/8	7/8	3/8	7/8	7/8	3/8	7/8	7/8
15,000	3/8	1 1/8	7/8	3/8	1 1/8	7/8	3/8	7/8	7/8	3/8	1 1/8	7/8
18,000	3/8	1 1/8	7/8	3/8	1 1/8	7/8	3/8	1 1/8	7/8	3/8	1 1/8	7/8
21,000	1/2	1 1/8	1 1/8	1/2	1 1/8	1 1/8	1/2	1 1/8	1 1/8	1/2	1 1/8	1 1/8
24,000	1/2	1 3/8	1 1/8	1/2	1 1/8	1 1/8	1/2	1 1/8	1 1/8	1/2	1 1/8	1 1/8
30,000	5/8	1 3/8	1 1/8	1/2	1 3/8	1 1/8	1/2	1 3/8	1 1/8	5/8	1 3/8	1 1/8
36,000	5/8	1 3/8	1 1/8	5/8	1 3/8	1 1/8	1/2	1 3/8	1 1/8	5/8	1 3/8	1 1/8
42,000	5/8	1 3/8	1 3/8	5/8	1 3/8	1 3/8	5/8	1 3/8	1 3/8	5/8	1 3/8	1 3/8
48,000	5/8	1 5/8	1 3/8	5/8	1 5/8	1 3/8	5/8	1 3/8	1 3/8	5/8	1 5/8	1 3/8
54,000	5/8	1 5/8	1 3/8	5/8	1 5/8	1 3/8	5/8	1 5/8	1 3/8	5/8	1 5/8	1 3/8
60,000	7/8	1 5/8	1 3/8	5/8	1 5/8	1 3/8	5/8	1 5/8	1 3/8	5/8	1 5/8	1 3/8
72,000	7/8	2 1/8	1 5/8	7/8	1 5/8	1 5/8	7/8	1 5/8	1 5/8	7/8	2 1/8	1 5/8
96,000	7/8	2 1/8	1 5/8	7/8	2 1/8	1 5/8	7/8	2 1/8	1 5/8	7/8	2 1/8	1 5/8
108,000	7/8	2 5/8	2 1/8	7/8	2 1/8	1 5/8	7/8	2 1/8	1 5/8	7/8	2 1/8	1 5/8
120,000	1 1/8	2 5/8	2 1/8	7/8	2 1/8	2 1/8	7/8	2 1/8	2 1/8	1 1/8	2 1/8	2 1/8
150,000	1 1/8	2 5/8	2 1/8	1 1/8	2 1/8	2 1/8	1 1/8	2 1/8	2 1/8	1 1/8	2 5/8	2 1/8
180,000	1 1/8	3 1/8	2 1/8	1 1/8	2 5/8	2 1/8	1 1/8	2 5/8	2 1/8	1 1/8	2 5/8	2 1/8
210,000	1 3/8	3 1/8	2 5/8	1 1/8	2 5/8	2 1/8	1 1/8	2 5/8	2 1/8	1 3/8	2 5/8	2 1/8
240,000	1 3/8	3 1/8	2 5/8	1 3/8	3 1/8	2 5/8	1 3/8	2 5/8	2 1/8	1 3/8	3 1/8	2 5/8
300,000	1 3/8	3 5/8	2 5/8	1 3/8	3 1/8	2 5/8	1 3/8	3 1/8	2 5/8	1 3/8	3 1/8	2 5/8
360,000	1 3/8	3 5/8	3 1/8	1 1/8	3 1/8	2 5/8	1 3/8	3 1/8	2 5/8	1 5/8	3 1/8	2 5/8
420,000	1 5/8	3 5/8	3 1/8	1 3/8	3 1/8	3 1/8	1 5/8	3 1/8	2 5/8	1 5/8	3 5/8	3 1/8
480,000	1 5/8	4 1/8	3 1/8	1 5/8	3 5/8	3 1/8	1 5/8	3 5/8	3 1/8	1 5/8	3 5/8	3 1/8
540,000	1 5/8	4 1/8	3 1/8	1 5/8	3 5/8	3 1/8	1 3/8	3 5/8	3 1/8	1 5/8	3 5/8	3 1/8
600,000	1 5/8	4 1/8	3 1/8	1 5/8	4 1/8	3 1/8	1 5/8	3 5/8	3 1/8	1 5/8	4 1/8	3 1/8

To convert Btu per hour to tons of refrigeration—divide by 12,000
Suction temperature, condensing medium, compressor design, and many other factors determine horsepower required for a ton of refrigerating capacity. Consult ASHRAE Handbook.
Mueller Brass

Figure 11-8   Suction line P-trap. (*Courtesy of Mueller Brass.*)

well in some instances. In fact, 50-50 solder will deteriorate in some refrigerated food-storage compartments where normally wet refrigerant lines and high carbon dioxide content are present. For this reason, No. 95 solder is recommended. It has 95 percent tin and 5 percent antimony. Number 122 solder (45 percent silver brazing alloy) is used for joints in refrigerant lines where 50-50 solder may deteriorate.

Suction line P-traps

For years, the P-trap was made by forming two or more fittings. It has now become available in one piece (see Fig. 11-8). The newer one-piece P-trap promotes efficient oil migration in refrigeration systems. This is increasingly important today. Many large food markets place their compressors and condensers on balconies or mezzanines. Such remote condensing units are likely to have long horizontal suction lines or vertical risers exceeding 3 ft in height in the suction line. In such cases, the oil concentration in the circulating refrigerant may be expected to be above 0.6 percent. A low vapor velocity may be encountered. This results in unsatisfactory oil return to the compressor. Tests have proven that with a P-trap installed, vapor velocity can fall as low as 160 ft/min and satisfactory oil return can still be achieved. The P-trap drains the oil from the horizontal runs approaching the risers. This oil, in turn, migrates up through the riser, to the compressor in one of three different forms: as a rippling oil film, mist, or transparent colloidal dispersion in the vaporized refrigerant. The method of oil migration depends upon the vapor velocity in the suction line.

Compressor valves

There are three types of compressor valves:

- Adjustable
- Double port
- Single port

Open and semihermetic compressors are usually fitted with compressor service valves, one each at the suction and discharge ports.

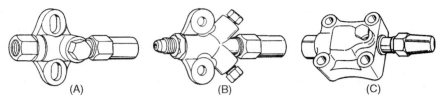

Figure 11-9 Compressor valves: (*a*) adjustable compression valve, (*b*) double-port compressor valve, and (*c*) single-port compressor valve. (*Courtesy of Mueller Brass.*)

The service valve has no operating function. Nevertheless, it is indispensable when service is to be performed on any part of the refrigeration system (see Fig. 11-9).

Compressor service valves are back-seating. They are constructed so that the stem forms a seal against a seat whether the stem is full-forward or full-backward. Valve packing is depended upon only when the stem is in the intermediate position. In one style of construction, the front seat including the one connection is threaded. Silver is brazed into the body after the stem has been assembled. When the valve is full-open (normal position when the unit is running) the gage and charging port plug or cap may be removed without loss of refrigerant. A charging line or pressure gage may be attached to this side port. It is also possible to repack the valve without interruption of service.

Line valves

Line valves are essential components of refrigerant systems (see Fig. 11-10). Installed in key locations, line valves make it possible to isolate any portion of a system or, in a multiple hookup, to separate one system from the rest. Local codes frequently specify the location of line valves in commercial- and industrial-refrigeration and air-conditioning systems. There are two types of line valves:

- Packed
- Pack-less

They must be designed to prevent refrigerant leakage. Since refrigerants are difficult to retain, packed valves are usually equipped with seal caps. Some seal caps are designed to be removed and used as wrenches for operating the valves.

In large packed valves, such as that shown in Fig. 11-11, O-rings are used as seals between the bonnets and valve bodies. They are available in either straight-through (7/8 to 4-1/8 in.) or angle-type (7-7/8 through 3-1/8 in.) construction.

The pack-less design is often preferred for smaller valves. The pack-less-type valve is used to good advantage on charging boards. The valves

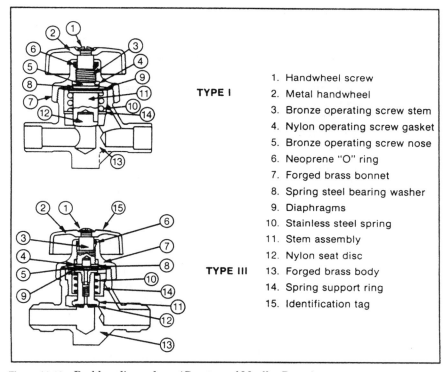

TYPE I

1. Handwheel screw
2. Metal handwheel
3. Bronze operating screw stem
4. Nylon operating screw gasket
5. Bronze operating screw nose
6. Neoprene "O" ring
7. Forged brass bonnet
8. Spring steel bearing washer
9. Diaphragms
10. Stainless steel spring
11. Stem assembly
12. Nylon seat disc

TYPE III

13. Forged brass body
14. Spring support ring
15. Identification tag

Figure 11-10   Packless-line valves. (*Courtesy of Mueller Brass.*)

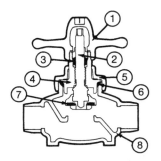

1. Cast bronze wing-seal cap
2. Bronze stem
3. Molded stem packing
4. Forged brass bonnet in sizes over 1⅛

5. Forged brass union collar
6. Neoprene "O" ring
7. Nylon seat disc
8. Cast bronze body

Figure 11-11   Packed-line valve. (*Courtesy of Mueller Brass.*)

contain triple diaphragms, one of phosphor bronze and two of stainless steel. These valves must be frost-proof. They must be designed for use where condensation is likely to occur. During the off cycle, condensation may seep down the stem of a non-frost-proof valve into the bonnet. There, it will alternately freeze and thaw. The eventual buildup of ice against the diaphragms may close the valve. Another factor that should be considered is whether or not the valve has back-seating. This prevents all pressure pulsations while the valve is open. The back-seating should allow inspection of the diaphragms without shutting down the system.

Driers, Line Strainers, and Filters

Of the three items to be examined and understood, one of the first to be considered is the drier.

Driers

Most authorities agree that moisture is the most detrimental material in a refrigeration system. A unit can stand only minute amounts of water. For this reason, most refrigeration and air-conditioning systems, both field- and factory-assembled, contain driers (see Fig. 11-12).

Moisture.   Moisture or water is always present in refrigeration systems. Acceptable limits vary from one unit to another and from one refrigerant to another. Moisture is harmful even if freeze-ups do not occur. Moisture is an important factor in the formation of acids, sludge, and corrosion. To be safe, keep the moisture level as low as possible.

Moisture will react with today's halogen-type refrigerants to form harmful hydrochloric and hydrofluoric acids within the system. To minimize the possibility of freeze-up or corrosion, the following maximum safe limits of moisture should be observed: (Note: These are *halogen-type* refrigerants.)

Refrigerant 12   15 parts per million (15 ppm)

Refrigerant 22   60 parts per million (60 ppm)

Refrigerant 502   30 parts per million (30 ppm)

If the moisture exceeds these figures, corrosion is possible. Also, excess water may freeze at the metering device if the system operates below

Figure 11-12   Filter drier. (*Courtesy of Mueller Brass.*)

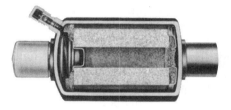

Figure 11-13 Suction-line filter drier. (*Courtesy of Mueller Brass.*)

32°F (0°C). Freeze-ups do not occur in air-conditioning systems where evaporator temperatures are normally above 40°F (4.4°C).

A drier charged with a moisture-removing substance and installed in the liquid line is the most practical way to remove moisture. With a drier of the proper size, excess water is stored in the drier. Here, it can neither react with the refrigerant nor travel through the system.

Many materials have been tried as desiccants, or drying agents. Today, the desiccant materials most commonly used are

- Silica gel
- Activated alumina
- Calcium sulfate
- Zeolite-type materials

These are known as molecular sieves and micro traps. The total drier design considers not only drying and filtering, but also maintaining maximum refrigerant flow. Filter-driers must allow free flow of refrigerant. They must also prevent fine particles of the adsorbent or other foreign matter from passing through to the metering device, usually located downstream from the drier (see Fig. 11-13).

Dirt. Dirt, sludge, flux, and metallic particles are frequently found in refrigeration systems. Numerous metallic contaminants—cast iron dust, rust, and scale, plus steel, copper, and brass filings—can damage cylinder walls and bearings. They can plug capillary tubes and thermostatic-expansion valve screens. These contaminants are catalytic and contribute to decomposition of the refrigerant-oil mixture at high temperatures.

Acids. By themselves, Refrigerants 12 and 22 are very stable—even when heated to a high temperature. However, under some conditions, reactions occur that can result in the formation of acids. For example, at elevated temperatures, Refrigerant 12 will react with the oil to form hydrochloric and hydrofluoric acids. These acids are usually present as a gas in the system and are highly corrosive. Where an "acid acceptor" such as electrical-insulation paper is present, Refrigerant 22 will decompose at high temperatures to form hydrochloric acid. The reaction of refrigerants with water may cause hydrolysis and the formation of hydrochloric and hydrofluoric acids. In ordinary usage this reaction is negligible. However, in a very wet system operating at abnormally high temperatures, some hydrolysis may occur. All of these reactions are increased by elevated temperature and are catalytic in effect. They result in the formation of corrosive compounds.

Another source of acidity in refrigeration systems is the organic acid formed from oil breakdown. Appreciable amounts of organic acid are found in the majority of oil samples analyzed in the laboratory. These acids will also corrode the metals in a system. Therefore, they must be removed.

Acid may be neutralized by the introduction of an alkali, but the chemical combination of the two creates further hazards. They release additional moisture and form a salt. Both of these are detrimental to the system.

Sludge and varnish.   Utmost care must be taken in the design and fabrication of a system. Nonetheless, in operation, unusually high discharge temperatures will cause the oil to break down and form sludge and varnish.

Temperatures may vary in different makes of compressors and under different operating conditions. Temperatures of 265°F (130°C) are not unusual at the discharge valve under normal operation. Temperatures well above 300°F (149°C) frequently occur under unusual conditions. Common causes of high temperatures in refrigeration systems are dirty condensers, noncondensable gases in the condenser, high-compression ratio, high superheat of suction gas returned to the compressor, and fan failure on forced convection condensers.

In addition to high-discharge temperatures, certain catalytic metals contribute to oil-refrigerant mixture breakdown. The most significant of these is iron. It is used in all systems and is an active catalyst. Copper is a catalyst also, but its action is slower.

However, the end result is the same. The reaction produces sludge and other corrosive materials that will hinder the normal operation of compressor valves and control devices. In addition, air in a system will also accelerate oil deterioration.

Line strainers and filters

It is impossible to keep all foreign matter out of factory or field-assembled refrigeration systems. Core sand from the compressor casting, brazing oxides in piping or tubing, chips from cutting or baring, sawdust, and dirt are found in most refrigeration systems. This is especially so with field-assembled installations. Tubing, for example, may have been exposed to air carrying dirt for several days. Cleanliness is difficult to maintain in the field. All tubing to be used for refrigeration applications should be protected by capping or sealing. It should be recapped or sealed after each use.

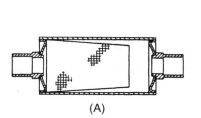
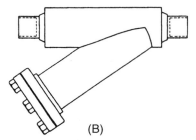

Figure 11-14   Strainers: (*a*) noncleanable strainer and (*b*) Y-type line strainer. (*Courtesy of Sporlan Valve.*)

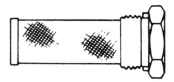

Figure 11-15   Strainer-filter, removable metal screen. (*Courtesy of Mueller Brass.*)

Moving through the system with the flowing refrigerant, particles of foreign matter may score critical moving parts or clog orifices. To prevent such damage, strainers or filters are frequently installed in the system (see Fig. 11-14). A filter-drier placed in the liquid line ahead of the metering device is the normal precaution. However, on multiple installations, it is usual to install a strainer upstream of each metering device just ahead of key valves and controls. To protect the compressor, most engineers also specify filters for the suction line.

The simplest strainer consists of a set of metal screens of the proper mesh (see Fig. 11-15). Adding felt pads or asbestos cloth creates a very effective filter, rather than a mere strainer. A cellulose-fiber core as used in Fig. 11-13 is also very effective for this purpose.

Suction-line strainers and filters are designed with sufficient flow capacity to prevent excessive pressure drop. Since determining the need for cleaning or replacement of a suction-line filter is related to pressure drop, some designs are offered with pressure taps. These permit pressure gage installation to determine the degree of pressure drop.

Strainers are made in several designs. They are also supplied in a wide range of sizes. Screen areas are large. In cartridge-type strainers, provision is made for removal of the screens or filters (see Fig. 11-16).

Liquid Indicators

Liquid indicators are inserted in a refrigerant line to indicate the amount of refrigerant in a system (see Fig. 11-17). Proper operation of a refrigeration system depends upon there being the correct amount of refrigerant in the unit. Looking through the window of a liquid indicator is

Figure 11-16   A strainer and its replacement-type filter. (*Courtesy of Sporlan Valve.*)

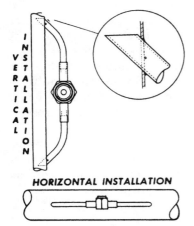

HORIZONTAL INSTALLATION

Figure 11-17   Liquid-line indicator installation. (*Courtesy of Sporlan Valve.*)

the simplest way to determine whether there is a refrigerant shortage. A shortage of refrigerant may be due to a leak in the system or failure to charge enough refrigerant into a unit after field service.

Liquid indicators normally disclose a shortage of refrigerant by the appearance of bubbles. Some use a special assembly that shows by the appearance of the word "Full" that there is sufficient refrigerant at that point in the system.

Liquid indicators are manufactured in single-port, double-port, and straight-through types. In the single- and double-port indicators, an internal compression bushing seals the glass firmly against the body. Assemblies are furnished with a protective dust cap or seal cap. To observe the liquid stream, it is necessary to remove the cap (see Fig. 11-18).

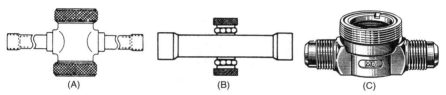

Figure 11-18   Liquid indicators: (*a*) and, (*b*) double port; and (*c*) single port. (*Courtesy of Mueller Brass.*)

A relatively new addition to the function of the liquid indicator is that of moisture detection. Special materials used in the ports of liquid-moisture indicators change color to indicate excessive moisture in the system.

Indicators with solder-type ends but without extended ends are normally furnished disassembled so the heating required for soldering will not damage glass or gaskets. In addition, single- and double-port types are supplied and assembled with extended ends. These make it possible to solder without damaging the indicator, as long as normal precautions are observed.

Construction

The indicator is a porous filter paper impregnated with a chemical salt that is sensitive to moisture. The salt changes color according to the moisture content (relative saturation) of the refrigerant. The indicator changes color below moisture levels generally accepted as a safe operation range. This device is not suitable for use with ammonia or sulfur dioxide. However, it does have a full application with Refrigerants 11, 12, 22, 113, 114, 500, and 502 (see Table 11-2). The indicator should be installed after the filter-drier and ahead of the expansion device. Prior to installation, the indicator will be yellow, indicating a wet condition. This is a normal situation, since the relative humidity of the air in contact with the element is above 0.5 percent. This does not affect the operation or calibration of the indicator. As soon as it is installed in a system, the indicator element will begin to change according to the moisture content in the refrigerant. The action of the indicator element is completely reversible. The element will change color as often as the moisture content of the system varies. Some change may take place rapidly at the start-up of a new system or after replacement of a drier on existing installations. However, the equipment should be operated for about 12 hours to allow the system to reach equilibrium before deciding if the drier needs to be changed.

Installation

Indicators with 1/4- through 1-1/8-in. ODF (outside diameter flanged) connections should not be disassembled in the field for brazing or any

TABLE 11-2 Moisture Content (In Parts Per Million) (Courtesy of Sporlan Valve.)

Unit shows	Liquid line temp. (°F)	Refrigerants 11 & 12			Refrigerant 22			Refrigerant 500			Refrigerants 502, 113, & 114		
		75°	**100°**	125°	75°	**100°**	125°	75°	**100°**	125°	75°	**100°**	125°
Green DRY		Below 5	Below **10**	Below 20	Below 30	Below **45**	Below 60	Below 40	Below **60**	Below 100	Below 10	Below **20**	Below 30
Chartreuse CAUTION		5–15	**10–30**	20–50	30–90	**45–130**	60–180	40–90	**60–150**	100–230	10–45	**20–65**	30–110
Yellow WET		Above 15	Above **30**	Above 50	Above 90	Above **130**	Above 180	Above 90	Above **150**	Above 230	Above 45	Above **65**	Above 110

Sporlan Valve

BOLD figures are for the average design conditions of refrigerant liquid lines operating at 100° F. Since the actual temperature is not critical, a satisfactory estimate can be made by comparing it to body temperature. If it feels cool to touch, use 75° F. If it feels warm, use 125° F column figures.

The unit calibration information given above is based on detailed experimental data for Refrigerants 12, 22, 500, 502, and 113. The calibration information on other refrigerants and solvents was obtained from a comparison of their properties with 12, 22, 500, 502, and 113. For the less common liquids, the following moisture calibration is suggested.

Refrigerant 13.........use ....“12” calibration   Perchloroethylene.........use ....“113” calibration
Refrigerant 21.........use ....“22” calibration   Carbon Tetrachloride.........use.......“12” calibration
Trichloroethylene.......use.......“22” calibration   Propane or Butane.........use.......“500” calibration

AIR TEST—Recent tests on AIR show that the unit changes color in the range of 0.5 to 2.0%. In ordinary air lines this means that the unit will change color at dew points in the range of −40 to −60° F.

other purpose. The long fittings on sweat models are copper-plated steel and do not conduct heat as readily as copper fittings.

On indicators with $1^3/_8$-, $1^5/_8$-, and $2^1/_8$-in. ODF connections, the indicator cartridge must be removed from the brass-saddle fitting before brazing the indicator in the main liquid line. It is shipped hand tight for easy removal.

Bypass installations

On systems having liquid lines larger than $2^1/_8$-in. OD, the indicator should be installed in a bypass line. During the operating cycle, this will provide sufficient flow to obtain a satisfactory reading for both moisture and liquid indication.

Best results will be obtained if the bypass line is parallel to the main liquid line and the take-off and return tubes project into the main liquid line at a 45° angle. Preformed 1/4 - and 3/8-in. tubing is available. It can be used with either flare or sweat-type indicators.

Excess oil and the indicator

When a system is circulating an excessive amount of oil. The indicator may become saturated. This causes the indicator to appear brown or translucent and lose its ability to change color. However, this does not damage the indicator. Let the indicator unit remain in the line. The circulating refrigerant will remove the excess oil and the indicator element will return to its proper color.

Alcohol

Do not install the color-changing indicator in a system that has methyl alcohol or a similar liquid dehydrating agent. Remove the alcohol by using a filter and then install the indicator. Otherwise, the alcohol will damage the color indicator.

Leak detectors

Die-type visual leak detectors will also mask the color-changing indicator. Here again, use a filter to remove all leak-detector color from the system before installing the indicator.

Liquid water

On occasions, it is possible for large quantities of water to enter a refrigeration system. An example would be a broken tube in a water-cooled condenser. If the free water contacts the indicator element, the element will be damaged. All moisture indicators are made of a chemical salt. These salts must be soluble in water to change color. If excessive water is present, the salts will dissolve. Permanent damage to the indicator will result. The indicator may remain yellow, or even turn white.

Hermetic-motor burnouts

After a hermetic-motor burnout, install a filter to remove the acid and sludge contamination. When the system has operated for 48 hours, replace the filter. At the same time, install the color indicator for moisture.

The acid formed by the burnout may damage the indicator element of the color-changing unit. Thus, it should be installed only after the greater percentage of contaminants has been removed.

Hardware and fittings

In assembling a unit in the factory or the field, strict standards of quality must be observed. Cleanliness is very important. The cleanliness of a part can determine the efficiency of a piece of equipment. Figure 11-19 illustrates some of the hardware and fittings.

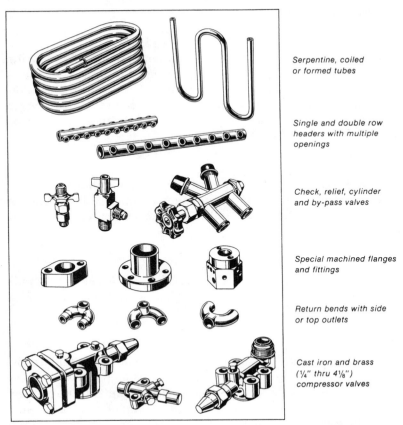

Serpentine, coiled or formed tubes

Single and double row headers with multiple openings

Check, relief, cylinder and by-pass valves

Special machined flanges and fittings

Return bends with side or top outlets

Cast iron and brass (¼" thru 4⅛") compressor valves

Figure 11-19 Hardware and fittings for refrigeration and air-conditioning installation. (*Courtesy of Mueller Brass.*)

Thermostatic-Expansion Valve

Several different valves are used to control the flow of refrigerants. All refrigerants are relatively expensive. They will leak through fittings and tubing capable of retaining water at high pressures. A leak results in the loss of expensive refrigerant and in possible product loss, such as of frozen food. For this reason, all refrigerant lines and fittings must be absolutely seepage-proof.

Proper fittings and controls also have a bearing on the efficiency and capacity of a refrigerating machine. The capacity of a condensing unit depends, among other things, upon the suction pressure at which the unit operates. Normally, the higher the suction pressure, the greater the efficiency of the compressor.

Suction pressure at the compressor is governed by the design of the evaporator. The desired temperature in the medium being cooled and the pressure drop in the suction line from the evaporator to the compressor also govern the design pressure. This pressure drop in the suction line can be kept to a minimum by use of ample line sizes, fittings, and accessories designed to eliminate restrictions. Pressure drop is also a factor in liquid lines between the receiver and the metering device. Excessive pressure drop will result in "flashing" or partial vaporization of the liquid refrigerant before it reaches the metering device. The metering device is designed to handle liquid. It will not function properly if fed a mixture of vapor and liquid. Here, valves play an important role in controlling and metering the flow of liquid in the system.

The thermostatic-expansion valve (TEV) uses the fluctuations of the pressure of the saturated refrigerant sealed inside the power element to control the flow of refrigerant through the valve (see Fig. 11-20).

Basically, thermostatic-expansion valve operation is determined by the following three fundamental pressures:

- Bulb pressure on one side of the diaphragm tends to open the valve.

- Evaporator pressure on the opposite side of the diaphragm tends to close the valve.

- Spring pressure is applied to the pin carrier and is transmitted through the push rods to the evaporator side of the diaphragm. This assists in closing the valve.

When the valve is modulating, bulb pressure is balanced by the evaporator pressure and spring pressure. When the same refrigerant is used in the thermostatic element and refrigeration system, each will exert the same pressure if their temperatures are identical. After evaporation of the liquid refrigerant in the evaporator, the suction gas is superheated. Its temperature will increase. However, the evaporator pressure, neglecting

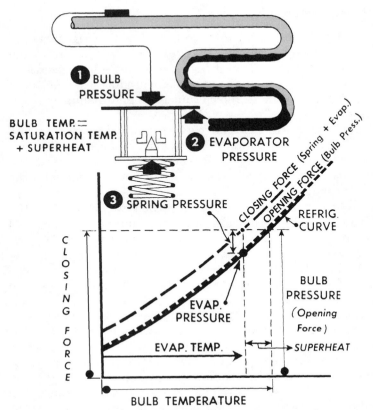

Figure 11-20   Basic thermostatic-expansion valve operation. (*Courtesy of Virginia Chemicals.*)

pressure drop, is unchanged. This warmer vapor flowing through the suction line increases the bulb temperature. Since the bulb contains both vapor and liquid refrigerant, its temperature and pressure increase. This higher bulb pressure acting on the top (bulb side) of the diaphragm is greater than the opposing evaporator pressure and spring pressure, which causes the valve pin to be moved away from the seat. The valve is opened until the spring pressure—combined with the evaporator pressure—is sufficient to balance the bulb pressure (see Fig. 11-20).

If the valve does not feed enough refrigerant, the evaporator pressure drops or the bulb temperature is increased by the warmer vapor leaving the evaporator (or both). The valve then opens. This admits more refrigerant until the three pressures are again in balance. Conversely, if the valve feeds too much refrigerant, the bulb temperature is decreased, or the evaporator pressure increases (or both). The spring pressure tends to close the valve until the three pressures are in balance.

With an increase in evaporator load, the liquid refrigerant evaporates at a faster rate and increases the evaporator pressure. The higher evaporator pressure results in a higher evaporator temperature and a correspondingly higher bulb temperature. The additional evaporator pressure (temperature) acts on the bottom of the diaphragm. The additional bulb pressure (temperature) acts on the top of the diaphragm. Thus. the two pressure increases on the diaphragm cancel each other. The valve easily adjusts to the new load condition with a negligible chance in superheat.

Valve location

Thermostatic-expansion valves may be mounted in any position. However, they should be installed as close to the evaporator inlet as possible. If a refrigerant distributor is used, mount the distributor directly to the valve outlet for best performance. If a hand valve is located on the outlet side of the thermostatic-expansion valve, it should have a full-sized port. No restrictions should appear between the thermostatic-expansion valve and the evaporator, except a refrigerant distributor if one is used.

When the evaporator and thermostatic-expansion valve are located above the receiver, there is a static pressure loss in the liquid line. This is due to the weight of the column of liquid refrigerant. This weight may be interpreted in terms of pressure loss in pounds per square inch (see Table 11-3). If the vertical lift is great enough, vapor, or flash gas, will form in the liquid line. This greatly reduces the capacity of the thermostatic-expansion valve. When an appreciable vertical lift is unavoidable, precautions should be taken to prevent the accompanying pressure loss from producing liquid-line vapor. This can be accomplished by providing enough subcooling to the liquid refrigerant, either in the condenser or after the liquid leaves the receiver. Subcooling is found by subtracting the actual liquid temperature from the condensing temperature (corresponding to the condensing pressure). The amount of subcooling necessary to prevent vapor formation in the liquid line is usually available in a table (see Table 11-4).

TABLE 11-3   Vertical Lift and Pressure Drop (*Courtesy of Sporlan Valve.*)

	Vertical Lift (°F)					Average pressure drop across distributor
	20	40	60	80	100	
Refrigerant	Static pressure loss (psi)					
12	11	22	33	44	55	25 psi
22	10	20	30	40	50	35 psi
500	10	19	29	39	49	25 psi
502	10	21	31	41	52	35 psi
717 (Ammonia)	5	10	15	20	25	40 psi

Sporlan Valve

TABLE 11-4   Pressure Loss and Required Subcooling for 100 and 130°F Condensing of Refrigerations (*Courtesy of Sporlan Valve.*)

Refrigerant	100°F (37.8°C) condensing						130°F (54.4°C) condensing					
	Pressure loss (psi)						Pressure loss (psi)					
	5	10	20	30	40	50	5	10	20	30	40	50
	Required subcooling (zF)						Required subcooling (°F)					
12	3	6	12	18	25	33	3	5	9	14	18	23
22	2	4	8	11	15	19	2	4	6	9	12	14
500	3	5	10	15	21	27	2	4	8	11	15	19
502	2	3	7	10	14	18	1	3	5	8	11	13
717 (Ammonia)	2	4	7	10	14	17	2	3	5	7	10	12

Sporlan Valve

Caution: Ammonia valves should never be permitted to operate with vapor in the liquid line. This causes severe pin and seat erosion. It also will drastically reduce the life of the valve.

Bulb location

The location of the bulb is extremely important. In some cases, it determines the success or failure of the refrigerating plant. For satisfactory expansion-valve control, good *thermal contact* between the bulb and suction line is essential. The bulb should be securely fastened with two bulb straps to a clean, straight section of the suction line.

Application of the bulb to a horizontal run of suction line is preferred. If a vertical installation cannot be avoided, the bulb should be mounted so that the capillary tubing comes out at the top. On suction lines with larger OD, the surface temperature may vary slightly around the circumference of the line. On these lines, it is generally recommended that the bulb be installed at a point midway on the side of the horizontal line and parallel to the direction of flow. On smaller lines the bulb may be mounted at any point around the circumference. However, locating the bulb on the bottom of the line is not recommended, since an oil-refrigerant mixture is generally present at that point. Certain conditions peculiar to a particular system may require a different bulb location than that normally recommended. In these cases, the proper bulb location may be determined by trial. Accepted principles of good suction-line piping should be followed to provide a bulb location that will give the best possible valve-control. Never locate the bulb in a trap or pocket in the suction line. Liquid

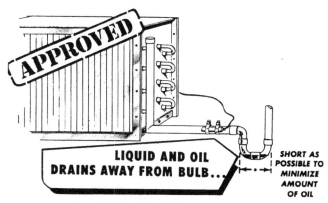

Figure 11-21 Installation of TEV with the compressor above the evaporator. (*Courtesy of Virginia Chemicals.*)

refrigerant or a mixture of liquid refrigerant and oil boiling out of the trap will falsely influence the temperature of the bulb and result in poor valve-control.

Recommended suction-line piping includes a horizontal line leaving the evaporator to which the thermostatic-expansion valve bulb is attached. This line is pitched slightly downward. When a vertical riser follows, a short trap is placed immediately ahead of the vertical line (see Fig. 11-21). The trap will collect any liquid refrigerant or oil passing through the suction line and prevent it from influencing the bulb temperature.

On multiple-evaporator installations the piping should be arranged so that the flow from any valve cannot affect the bulb of another. Approved piping practices, including the proper use of traps, ensure individual control for each valve without the influence of refrigerant and oil flow from other evaporators (see Fig. 11-22).

For recommended suction-line piping when the evaporator is located above the compressor see Fig. 11-23. The vertical riser extending to the height of the evaporator prevents refrigerant from draining by gravity into the compressor during the off-cycle. When a pump-down control is used, the suction line may turn down without a trap.

In commercial and low-temperature applications, the bulb should be the same as the evaporator temperature during the off cycle. This will ensure tight closing of the valve when the compressor stops. If bulb insulation is used on lines operating below 32°F (0°C), use non-water-absorbing insulation to prevent water from freezing around the bulb.

In brine tanks and water coolers the bulb should be below the liquid surface. Here, it will be at the same temperature as the evaporator during the off cycle. A solenoid valve must be used ahead of the thermostatic-expansion valve.

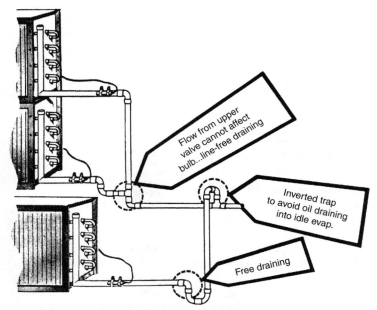

Figure 11-22   Installation of the TEV with multiple evaporators, above and below main suction line. (*Courtesy of Virginia Chemical.*)

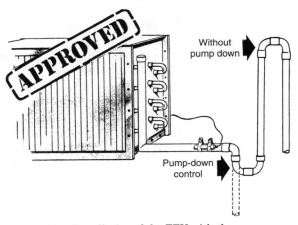

Figure 11-23   Installation of the TEV with the compressor below the evaporator. (*Courtesy of Virginia Chemicals.*)

Some air-conditioning applications have thermostatic-expansion valves equipped with charged elements. Here, the bulb may be located inside or outside the cooled space or duct. The valve body should not be located in the air stream leaving the evaporator. Avoid locating the bulb in the return air stream unless the bulb is well insulated.

External equalizer

As the evaporating temperature drops, the maximum pressure drop that can be tolerated between the valve outlet and the bulb location without serious capacity loss for an internally equalized valve also decreases. This is shown in Table 11-4. There are, of course, applications that may satisfactorily employ the internal equalizer when higher pressure drop is present. This should usually be verified by laboratory tests. The general recommendations given in Table 11-4 are suitable for most field-installed systems. Use the external equalizer when pressure drop between the outlet and bulb locations exceeds values shown in Table 11-4. When the expansion valve is equipped with an external equalizer, it must be connected. Never cap an external equalizer. The valve may flood, starve, or regulate erratically. There is no operational disadvantage in using an external equalizer, even if the evaporator has a low pressure drop. [Note: The external equalizer must be used on evaporators that use a pressure-drop-type refrigerant distributor (see Fig. 11-24).] Generally, the external equalizer connection is in the suction line immediately downstream of the bulb (see Fig. 11-24). However, equipment manufacturers sometimes select other locations that are compatible with their specific design requirements.

Field service

The thermostatic-expansion valve is erroneously considered by some to be a complex device. As a result, many valves are needlessly replaced when the cause of the system malfunction is not immediately recognized.

Actually, the thermostatic-expansion valve performs only one very simple function. It keeps the evaporator supplied with enough refrigerant to satisfy all load conditions. It is not a temperature control, suction pressure control, a control to vary the compressor running time, or a humidity control.

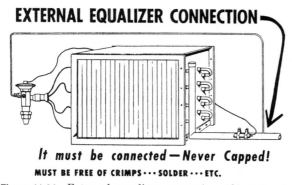

EXTERNAL EQUALIZER CONNECTION

It must be connected — Never Capped!

MUST BE FREE OF CRIMPS ··· SOLDER ··· ETC.

Figure 11-24   External equalizer connection. (*Courtesy of Virginia Chemicals.*)

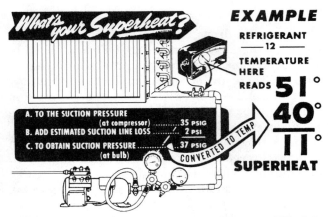

Figure 11-25 How to figure superheat (*Courtesy of Virginia Chemicals.*)

The effectiveness of the valve's performance is easily determined by measuring the superheat (see Fig. 11-25). Observing the frost on the suction line or considering only the suction pressure may be misleading. Checking the superheat is the first step in a simple and systematic analysis of thermostatic-expansion-valve performance.

If insufficient refrigerant is being fed to the evaporator the superheat will be high. If too much refrigerant is being fed to the evaporator the superheat will be low. Although these symptoms may be attributed to improper thermostatic-expansion-valve control. more frequently the origin of the trouble lies elsewhere.

Crankcase Pressure-Regulating Valves

Crankcase pressure-regulating valves are designed to prevent overloading of the compressor motor. They limit the crankcase pressure during and after a defrost cycle or after a normal shutdown period. When properly installed in the suction line, these valves automatically throttle the vapor flow from the evaporator until the compressor can handle the load. They are available in the range of 0 to 60 psig.

Operation of the valve

Crankcase pressure-regulating valves (CROs) are sometimes called suction pressure-regulating valves. They are sensitive only to their outlet pressure. This would be the compressor crankcase or suction pressure. To indicate this trait, the designation describes the operation as close on rise of outlet pressure (CRO). As shown in Fig. 11-26, the inlet pressure is exerted on the underside of the bellows and on top of the seat disc. Since the effective area of the bellows is equal to the area of the port, the inlet

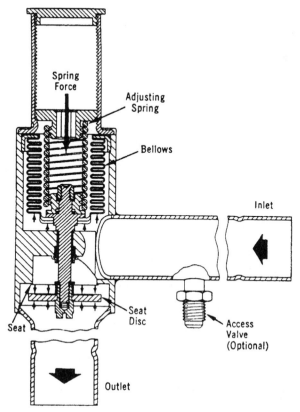

Figure 11-26   Crankcase pressure-regulating valve. (*Courtesy of Sporlan Valve.*)

pressure cancels out and does not affect valve operation. The valve outlet pressure acting on the bottom of the disc exerts a force in the closing direction. This force is opposed by the adjustable spring force. These are the operating forces of the CRO. The CROs pressure setting is determined by the spring force. Thus, by increasing the spring force, the valve setting or the pressure at which the valve will close is increased.

As long as the valve outlet pressure is greater than the valve pressure setting, the valve will remain closed. As the outlet pressure is reduced, the valve will open and pass refrigerant vapor into the compressor. Further reduction of the outlet pressure will allow the valve to open to its rated position, where the rated pressure drop will exist across the valve port. An increase in the outlet pressure will cause the valve to throttle until the pressure setting is reached.

The operation of a valve of this type is improved by an antichatter device built into the valve. Without this device, the CRO would be susceptible to

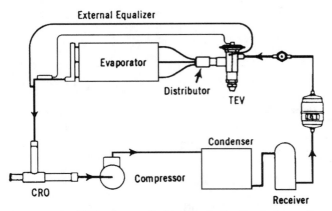

Figure 11-27 CRO valve applied in the suction line between the evaporator and the compressor. (*Courtesy of Sporlan Valve.*)

compressor pulsations that greatly reduce the life of a bellows. This feature allows the CRO to function at low-load conditions without any chattering or other operational difficulties.

Valve location

As Fig. 11-27 indicates, the CRO valve is applied in the suction line between the evaporator and the compressor. Normally, the CRO is installed downstream of any other controls or accessories. However, on some applications it may be advisable or necessary to locate other system components, such as an accumulator, downstream of the CRO. This is satisfactory as long as the CRO valve is applied only as a CRO valve. CRO valves are designed for application in the suction line only. They should not be applied in hot-gas bypass lines or any other refrigerant line of a system.

Strainer

Just as with any refrigerant flow-control device, the need for an inlet strainer is a function of system cleanliness and proper installation procedures (see Fig. 11-28). When the strainer is used, the tubing is inserted in the valve connection up to the tubing stop. Thus, the strainer has been

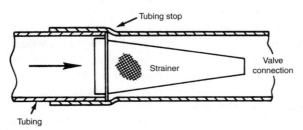

Figure 11-28 Strainer for cleanliness. (*Courtesy of Sporlan Valve.*)

locked in place. Moisture and particles too small for the inlet strainer are harmful to the system and must be removed. Therefore, it is recommended that a filter-drier be installed according to the application recommendations.

Brazing procedures

When installing CROs with solder connections, the internal parts must be protected by wrapping the valve with a *wet* cloth to keep the body temperature below 250°F (121°C). The tip of the torch should be large enough to avoid prolonged heating of the connections. Overheating can also be minimized by directing the flame away from the valve body.

Test and operating pressures

Excessive leak testing or operating pressures may damage these valves by reducing the life of the bellows. For leak detection, an inert gas such as nitrogen or carbon dioxide may be added to an idle system to supplement the refrigerant pressure.

Caution: Inert gas must be added to the system carefully. Use a pressure regulator. Unregulated gas pressure can seriously damage the system and endanger human life. Never use oxygen or explosive gases. The values will withstand 200 to 300 psig. However, check the manufacturer's recommendations first.

Adjusting the pressure

The standard setting by the factory for CROs in the 0/60 psig range is 30 psig. Since these valves are adjustable, the setting may be altered to suit the specific system requirements. CROs should be adjusted at start-up when the pressure in the evaporator is above the desired setting. The final valve setting should be below the maximum suction pressure recommended by the compressor or unit manufacturer.

The main purpose of the CRO is to prevent the compressor motor from overloading due to high suction pressure. Thus, it is important to arrive at the correct pressure setting. The best way to see if the motor is overloaded is to check the current draw at start-up or after a defrost cycle. If overloading is evident, a suction gage should be put on the compressor. The CRO setting may be too high and may have to be adjusted. If the compressor is overloaded and the CRO valve is to be reset, the following procedure should be followed.

The unit should be shut off long enough for the system pressure to equalize. Observe the suction pressure as the unit is started, since this is the pressure the valve is controlling. If the setting is to be decreased, slowly adjust the valve in a counterclockwise direction approximately one-quarter turn for each 1 psi pressure change required. After a few moments of operation, the unit should be cycled off and the system pressure allowed to equalize again. Observe the suction pressure (valve setting) as the unit is started up. If the setting is still too high, the adjustment should be repeated. The proper size hex wrench is used to adjust these valves. A clockwise rotation increases the valve setting, while a counterclockwise rotation decreases the setting.

When CROs are installed in parallel, each should be adjusted the same amount. If one valve has been adjusted more than the other, best performance will occur if both are adjusted all the way in before resetting them an equal amount.

Service

Since CRO valves are hermetic and cannot be disassembled for inspection and cleaning, they are usually replaced if inoperative. If a CRO fails to open, close properly, or will not adjust, solder or other foreign material is probably lodged in the port. It is sometimes possible to dislodge these materials by turning the adjustment nut all the way in, with the system running. If the CRO develops a refrigerant leak around the spring housing, it probably has been overheated during installation or the bellows has failed due to severe compressor pulsations. In either case, the valve must be replaced.

Evaporator Pressure-Regulating Valves

Evaporator pressure-regulating valves offer an efficient means of balancing the system capacity and the load requirements during periods of low loads. They are also able to maintain different evaporator conditions on multitemperature systems. The main function of this valve is to prevent the evaporator pressure from falling below a predetermined value at which the valve has been set.

Control of evaporator pressure by cycling the compressor with a thermostat or some other method is quite adequate on most refrigeration systems. Control of the evaporator pressure also controls the saturation temperature. As the load drops off, the evaporating pressure starts to decrease and the system performance falls off. These valves automatically throttle the vapor flow from the evaporator. This maintains the desired minimum evaporator pressure. As the load increases, the evaporating pressure will increase above the valve setting and the valve will open further.

Operation

For any pressure sensitive valve to modulate to a more closed or open position, a change in operating pressure is required. The unit change

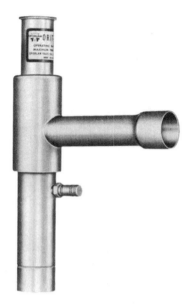

Figure 11-29   Evaporator pressure-regulating valve. (*Courtesy of Sporlan Valve.*)

in the valve stroke for a given change in the operating pressure is called the valve gradient. Every valve has a specific gradient designed into it for the best possible operation. Valve sensitivity and the valve's capacity rating are functions of the valve gradient. Thus, a relatively sensitive valve is needed when a great change in the evaporating temperature cannot be tolerated. Therefore, the valves have nominal ratings based on the 8-psi evaporator-pressure change, rather than a full stroke.

Evaporator pressure-regulator valves respond only to variations in their inlet pressure (evaporator pressure). Thus, the designation for evaporator pressure-regulating valves is ORI (opens on the rise of the inlet pressure) (see Fig. 11-29).

Pressure at the outlet is exerted on the underside of the bellows and on top of the seat disc. The effective area of the bellows is equal to the area of the port. Thus, the outlet pressure cancels out and the inlet pressure acting on the bottom of the seat disc opposes the adjustable spring force. These two forces are the operating forces of the ORIT. (The "T" added to the valve designation indicates an access valve on the inlet connection.) When the evaporator load changes, the ORIT opens or closes in response to the change in evaporator pressure. An increase in inlet pressure above the valve setting tends to open the valves. If the load drops, less refrigerant is boiled off in the evaporator and evaporator pressure will decrease. The decrease in evaporator pressure tends to move the ORIT to a more closed position. This, in turn, keeps the evaporator pressure up. The result is that the evaporator pressure changes as the load changes. The operation of a valve of this type is improved by an

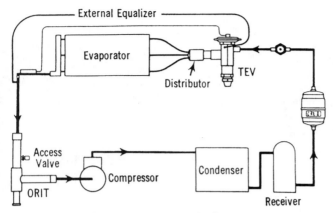

Figure 11-30 Valve location in a single-evaporator system. (*Courtesy of Sporlan Valve.*)

antichatter device built into the valve. Without this device, the OBIT would be susceptible to compressor pulsations that can reduce the life of a bellows. This antichatter feature allows the ORIT to function at low-load conditions without chattering or other operating difficulties.

Type of system

The proper application of the evaporator pressure-regulating valve involves the consideration of several system factors.

One type of system is a single evaporator type, such as a water chiller. Here, the valve is used to prevent freeze-up at light loads (see Fig. 11-30).

Another type of system is a multitemperature-refrigeration system with evaporators operating at different temperatures (see Fig. 11-31). A valve may be required on one or more of the evaporators to maintain pressures higher than that of the common suction line. For example, if evaporator A in Fig. 11-31 is designed for 35°F (1.7°C) (72.6 psig on Refrigerant 502), evaporator B for 32°F (0°C) (68.2 psig on the same

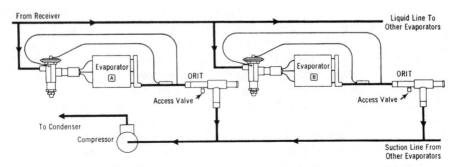

Figure 11-31 Multitemperature refrigeration system. (*Courtesy of Sporlan Valve.*)

refrigerant), and other evaporators for 25°F (–3.9°C) (58.7 psig), the valves (ORIT) are used to maintain a pressure of 72.6 psig in evaporator A and 68.2 psig in evaporator B. However, some multitemperature systems may require an OBIT on each evaporator, depending on the type of product being refrigerated.

Valve location

ORITs must be installed upstream of any other suction-line controls or accessories. These valves may be installed in the position most suited to the application. However, these valves should be located so that they do not act as an oil trap or so that solder cannot run into the internal parts during brazing in the suction line. Since these valves are hermetic, they cannot be disassembled to remove solder trapped in the internal parts. Installation of a filter-drier and a strainer may be worth the expense to keep the system clean and operational. Brazing procedures are the same as for other valves of this type. The valve core of the access valve is shipped in an envelope attached to the access valve. If the access valve connection is to be used as a reusable pressure tap to check the valve setting, the OBIT must be brazed in before the core is installed. This protects the synthetic material of the core. If the access valve is to be used as a permanent pressure tap, the core and access valve cap may be discarded.

Test and operating pressures

As with other pressure valves, it is possible to introduce nitrogen or carbon dioxide in an idle system to check for correct pressure settings.

The usual precautions for working with gases apply here. The standard factory setting for the 0/50 psig range is 30 psi. For the 30/100 psig range, it is 60 psig. Since these valves are adjustable, the setting may be altered to suit the system.

The main purpose of an OBIT valve is to keep the evaporator pressure above some given point at minimum-load conditions. The valves are selected on the basis of the pressure drop at full-load conditions. Nevertheless, they should be adjusted to maintain the minimum allowable evaporator pressure under the actual minimum-load conditions.

These valves can be adjusted by removing the cap and turning the adjustment screw with a hex wrench of the proper size. A clockwise rotation increases the valve setting, while a counterclockwise rotation decreases the setting. To obtain the desired setting, a pressure gage should be utilized on the inlet side of the valve. Thus. the effects of any adjustments can be observed.

When these valves are installed in parallel, each should be adjusted the same amount. If one valve has been adjusted more than the other, the best performance will occur if both are adjusted all the way in before resetting them to an equal amount.

Service

These valves are hermetic and cannot be disassembled for inspection and cleaning. They usually must be replaced if found defective or inoperative. It is possible sometimes to adjust the valve until the obstruction is dislodged. This usually works best when the system is running. If it leaks around the spring housing, it will have to be replaced. The bellows have been permanently damaged.

Head-Pressure Control Valves

Design of air-conditioning and refrigeration systems using air-cooled condensing units involves two main problems that must be solved if the system is to be operated reliably and economically. These problems are high-ambient and low-ambient operation. If the condensing unit is properly sized, it will operate satisfactorily during extreme-ambient temperatures. However, most units will be required to operate at ambient temperatures below their design dry-bulb temperature during most of the year. Thus, the solution to low-ambient operation is more complex.

Without good head-pressure control during low-ambient operation, the system can have running-cycle and off-cycle problems. Two running-cycle problems are of prime concern:

- The pressure differential across the thermostatic-expansion valve port affects the rate of refrigerant flow. Thus, low head pressure generally causes insufficient refrigerant to be fed to the evaporator.

- Any system using hot gas for defrost or compressor-capacity control must have a normal head pressure to operate properly. In either case, failure to have sufficient head pressure will result in low suction pressure and/or iced evaporator coils.

The primary off-cycle problem is the possible inability to get the system on-the-line if the refrigerant has migrated to the condenser. The evaporator pressure may not build up to the cut-in point of the low pressure control. The compressor cannot start, even though refrigeration is required. Even if the evaporator pressure builds up to the cut-in setting, insufficient flow through the TEV will cause a low suction pressure, which results in compressor cycling.

There are nonadjustable and adjustable methods of head-pressure control by valves. Each method uses two valves designed specifically for this type of application. Low-ambient conditions are encountered during fall-winter-spring operation on air-cooled systems, with the resultant drop in condensing pressure. Then, the valve's purpose is to hold back enough of the condensed liquid refrigerant to make part of the condenser surface inactive. This reduction of active condensing surface raises condensing pressure and sufficient liquid-line pressure for normal system operation.

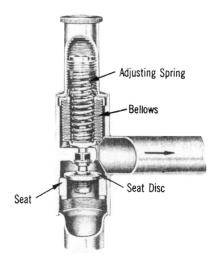

Seat

Seat Disc

Figure 11-32   Head-pressure control valve. (*Courtesy of Sporlan Valve.*)

Operation

The ORI head pressure-control valve is an inlet pressure regulating valve. It responds to changes in condensing pressure only. The valve designation stands for opens on rise of inlet pressure. As shown in Fig. 11-32, the outlet pressure is exerted on the underside of the bellows and on top of the seat disc. Since the effective area of the bellows is equal to the area of the port, the outlet pressure cancels out. The inlet pressure acting on the bottom of the seat disc opposes the adjusting spring force. These two forces are the operating forces of the ORI.

When the outdoor ambient temperature changes. the ORI opens or closes in response to the change in condensing pressure. An increase in inlet pressure above the valve setting tends to open the valve. If the ambient temperature drops, the condenser capacity is increased and the condensing pressure drops off. This causes the ORI to start to close or assume a throttling position.

ORO valve operation

The ORO head pressure-control valve is an outlet pressure–regulating valve that responds to changes in receiver pressure. The valve designation stands for opens on rise of outlet pressure (see Fig. 11-33). The inlet and outlet pressures are exerted on the underside of the seat disc in an opening direction. Since the area of the port is small in relationship to the diaphragm area, the inlet pressure has little direct effect on the operation of the valve. The outlet or receiver pressure is the control pressure. The force on top of the diaphragm that opposes the control pressure is due to the air charge in the element. These two forces are the operating forces of the ORO.

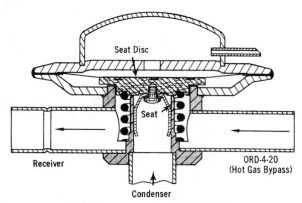

Figure 11-33 Head-pressure control valve that opens on rise of outlet pressure (ORO). (*Courtesy of Sporlan Valve.*)

When the outdoor ambient temperature changes, the condensing pressure changes. This causes the receiver pressure to fluctuate accordingly. As the receiver pressure decreases, the ORO throttles the flow of liquid from the condenser. As the receiver pressure increases, the valve modulates in an opening direction to maintain a nearly constant pressure in the receiver. Since the ambient temperature of the element affects the valve pressure setting, the control pressure may change slightly when the ambient temperature changes. However, the valve and element temperature remain fairly constant.

ORD valve operation

The ORD valve is a pressure *differential* valve. It responds to changes in the pressure difference across the valve (see Fig. 11-34). The valve designation stands for opens on rise of differential pressure. Therefore, the ORD is dependent on some other control valve or action for its operation. In this respect, it is used with either the ORI or ORO for head-pressure control.

As either the ORI or ORO valve starts to throttle the flow of liquid refrigerant from the condenser, a pressure differential is created across the ORD. When the differential reaches 20 psi, the ORD starts to open

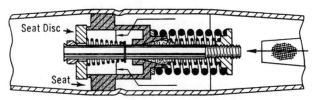

Figure 11-34 Head-pressure control valve that opens on rise of differential across the valve (ORD). (*Courtesy of Sporlan Valve.*)

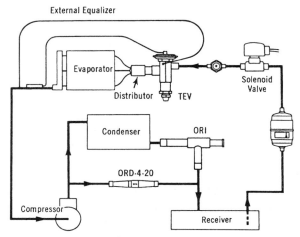

Figure 11-35 Adjustable ORI/ORD system. (*Courtesy of Sporlan Valve.*)

and bypasses hot gas to the liquid drain-line. As the differential increases, the ORD opens further until its full stroke is reached at a differential of 30 psi. Due to its function in the control of head pressure, the full stroke can be utilized in selecting the ORD. While the capacity of the ORD increases as the pressure differential increases, the rating point at 30 psi is considered a satisfactory maximum value.

The standard pressure setting for the ORD is 20 psig. For systems where the condenser pressure drop is higher than 10 or 12 psi, an ORD with a higher setting can be ordered.

Head-pressure control can be improved with an arrangement such as that shown in Fig. 11-35. In this operation, a constant receiver pressure is maintained for normal system operation. The ORI is adjustable over a nominal range of 100 to 225 psig. Thus, the desired pressure can be maintained for all of the commonly used refrigerants—12, 22, and 502 as a well as the latest alternatives.

The ORI is located in the liquid-drain line between the condenser and the receiver. The ORD is located in a hot-gas line bypassing the condenser. During periods of low ambient temperature, the condensing pressure falls until it approaches the setting of the ORI valve. The ORI then throttles, restricting the flow of liquid from the condenser. This causes refrigerant to back up in the condenser, thus reducing the active condenser surface. This raises the condensing pressure. Since it is really receiver pressure that needs to be maintained, the bypass line with the ORD is required.

The ORD opens after the ORI has offered enough restriction to cause the differential between condensing pressure and receiver pressure to exceed 20 psi. The hot gas flowing through the ORD heats up the cold liquid being passed through the ORI. Thus, the liquid reaches the receiver warm and with sufficient pressure to assure proper expansion-valve operation. As long as sufficient refrigerant charge is in the system, the two valves modulate the flow automatically to maintain proper receiver pressure regardless of outside ambient temperature.

Installation

To ensure proper performance, head pressure-control valves must be selected and applied correctly. These valves can be installed in either horizontal or vertical lines. if possible, the valves should be oriented so solder cannot run into the internal parts during brazing. Care should be taken to install the valves with the flow in the proper direction. The ORI and ORO valves *cannot* be installed in the discharge line for any reason.

In most cases the valves are located at the condensing unit. When the condenser is remote from the compressor, the usual location is near the compressor. In all cases it is important that some precautions be taken in mounting the valves. While the heaviest valve is approximately 2.5 lb (1.14 kg) in weight, it is suggested that they be adequately supported to prevent excessive stress on the connections. Since discharge lines are a possible source of vibrations that result from discharge-gas pulses and inertia forces associated with the moving parts, fatigue in tubing, fittings, and connections may result. Pulsations are best handled by placing a good muffler as close to the compressor as possible.

Vibrations from moving parts of the compressor are best isolated by flexible loops or coils (discharge lines or smaller) or flexible metal hoses for larger lines. For best results, the hoses should be installed as close to the compressor shut-off valves as possible. The hoses should be mounted horizontally and parallel to the crankshaft or vertically. The hoses should *never* be mounted horizontally and 90° from the crankshaft. A rigid brace should be placed on the outlet end of the hose. This brace will prevent vibrations beyond the hose.

Brazing procedures

Any of the commonly used brazing alloys for high-side usage are satisfactory. It is very important that the internal parts be protected by wrapping the valve with a wet cloth to keep the body temperature below 250°F (121°C). Also, when using high-temperature solders, the torch tip should be large enough to avoid prolonged heating of the copper connections. Always direct the flame away from the valve body.

Test and operating pressures

Excessive leak testing or operating pressures may damage these valves and reduce the life of the operating members. For leak detection, an inert dry gas such as nitrogen or carbon dioxide can be added to an idle system to supplement the refrigerant pressure. Remove the cap and adjust the adjustment screw with the proper wrench. Check the manufacturer's recommended pressures before making adjustments.

Refrigerant and charging procedures require that enough refrigerant be available for flooding the condenser at the lowest expected ambient temperature. There must still be enough charge in the system for proper operation. A shortage of refrigerant will cause hot gas to enter the liquid line and the expansion valve. Refrigeration will cease.

The receiver must have sufficient capacity to hold at least all of the excess liquid refrigerant in the system. This is because such refrigerant will be returned to the receiver when high-ambient conditions prevail. If the receiver is too small, liquid refrigerant will be held back in the condenser during high-ambient condition. Excessively high discharge pressures will be experienced.

> Caution: All receivers must utilize a pressure relief value or device according to the applicable standards or codes.

Follow the manufacturer's recommendations for charging the system. Procedures may vary with different valve manufacturers.

Service

There are several possible causes for system malfunction with "refrigerant side" head-pressure control. These may be difficult to isolate from each other. As with any form of system troubleshooting, it is necessary to know the existing operating temperatures and pressures before system problems can be determined. Once the malfunction is established, it is easier to pinpoint the cause and then take suitable action. Table 11-5 lists the most common malfunctions, the possible causes, and the remedies.

Nonadjustable ORO/ORD system operation

The nonadjustable ORO head pressure-control valve and the ORD pressure-differential valve offer the most economical system of refrigerant side head-pressure control. Just as the ORI/ORD system simplified this type of control, the ORO/ORD system offers the capability of locating the condenser and receiver on the same elevation (see Fig. 11-36). By making these two valves available either separately or brazed together, there is added flexibility in the piping layout. The operation of the ORO/ORD system is such that a nearly constant receiver pressure is maintained for normal operation. As the temperature of the ORO element decreases, the pressure setting decreases accordingly. However, by running the bypassed hot gas through the ORO the element temperature is

TABLE 11-5   Troubleshooting Head-Pressure Control Valves (*Courtesy of Sporlan Valves.*)

Malfunction—low head pressure

Possible cause	Remedy
1. Insufficient refrigerant charge to adequately flood condenser.	1. Add charge.
2. Low-pressure setting on ORI.	2. Increase setting.
3. ORI fails to close due to foreign material in valve.	3. Turn adjustment out so material passes through valve. If unsuccessful, replace ORI.
4. ORI fails to adjust properly	4. See 3 above.
5. Wrong setting on ORO (e.g., 100 psig on Refrigerant 22 or 502 system).	5. Replace ORO with valve with correct setting.
6. ORO fails to close due to: a. Foreign material in valve.	6. See below: a. Cause ORO to open by raising condensing/receiver pressure above valve setting by cycling condenser fan. If foreign material does not pass through valve, replace ORO.
b. Loss of air charge in element.	b. Replace ORO.
7. ORD fails to open (on ORI/ORD system only) due to: a. Less than 20 psi pressure drop across ORD. b. Internal parts damaged by overheating when installed.	7. See below: a. Check ORI causes/remedies above: 2, 3, or 4. b. Replace ORD.
8. Refrigerant leak at adjustment housing of ORI.	8. Replace ORI.

Malfunction—high head pressure

1. Dirty condenser coil.	1. Clean coil.
2. Air on condenser blocked off.	2. Clear area around unit.
3. Too much refrigerant charge.	3. Remove change until proper head pressure is maintained.
4. Undersized receiver.	4. Check receiver capacity against refrigerant required to maintain desired head pressure.
5. Noncondensibles (air) in system.	5. Purge from system.
6. High-pressure setting on ORI.	6. Decrease setting.
7. ORI or ORO restricted due to inlet strainer being plugged.	7. Open inlet connection to clean strainer.
8. ORI fails to adjust properly or to open due to foreign material in valve.	8. Turn adjustment out so material passes through valve. If unsuccessful, replace ORI.
9. Wrong setting on ORO (e.g., 180 psig on Refrigerant 12 system).	9. Replace ORO with valve with correct setting.
10. ORD fails to open due to internal parts being damaged by overheating when installed (only when used with ORO).	10. Replace ORD.
11. ORD bypassing hot gas when not required due to: a. Internal parts damaged by overheating when installed. b. Pressure drop across condenser coil, ORI or ORO, and connecting piping above 14 psi.	11. See below: a. Replace ORD. b. Reduce pressure drop (e.g., use larger ORI or ORO valves in parallel) or order ORD-4 with higher setting.

Sporlan Valve

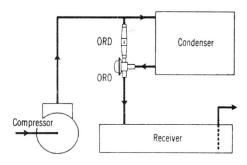

Figure 11-36 The ORO is located in the liquid drain line between the condenser and the receiver. (*Courtesy of Sporlan Valve.*)

adequately maintained so the ORO/ORD system functions well to ambient temperatures of −40°F (−40°C) and below. This third connection on the ORO also eliminates the need for a tee connection in the liquid-drain line.

Note that in Fig. 11-36 the ORO is located in the liquid-drain line between the condenser and the receiver, while the ORD is located in a hot-gas line bypassing the condenser. Other than the fact that the ORO operates in response to its outlet pressure (receiver pressure), the ORO/ORD operates in the same basic manner as the ORI/ORD system previously explained.

Discharge-Bypass Valves

On many air-conditioning and refrigeration systems it is desirable to limit the minimum evaporating pressure. This is so especially during periods of low load either to prevent coil icing or to avoid operating the compressor at lower suction pressure than it was designed for. Various methods of operation have been designed to achieve the result—integral cylinder unloading, gas engines with variable speed control, or multiple smaller systems. Compressor cylinder unloading is used extensively on larger systems. However, it is too costly on small equipment, usually 10 hp or below. Cycling the compressor with a low pressure-cutout control has had widespread usage, but is being reevaluated for three reasons:

- On-off control on air-conditioning systems is uncomfortable and does a poor job of humidity control.

- Compressor cycling reduces equipment life.

- In most cases, compressor cycling is uneconomical because of peak load demand charges.

One solution to the problem is to bypass a portion of the hot discharge gas directly into the low side. This is done by the modulating-control valve—commonly called a discharge-bypass valve (DBV). This valve, which opens on a decrease in suction pressure, can be set to maintain automatically a desired minimum evaporating pressure, regardless of the decrease in evaporator load.

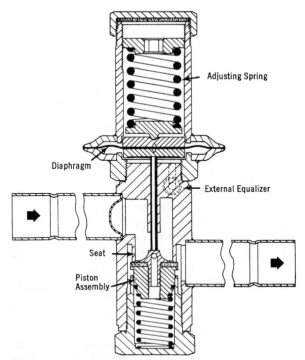

Figure 11-37   Discharge-bypass valve. (*Courtesy of Sporlan Valve.*)

Operation

Discharge-bypass valves (DBV) respond to changes in downstream or suction pressure (see Fig. 11-37). When the evaporating pressure is above the valve setting, the valve remains closed. As the suction pressure drops below the valve setting, the valve responds and begins to open. As with all modulating-type valves, the size of the opening is proportional to the change in the variable being controlled. In this case, the variable is the suction pressure. As the suction pressure drops, the valve opens further until the limit of the valve stroke is reached. However, on normal applications there is not sufficient pressure change to open these valves to the limit of their stroke. The amount of pressure change available from the point at which it is desired to have the valve closed to the point at which it is to be open varies widely with the refrigerant used and the evaporating temperature. For this reason, DBVs are rated on the basis of allowable evaporator temperature change from closed position to rated opening. A 6°F (3.3°C) change is considered normal for most applications and is the basis of capacity ratings.

Application

DBVs provide an economical method of compressor capacity control in place of cylinder unloaders or of handling unloading requirements below the last step of cylinder unloading.

On air-conditioning systems, the minimum allowable evaporating temperature that will avoid coil icing depends on evaporator design. The amount of air passing over the coil also determines the allowable evaporator minimum temperature. The refrigerant temperature may be below 32°F (0°C). However, coil icing will not usually occur with high air velocities, since the external-surface temperature of the tube will be above 32°F (0°C). For most air-conditioning systems the minimum evaporating temperature should be 26 to 28°F (−3.3 to −2.2°C). DBVs are set in the factory. They start to open at an evaporating pressure equivalent to 32°F (0°C) saturation temperature. Therefore, evaporating temperature of 26°F (−3.3°C) is their rated capacity. However, since they are adjustable, these valves can be set to open at a higher evaporating temperature.

On refrigeration systems, discharge-bypass valves are used to prevent the suction pressure from going below the minimum value recommended by the compressor manufacturer. A typical application would be a low-temperature compressor designed for operation at a minimum evaporating temperature on Refrigerant 22 of −40°F (−40°C). The required evaporating temperature at normal load conditions is −30°F (−34°C). A discharge-bypass valve would be selected that would start to open at the pressure equivalent to −34°F (−36°C) and bypass enough hot gas at −40°F (−40°C) to prevent a further decrease in suction pressure. Valve settings are according to manufacturer's recommendations.

The discharge-bypass valve is applied in a branch line off the discharge line as close to the compressor as possible. The bypassed vapor can enter the low side at one of the following locations:

- To evaporator inlet with distributor
- To evaporator inlet without distributor
- To suction line

Figure 11-38 shows the bypass to evaporator inlet with a distributor. The primary advantage of this method is that the system thermostatic-expansion valve will respond to the increased superheat of the vapor leaving the evaporator and will provide the liquid required for de-superheating. The evaporator also serves as an excellent mixing chamber for the bypassed hot gas and the liquid-vapor mixture from the expansion valve. This ensures that dry vapor reaches the compressor suction. Oil return from the evaporator is also improved, since the velocity in the evaporator is kept high by the hot gas.

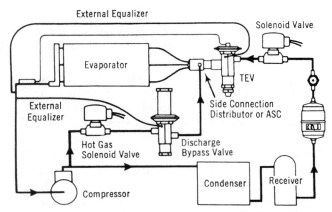

Figure 11-38   Connection arrangement for a discharge-bypass valve. (*Courtesy of Sporlan Valve.*)

Externally Equalized Bypass Valves

The primary function of the DBV is to maintain suction pressure. Thus, the compressor suction pressure is the control pressure. It must be exerted on the underside of the valve diaphragm. When the DBV is applied as shown in Fig. 11-38, where there is an appreciable pressure drop between the valve outlet and the compressor suction, the externally equalized valve must be used. This is true because when the valve opens, a sudden rise in pressure occurs at the valve outlet. This creates a false control pressure, which would cause the internally equalized valve to close.

Many refrigeration systems and water chillers do not use refrigerant distributors but may require some method of compressor capacity control. This type of application provides the advantages discussed earlier.

Bypass to Evaporator Inlet
without Distributor

On many applications, it may be necessary to bypass directly into the suction line. This is generally true of systems with multievaporators or remote-condensing units. It may also be true for existing systems where it is easier to connect the suction line than the evaporator inlet. The latter situation involves systems fed by TEVs or capillary tubes. When hot gas is bypassed, temperature starts to increase. This can cause breakdown of the oil and refrigerant, possibly resulting in a compressor burnout. On close-coupled systems, this can be eliminated by locating the main expansion-valve bulb downstream of the bypass connection, as shown in Fig. 11-39.

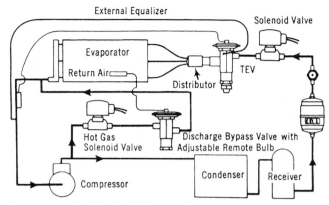

Figure 11-39   Application of a hot-gas bypass to an existing system with only minor piping changes. (*Courtesy of Sporlan Valve.*)

Installation

Bypass valves can be installed in horizontal or vertical lines, whichever best suits the application and permits easy accessibility to the valves. However, consideration should be given to locating these valves so that they do not act as oil traps. Also solder must not run into the internal parts during brazing.

The DBV should always be installed at the condensing unit rather than at the evaporator section. This will ensure the rated bypass capacity of the DBV. It will also eliminate the possibility of hot gas condensing in the bypass line. This is especially true on remote systems.

When externally equalized lines are used, the equalizer connection must be connected to the suction line where it will sense the desired operating pressure. See Fig. 11-40

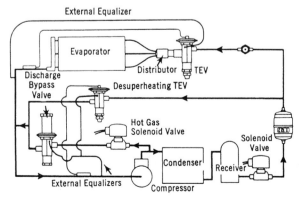

Figure 11-40   Externally equalized discharge-bypass valve. (*Courtesy of Sporlan Valve.*)

Since the DBV is applied in a bypass line between the discharge line and the low side of a system, the valve is subjected to compressor vibrations. Unless the valve, connecting fittings, and tubing are properly isolated from the vibrations, fatigue failures may occur. While the heaviest valve weights only 3.5 lb (1.6 kg), it should be adequately supported to prevent excessive stress on the connections..

If the remote-bulb type bypass valve is used, the bulb must be located in a fairly constant ambient temperature because the element-bulb assembly is air charged. These valves are set at the factory in an 80°F (27°C) ambient temperature. Thus, any appreciable variation from this temperature will cause the pressure setting to vary from the factory setting. For a nonadjustable valve, the remote bulb may be located in an ambient 80°F ±10°F (27°C ± 5.5°C) The adjustable remote bulb model can be adjusted to operate in a temperature of 80°F ± 30°F (27°C ± 16.7°C). On many units the manufacturer will have altered the pressure setting to compensate for an ambient temperature appreciably different than 80°F (27°C). Therefore on some units it may be necessary to consult with the equipment manufacturer for the proper opening pressure setting of the bypass valve

There are numerous places on a system where the remote bulb can be located. Two possible locations are the return air stream and a structural member of the unit, if it is located in a conditioned space. Other locations, where the temperature is fairly constant but different than 80°F (27°C) are also available. These include the return water line on a chiller, the compressor suction line, or the main liquid line. As previously mentioned, the setting may have been altered.

A bulb strap with bolts and nuts is usually supplied with each remote-bulb-type DBV. This strap is for the use in fastening the bulb in place.

Special Considerations

If a DBV is applied on a system with an evaporator pressure regulating valve (ORIT or other type), the DBV may bypass into either the evaporator inlet or the suction line. The bypass will depend on the specific system. Valve function and the best piping method to protect the compressor should be deciding factors.

If the DBV is required on a system with a crankcase pressure-regulating valve (CRO or other type), the bypass valve can bypass to the low side of the evaporator inlet or the suction line without difficulties. The only decision necessary is whether an internally or externally equalized valve is required. This depends on where the hot gas enters the low side. The pressure setting of the DBV must be lower than the CRO setting for each valve to function properly.

The hot gas solenoid valve is to be located upstream of the bypass valve. If the solenoid valve is installed downstream of the DBV, the oil and/or liquid refrigerant may be trapped between the two valves. Depending on the ambient temperature surrounding the valves and piping, this could be dangerous.

The hot-gas solenoid is sometimes used for protection against high superheat conditions because the compressor does not have an integral temperature protection device. If this is done, the solenoid valve is wired in series with a bimetal thermostat fastened to the discharge line close to the compressor.

Testing and Operating Pressures

Excessive leak testing or operating pressures may damage these valves and reduce the life of the operating members. Since a high-side test pressure differential of approximately 350 psi or higher will force the DBV open, the maximum allowable test pressure for DBV are the same as for the high and low side of the system. If greater high-side test pressures than those given in the manufacturer's specifications are to be encountered, some method of isolating the DBV from these high pressure must be found.

Valve settings and adjustment must be done according to the manufacturer's recommendations. Proper instrumentation must be used to determine exactly when these valves are open.

Hot Gas

Hot gas may be required for other systems functions besides bypass capacity control. Hot gas may be needed for defrost and head pressure control. Normally, these functions will not interfere with each other. However, compressor cycling on low suction pressure may be experienced on system start-up when the discharge-bypass valve is operating and other functions require the hot gas also. For example, the head-pressure control requires hot gas to pressurize the receiver and liquid-line to get the thermostatic-expansion valve operating properly. In this case, the discharge-bypass valve should be prevented from functioning by keeping the hot-gas solenoid valve closed until adequate liquid line or suction pressure is obtained.

Malfunctions

There are several reasons for system malfunctions. Possible causes of trouble when hot-gas bypass for capacity control is used are listed in Table 11-6.

Valves are coded by the manufacturer. The part numbers given in Table 11-6 are those of the Sporlan Valve Company. Note that each letter and number has a meaning. The coded part numbers in Fig. 11-41 are given as examples. Similar codes are used by other valve manufacturers. To be

TABLE 11-6 Troubleshooting Discharge-Bypass Valves (*Courtesy of Sporlan Valve.*)

Fully adjustable models—ADR type

Valve type*	Malfunction	Cause	Remedy
ADRS-2 ADRSE-2 ADRP-3 ADRPE-3	Failure to open	1. Dirt or foreign material in valve.	1. Disassemble valve and clean.
	Failure to close	1. Dirt or foreign material in valve. 2. Diaphragm failure. 3. Equalizer passageway plugged. 4. External equalizer not connected or equalizer line pinched shut.	1. Disassemble valve and clean. 2. Replace element only. 3. Disassemble valve and clean. 4. Connect or replace equalizer line.
ADRH-6 ADRHE-6	Failure to open	1. Dirt or foreign material in valve. 2. Equalizer passageway plugged. 3. External equalizer not connected or equalizer line pinched shut.	1. Disassemble valve and clean. 2. Disassemble valve and clean. 3. Connect or replace equalizer line.
	Failure to close	1. Dirt or foreign material in valve. 2. Diaphragm failure.	1. Disassemble valve and clean. 2. Replace element only.

"Limited" adjustable models—DR–AR type

Valve type*	Malfunction	Cause	Remedy
DRP-3-AR DRPE-3-AR	Failure to open	1. Dirt or foreign material in valve. 2. Diaphragm failure. 3. Air charge in element lost.	1. Disassemble valve and clean. 2. Replace element only. 3. Replace element only.
	Failure to close	1. Dirt or foreign material in valve. 2. Equalizer passageway plugged. 3. External equalizer not connected or equalizer line pinched shut.	1. Disassemble valve and clean. 2. Disassemble valve and clean. 3. Connect or replace equalizer line.
DRH-6-AR DRHE-6-AR	Failure to open	1. Dirt or foreign material in valve. 2. Diaphragm failure. 3. Equalizer passageway plugged. 4. External equalizer not connected or equalizer line pinched shut.	1. Disassemble valve and clean. 2. Replace element only. 3. Disassemble valve and clean. 4. Connect or replace equalizer line.

Model	Condition	Cause	Remedy
	Failure to close	5. Air charge in element lost.	5. Replace element only.
		1. Dirt or foreign material in valve.	1. Disassemble valve and clean.

Nonadjustable models—remote bulb and dome type

Model	Condition	Cause	Remedy
DRS-2 DRSE-2	Failure to open	1. Dirt or foreign material in valve.	1. Disassemble valve and clean.
		2. Diaphragm failure.	2. Replace element only.
		3. Air charge in element lost.	3. Replace element only.
DRP-3 DRPE-3	Failure to close	1. Dirt or foreign material in valve.	1. Disassemble valve and clean.
		2. Equalizer passageway plugged.	2. Disassemble valve and clean.
		3. External equalizer not connected or equalizer line pinched shut.	3. Connect or replace equalizer line.
DRH-6 DRHE-6	Failure to open	1. Dirt or foreign material in valve.	1. Disassemble valve and clean.
		2. Diaphragm failure.	2. Replace element only.
		3. Equalizer passageway plugged.	3. Disassemble valve and clean.
		4. External equalizer not connected or equalizer line pinched shut.	4. Connect or replace equalizer line.
		5. Air charge in element lost.	5. Replace element only.
	Failure to close	1. Dirt or foreign material in valve.	1. Disassemble valve and clean.

*The model numbers are for Sporlan valves.

453

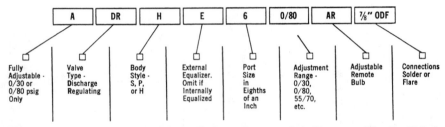

Figure 11-41   Codes used to identify the discharge-bypass valve. (*Courtesy of Sporlan Valve.*)

informed of such codes, you will need the manufacturers' bulletins. A good file of such bulletins will enable you to quickly identify the various valve problems. These can be obtained on the internet. Just use the manufacturer's name and "dot com" and you will usually open to their Web site. From their home page you will be able to navigate their various departments and offerings.

Level Control Valves

Capillary tubes and float valves are used to control the refrigerant in a system.

Capillary tubes

Capillary tubes are used to control pressure and temperature in a refrigeration unit. They are most commonly used in domestic refrigeration, milk coolers, ice-cream cabinets, and smaller units. Commercial refrigeration units use other devices. The capillary tube consists of a tube with a very small diameter. The length of the tube depends on the size of the unit to be served, the refrigerant used, and other physical considerations. To effect the necessary heat exchange, this tube is usually soldered to the suction line between the condenser and the evaporator. The capillary tube acts as a constant throttle or restrictor on the refrigerant. Its length and diameter offer sufficient frictional resistance to the flow of refrigerant to build up the head pressure needed to condense the gas.

If the condenser and evaporator were simply connected by a large tube, the pressure would rapidly adjust itself to the same value in both of them. A small diameter water pipe will hold back water, allowing a pressure to be built up behind the water column, but with a small rate of flow. Similarly, the small-diameter capillary tube holds back the liquid refrigerant. This enables a high pressure to be built up in the condenser during the operation of the compressor. At the same time, this permits the refrigerant to flow slowly into the evaporator (see Fig. 11-42). A filter drier is usually inserted between the condenser and the capillary

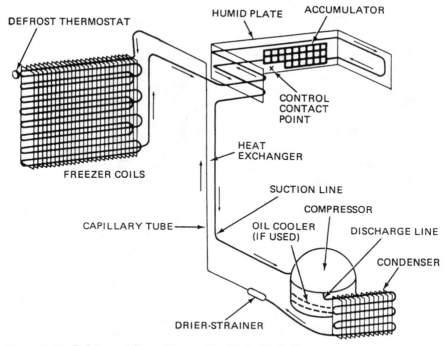

Figure 11-42 Refrigerant flow with a capillary tube in the line.

tube. This is necessary because the line or tube is so small that it is easily clogged.

Capillary tubes may be cleaned and unplugged by the method suggested in Chap. 1. Replacement should be performed in the shop after discharging the unit. In replacing the capillary tube, make sure that the same length of tube is used. The bore or inner diameter should be exactly the same as the old tube. It is easy to check with a proper tool.

Float valve

A hollow float is sometimes used to control the level of refrigerant (see Fig. 11-43). The float is fastened to a lever arm. The arm is pivoted at a given point and connected to a needle that seats at the valve opening. If there is no liquid in the evaporator, the ball-lever arm rests on a stop and the needle is not seated, thus leaving he valve open. Once liquid refrigerant under pressure from the compressor enters the float chamber, the float rises with the liquid level until, at a predetermined level, the needle closes the needle-valve opening.

In some large size plants where Freon-12 is used as a refrigerant, multiple ports are provided for handling the larger quantities of liquid.

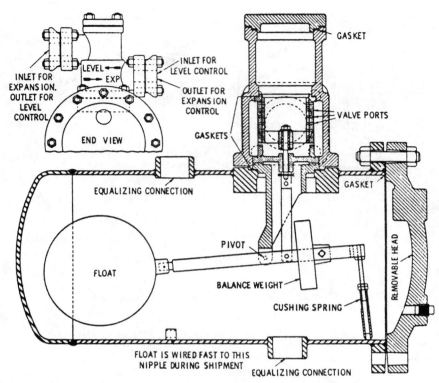

Figure11-43   Interior construction of a typical float valve. (*Courtesy of Frick.*)

Installation.   The following precautions must be observed before installation of a float valve:

- Most float controls are designed for a maximum differential pressure of 200 lb.

- If the pressure will exceed 190 psi, there are stems and orifices of special size available for low-temperature use.

- In any application, keep the bottom equalizing line above the bottom of the evaporator to avoid oil logging.

- Make sure there are no traps in the equalizing line.

- The stems of a globe valve must be in a horizontal plane.

- Refrigerant flow must be kept to less than 100 ft/min where a bottom float equalizing connection is made to the header or accumulator return. That means the header and accumulator pipe must be properly sized.

- Accumulators of a small diameter with a velocity of over 50 fpm are not suitable for accurate float application. However, the float may control

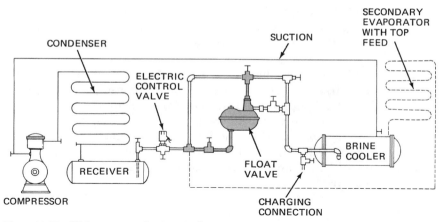

Figure 11-44   High-pressure float-control system.

within wider limits with higher velocities. The top equalizing connection must be connected to a point of practically zero gas velocity.

- In automatic plants, always provide a solenoid valve in the liquid line ahead of the float control. This solenoid valve is to close either when the temperatures are satisfactory or when the compressor stops.

Figure 11-44 illustrates the connections for a high-pressure float control. There have been new developments in the control of liquid level since the early days of refrigeration.

Level-Master Control

The level-master control is a positive liquid-level-control device suitable for application to all flooded evaporators (see Fig. 11-45). The level-master control is a standard thermostatic-expansion valve with a level-master element. The combination provides a simple, economical, and highly effective liquid-level control. The bulb of the conventional thermostatic

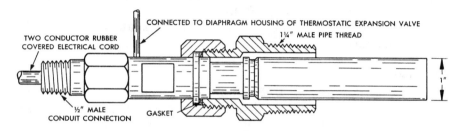

LEVEL MASTER ELEMENT WITH ½" MALE CONDUIT CONNECTION

Figure 11-45   Level-master control. (*Courtesy of Sporlan Valve.*)

element has been modified to an insert-type bulb that incorporates a low-wattage heater. A 15-W heater is supplied as standard. For applications below –60°F (–51°C) evaporating temperature, a special 25-W heater is needed.

The insert bulb is installed in the accumulator or surge drum at the point of the desired liquid level. As the level at the insert bulb drops, the electrically added heat increases the pressure within the thermostatic element and opens the valve. As the liquid level at the bulb rises, the electrical input is balanced by the heat transfer from the bulb to the liquid refrigerant. The level-master control either modulates or eventually shuts off. The evaporator pressure and spring assist in providing a positive closure.

Installation

The level-master control is applicable to any system that has been specifically designed for flooded operation. The valve is usually connected to feed into the surge drum above the liquid-level. It can feed into the liquid leg or coil header.

The insert bulb can be installed directly in the shell, surge drum, or liquid leg on new or existing installations. Existing float systems can be easily converted by installing the level-master control insert bulb in the float chamber.

Electrical connections

The heater is provided with a two-wire neoprene-covered cord 2 ft in length. It runs through a moisture-proof grommet and a 1/2-in. male conduit connection affixed to the insert-bulb assembly (see Fig. 11-46).

The heater circuit must be interrupted when refrigeration is not required. Wire the heater in parallel with the holding coil of the compressor line starter or solenoid valve—not in series.

Hand valves

On some installations, the valve is isolated from the surge drum by a hand valve. A 2- to 3-lb pressure drop from the valve outlet to the bulb location is likely. For such installations, an externally equalized valve is recommended.

Oil return

All reciprocating compressors will allow some oil to pass into the discharge line along with the discharge gas. Mechanical oil separators are used extensively. However, they are never completely effective. The untrapped oil passes through the condenser, liquid line, expansion device, and into the evaporator.

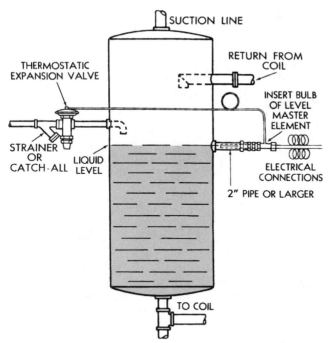

Figure 11-46 Installation of the level-master control. (*Courtesy of Sporlan Valve.*)

In a properly designed direct-expansion system, the refrigerant velocity in the evaporator tubes and the suction line is sufficiently high to ensure a continuous return of oil to the compressor crankcase. However, this is not characteristic of flooded systems. Here, the surge drum is designed for a relatively low vapor velocity. This prevents entrainment of liquid refrigerant droplets and consequent carryover into the suction line. This design also prevents the return of any oil from the low side in the normal manner.

If oil is allowed to concentrate at the insert-bulb location of the level-master control, overfeeding with possible flood-back can occur. The tendency to overfeed is due to the fact that the oil does not convey the heat from the low-wattage heater element away from the bulb as rapidly as does pure liquid refrigerant. The bulb pressure is higher than normal and the valve remains in the open or partially open position.

Oil and ammonia systems

For all practical purposes, liquid ammonia and oil are immiscible (not capable of being mixed). Since the density of oil is greater than that of ammonia, it will fall to the bottom of any vessel containing such a mixture

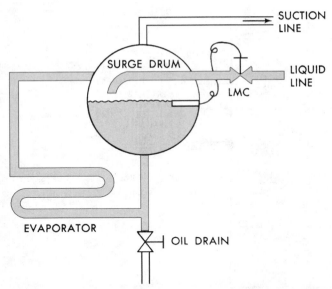

Figure 11-47   Location of LM in liquid line. (*Courtesy of Sporlan Valve.*)

if the mixture is relatively placid. Therefore. the removal of oil from an ammonia system is a comparatively simple task. Generally, on systems equipped with a surge drum, the liquid leg is extended downward below the point where the liquid is fed off to the evaporator. A drain valve is provided to allow periodic manual draining (see Fig. 11-47).

For flooded chillers that do not use a surge drum, a sump with a drain valve is usually provided at the bottom of the chiller shell. These methods are quite satisfactory, except possibly on some low-temperature systems. Here, the drain leg or sump generally must be warmed prior to attempting to draw off the oil. The trapped oil becomes quite viscous at lower temperatures.

If oil is not drained from a flooded ammonia system, a reduction in the evaporator heat-transfer rate can occur due to an increase in the refrigerant film resistance. Difficulty in maintaining the proper liquid level with any type of flooded control can also be expected.

With a float valve, you can expect the liquid level in the evaporator to increase with high concentration of oil in a remote float chamber. If a level-master control is used with the insert bulb installed in a remote chamber, oil concentration at the bulb can cause overfeeding with possible flood-back. The lower or liquid-balance line must be free of traps and be free-draining into the surge drum or chiller, as shown in Fig. 11-48. The oil drain leg or sump must be located at the lowest point in the low side.

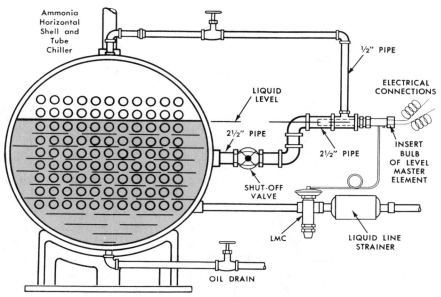

Figure 11-48 Level-master control with the bulb inserted in a remote chamber. (*Courtesy of Sporlan Valve.*)

Oil and halocarbon systems

With halocarbon systems (Refrigerants 12, 22, 502, etc.) the oil and refrigerant are miscible (capable of being mixed) under certain conditions. Oil is quite soluble in liquid Refrigerant 12 and partially so in liquid Refrigerant 22 and 502. For example, for a 5% (by weight) solution of a typical *napthenic* (a petroleum-based oil) oil in liquid refrigerant, the oil will remain in solution down to about −75°F (−59°C) for Refrigerant 12, down to about 0°F (−18°C) for Refrigerant 22, and down to about 20°F (−7°C) for Refrigerant 502. Depending upon the type of oil and the percentage of oil present, these figures can vary. However, based on the foregoing, we can assume that for the majority of Refrigerant-12 systems the oil and refrigerant are completely miscible at all temperatures normally encountered. However, at temperatures below 0°F (−18°C) with Refrigerant 22 and a 5% oil concentration and temperatures below 20°F (−7°C) with Refrigerant 502 and a 5% oil concentration, a liquid phase separation occurs. An oil-rich solution will appear at the top and a refrigerant-rich solution will lay at the bottom of any relatively placid remote bulb chamber.

Oil in a halocarbon-flooded evaporator can produce many results. Oil as a contaminant will raise the boiling point of the liquid refrigerant. For example, with Refrigerant 12, the boiling point increases approximately 1°F [(0.56°C)] for each 5% of oil (by weight) in solution. As in an ammonia system, oil can foul the heat-transfer surface with a consequent loss in system capacity. Oil can produce foaming and possible carryover of liquid into the suction line. Oil can also affect the liquid-level control. With a float valve you can normally expect the liquid level in the evaporator to decrease with increasing concentrations of oil in the float chamber. This is due to the difference in density between the lighter oil in the chamber and the lower balance leg and the heavier refrigerant/oil mixture in the evaporator. A lower column of dense mixture in the evaporator will balance a higher column of oil in the remote chamber and piping. This is similar to a "U" tube manometer with a different fluid in each leg.

With the level-master control, the heat-transfer rate at the bulb is decreased, producing overfeeding and possible flood-back. What can be done? First of all, the oil concentration must be kept as low as possible in the evaporator, surge drum, and remote insert bulb chamber (if one is used). With Refrigerant 12, since the oil/refrigerant mixture is homogenous, it can be drained from almost any location in the chiller, surge drum, or remote chamber that is below the liquid level. With Refrigerants 22 and 502, the drain must be located at or slightly below the surface of the liquid, since the oil-rich layer is at the top. There are many types of oil-return devices:

- Direct drain into the suction line
- Drain through a high-pressure, liquid-warmed heat exchanger
- Drain through a heat exchanger with the heat supplied by an electric heater

Draining directly into the suction line, as shown in Fig. 11-49, is the simplest method. However, the hazard of possible flood-back to the compressor remains.

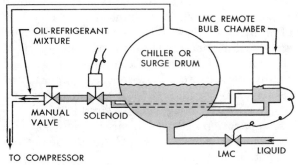

Figure 11-49 Direct drain of oil to the suction line is one of the three ways to recover oil in flooded systems. Heat from the environment or a liquid-suction heat exchanger is required to vaporize the liquid refrigerant so drained. Vapor velocity carries oil back to the compressor. (*Courtesy of Sporlan Valve.*)

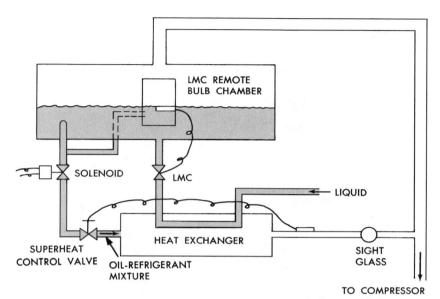

Figure 11-50   Oil return by draining oil-refrigerant mixture through a heat exchanger is shown here. Heat in incoming liquid vaporizes refrigerant to prevent return of liquid to the compressor. Liquid feed is controlled by a thermostatic- or hand-expansion valve. (*Courtesy of Sporlan Valve.*)

Draining through a heat exchanger, as indicated in Fig. 11-50, is a popular method. The liquid refrigerant flood-back problems are minimized by using the warm liquid to vaporize the liquid refrigerant in the oil/refrigerant mixture.

The use of a heat exchanger with an insert electric heater, as shown in Fig. 11-51, is a variation of the preceding method.

In all of the return arrangements discussed, a solenoid valve should be installed in the drain line and arranged to close when the compressor is

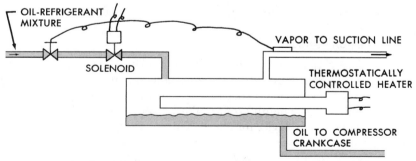

Figure 11-51   An electric heater may also be added to separate oil and refrigerant. This system is similar to that shown in Fig. 10-49, except that the heat required for vaporization is added electrically. (*Courtesy of Sporlan Valve.*)

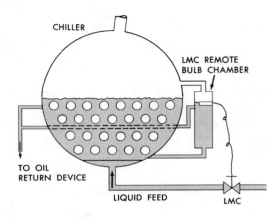

Figure 11-52 Level-master control inserted in remote chamber. (*Courtesy of Sporlan Valve.*)

not in operation. Otherwise, liquid refrigerant could drain from the low side into the compressor crankcase during the off-cycle.

If the insert bulb is installed directly into the surge drum or chiller, oil return is necessary only from this point. However, the insert bulb is sometimes located in a remote chamber that is tied to the surge drum or chiller with liquid- and gas-balance lines. Then oil return should be made from both locations, as shown in Figs. 11-49, 11-50, and 11-52.

Conclusions

The problem of returning oil from a flooded system is not highly complex. There are undoubtedly other methods in use today that are comparable to those outlined here. Regardless of how it is accomplished, oil return must be provided for proper operation of any flooded system. This is necessary not only with the level-master control, but also with a float or other type of level-control device.

Other Types of Valves

There are check valves, water valves, and receiver valves.

Service valves on sealed units

Hermetic refrigeration systems, also called sealed units, normally have no service valves on the compressor. Instead, a charging plug or valve maybe mounted on the compressor. A special tool is needed to operate the charging device, which varies on different makes. A service engineer needs the correct valve-operating device for a unit. Thus, a kit is made that contains adapters and wrench ends to fit many makes of sealed units.

Essentially, the device is a body with a union connection and provisions for charging line- and pressure-gage connections. The stem may be turned or pushed in or out of the body as required. Fig. 11-53 shows a line-piercing

Figure 11-53 Line-piercing valve. (*Courtesy of Mueller Brass.*)

valve. They are used for charging, testing, or purging those hermetic units not provided with a charging plug or valve. These valves may be permanently attached to the line without danger of refrigerant loss.

Water valves

Manually operated valves are installed on water circuits associated with refrigeration systems—either on cooling towers or in secondary brine circuits. They are installed for convenience in servicing and for flexibility in operating conditions. These valves make it possible to recircuit, bypass, or shut off water flow as desired (see Fig. 11-54).

These manually operated shutoff or flow-control valves are available in a wide variety of styles and sizes. Valve stems and body seats are accurately machined to close tolerances, ensuring easy and positive shutoff. They are made of nonporous cast bronze.

There are three main types (see Fig. 11-54):

- Stop valves
- Globe valves
- Gate valves

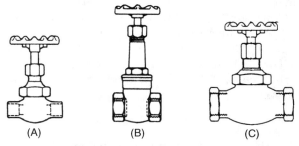

Figure 11-54 Water valves: (*a*) stop valve, (*b*) gate valve, and (C) glove valve. (*Courtesy of Mueller Brass.*)

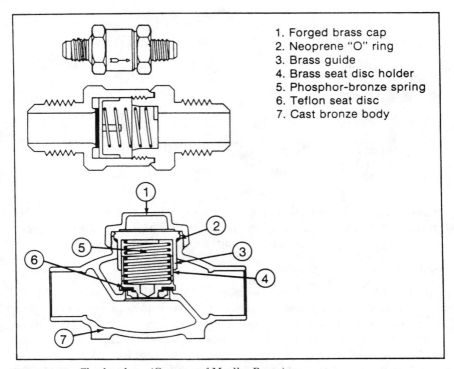

1. Forged brass cap
2. Neoprene "O" ring
3. Brass guide
4. Brass seat disc holder
5. Phosphor-bronze spring
6. Teflon seat disc
7. Cast bronze body

Figure 11-55   Check valves. (*Courtesy of Mueller Brass.*)

Check valves

Some refrigeration systems are designed in which the refrigerant liquid or vapor flows to several components, but must never flow back through a given line. A check valve is needed in such installations. As its name implies, a check valve checks or prevents the flow of refrigerant in one direction, while allowing free flow in the other direction. For example, two evaporators might be controlled by a single condensing system. In this case, a check valve should be placed in the line from the lower-temperature evaporator to prevent the suction gas from the higher-temperature evaporator from entering the lower-temperature evaporator (see Fig. 11-55).

Check valves are designed to eliminate chattering and to give maximum refrigerant flow when the unit is operating. If the spring tension is sufficient to overcome the weight of the valve disc, the check valve may be mounted in any position.

Receiver valves

Receivers may be fitted with two valves—an inlet valve and an outlet valve. The outlet valve may have the inlet in the form of an ordinary

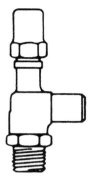

Figure 11-56   Receiver-angle valve. (*Courtesy of Mueller Brass.*)

connection, such as an elbow. An inlet valve permits closing the receiver should a leak develop between the compressor and the receiver. The receiver outlet valve is important when the system is "pumped down," when for reasons of service all the refrigerant is conveyed to the receiver for temporary storage (see Fig. 11-56).

Accumulators

Accumulators have been used for years on original equipment. More recently they have been field installed. The significance with respect to accumulator and system performance has never been clarified. Engineers have been forced to evaluate each model in terms of the system on which it is to be applied. Application in the field has been primarily based on choosing a model with fittings that will accommodate the suction line and be large enough to hold about half of the refrigerant charge.

There is no standard rating system for accumulators. The accuracy of rating data becomes a function of the type of equipment used to determine the ratings. Some data is now available to serve as a guide to those checking the use of an accumulator.

Purpose

The purpose of an accumulator is to prevent compressor damage due to slugging of refrigerant and oil. They provide a positive oil return at all rated conditions. They are designed to operate at −40°F (−40°C) evaporator temperature. Pressure drop is low across them. They act as a suction muffler. They can take suction-gas temperatures as low as 10°F (−12.2°C) at the accumulator. Most of them can withstand a working pressure of 300 psi and have fusible relief devices.

Compressors are designed to compress vapors, not liquids. Many systems, especially low-temperature systems, are subject to the return of excessive quantities of liquid refrigerant. This returned refrigerant dilutes the oil and washes out bearings. In some cases, it causes complete loss of

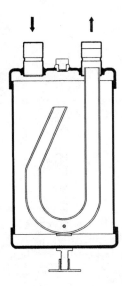

Figure 11-57 Suction-line accumulator. (Virginia Chemicals.)

oil in the crankcase. This results in broken valve reeds, and damage to pistons, rods, crankcase, and other moving parts. The accumulator acts as a reservoir to hold temporarily the excess oil-refrigerant mixture and return it at a rate the compressor can safely handle. Figure 11-57 shows the interior view.

Rating data

The refrigerant-holding capacity of the accumulator is based on an average condition of 65 percent fill under running conditions.

Refrigerant-holding capacity. It is obvious that directly on startup or after long off-cycles the amount held may fluctuate from empty to nearly full.

Minimum evaporator temperature and also of suction gas at the accumulator

The oil-refrigerant mixture in the suction line has been studied over the range of $-50°F$ to $+40°F$ ($-46°C$ to $+4°C$). The value of $-40°F$ ($-40°C$) was chosen as a minimum evaporator temperature because it appears adequate for commercial refrigeration. Yet, it is conservative enough to provide a margin of safety. More important is the requirement that the temperature of the suction gas at the accumulator be $10°F$ ($-12°C$) or higher. Particularly with refrigerants such as Freon-502 in the low temperature range up to $0°F$ ($-17.8°C$), the oil and refrigerant separate into two layers, with the upper layer being the oil-rich layer. At these low temperatures, the oil-rich layer can become so viscous that it will not flow. When the refrigerant below the heavy oil layer leaves the accumulator, the very thick oil settles over the oil-return port and stops all oil return. This condition will occur regardless of accumulator design. If temperatures below 10°F (−12°C) at the accumulator are to be used, auxiliary heat must be added to keep the oil fluid.

Maximum recommended actual tonnage is based on pressure drop through the accumulator equivalent to an effect of 1°F (0.56°C) on evaporator temperature.

Minimum recommended actual tonnage is based on the minimum flow through the accumulator necessary to ensure positive oil return.

For operating conditions outside the manufacturer's published ratings, contact the manufacturer for recommendations.

Installation of the Accumulator

Locate the accumulator as close to the compressor as possible. In systems employing reverse cycle, the accumulator must be installed between the reversing valve and the compressor. Proper inlet (from the evaporator) and outlet (to the compressor) must be observed. The accumulator must be installed vertically. Proper sizing of an accumulator may not necessarily result in the accumulator connections matching the suction-line size. This new technology must replace the dangerous and outmoded practice of matching the accumulator connections to the suction-line size. To accommodate mismatches, bushing down may be required.

The accumulator should not be installed in a bypass line or in suction lines that experience other than total refrigerant flow.

When installing an accumulator with solder connections, direct the torch away from the top access plug to prevent possible damage to the O-ring seal. When installing a model equipped with a fusible plug, a dummy plug should be inserted in place of the fusible plug until all brazing or soldering is complete.

12

Complying with the Section 608 Refrigerant-Recycling Rule

This chapter provides an overview of the refrigerant-recycling requirements of Section 608 of the Clean Air Act of 1990, as amended (CAA), including final regulations published on May 14, 1993 (58 FR 28660), August 19, 1994 (59 FR 42950), November 9, 1994 (59 FR 55912), and July 24, 2003 (68 FR 43786). The chapter also describes the prohibition on venting that became effective on July 1, 1992.

Overview

Under Section 608 of the CAA, EPA has established regulations (40 CFR Part 82, Subpart F) that

- Require service practices that maximize recycling of ozone-depleting compounds [both chlorofluorocarbons (CFCs) and hydrochlorofluoro-carbons (HCFCs) and their blends] during the servicing and disposal of air-conditioning and refrigeration equipment.

- Set certification requirements for recycling and recovery equipment, technicians, and reclaimers.

- Restrict the sale of refrigerant to certified technicians.

- Require persons servicing or disposing of air-conditioning and refrigeration equipment to certify to EPA that they have acquired recycling or recovery equipment and are complying with the requirements of the rule.

- Require the repair of substantial leaks in air-conditioning and refrigeration equipment with a charge of greater than 50 lb.

- Establish safe disposal requirements to ensure removal of refrigerants from goods that enter the waste stream with the charge intact (e.g., motor vehicle air conditioners, home refrigerators, and room air conditioners).

The Prohibition on Venting

Effective July 1, 1992, Section 608 of the Act prohibited individuals from knowingly venting ozone-depleting compounds (generally CFCs and HCFCs) used as refrigerants into the atmosphere while maintaining, servicing, repairing, or disposing of air-conditioning or refrigeration equipment (appliances). Only four types of releases are permitted under the prohibition:

1. "De minimis" quantities of refrigerant released in the course of making good faith attempts to recapture and recycle or safely dispose of refrigerant.

2. Refrigerants emitted in the course of normal operation of air-conditioning and refrigeration equipment (as opposed to during the maintenance, servicing, repair, or disposal of this equipment) such as from mechanical purging and leaks. However, EPA requires the repair of leaks above a certain size in large equipment (see Refrigerant Leaks).

3. Releases of CFCs or HCFCs *that are not used as refrigerants.* For instance, mixtures of nitrogen and R-22 that are used as holding charges or as leak-test gases may be released, because in these cases, the ozone-depleting compound is not used as a refrigerant. However, a technician may not avoid recovering refrigerant by adding nitrogen to a charged system. Before nitrogen is added, the system must be evacuated to the appropriate level as shown in Table 12-1. Otherwise, the CFC or HCFC vented along with the nitrogen will be considered a refrigerant. Similarly, *pure* CFCs or HCFCs released from appliances will be presumed to be refrigerants, and their release will be considered a violation of the prohibition on venting.

4. Small releases of refrigerant that result from purging hoses or from connecting or disconnecting hoses to charge or service appliances will not be considered violations of the prohibition on venting. However, recovery and recycling equipment manufactured after November 15, 1993 must be equipped with low-loss fittings.

Regulatory Requirements

Service practice requirements

Evacuation requirements. Since July 13, 1993, technicians have been required to evacuate air-conditioning and refrigeration equipment to

TABLE 12-1 Ozone Layer Depletion

Ozone Layer Depletion—Regulatory Programs

Recent Additions | Contact Us   **Search:** ○ All EPA ⦿ This Area [   ]
[Go]

Ozone Depletion Home   You are here: EPA Home   Ozone Layer Depletion   Regulatory Programs   Stationary AC
{Page N ame}

Regulatory Programs
Home

{Page N ame}

Phaseout

Exemptions:
  Methyl bromide
  Essential uses
  (MDIs)
  Lab uses

Stationary
  Refrigeration and A/C

Auto A/C

Halons

Nonessential Products

Labeling

Enforcement

Reporting

Text version of this table (use if the table is a jumble)

REQUIRED LEVELS OF EVACUATION FOR APPLIANCES EXCEPT FOR SMALL APPLIANCES, MVACS, AND MVAC-LIKE APPLIANCES

Type of Appliance	Inches of Mercury Vacuum* Using Equipment Manufactured:	
	Before Nov. 15, 1993	On or after Nov. 15, 1993
HCFC-22 appliance** normally containing less than 200 pounds of refrigerant	0	0
HCFC-22 appliance** normally containing 200 pounds or more of refrigerant	4	10
Other high-pressure appliance** normally containing less than 200 pounds of refrigerant (CFC-12, -500, -502, -114)	4	10
Other high-pressure appliance** normally containing 200 pounds or more of refrigerant (CFC-12, -500, -502, -114)	4	15
Very High-Pressure Appliance (CFC-13, -503)	0	0
Low-Pressure Appliance (CFC-11, HCFC-123)	25	25 mm Hg absolute

* Relative to standard atmospheric pressure of 29.9" Hg
** Or isolated component of such an appliance

http://www.epa.gov/ozone/title6/608/608evtab.html

established vacuum levels when opening the equipment. If the technician's recovery or recycling equipment was manufactured any time before November 15, 1993, the air-conditioning and refrigeration equipment must be evacuated to the levels described in the first column of Table 12-1. If the technician's recovery or recycling equipment was manufactured on or after November 15, 1993, the air-conditioning and refrigeration equipment must be evacuated to the levels described in the second column of Table 12-1, and the recovery or recycling equipment must have been certified by an EPA-approved equipment-testing organization. Persons who simply add refrigerant to (top-off) appliances are not required to evacuate the systems.

Technicians repairing small appliances, such as household refrigerators, window air conditioners, and water coolers, must recover:

80 percent of the refrigerant when

■ The compressor in the appliance is not operating

or

90 percent of the refrigerant when

■ The technician uses recovery or recycling equipment manufactured after November 15, and

■ The compressor in the appliance is operating

In order to ensure that they are recovering the correct percentage of refrigerant, technicians must use the recovery equipment according to the directions of its manufacturer. Technicians may also satisfy recovery requirements by evacuating the small appliance to 4 in. of mercury vacuum.

Exceptions to evacuation requirements. EPA has established limited exceptions to its evacuation requirements for (*a*) repairs to leaky equipment and (*b*) repairs that are not major and that are not followed by an evacuation of the equipment to the environment.

If, due to leaks, evacuation to the levels indicated in Table 12-1 is not attainable, or would substantially contaminate the refrigerant being recovered, persons opening the appliance must

■ Isolate leak from nonleaking components wherever possible.

■ Evacuate nonleaking components to the levels shown in Table 12-1.

■ Evacuate leaking components to the lowest level that can be attained without substantially contaminating the refrigerant. This level cannot exceed 0 psig.

If evacuation of the equipment to the environment is not to be performed when repairs are complete, *and* if the repair is not major, then the appliance must

■ Be evacuated to at least 0 psig before it is opened if it is a high- or very high pressure appliance, or

■ Be pressurized to 0 psig before it is opened if it is a low-pressure appliance. Methods that require subsequent purging (e.g., nitrogen) *cannot* be used except with appliances containing R-113

Reclamation requirement. EPA has also established that refrigerant recovered and/or recycled can be returned to the same system or other systems owned by the same person without restriction. If refrigerant changes ownership, however, that refrigerant must be reclaimed (i.e., cleaned to the ARI 700-1993 standard of purity and chemically analyzed to verify that it meets this standard) unless the refrigerant was used only in a motor vehicle air conditioner (MVAC) or MVAC-like appliance and will be used in the same type of appliance. (Refrigerant used in MVACs and MVAC-like appliances is subject to the purity requirements of the MVAC regulations at 40 CFR Part 82 Subpart B.) EPA updates the list of reclaimers as new companies are added.

Equipment Certification

The agency has established a certification program for refrigerant recovery and recycling equipment. Under the program, EPA requires that manufacturers or importers of refrigerant recovery and recycling equipment manufactured on or after November 15, 1993, have their equipment tested by an EPA-approved testing organization to ensure that it meets EPA requirements. Equipment intended for use with air-conditioning and refrigeration appliances must be tested under EPA requirements based upon the ARI 740 test protocol (i.e., EPA Apps. B and B1 to 40 CFR 82 subpart F). Recycling and recovery equipment intended for use with small appliances must be tested under EPA App. C or alternatively under requirements based upon the ARI 740 test protocol (i.e., Apps. B and B1 to 40 CFR 82 subpart F).

The agency requires recovery efficiency standards that vary depending on the size and type of air-conditioning or refrigeration equipment being serviced. For recovery and recycling equipment intended for use with air-conditioning and refrigeration equipment besides small appliances, these standards are the same as those in the second column of Table 12-1. Recovery equipment intended for use with small appliances must be able to recover 90 percent of the refrigerant in the small appliance when the small appliance compressor is operating and 80 percent of the refrigerant in the small appliance when the compressor is not operating.

EPA has approved both the Air-Conditioning and Refrigeration Institute (ARI) and Underwriters Laboratories (UL) to certify recycling and recovery equipment. Certified equipment can be identified by a label reading "This equipment has been certified by ARI/UL to meet EPA's minimum requirements for recycling and/or recovery equipment intended for use with (appropriate category of appliance—e.g., small appliances, HCFC appliances containing less than 200 lb of refrigerant, all high-pressure appliances)." Lists of certified equipment may be obtained by contacting ARI at 703-524-8800.

Equipment Grandfathering

Equipment manufactured before November 15, 1993, including homemade equipment, may be grandfathered if it meets the standards in the first column of Table 12-1. Third-party testing is not required for equipment manufactured before November 15, 1993, but equipment manufactured on or after that date, including homemade equipment, must be tested by a third-party (see Equipment Certification).

Refrigerant Leaks

Owners of equipment with charges of greater than 50 lb are required to repair leaks in the equipment when those leaks together would result in the loss of more than a certain percentage of the equipment's charge over a year. For the commercial and industrial process refrigeration sectors, leaks must be repaired when the appliance leaks at a rate that would release 35 percent or more of the charge over a year. For all other sectors, including comfort cooling, leaks must be repaired when the appliance leaks at a rate that would release 15 percent or more of the charge over a year.

The trigger for repair requirements is the current leak rate rather than the total quantity of refrigerant lost. For instance, owners of a commercial refrigeration system containing 100 lb of charge must repair leaks if they find that the system has lost 10 lb of charge over the past month; although 10 lb represents only 10 percent of the system charge in this case, a leak rate of 10 lb per month would result in the release of over 100 percent of the charge over the year. To track leak rates, owners of air-conditioning and refrigeration equipment with more than 50 lb of charge must keep records of the quantity of refrigerant added to their equipment during servicing and maintenance procedures.

Owners are required to repair leaks within 30 days of discovery. This requirement is waived if, within 30 days of discovery, owners develop a 1-year retrofit or retirement plan for the leaking equipment. Owners of industrial process refrigeration equipment may qualify for additional time under certain circumstances. For example, if an industrial process shutdown is required to repair a leak, owners have 120 days to repair the leak. Owners of leaky industrial process refrigeration equipment should see the Compliance Assistance Guidance Document for Leak Repair (available from the hotline) for additional information concerning time extensions and pertinent recordkeeping and reporting requirements. EPA anticipates putting this document on the Web site, but does not have an estimated date for when that will happen. A longer fact sheet about leak repair is also available.

Technician Certification

EPA has established a technician certification program for persons ("technicians") who perform maintenance, service, repair, or disposal that could be reasonably expected to release refrigerants into the atmosphere. The definition of "technician" specifically includes and excludes certain activities as follows:

- Attaching and detaching hoses and gauges to and from the appliance to measure pressure within the appliance.

- Adding refrigerant to (e.g., "topping-off") or removing refrigerant from the appliance.

- Any other activity that violates the integrity of the MVAC-like appliances, and small appliances.

In addition, apprentices are exempt from certification requirements provided the apprentice is closely and continually supervised by a certified technician.

The agency has developed four types of certification:

1. For servicing small appliances (Type I).

2. For servicing or disposing of high- or very high pressure appliances, except small appliances and MVACs (Type II).

3. For servicing or disposing of low-pressure appliances (Type III).

4. For servicing all types of equipment (Universal).

Technicians are required to pass an EPA-approved test conducted by an EPA-approved certifying organization to become certified under the mandatory program. Section 608 technician certification credentials do not expire.

Refrigerant Sales Restrictions

Since November 14, 1994, the sale of refrigerant in any size container has been restricted to technicians certified either under the program described previously in the chapter under Technician Certification or under EPAs motor vehicle air-conditioning regulations. The sales restriction covers refrigerant contained in bulk containers (cylinders or drums) and precharged parts.

The restriction excludes refrigerant contained in refrigerators or air conditioners with fully assembled refrigerant circuits (such as household refrigerators, window air conditioners, and packaged air conditioners), pure HFC refrigerants (such as R-134a), and CFCs or HCFCs that are not intended for use as refrigerants. In addition, a restriction on sale of precharged split systems has been stayed (suspended) while EPA reconsiders this provision.

Under Section 609 of the Clean Air Act, sales of CFC-12 in containers smaller than 20 lb are restricted solely to technicians certified under EPA's motor vehicle air-conditioning regulations (i.e., Section 609 certified technicians). Technicians certified under EPA's stationary refrigeration and air-conditioning equipment (i.e., Section 608 certified technicians) may buy containers of CFC-12 larger than 20 lb.

Section 609 technicians are only allowed to purchase refrigerants that are suitable for use in motor vehicle air-conditioners. Effective September 22, 2003, EPA has restricted the sale of ozone-depleting refrigerants, approved for use in stationary refrigeration and air-conditioning equipment, to Section 608 certified technicians. Therefore, the sale of ozone-depleting refrigerants (such as HCFC-22) that are approved for use in stationary equipment but not for use in motor vehicle air-conditioners is restricted to Section 608 certified technicians.

More detailed information is available in an EPA fact sheet titled "The Refrigerant Sales Restriction."

Certification by Owners of Recycling and Recovery Equipment

EPA requires that persons servicing or disposing air-conditioning and refrigeration equipment certify to the appropriate EPA regional office that they have acquired (built, bought, or leased) recovery or recycling equipment and that they are complying with the applicable requirements of this rule. This certification must be signed by the owner of the equipment or another responsible officer and sent to the appropriate EPA regional office. Although owners of recycling and recovery equipment are required to list the number of trucks based at their shops, they do not need to have a piece of recycling or recovery equipment for every truck. Owners do not have to send in a new form each time they add recycling or recovery equipment to their inventory.

Reclaimer Certification

Reclaimers are required to return refrigerant to the purity level specified in ARI Standard 700-1993 (an industry-set purity standard) and to verify this purity using the laboratory protocol set forth in the same standard. In addition, reclaimers must release no more than 1.5 percent of the refrigerant during the reclamation process and must dispose of wastes properly. Reclaimers must certify to the Section 608 recycling program manager at EPA headquarters that they are complying with these requirements and that the information given is true and correct. Certification must also include the name and address of the reclaimer and a list of equipment used to reprocess and to analyze the refrigerant.

EPA encourages reclaimers to participate in a voluntary third-party reclaimer certification program operated by the ARI. The voluntary program offered by ARI involves quarterly testing of random samples of reclaimed refrigerant. Third-party certification can enhance the attractiveness of a reclaimer's product by providing an objective assessment of its purity. EPA maintains a list of approved reclaimers that is available from the hotline. In addition, a checklist helps prospective reclaimers provide appropriate information for EPA to review.

MVAC-Like Appliances

Some of the air conditioners that are covered by this rule are identical to motor vehicle air conditioners (MVACs), but they are not covered by the MVAC refrigerant-recycling rule (40 CFR Part 82, Subpart B) because they are used in vehicles that are not defined as *motor vehicles.* These air conditioners include many systems used in construction equipment, farm vehicles, boats, and airplanes. Like MVACs in cars and trucks, these air conditioners typically contain 2 or 3 lb of CFC-12 and use open-drive compressors to cool the passenger compartments of vehicles. (Vehicle air conditioners utilizing HCFC-22 are not included in this group and are therefore subject to the requirements outlined above for HCFC-22 equipment.) EPA is defining these air conditioners as *MVAC-like appliances* and is applying the MVAC rule's requirements for the certification and use of recycling and recovery equipment to them. That is, technicians servicing MVAC-like appliances must "properly use" recycling or recovery equipment that has been certified to meet the standards in App. A to 40 CFR Part 82, Subpart B. In addition, EPA is allowing technicians who service MVAC-like appliances to be certified by a certification program approved under the MVAC rule, if they wish.

More detailed information is presented in an EPA fact sheet titled "Servicing Farm and Heavy-Duty Equipment."

Safe Disposal Requirements

Under EPAs rule, equipment that is typically dismantled on-site before disposal (e.g., retail food refrigeration, central residential air-conditioning, chillers, and industrial process refrigeration) has to have the refrigerant recovered in accordance with EPAs requirements for servicing. However, equipment that typically enters the waste stream with the charge intact (e.g., motor vehicle air conditioners, household refrigerators and freezers, and room air conditioners) is subject to special safe disposal requirements.

Under these requirements, the final person in the disposal chain (e.g., a scrap metal recycler or landfill owner) is responsible for ensuring that refrigerant is recovered from equipment before the final disposal of the equipment. However, persons "upstream" can remove the refrigerant and provide documentation of its removal to the final person if this is more cost-effective. If the final person in the disposal chain (e.g., a scrap metal recycler or landfill owner) accepts appliances that no longer hold a refrigerant charge, that person is responsible for maintaining a signed statement from whom the appliance/s is being accepted. The signed statement must include the name and address of the person who recovered the refrigerant, and the date that the refrigerant was recovered, or a copy of a contract stating that the refrigerant will be removed prior to delivery. EPA does not mandate a sticker as a form of verification that the refrigerant has been removed prior to disposal of the appliance. Such stickers do not relieve the final disposer of their responsibility to recover any remaining refrigerant in the appliance, unless the sticker contains a signed statement that includes the name and address of the person who recovered the refrigerant, and the date that the refrigerant was recovered.

The equipment used to recover refrigerant from appliances prior to their final disposal must meet the same performance standards as equipment used prior to servicing, but it does not need to be tested by a laboratory. This means that self-built equipment is allowed as long as it meets the performance requirements. For MVACs and MVAC-like appliances, the performance requirement is 102 mm of mercury vacuum and for small appliances, the recovery equipment performance requirements are 90 percent efficiency when the appliance compressor is operational, and 80 percent efficiency when the appliance compressor is not operational.

Technician certification is not required for individuals removing refrigerant from appliances in the waste stream.

The safe disposal requirements went into effect on July 13, 1993. Equipment must be registered or certified with the EPA.

Major Recordkeeping Requirements

Technicians servicing appliances that contain 50 lb or more of refrigerant must provide the owner with an invoice that indicates the amount of refrigerant added to the appliance. Technicians must also keep a copy of their proof of certification at their place of business.

Owners of appliances that contain 50 lb or more of refrigerant must keep servicing records documenting the date and type of service, as well as the quantity of refrigerant added.

Wholesalers who sell CFC and HCFC refrigerants must retain invoices that indicate the name of the purchaser, the date of sale, and the quantity of refrigerant purchased.

Reclaimers must maintain records of the names and addresses of persons sending them material for reclamation and the quantity of material sent to them for reclamation. This information must be maintained on a transactional basis. Within 30 days of the end of the calendar year, reclaimers must report to EPA the total quantity of material sent to them that year for reclamation, the mass of refrigerant reclaimed that year, and the mass of waste products generated that year.

Hazardous Waste Disposal

If refrigerants are recycled or reclaimed, they are not considered hazardous under federal law. In addition, used oils contaminated with CFCs are not hazardous on the condition that.

- They are not mixed with other waste.
- They are subjected to CFC recycling or reclamation.
- They are not mixed with used oils from other sources.

Used oils that contain CFCs after the CFC reclamation procedure, however, are subject to specification limits for used-oil fuels if these oils are destined for burning.

Enforcement

EPA is performing random inspections, responding to tips, and pursuing potential cases against violators. Under the Act, EPA is authorized to assess fines of up to $32,500 per day for any violation of these regulations. Information on selected enforcement actions is available in the enforcement section.

If you wish to report a possible violation of the Clean Air Act, please file a complaint form or contact the ozone hotline at 800-296-1996.

Planning and Acting for the Future

Observing the refrigerant-recycling regulations for Section 608 is essential in order to conserve existing stocks of refrigerants, as well as to comply with Clean Air Act requirements. However, owners of equipment that contains CFC refrigerants should look beyond the immediate need to maintain existing equipment in working order. Owners are advised to begin planning for conversion or replacement of existing equipment with equipment that uses alternative refrigerants.

To assist owners, suppliers, technicians, and others involved in comfort chiller and commercial-refrigeration management, EPA has published a series of short fact sheets and expects to produce additional material. Copies of material produced by the EPA Stratospheric Protection Division are available from the Stratospheric Ozone Information Hotline (see hotline number below).

For Further Information

For further information concerning regulations related to stratospheric ozone protection, please call the Stratospheric Ozone Information Hotline: 1-800-296-1996. Lists of certified equipment may be obtained by contacting ARI at 703-524-8800 and UL at 708-272-8800 ext. 42371.

This information was provided by the EPA. Visit the Internet to get the complete document in it's original form. A number of Web sites are also in this chapter and provide more additional information important to technicians.

Figures 12-1 and 12-2, and Table 12-1 are examples of some of the information available on the Internet.

Form Approved
OMB No. 2060-0256
Expires: 07/31/2010

ENVIRONMENTAL PROTECTION AGENCY
REFRIGERANT RECOVERY OR RECYCLING DEVICE
ACQUISITION CERTIFICATION FORM

EPA regulations require establishments that service or dispose of refrigeration or air-conditioning equipment to certify that they have acquired recovery or recycling devices that meet EPA standards for such devices. To certify that you have acquired equipment, please complete this form according to the instructions and **mail it to the appropriate EPA Regional Office. BOTH THE INSTRUCTIONS AND MAILING ADDRESSES CAN BE FOUND ON THE REVERSE SIDE OF THIS FORM.**

PART 1: ESTABLISHMENT INFORMATION

Name of Establishment

(Area Code) Telephone Number

Number of Service Vehicles Based at Establishment

Street

City   State   Zip Code

County

PART 2: REGULATORY CLASSIFICATION

Identify the type of work performed by the establishment. **Check all boxes that apply.**

☐ Type A - Service small appliances
☐ Type B - Service refrigeration or air-conditioning equipment other that small appliances
☐ Type C- Dispose of small appliances
☐ Type D - Dispose of refrigeration or air-conditioning equipment other than small appliances

PART 3: DEVICE IDENTIFICATION

	Name of Device(s) Manufacturer	Model Number	Year	Serial Number (if any)	Check Box if Self-Contained
1.					☐
2.					☐
3.					☐
4.					☐
5.					☐

PART 4: CERTIFICATION SIGNATURE

I certify that the establishment in Part 1 has acquired the refrigerant recovery or recycling device(s) listed in Part 2, that the establishment is complying with Section 608 regulations, and that the information gives is true and correct.

Signature of Owner/Responsible Officer   Date   Name (Please Print)   Title

EPA FORM 7610-31

Figure 12-1   Environmental protection agency.

Form Approved
OMB No. 2060-0256
Expires: 07/31/2010

INSTRUCTIONS

Part 1: Please provide the name, address, and telephone number of the establishment where the refrigerant recovery or recycling device(s) is (are) located. Please complete one form for each location. State the number of vehicles based at this location that are used to transport technicians and equipment to and from service sites.

Part 2: Check the appropriate boxes for the type of work performed by technicians who are employees of the establishment. The term "small appliance" refers to any of the following products that are fully manufactured, charged, and hermetically sealed in a factory with five pounds or less of refrigerant: refrigerators, and freezers designed for home use, room air conditioners (including window air conditioners and packaged terminal air conditioners), packaged terminal heat pumps, dehumidifiers, under-the-counter ice makers, vending machines, and drinking water coolers.

Part 3: For each recovery or recycling device acquired, please list the name of the manufacturer of the device, and (if applicable) its model number and serial number.

If more that seven devices have been acquired, please fill out an additional form and attach it to this one. Recovery devices that are self-contained should be listed first and should be identified by checking the box in the last column on the right. Self-contained recovery equipment means refrigerant recovery or recycling equipment that is capable of removing te refrigerant from an appliance without the assistance of components contained in the appliance. On the other hand, system-dependent recovery equipment means refrigerant recovery equipment that requires the assistance of components contained in an appliance to remove the refrigerant from the appliance.

If the establishment has been listed as Type B and/or Type D in Part 2, then the first device listed in Part # must be a self-contained device and identifies as such by checking the box in the last column on the right.

If any of the devices are homemade, they should be identified by writing "homemade" in the column provided for listing the name of the device manufacturer. Type A or Type B establishments can use homemade devices manufactured before November 15, 1993. Type C or Type D establishments can use homemade manufactured anytime. If, however, a Type C or Type D establishment is using homemade equipment manufactured after November 15, 1993, then it must <u>not</u> use these devices for service jobs.

EPA FORM 7610-31   -i-

Figure 12-1 *(Continued)*

Form Approved
OMB No. 2060-0256
Expires: 07/31/2010

EPA REGIONAL OFFICES

Send your form to the EPA office listed under the state or territory in which the establishment is located.

Connecticut, Maine, Massachusetts, New Hampshire, Rhode Island, Vermont

> CAA 608 Enforcement Contact: EPA Region I; Mail Code SEA; JFK Federal Building; One Congress Street, Suite 1100; Boston, MA 02114-2023

New York, New Jersey, Puerto Rico, Virgin Islands

> CAA 608 Enforcement Contact: EPA Region II; Mail Code 2DECA-AC; 290 Broadway; New York, NY 10007-1866

Delaware, District of Columbia, Maryland, Pennsylvania, Virginia, West Virginia

> CAA 608 Enforcement Contact: EPA Region III-Wheeling Office; Mail Code 3AP12; 303 Methodist Building; 11th and Chapline Streets; Wheeling, WV 26003

Alabama, Florida, Georgia, Kentucky, Mississippi, North Carolina, South Carolina, Tennessee

> CAA 608 Enforcement Contact: EPA Region IV; Mail Code APT-AE; 61 Forsyth Street, SW; Atlanta, GA 30303-8960

Illinois, Indiana, Michigan, Minnesota, Ohio, Wisconsin

> CAA 608 Enforcement Contact: EPA Region V; Mail Code AE-17J; 77 West Jackson Blvd.; Chicago, IL 60604

Arkansas, Louisiana, New Mexico, Oklahoma, Texas

> CAA 608 Enforcement Contact: EPA Region VI; Mail Code 6EN-AA; 1445 Ross Ave., Suite 1200; Dallas, TX 75202

Iowa, Kansas, Missouri, Nebraska

> CAA 608 Enforcement Contact: EPA Region VII; Mail Code APCOARTD; 901 North Fifth Street; Kansas City, KS 66101

Colorado, Montana, North Dakota, South Dakota, Utah, Wyoming

> CAA 608 Enforcement Contact: EPA Region VIII; Mail Code 8ENF-T; 1595 Wynkoop Street, Denver, CO 80202-1129

American Samoa, Arizona, California, Guam, Hawaii, Nevada

> CAA 608 Enforcement Contact: EPA Region IX; Mail Code AIR-5; 75 Hawthorne Street; San Francisco, CA 94105

Alaska, Idaho, Oregon, Washington

> CAA 608 Enforcement Contact: EPA Region X; Mail Code OAQ-107; 1200 Sixth Ave.; Seattle, WA 98101

Figure 12-1 *(Continued)*

Form Approved
OMB No. 2060-0256
Expires: 07/31/2010

PUBLIC BURDEN

The purpose and need of this renewed collection request is to facilitate compliance with and enforcement of Section 608 of the Act by reducing emissions of class I and class II ozone-depleting refrigerants to the lowest achievable level during the service, maintenance, repair, and disposal of appliances. EPA has used and will continue to use these records and reports to ensure that refrigerant releases are minimized during the recovery and recycling of ozone-depleting refrigerants during the service, maintenance, repair, and disposal of appliances. Collection of this information is mandated by EPA regulations, in accordance with 40 CFR 82.162. This information is not shared with parties outside of the Federal government. EPA's confidentiality regulations (40 CFR 2.201 et seq.) assure computer data security, disclosure prevention, proper handling, proper storage, and proper disposal of the submitted information.

The public reporting and recordkeeping burden for this collection of information is estimated to average one (1) hour per response per respondent annually. Burden means the total time, effort, or financial resources expended by persons to generate, maintain, retain, or disclose or provide information to or for a Federal agency. This includes the time needed to review instructions; develop, acquire, install, and utilize technology and systems for the purposes of collecting, validating, and verifying information, processing and maintaining information, and disclosing and providing information; adjust the existing ways to comply with any previously applicable instructions and requirements; train personnel to be able to respond to a collection of information; search data sources; complete and review the collection of information; and transmit or otherwise disclose the information. An agency may not conduct or sponsor, and a person is not required to respond to, a collection of information unless it displays a currently valid OMB control number.

To comment on the Agency's need for this information, the accuracy of the provided burden estimates, and any suggested methods for minimizing respondent burden, including the use of automated collection techniques, EPA has established a public docket for this ICR under Docket ID No. OAR-2003-0018, which is available for public viewing at the Air and Radiation Docket and Information Center in the EPA Docket Center (EPA/DC), EPA West, Room B102, 1301 Constitution Ave., NW, Washington, DC. The EPA Docket Center Public Reading Room is open from 8:30 a.m. to 4:30 p.m., Monday through Friday, excluding legal holidays. The telephone number for the Reading Room is (202) 566-1744, and the telephone number for the OAR Docket is (202) 566-1742. An electronic version of the public docket is available through EPA Dockets (EDOCKET) at http://www.epa.gov/edocket. Use EDOCKET to submit or view public comments, access the index listing of the contents of the public docket, and to access those documents in the public docket that are available electronically. Once in the system, select "search," then key in the docket ID number identified above. Also, you can send comments to the Office of Information and Regulatory Affairs, Office of Management and Budget, 725 17th Street, NW, Washington, DC 20503, Attention: Desk Office for EPA. Please include the EPA Docket ID No. (OAR-2003-0018) and OMB control number (2060-0256) in any correspondence.

Figure 12-1 *(Continued)*

Technician Certification (Section 608): Steps For Replacing a Lost Card

STEP 1. Is your testing organization that issued your certification still in business? Check the list of certifying organizations that are still operating here.

> Yes, my organization is still operating.
> Go to that organization and get a replacement card. They are required to maintain records of people issued cards.
>
> No. - Go to 2 below.

STEP 2. Do you have documentation from your original testing organization that demonstrates successful completion of the Section 608 Technician Certification exam? Do you or a current or former employer have a copy of your lost card?

> Yes. I have documentation from my original testing organization.
> Go to the list of certifying organization that will replace cards. Send a copy of your documentation to one of the organizations (who have volunteered to make cards for people who can't get them from their certifying organization) on the list . They will issue you a new card and they will maintain a record of your certification.
>
> No. - Go to 3 below.

STEP 3. Is the record of your certification in our centralized files which were compiled from data submitted by certifying organizations that have gone out of business? Go to the list of certifying organizations that have closed.

> Yes, the record of my certification is in the data submitted by companies that have gone out of business.
> Download and complete the Assistance with Obtaining a Replacement Card form (PDF, 1 pp., 14K, about PDF). After completing the form mail or fax it to the Section 608 Technician Certification Program Manager. Once it is received we will contact you with information on how to obtain a replacement card.
>
> No. - Go to 4 below.

STEP 4. If you cannot answer "yes" to any of the steps above, EPA will not issue a replacement card. You will need to retake the Section 608 certification test. Please go to the Section 608 Technician Certification Programs page to find testing organizations which meet your needs.

Figure 12-2   Replacing a lost card.

Programming Thermostats

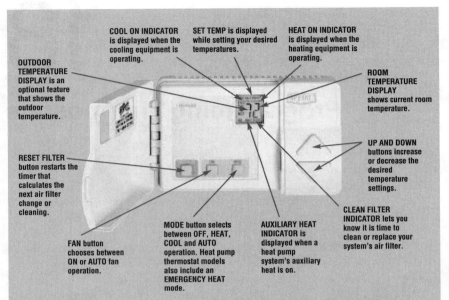

COOL ON INDICATOR is displayed when the cooling equipment is operating.

SET TEMP is displayed while setting your desired temperatures.

HEAT ON INDICATOR is displayed when the heating equipment is operating.

OUTDOOR TEMPERATURE DISPLAY is an optional feature that shows the outdoor temperature.

ROOM TEMPERATURE DISPLAY shows current room temperature.

RESET FILTER button restarts the timer that calculates the next air filter change or cleaning.

UP AND DOWN buttons increase or decrease the desired temperature settings.

CLEAN FILTER INDICATOR lets you know it is time to clean or replace your system's air filter.

MODE button selects between OFF, HEAT, COOL and AUTO operation. Heat pump thermostat models also include an EMERGENCY HEAT mode.

AUXILIARY HEAT INDICATOR is displayed when a heat pump system's auxiliary heat is on.

FAN button chooses between ON or AUTO fan operation.

NOTE: Not all messages displayed in above illustration will appear at once in any situation.

Comfort At Your Command.

This is no ordinary thermostat. Bryant listened to the needs of homeowners nationwide and delivered a product to meet those needs. The result is a thermostat that interacts with people as effectively as it does with your heating and cooling system. It's a simple, yet powerful control that puts comfort at your command.

Making Life Easier.

Take a few minutes to review the features and functions listed above. Bryant gives you control over your comfort with simple instructions, responsive push buttons and an easy-to-read backlit display. Once set, this thermostat reliably monitors indoor temperatures and responsively meets your comfort demands.

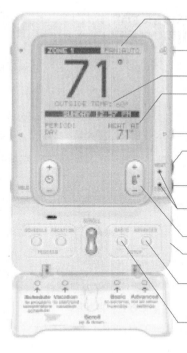

Fan Setting

Fan Button chooses "high," "medium," "low," or "auto" fan mode.

Outside Temperature

Desired Heating/Cooling Temperature

Right Button provides system status.

Heat Button selects heating operation.

Cool Button selects cooling operation.

Heat/Cool LEDs indicate heating or cooling operation.

Temp (+\-) Button

Off Button turns the system on and off.

Advanced Setup Button provides access to customizable features.

Basic Setup Button provides access to current day, time and desired humidity level.

HOW TO DETERMINE A MODEL NUMBER FOR TOTALINE, CARRIER, AND BRYANT THERMOSTATS

Gold Series

All of these thermostats have a door with Up and Down buttons on the right-hand side.

The **Standard Residential** stats have the following buttons from left to right:
Reset Filter – Fan - Mode
There are three different models in this group and the same Owners Manual covers all three.

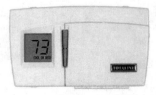

Old Version

New Version As Of 11/1/04

Totaline: P274-0100, 0200, 0300
Carrier: TSTATCCNAC01-B, NHP01-B, N2S01-B
Bryant: TSTATBBNAC01-B, NHP01-B, N2S01-B

Totaline: P274-0100-C, 0200-C, 0300-C
Carrier: TSTATCCNAC01-C, NHP01-C, N2S01-C
Bryant: TSTATBBNAC01-C, NHP01-C, N2S01-C

The **Programmable Residential** stats have the following buttons from left to right:
Top Row: **Copy Previous Day – Program – Mode**
Middle row: **Change Day – End – Fan**
Bottom Row: **Set Time/Temp – Reset Filter - Hold**
There are three different models in this group and the three same Owners Manual covers all three.

Totaline: P274-1100, 1200, 1300
Carrier: TSTATCCPAC01-B, PHP01-B, P2S01-B
Bryant: TSTATBBPAC01-B, PHP01-B, P2S01-B

In addition to these three, Carrier and Bryant also have two other models.
The Duel Fuel model has the same appearance and buttons as the others, so you would have the customer pull the stat apart in order to read the model # on the back of the circuit board. These stats open like a door from left to right.
Carrier: TSTATCCPDF01-B
Bryant: TSTATBBPDF01-B

The Thermidistat also has the same appearance, but the following buttons:
Top Row: **Copy Previous Day – Program – Mode**
Middle row: **Change Day – Humidity – Fan**
Bottom Row: **Set Time/Temp – Vacation – Hold/End**

Carrier: TSTATCCPDF01-B
Bryant: TSTATBBPDF01-B

The **Commercial** stat has the same appearance as the Residential stats, but has slightly different buttons.
Top Row: **Copy Previous Day – Program – Mode**
Middle row: **Change Day – End – Fan**
Bottom Row: **Set Time/Temp – Occupied <Reset Filter>Hold**

Signature Series

The **Standard Residential** stats do not have a door and contain an Up and Down button on the right side and two buttons that read **Mode** and **Fan** under the display.
There are two Carrier/Bryant models and four Totaline models. One manual is used for all models.

Totaline: P374-0000, 0100, 0200, 0300
Carrier: TSTATCCBAC01-B, TSTATCCBHP01-B
Bryant: TSTATBBBAC01-B, TSTATBBBHP01-B

The **Programmable Residential** stats have three different styles.
The first style has a door with three buttons on the outside. There are four different versions.

Version #1: **Fan – Emergency Heat – Backlight – Program – Set Clock – Mode**
These functions refer to the following models:
Totaline: P374-1000
Carrier: TSTATCCPS101
Bryant: TSTATBBPS101

Version #2: **Fan – Outside – Vacation – Program – Set Clock**
These functions refer to the following models:
Totaline: P374-1100
Carrier: TSTATCCPS701
Bryant: TSTATBBPS701

Version #3: **Fan – Outside – Vacation – Program – Set Clock**
While these functions are the same as Version #2, check to see if the customer has a duel fuel system ... i.e., A heat pump and a gas furnace.
Totaline Only: P374-1500

Version #4: "INTELLISTAT" **Fan – Outside – Humidity – Program – Set Clock - Mode**
Totaline Only: P374-1600

The second style is a flushmount or "flatstat". It has four buttons under the display, plus a raised round bubble where the sensor is enclosed. There are two different versions that both have the same buttons.
Mode – Fan – Up - Down

1-Day or Nonprogrammable:
Totaline: P374-1000FM
Carrier: TSTATCCPF101
Bryant: TSTATBBPF101

7-Day   Programmable:
Totaline: P374-1100FM
Carrier: TSTATCCPF701
Bryant: TSTATBBPF701

The third style has the Up/Down buttons on the right and six buttons under the display. There are two versions but the manual is the same for both. These are not available in Totaline branding.
Top Row: **Mode – Fan**
Bottom Row: **Program – Time/Temp – Day – Hold/End**

Carrier: TSTATCCSAC01 and TSTATCCSHP01
Bryant: TSTATBBSAC01 and TSTATBBSHP01

The **Commercial** stats have two different styles and several different versions.
The first style has a door with three buttons on the outside. There are four different versions.

Version #1: **Mode – Fan – Emerg. Heat – Backlight – Reset Filter**
Totaline: P374-2100
Carrier: 33CS071-01
Bryant: TSTATBB071-01

Version #2: **Mode – Fan – Holiday – Program – Set Clock**
Totaline: P374-2200 or P374-2200LA (Light Activated)
Carrier: 33CS220-01 or 33CS220-LA
Bryant: TSTATBB220-01 or TSTATBB220-LA

Version #3: **Mode – Fan – Holiday – Program – Set Clock**
While these buttons are the same as Version #2, there is a difference, please call.
Totaline: P374-2300 or P374-2300LA (Light Activated)
Carrier: 33CS250-01 or 33CS250-LA
Bryant: TSTATBB250-01 or TSTATBB250-LA

The second style is a flushmount or "flatstat." It has four buttons under the display, plus a raised round bubble where the sensor is enclosed. There are two different versions that both have the same buttons. This is not available in the Bryant branding.
Mode – Override – Up - Down

Standard:
Totaline: P374-2200FM
Carrier: 33CS220-FS

Delux:
Totaline: P374-2300FM
Carrier: 33CS250-FS

Star Series

The **Standard Residential** stats have two different styles.
The first style is a battery operated stat that has the Up/Down buttons plus slide switches on the bottom – and the side if a Heat Pump model (P474-0140).

Totaline: P474-0130   Totaline Only: P474-0140
Carrier: TSTATCCNQ001
Bryant: TSTATBBNQ001

The second style has two buttons on each side of the display. There are two different versions, but they both have the same buttons. Have the consumer pull the stat off of the backplate. The part # is located on the circuit board.
Left side: **Mode – Fan**   Right side: **Up - Down**

Totaline: P474-0100 or P474-0220
Carrier: TSTATCCNB001 or TSTATCCN2W01
Bryant: T STATBBNB001 or TSTATBBN2W01

The **Programmable Residential** stats have four different styles.
The first style has two buttons on each side of the display. It looks exactly like the above picture, but is 1-day programmable. Have the customer pull the body of the stat off of the back plate. The part # is located on the circuit board.

Totaline: P474-1010
Carrier: TSTATCCBP101
Bryant: TSTATBBPB101

The second style is a battery operated stat that has the Up/Down buttons plus slide switches on the bottom and the side.

Totaline: P474-1035
Carrier: TSTATCCPQ501
Bryant: TSTATBBPQ501

The third style has four buttons all located under the display. There are two different versions so the customer needs to pull the body of the stat off of the back plate. The part # is printed on the circuit board. **Mode – Fan – Down - Up**

Totaline: P474-1020 or P474-1050
Carrier: TSTATCCP2W01 or TSTATCCPB501
Bryant: TSTATBBP2W01 or TSTATBBPB501

There are also two Commercial versions of this stat. One version has the same buttons as the Residential model. Again, the customer needs to pull the body of the stat off to determine the model #. The second version has one different button: **Mode – Override – Down- Up**

Totaline: P474-2050 and P474-2150
Carrier: 33CSN2-WC and 33CSSP2-WC
Bryant: Not Available

The fourth style is the wireless stat. It consists of two parts, the transmitter and the receiver. The transmitter has four buttons located vertically under the display: **Up – Down – Mode – Fan**

Totaline: P474-1100RF and P474-1100REC
Carrier: TSTATCCPRF01 and TSTATCCREC01
Bryant: TSTATBBPRF01 and TSTATBBREC01

The Commercial version of the wireless stat looks exactly the same, but one button changes:
Up – Down – Mode – Override

Totaline: P474-2300RF and P474-2300REC
Carrier: 33CS250-RC and 33CS250-RE
Bryant: Not Available

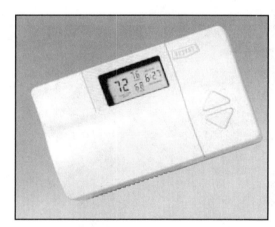

Thermidistat Control™ Homeowner's Guide

Simple, Energy Efficient Control.

As the owner of a Bryant Thermidistat Control, you can look forward to years of reliable, energy-efficient indoor comfort. This smart, easy-to-use control center combines computer-like intelligence with simple operation in a streamlined design. It maximizes the performance of your Bryant indoor comfort system so you and your family can enjoy consistent indoor temperatures enhanced by proper humidity control. The precision performance of our Thermidistat Control keeps you comfortable while conserving energy. Just follow the simple instructions outlined in this manual and let Bryant's Thermidistat Control deliver the simple, worry-free comfort and extra energy savings that you deserve.

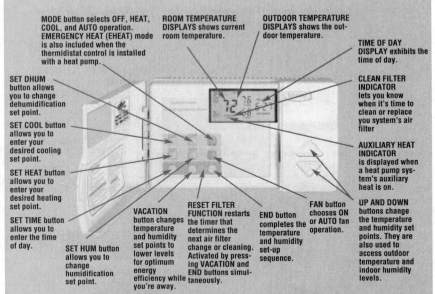

MODE button selects OFF, HEAT, COOL, and AUTO operation. EMERGENCY HEAT (EHEAT) mode is also included when the thermidistat control is installed with a heat pump.

ROOM TEMPERATURE DISPLAYS shows current room temperature.

OUTDOOR TEMPERATURE DISPLAYS shows the outdoor temperature.

TIME OF DAY DISPLAY exhibits the time of day.

SET DHUM button allows you to change dehumidification set point.

CLEAN FILTER INDICATOR lets you know when it's time to clean or replace you system's air filter

SET COOL button allows you to enter your desired cooling set point.

AUXILIARY HEAT INDICATOR is displayed when a heat pump system's auxiliary heat is on.

SET HEAT button allows you to enter your desired heating set point.

SET TIME button allows you to enter the time of day.

VACATION button changes temperature and humidity set points to lower levels for optimum energy efficiency while you're away.

RESET FILTER FUNCTION restarts the timer that determines the next air filter change or cleaning. Activated by pressing VACATION and END buttons simultaneously.

END button completes the temperature and humidity set-up sequence.

FAN button chooses ON or AUTO fan operation.

UP AND DOWN buttons change the temperature and humidity set points. They are also used to access outdoor temperature and indoor humidity levels.

SET HUM button allows you to change humidification set point.

NOTE: Not all messages displayed in above illustration will appear at once in any situation.

Year-Round Comfort.

Bryant's Thermidistat Control provides enhanced year-round comfort. By providing both temperature and humidity control, this simple yet powerful device lets you take the steam out of summer, the static out of winter and helps eliminate that annoying moisture build-up on your windows when it's cold outside. Comfort has never been quite this easy.

Making Life Easier.

Bryant puts your family's comfort at your fingertips with simple instructions, responsive push buttons an an easy-to-read backlit LCD display. Once set, the Thermidistat Control accurately monitors indoor conditions and reliably meets your needs for consistent, energy-efficient indoor comfort.

TABLE OF CONTENTS i

SETTING UP THE THERMIDISTAT CONTROL 1

Setting your temperature set points

Setting the desired heating set point

SET HEAT
1 🔲 Press the SET HEAT button.

The word HEAT flashes on the display.

2 △▽ Press the UP or DOWN
button until the correct
heating set point is displayed.

3 *END* ◯ Press the END button to exit.

Setting the desired cooling set point

SET COOL
1 🔲 Press the SET COOL button.

The word COOL flashes on the display.

SETTING UP THE THERMIDISTAT CONTROL 2

2 △▽ Press the UP or DOWN
button until the correct
cooling set point is displayed.

3 *END* ◯ Press the END button to exit.

*NOTE: Your COOL setting must be at least
2° higher than your HEAT setting. Your
Thermidistat Control will automatically
change your previously set temperature to
maintain that 2° difference.*

Setting the current time

SET TIME
1 ◯ Press the SET TIME button.

The word TIME flashes on the display.

2 △▽ Press the UP or DOWN
button until the correct
time is displayed. To quickly advance to

SETTING UP THE THERMIDISTAT CONTROL   3

the proper time, press and hold the UP or DOWN button. Be sure that AM or PM is properly selected.

3 ◯ *END*   When the correct time appears on the display, press the END button.

NOTE: If you choose not to press the END button, the word TIME will stop flashing after 15 seconds.

NOTE: If you live in a Daylight Savings Time area, you may advance the time 1 hour in the spring by simultaneously pressing the SET TIME button and the UP button. In the fall, you may set the time back 1 hour by simultaneously pressing the SET TIME button and the DOWN button.

Setting the mode operation

1 ◯ *MODE*   Press the MODE button.

SETTING UP THE THERMIDISTAT CONTROL   4

2 When the MODE button is pressed, the words

OFF,

HEAT,

COOL,

and AUTO
will rotate on the display.
If you have a heat pump, the display will show OFF, HEAT, COOL, AUTO, and EHEAT.

SETTING UP THE THERMIDISTAT CONTROL   5

NOTE: In AUTO mode, your system will heat OR cool as needed to reach your temperature settings. AUTO mode may be disabled.

3 ◯ *MODE*   Continue to press the MODE button until you reach the desired setting.

Setting the fan operation

1 ◯ *FAN*   Press the FAN button to switch between ON and AUTO fan settings.

When the ON mode is selected, the fan runs continuously for improved air circulation.

SETTING UP THE THERMIDISTAT CONTROL   6

NOTE: If the indoor humidity level is above the dehumidify setting, the fan will shut off for 5 minutes after the air conditioner shuts off. During this period, a triangle next to the word ON will flash.

When the AUTO mode is selected, the fan runs only as needed to maintain your preferred indoor temperature settings.

Setting the humidification set point

SET HUM

1   Press the SET HUM button to enter the humidify mode.

The current indoor humidity (large number) and humidify set point (small number) are displayed along with the humidify indicator (hu).

SETTING UP THE THERMIDISTAT CONTROL   7

2   With the humidify indicator (hu) displayed, press the UP or DOWN button to adjust the humidify set point. Humidity levels can be set from 10% to 45%.

Or, to turn humidification off, press the MODE button until "OF" appears on the display. (See suggested settings on page 27.)

MODE

3   Press the MODE button to select between:

a. FAN indicator displayed — fan and humidifier on every time humidification is needed.

NOTE: With the FAN indicator displayed, you will realize maximum humidification, but the air may feel cold because of the heat.

SETTING UP THE THERMIDISTAT CONTROL   8

source is not always on. Without the FAN indicator displayed, you will conserve water and electricity with adequate humidification.
b. AUTO indicator displayed — humidify setting automatically changes according to outdoor weather. This reduces the chance of moisture buildup on windows in colder weather.

NOTE: This feature requires the use of an outdoor air temperature sensor.

c. AUTO and FAN displayed — combines the features of a and b.

d. OF displayed — humidify function is turned off.

SETTING UP THE THERMIDISTAT CONTROL   9

e. Humidify setting only displayed —
humidify setting does not change
according to outdoor temperature.

$$60 \; 45 \; hu$$

4 ○ *END* Press the END button to
exit the humidify mode.

Setting the dehumidification set point
*NOTE: This function is for use with
variable-speed equipment only.*

1 ▣ *SET DHUM* Press the SET DHUM
button to enter the dehumidify mode.

$$60 \; ^{65} \; d\,hu$$

The current indoor humidity (large number)
and dehumidify set point (small number)
are displayed along with the dehumidify
indicator (dhu).

SETTING UP THE THERMIDISTAT CONTROL   10

2 △▽ With the dehumidify
indicator (dhu) displayed,
press the UP or DOWN button to adjust
the dehumidify set point. Dehumidify
levels can be set from 50% to 90%. Or, to
turn dehumidification off, press the MODE
button until "OF" appears on the display.

$$60 \; ^{OF} \; d\,hu$$

(See suggested settings on page 26.)

3 ○ *END* Press the END button to exit.

**Setting the "cool to dehumidify"
function**
This setting allows a standard comfort
system to provide moderate dehumidification by running the air conditioner.
The function can also be used with
variable-speed equipment.

*NOTE: While in the "cool to dehumidify"
mode, the indoor air temperature will not
drop more than 3° below the cooling set
point with a dehumidification demand.*

SETTING UP THE THERMIDISTAT CONTROL   11

1 ▣ *SET DHUM* Press the SET DHUM
button. "dhu" is displayed.

2 △▽ Press the UP or DOWN
button to raise or lower the
dehumidify set point.
Dehumidification can be set from 50%
to 90%.

3 ○ *MODE* Press the MODE button
until the COOL icon is displayed.

$$_{COOL}\,60 \; ^{65} \; d\,hu$$

4 ○ *END* Press the END button to exit.

OPERATING THE THERMIDISTAT CONTROL 12

Checking current temperature

The Thermidistat Control will display the current temperature.

To view your current temperature set points, press the UP or DOWN button once. The heating and cooling set points will be displayed.

Checking the outdoor temperature and indoor humidity

1 Press the UP and DOWN buttons simultaneously.

2

The outdoor temperature will appear on the display. Then, the indoor humidity will be displayed.

OPERATING THE THERMIDISTAT CONTROL 13

NOTE: If two dashes (--) appear, your Thermidistat Control does not include the outdoor air temperature sensor or the sensor is not working properly. Check with your dealer if you are unsure.

Checking current humidification and dehumidification set points

1 *SET HUM* ▢ Press the SET HUM button.

The current indoor humidity (large number) is displayed along with the humidify set point (small number).

```
60  45  hu
```

2 *END* ◯ Press the END button.

OPERATING THE THERMIDISTAT CONTROL 14

3 *SET DHUM* ▢ Press the SET DHUM button.
The current indoor humidity (large number) is displayed along with the dehumidify set point (small number).

```
60  65  dhu
```

4 *END* ◯ Press the END button.

Clean filter feature

Your Thermidistat Control reminds you when it's time to change or clean your filter by displaying the CLEAN FILTER indicator.

1 *VACATION* ◯ Press the VACATION and
END ◯ END buttons simultaneously after you have changed or cleaned your filter to restart the timer.

OPERATING THE THERMIDISTAT CONTROL   **15**

Vacation feature setup

The vacation feature allows a separate set of temperature and humidity set points to be stored for vacation and recalled with a single button press.

The vacation feature is preprogrammed for you with vacation settings for temperature and humidity. (Heat 60°, cool 85°, hu 10%, dhu 75%) If these are okay, skip ahead to "vacation feature operation." If you wish to enter new settings, continue with this section.

VACATION

1 ◯ Press the VACATION button to display the vacation temperature settings.

The OUT indicator is displayed.

OPERATING THE THERMIDISTAT CONTROL   **16**

2 To change the cooling set point:

SET COOL

▢ a. Press the SET COOL button. COOL will flash on the display.

△ b. Press the UP or DOWN
▽ button to adjust the setting.

END
◯ c. Press the END button to end.

3 To change the heating set point:

SET HEAT

▢ a. Press the SET HEAT button. HEAT will flash on the display.

△ b. Press the UP or DOWN
▽ button to adjust the setting.

END
◯ c. Press the END button to end.

OPERATING THE THERMIDISTAT CONTROL   **17**

4 To change the dehumidification set point:

SET DHUM

 a. Press the SET DHUM button. The "dhu" indicator will be displayed.

△ b. Press the UP or DOWN
▽ button to adjust the setting.

MODE
◯ c. Press the MODE button to choose the dehumidification mode.

END
◯ d. Press the END button to end.

OPERATING THE THERMIDISTAT CONTROL   18

5 To change the humidification set point:

SET HUM

■ a. Press the SET HUM button. The "hu" indicator will be displayed.

```
 ┌──────────────────┐
 │      OUT          │
 │  60  ₁0  hu      │
 │                   │
 └──────────────────┘
```

△
▽ b. Press the UP or DOWN button to adjust the setting.

MODE
○ c. Press the MODE button to turn the humidification feature off (OF).

END
○ d. Press the END button to end.

NOTE: In dehumidify, you may enter a set point, choose "COOL" to dehumidify, or turn dehumidification off (OF). In humidify, you may enter a setting or turn humidification off (OF).

OPERATING THE THERMIDISTAT CONTROL   19

Vacation feature operation

VACATION
1 ○ Press the VACATION button when you are ready to leave. Be sure you have properly selected the mode (HEAT, COOL, AUTO).

```
 ┌──────────────────────┐
 │ Mode   OUT            │
 │ AUTO  72 85  AM      │
 │ Fan     COOL         │
 │ AUTO  60  6:00       │
 │        HEAT          │
 └──────────────────────┘
```

The OUT indicator is displayed, and your system will automatically follow your vacation temperature and humidity settings.

VACATION
2 ○ Press the VACATION button when you return to resume normal operation.

AUTO CHANGEOVER   20

Your Thermidistat Control provides complete, automatic control over heating and cooling with auto changeover. Auto changeover means your system will automatically heat or cool as needed to maintain your temperature set points.

Auto changeover makes life easier because you no longer have to manually switch the thermostat between heating or cooling operation. Just set your heating and cooling set points and let the Thermidistat Control do the rest!

NOTE: If Auto Changeover mode is not necessary in your area of the country, your installer may disable the AUTO mode.

WHAT IF... 21

AUXILIARY HEAT indicator is displayed...

The AUXILIARY HEAT indicator appears on the heat pump version of the Thermidistat Control only. It is displayed when your system is operating on auxiliary heat.

NOTE: This indicator does not reflect a problem with your system.

CLEAN FILTER indicator is displayed ...

The CLEAN FILTER indicator tells you when to clean or replace your system's air filter. Press the VACATION and END buttons simultaneously after cleaning or replacing the filter to turn off the indicator and restart the timer.

NOTE: This indicator does not reflect a problem with your system.

WHAT IF... 22

OUT Indicator is displayed ...

The OUT indicator reminds you that your system is in vacation mode. This function automatically adjusts the temperature and humidity settings to levels appropriate for when you're away. Press the VACATION button to resume normal system operation.

NOTE: This indicator does not reflect a problem with your system.

EQUIPMENT ON Indicator is displayed ...

When the cooling equipment is operating, the word COOL preceded by a small triangle is displayed below the cooling set point. When the heating equipment is operating, the word HEAT preceded by a small triangle is displayed below the heating set point. If the equipment turn on is being delayed, the triangle and the word will flash.

NOTE: This indicator does not reflect a problem with your system.

WHAT IF ... 23

You have a power outage ...

An internal power source eliminates the need to re enter your settings into the Thermidistat Control after power outages. The comfort settings you have entered will be maintained indefinitely. The clock will run for 8 hours.

You have a system error message ...

The display may appear as follows:

--, E3, E4, E5, or E6

-- indicates a problem with the indoor air-temperature sensor

E3 indicates a problem with the outdoor air-temperature sensor

E4, E5, or E6 indicates an internal failure.

WARRANTY 24

This Thermidistat Control includes a 1-year limited warranty. For detailed warranty information, please refer to the All Product Limited Warranty Card included in your information packet. This Thermidistat Control is also eligible for manufacturer's extended system warranties. Ask your dealer for details on extended warranties for longer-term protection.

COMMON TERMS AND WHERE TO FIND THEM 25

Auxiliary Heat ....Pg. 21

Most heat pump systems require a supplemental heating source, called auxiliary heat, to maintain your comfort when outdoor temperatures fall significantly. Your Thermidistat Control lets you know when your home is being warmed with supplemental heat.

Clean Filter ....Pg. 14

Your system's air filter will require regular cleaning to reduce the dirt and dust in the system and your indoor air. The CLEAN FILTER indicator lets you know when it's time to clean the filter.

COMMON TERMS AND WHERE TO FIND THEM 26

Dehumidification Set Point ....Pg. 9

The amount of moisture to be removed from your home. You can check your actual humidity level and your desired dehumidification set point by pressing the SET DHUM button.

Suggested settings: 50% – 60% suggested depending on installation, area of the country, and your heating and cooling equipment.

Emergency Heat ....Pg. 4

This indicates that auxiliary heat is being used without the heat pump.

End ....Pg. 1

The END button returns the Thermidistat Control to normal operation.

COMMON TERMS AND WHERE TO FIND THEM 27

Fan ....Pg. 5

Your system's fan can run continuously or only as called for during heating or cooling. Continuous operation helps with air circulation and cleaning. Automatic operation provides energy savings. Press the FAN button to make your choice.

Humidification Set Point ....Pg. 6

The amount of moisture desired in your home to be supplied by the humidifier. You can check the actual humidity level and your desired humidification set point by pressing the SET HUM button.
Suggested settings:

Outdoor Temp (F)	-20	-10	0	10	20
Suggested Hum Set Point	15	20	25	30	35

COMMON TERMS AND WHERE TO FIND THEM 28

Mode ....Pgs. 3-5

Mode refers to the type of operation your system is set up to perform. Mode settings include: OFF, HEAT, COOL, and AUTO. Heat pump systems also include EMERGENCY HEAT (EHEAT).

Outdoor Temperature ....Pg. 12

Your Thermidistat Control not only measures the indoor temperature, but it may also be equipped to measure and display the outdoor temperature as well. Press the UP and DOWN buttons simultaneously to read the outdoor temperature display.

COMMON TERMS AND WHERE TO FIND THEM 29

Power Outage ....Pg. 22

Complete loss of electricity. Your Thermidistat Control has an internal power source that allows the clock to continue to run for 8 hours or more without electricity. Settings are stored indefinitely without the aid of batteries.

Reset Filter ....Pg. 14

The reset filter function turns off the CLEAN FILTER indicator and restarts the timer. Press the VACATION and END buttons simultaneously after you've cleaned and replaced the system's air filter.

COMMON TERMS AND WHERE TO FIND THEM 30

Set Time ....Pg. 2

This function allows you to set the proper time. Press the SET TIME button to activate.

Temperature Sensor ....Pg. 12

Temperature sensors measure the current indoor or outdoor temperatures which are displayed on the Thermidistat Control.

Temperature Set Points ....Pg. 1

These are the desired heating and cooling set points entered into the Thermidistat Control. The actual room temperature will automatically be displayed, but you can check the desired temperature for the current mode by pressing either the UP or DOWN button.

Heating & Cooling Systems
Since 1904

7310 West Morris Street, Indianapolis, IN 46231

 Printed on recycled paper.

Copyright 1997 Bryant Heating and Cooling Systems
Form: OM17-25   Replaces: OM17-22   Printed in the U.S.A.   9-98   Catalog No. 13TS-TA11

COMMON TERMS AND WHERE TO FIND THEM   31

14

Control Devices

There are a number of devices needed to control the refrigeration process. Electricity and electronics are the prevalent type of control devices used in this field. Some of the devices, earlier models, usually, were controlled by hydraulics, pneumatics, or mechanicals.

Many types of controls are available for use on air-conditioning, refrigeration, and heating equipment. They come in many sizes and shapes and do the job well for a period of time, but they all require periodic inspections, repairs, and replacement.

Power Relays

One of the control devices is the power relay. It is one of the most often used controls for controlling compressors in refrigeration and air-conditioning. The power relay is also referred to as the *main conductor*. It is used to apply the main line voltage to the motor circuit. The coil of the relay is usually operated by voltages lower than the line provides. This means that it uses a transformer for the lower control-voltages (see Fig. 14-1).

Magnetic contactors are normally used for starting polyphase motors, either squirrel cage or single phase. Contactors may be connected at any convenient point in the main circuit between the fuses and the motor. Small control wires (using low voltage) may be run between the contactor and the point of control.

Motor-start relays

Relays are a necessary part of many control and pilot-light circuits. They are similar in design to contactors, but are generally lighter in construction so they carry smaller currents.

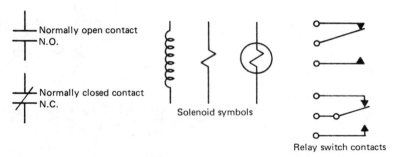

Figure 14-1   Symbols for the main contactor or power relay.

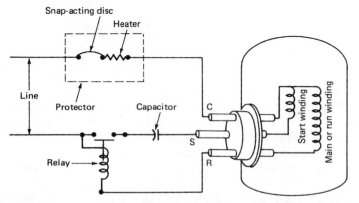

Figure 14-2   Capacitor-start, induction-run motor for a compressor with the potential relay used to take out the start winding once the motor comes up to speed.

Compressors used for household refrigerators, freezers, dehumidifiers, vending machines, and water coolers have the capacitor-start, induction-run type of motor. This type of compressor may have a circuit that resembles Fig. 14-2. When the compressor is turned on by the thermostat demanding action, the relay is closed and the start winding is in the circuit. Once the motor comes up to about 75 percent of rated speed, there is enough current flow through the relay coil to cause it to energize, and it pulls the contacts of the relay open, thereby taking the start capacitor and start winding out of the circuit. This allows the motor to run with one winding as designed.

Figure 14-3 shows the current type of relay. This is generally used with small refrigeration compressors up to 3/4 hp. Figure 14-4 shows the potential type of relay. This is generally used with large commercial air-conditioning compressors up to 5 hp.

Protection of the motor against prolonged overload is accomplished by time limit overload relays. They are operative during the starting period

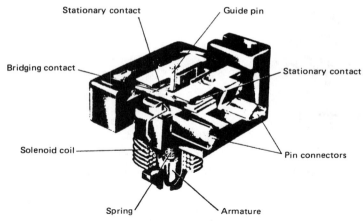

Figure 14-3 Current relay. (*Courtesy of Tecumseh.*)

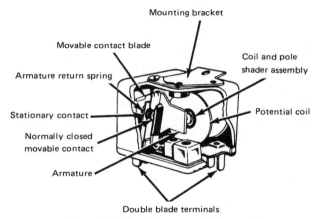

Figure 14-4 Potential relay. (*Courtesy of Tecumseh.*)

and running period. Relay action is delayed long enough to take care of the heavy starting currents and momentary overloads without tripping.

Relays with more than one contact

Some power relays are made with more than one set of contacts. They are used to cause a sequence of events to take place. The contacts can be wired into a circuit that controls functions other than the on-off operation of the compressor motor (see Fig. 14-5).

Thermal overload protectors

Motors for commercial units are protected by a bimetallic switch. The switch is operated on the heat principle. This is a built-in motor overload

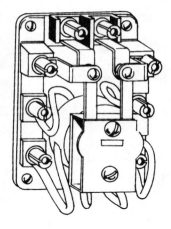

Figure 14-5 Relay with more than one set of contacts.

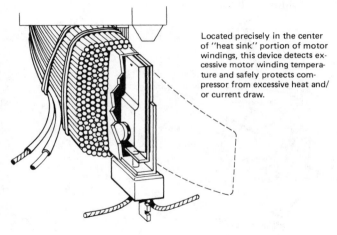

Located precisely in the center of "heat sink" portion of motor windings, this device detects excessive motor winding temperature and safely protects compressor from excessive heat and/or current draw.

Figure 14-6 Motor protector inserted in the windings of the compressor. (*Courtesy of Tecumseh.*)

protector (see Fig. 14-6). It limits the motor-winding temperature to a safe value. In its simplest form, the switch or motor protector consists essentially of a bimetal switch mechanism that is permanently mounted and connected in series with the motor circuit (see Fig. 14-7). Figure 14-8 shows the external line-break overload

Time-delay relays

In time-delay relays, bimetallic strips are heated with an electrical resistance mounted near or around them. The strips expand when heated. When they expand, they make contact and complete the circuit with their contacts closed (see Fig. 14-9). The time delay can be adjusted

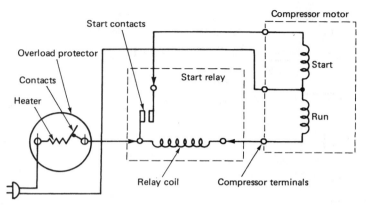

Figure 14-7 Domestic refrigerator circuit showing the start contacts and relay coil, as well as the overload protector.

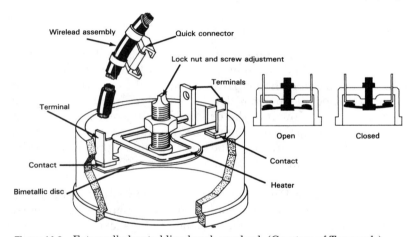

Figure 14-8 Externally located line-break overload. (*Courtesy of Tecumseh.*)

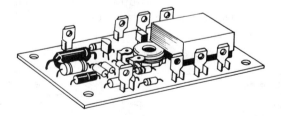

Figure 14-9 Time-delay relay.

by the resistance of the heater unit. This type of unit is different from that shown as a protector in Fig. 14-7. The heating element in Fig. 14-7 causes the circuit to open and protect the motor. The time-delay relay is used to make sure that certain things take place within the refrigeration cycle before another is commenced.

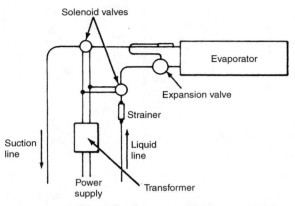

Figure 14-10   Solenoid valves connected in the suction and liquid evaporator lines of a refrigeration system.

Solenoids

Solenoid valves are used in many heating and cooling applications. They are electrically operated. A solenoid valve, when connected as in Fig 14-10, remains open when current is supplied to it. It closes when the current is turned off. In general, solenoid valves are used to control the liquid-refrigerant flow into the expansion valve or the refrigerant gas flow from the evaporator when it or the fixture it is controlling reaches the desired temperature. The most common application of the solenoid valve is in the liquid line, and it operates with a thermostat (see Fig. 14-11).

The solenoid shown in Fig. 14-12 controls the flow of natural gas in a hot-air furnace. Note how the coil is wound around the plunger. The plunger is the core of the solenoid. It has a tendency to be sucked into the coil whenever the coil is energized by current flowing through it. The electromagnetic effect causes the plunger to be attracted upward into the coil area. When the plunger is moved upward by the pull of the electromagnet,

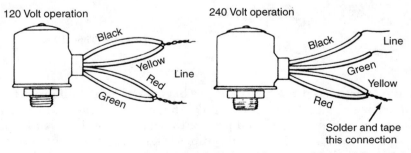

Figure 14-11   Solenoid valves. Note color-coded wires. (*Courtesy of General Controls.*)

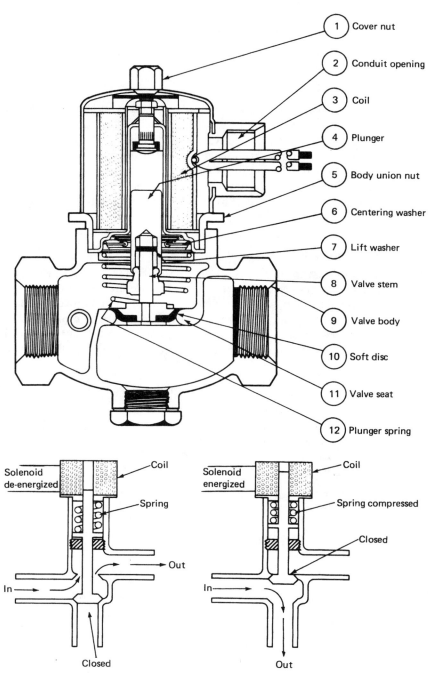

1. Cover nut
2. Conduit opening
3. Coil
4. Plunger
5. Body union nut
6. Centering washer
7. Lift washer
8. Valve stem
9. Valve body
10. Soft disc
11. Valve seat
12. Plunger spring

Solenoid de-energized
Coil
Spring
Out
In
Closed

Solenoid energized
Coil
Spring compressed
Closed
In
Out

Figure 14-12 Solenoid used for controlling natural gas flow to a furnace. (*Courtesy of Honeywell.*)

the soft disc (No. 10) is pulled upward, allowing gas to flow through the valve. This basic technique is used to control water, gasoline, oil, or any other liquid or gas.

Thermostats

Temperature control by using thermostats is common to both heating and cooling equipment. Thermostats are used to control heating circuits that cause furnaces and boilers to operate and provide heat. Thermostats are also used to control cooling equipment and refrigeration units. Each of these purposes may have its own specially designed thermostat or may use the same one. For instance, in the home you use the same thermostat to control the furnace and the air-conditioning unit.

Bellows-type thermostat

On modern condensing units, low-pressure control switches are largely superseded by thermostatic-control switches. A thermostatic control consists of three main parts: a bulb, a capillary tube, and a power element or switch. The bulb is attached to the evaporator in a manner that ensures contact with the evaporator. It may contain a volatile liquid, such as a refrigerant. The bulb is connected to the power element by means of a small capillary tube (see Fig. 14-13).

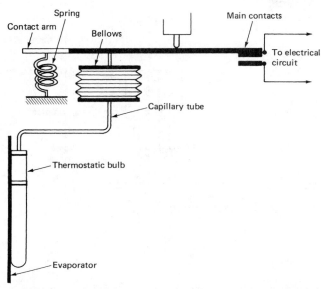

Figure 14-13   Bellows-type switch.

Operation of the bellows is provided by a change in temperature. Or the operation of the thermostatic-control switch is such that, as the evaporator temperature increases, the bulb temperature also increases. This raises the pressure of the thermostatic-liquid vapor. This, in turn, causes the bellows to expand and actuate an electrical contact. The contact closes the motor circuit, and the motor and compressor start operating. As the evaporator temperature decreases, the bulb becomes colder and the pressure decreases to the point where the bellows contracts sufficiently to open the electrical contacts, thus turning off the motor circuits. In this manner, the condensing unit is entirely automatic. Thus, it is able to produce exactly the amount of refrigeration needed to meet any normal operating condition.

Bimetallic-type thermostat

Temperature changes can cause a bimetallic strip to expand or contract in step with changes in temperature. These thermostats are designed for the control of heating and cooling in air-conditioning units, refrigeration storage rooms, greenhouses, fan coils, blast coils, and similar units. This is the type used in most homes for control of the central air-conditioning and central-heating system.

Figure 14-14 shows how the bimetallic strip thermostat works. Two metals, each having a different coefficient of expansion, are welded together to form a bimetallic unit or blade. With the blade securely anchored at one end, a circuit is formed and the contact points are closed. This allows the passage of an electric current through the closed points. Because an electric current provides heat in its passage through the bimetallic blade, the metals in the blade begin to expand. However, they expand at a different rate. The metals in the blade are so arranged that the one with a greater coefficient of expansion is placed at the bottom of the unit. After a certain time, the operating temperature is reached and the contact points become separated. This disconnects the device from its power source.

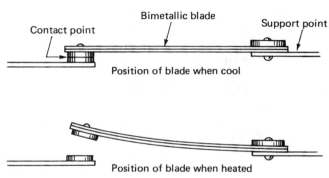

Figure 14-14   Bimetallic strip used in a thermostat.

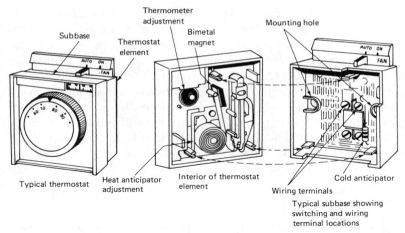

Figure 14-15   Modern thermostat for heating and cooling of a house. (*Courtesy of General Controls.*)

After a short period, the contact blade will again become sufficiently cooled to cause the contact point to join, thus reestablishing the circuit and permitting the current again to actuate the circuit. The cycle is repeated over and over again. In this way, the bimetallic thermostat prevents the temperature from rising too high or dropping too low.

Heating and cooling thermostats

Some thermostats can be used for both heating and cooling. The thermostat shown in Fig. 14-15 is such a device. The basic thermostat element has a permanently sealed, magnetic single pole double throw (SPDT) switch. The thermostat element plugs into the subbase and contains the heat anticipation, the magnetic switching, and a room temperature thermometer. The subbase unit contains fixed cool anticipation and circuitry. This thermostat is used with 24 V AC. In this case, the thermostatic element (bimetal) does not make direct contact with the electrical circuit. Instead, the expansion of the bimetal causes the magnet to move. This, in turn, causes the switch to close or open. Figure 14-16 shows that the bimetal is not in the electrical circuit.

Mercury contacts.   Some thermostats use the expanding bimetal arrangement to cause a tube of mercury to move. As the mercury moves in the tube, it comes in contact with two wires inserted into the glass tube. When the mercury comes in contact with the two wires, it completes the electrical circuit. This type of thermostat needs to be so arranged that the tube of mercury is pivoted and can be moved by the expanding or contracting bimetal strip, which exerts or releases pressure on the tube of mercury.

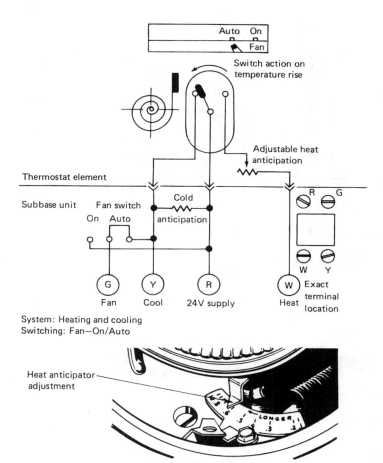

Figure 14-16 Wiring diagram for a thermostat.

Thermostats used in home air-conditioning and heating systems are now, equipped with mercury contacts (see Fig. 14-17). They are made so that the mercury contacts two wires that control the air-conditioning in one position and two wires that control the heating system in the other position (see Fig. 14-18).

The advantage of the mercury bulb type of switch is the elimination of switch contact points. Contact points are in need of constant attention. In most cases the dust from the air will eventually cause them to function improperly. It is necessary to clean the points by running a piece of clean paper through them to remove the dust particles and arcing residue. Since the mercury type is sealed and the arcing created on make and break of the circuit simply causes the mercury to vaporize slightly and then return to a liquid state, it provides a trouble-free switching operation.

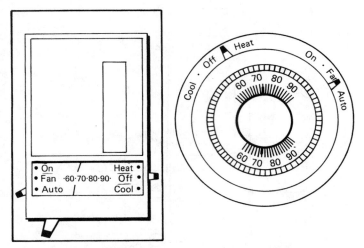

Figure 14-17   Thermostat for air-conditioning and heating.

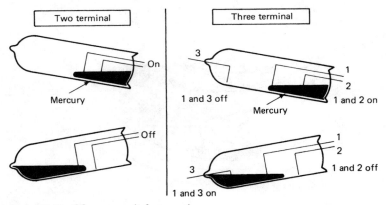

Figure 14-18   Mercury-switch operation.

Microprocessor thermostats

Semiconductor technology has produced another means of more accurately controlling air-conditioning and heating systems to provide better regulated temperatures in the home, office, and business. The microprocessor makes use of the semiconductor chip or integrated circuit discussed earlier in Chap. 19. All the external connections are the same as for any other type of thermostat. Only the internal circuitry has changed to provide a better regulated temperature and a variety of operations that allow you to set it for any energy-saving program desired (see Fig. 14-19). Unless a battery is included, it does not retain the program in most instances, and the clock, if there is one on the unit, has to be reset each time the power goes off.

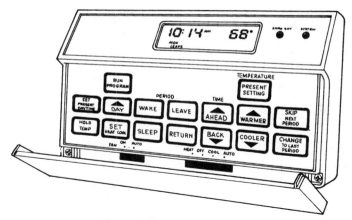

Figure 14-19 Microprocessor used for air-conditioning and furnace control in a home.

Thermostat adjustments

In Fig. 14-16, a cold-anticipation and a heat-anticipation adjustment are placed in the thermostat circuit. The heat-anticipation control is placed in series with the switch. The cold-anticipation resistor is placed in shunt or parallel with the switch. Thus, when the switch is closed the shunt is shorted out.

Heat anticipator.   The reason for the heat anticipator is to limit the degree of swing between turning on the furnace and the temperature of the room. It is a resistance heater element that is inserted in series with the thermostat line that runs to the heat-contactor coil. Once the thermostat contacts are closed current flows through the resistor. This causes it to heat up. The heat generated by the resistor causes the thermostat to open slightly before the desired room temperature is reached by the heating system. This allows the heat in the plenum of the furnace to continue to heat the room. Thus, the resistor aids the thermostat in anticipating the amount of heat that will be provided to the room by using the heat already produced in the plenum.

Cold anticipator.   The cold anticipator is a fixed resistor and is not adjustable. It heats the bimetallic coil that operates the points whenever the air-conditioner compressor is not on. When the compressor is on, the resistor is shorted out by the thermostat points being closed. The heating of the coil while the points are open causes it to close a little earlier than if it waited for the room to heat up sufficiently to cause it to turn on. This way the heat produced by the anticipator resistor causes the compressor to turn on a little before the thermostat would have

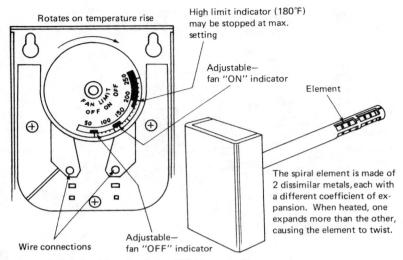

Figure 14-20   Combination fan and limit controller.

normally told it to do so. By turning it on before the room has reached the selected temperature, the anticipator causes the temperature swing in the room to be reduced and makes it more comfortable.

Limit Switches

Many types of switches are used to limit the amount of heat produced in a furnace. The upper limit has to be controlled so that the furnace does not cause fires by overheating. Limit switches take various forms depending on the manufacturer. However, Fig. 14-20 shows a typical switch and how it works. This is a combination of fan and limit controller that combines the functions of a fan controller and a limit controller in a single unit. One sensing element is used for both controls.

Combination controllers are wired in much the same way as individual controls. These combined controls can be used on line voltage, low voltage, or self-energizing milli-volt systems.

Figure 14-21 shows the fluid-filled type of capillary tube used in a limit switch. The one shown in Fig. 14-20 is the bimetal type that twists as it heats up, causing the control unit to move. These limit switches are placed in the plenum of the furnace to control when the fan goes on and off; when the plenum has reached the desired temperature, it turns off the solenoid and shuts off the flow of natural gas to the burner. Limit switches of a slightly different configuration are also used for electrical strip heaters. They may also be of the low-voltage (24-V) or line-voltage type.

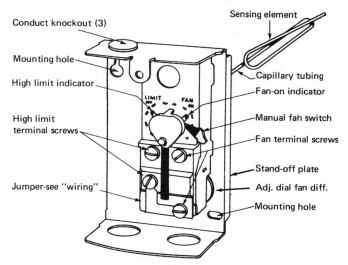

Figure 14-21   Combination fan and limit controller.

Pressure Control Switches

One safety feature for air-conditioning units with a compressor and condenser is a pressure-controlled switch. This switch is wired into the circuit to protect the system in case the system develops a leak. If a leak develops, it is possible to draw in moisture and air and damage the whole system. If the pressure builds too high, it can cause a rupture of any of the joints or weaker points in the system.

A low-voltage (24-V) relay is wired into the 240-V line that supplies the compressor motor. The relay contacts are wired into the supply line for the motor (see Fig. 14-22). The solenoid of the relay is wired in series with two pressure-operated switches. If the pressure builds too high, the

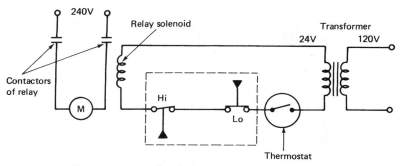

Figure 14-22   Pressure-operated switches control the compressor.

high switch will open and cause the solenoid to de-energize. If this happens, it causes the contacts of the relay to open. This removes power from the compressor motor. If the low-pressure switch opens, it will do the same thing. This way the compressor is protected from both high- and low-pressure causing damage to the system.

Both manual and automatic controls are available. Automatic controls reset when the pressure is stabilized in the system. If it is not stabilized, it will again turn the system off and keep recycling, until it reaches the design pressure.

Water Tower Controls

Temperature controls for refrigerating service are designed to maintain adequate head pressure with evaporative condensers and cooling towers. Low refrigerant head pressure, caused by abnormally low cooling water temperature, reduces the capacity of the refrigeration system.

Two systems of control for mechanical and atmospheric draft towers and evaporative condensers are shown in Figs. 14-23 and 14-24. The control opens the contacts when the temperature drops. These contacts are wired in series with the fan motor. Or they can be wired to the pilot of a fan-motor controller. Opening the contacts stops the fan when the cooling water temperature falls to a predetermined minimum value. This value corresponds to the minimum head pressure for proper operation. In the control system shown in Fig. 14-24, the contacts close on a temperature drop and are wired in series with a normally closed

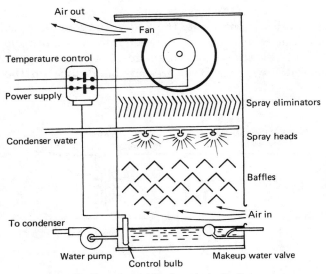

Figure 14-23   Cooling tower with forced-air draft.

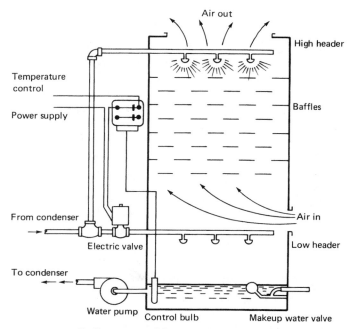

Air out

High header

Temperature
control

Baffles

Power supply

Air in

From condenser

Electric valve

Low header

To condenser

Water pump

Control bulb

Makeup water valve

Figure 14-24 Cooling tower with atmospheric draft cooling.

motorized valve or a solenoid valve. The contacts open the valve when low cooling temperature occurs. The cooling water then flows through a low header in the atmospheric tower. This reduces its cooling effect and the head pressure increases.

Float switches are used to control the level of water in the cooling tower. Automatic float switches provide automatic control for motors operating tank or sump pumps. They are built in several styles and can be supplied with several types of accessories that provide rod or chain operation and either wall or floor mounting. A sensor system may also be used. There are hundreds of sensor types. They usually sense the level of water by using two probes. When the water contacts the probes, it causes a small electrical current (at low voltages) to flow and energize a solenoid or relay that in turn causes the water to be turned off. When the level of water is below the two probes and a complete circuit is not available, the normally closed relay contacts are closed by de-energizing the relay. This causes the water solenoid to be energized. This allows makeup water to flow into the cooling tower until it reaches the point where the probes are immersed in water and the cycle is repeated.

15

Insulating Pipes and Tubing

Insulation

Insulation is needed to prevent the penetration of heat through a wall or air hole into a cooled space. There are several insulation materials, such as wood, plastic, concrete, and brick. Each has its application. However, more effective materials are constantly being developed and made available. Such types of insulation include

- Sheet
- Tubing
- Pipe

Piping insulation can be broken down into cork, rock cork, wool felt with waterproof jacket, and hair felt with waterproof jacket.

Sheet insulation

Vascocel is an expanded, closed cell, sponge rubber that is made in a continuous sheet form 36 in. wide. It comes in a wide range of thickness (3/8, 1/2, and 3/4 in.). This material is designed primarily for insulating oversize pipes, large tanks and vessels, and other similar medium- and low-temperature areas. Because of its availability on continuous rolls, this material lends itself ideally to application on large air ducts and irregular shapes.

This material is similar to its companion product, Vascocel tubing. It may be cut and worked with ordinary hand tools such as scissors or a knife. The sheet stock is easily applied to clean, dry surfaces with an adhesive (see Fig. 15-1). The *k* factor (heat transfer coefficient) of this

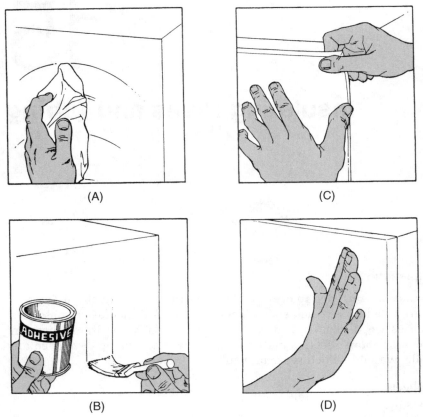

(A)

(C)

(B)

(D)

Figure 15-1 Installing sheet insulation. (*a*) Prepare the surfaces for application of the sheet insulation by wiping with a soft, dry cloth to remove any dust or foreign matter. Use a solvent to remove grease or oil. (*b*) Apply the adhesive in a thin, even coat to the surface to be insulated. (*c*) Position the sheet of insulation over the surface and then simply smooth it in place. The adhesive is a contact type. The sheet must be correctly positioned before it contacts the surface. (*d*) Check for adhesion of ends and edges. The surface can be painted.

material is 0.23. It has some advantages over other materials. It is resistant to water penetration, water absorption, and physical abrasion.

Tubing insulation

Insulation tape is a special synthetic rubber and cork compound designed to prevent condensation on pipes and tubing. It is usually soft and pliable. Thus, it can be molded to fit around fittings and connections. There are many uses for this type of insulation. It can be used on hot or cold pipes or tubing. It is used in residential buildings, air-conditioning units, and commercial installations. It comes in 2-in-wide rolls that are

Figure 15-2 Foam insulation tape.
(*Courtesy of Virginia Chemical.*)

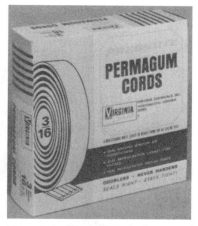

Figure 15-3 Slugs of insulation material and cords are workable into locations where sheet material cannot fit.
(*Courtesy of Virginia Chemicals.*)

30 ft long. The tape is thick, if stored or used in temperatures under 90°F (32.2°C), the lifetime is indefinite.

Foam insulation tape is made specifically for wrapping cold pipes to prevent pipe sweat (see Fig. 15-2). It can be used to hold down heat loss on hot pipes below 180°F (82.2°C). It can be cut in pieces and easily molded around fittings and valves. It adheres to itself and clean metal surfaces. It is wrapped over pipes with about 1/4-in. overlap on each successive lap. Remember one precaution: never wrap two or more parallel runs of tubing or pipe together, leaving air voids under the tape. Fill the voids between the pipes with Permagum before wrapping. This will prevent moisture from collecting in the air spaces. This foam insulation tape has a unicellular composition. The k factor is 0.26 at 75°F (23.9°C).

Permagum is a nonhardening, water-resistant sealing compound. It is formulated to be nonstaining, nonbleeding, and to have excellent adhesion to most clean surfaces. It comes in containers in either slugs or cords (see Fig. 15-3).

This sealer is used to seal metal-to-metal joints in air conditioners, freezers, and coolers. It can seal metal-to-wood joints and set plastic and glass windows in wood or metal frames. It can be used to seal electrical or wire entries in air-conditioning installations or in freezers. It can be worked into various spaces. It comes with a paper backing so that it will not stick to itself.

Extrusions are simple to apply. Unroll the desired length and smooth it into place. It is soft and pliable. The bulk slug material can be formed and applied by hand or with tools such as a putty knife.

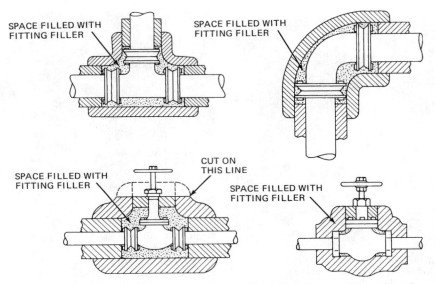

Figure 15-4 Pipe fittings covered with cork-type insulation. On one of the valves the top section can be removed if the packing needs replacing.

Sealing compounds are sometimes needed to seal a joint or an entry location. These compounds can be purchased in small units in white, nonstaining compositions.

Pipe insulation

Pipe fittings are insulated for a number of reasons. Methods of insulating three different fittings are shown in Fig. 15-4. In most cases it is advisable to clean all joints and waterproof them with cement. A mixture of hot crude paraffin and granulated cork can be used to fill the cracks around the fittings.

Figure 15-5 shows a piece of rock cork insulation. It is molded from a mixture of rock wool and waterproof binder. Rock wool is made from limestone that has been melted at about 3000°F (1649°C). It is then

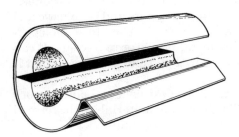

Figure 15-5 Pipe insulation.

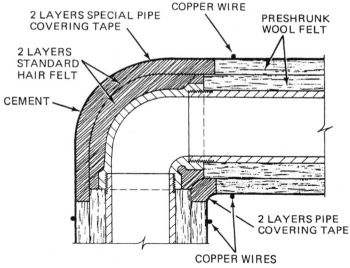

COPPER WIRE

2 LAYERS SPECIAL PIPE
COVERING TAPE

PRESHRUNK
WOOL FELT

2 LAYERS
STANDARD
HAIR FELT

CEMENT

2 LAYERS PIPE
COVERING TAPE

COPPER WIRES

Figure 15-6 Insulated pipe fittings.

blown into fibers by high-pressure steam. Asphaltum is the binder used to hold it into a moldable form. This insulation has approximately the same insulation qualities as cork. It can be made waterproof when coated with asphalt. Some more modern materials have been developed to give the same or better insulation qualities. The Vascocel tubing can be used in the insulation of pipes. Pipe wraps are available to give good insulation and prevent dripping, heat loss, or heat gain.

Figure 15-6 shows a fitting insulated with preshrunk wool felt. This is a built-up thickness of pipe covering made of two layers of hair felt. The inside portion is covered with plastic cement before the insulation material is applied. After the application, waterproof tape and plastic cement should be added for protection against moisture infiltration. This type of insulation is used primarily on pipes located inside a building. If the pipe is located outside, another type of insulation should be used.

Refrigeration Piping

The use of various materials for insulation purposes in the refrigeration field over the years has resulted in some equipment still operational today. It is this equipment that service people are most often called to repair or maintain. It is therefore necessary for the present day repair-person to be acquainted with the older types of insulations that may be encountered during the workday.

The success of any refrigeration plant depends largely on the proper design of the refrigeration piping and a thorough understanding of the necessary accessories and their functions in the system. In sizing refrigerant lines, it is necessary to consider the optimum sizes with respect to economics, friction losses, and oil return. It is desirable to have line sizes as small as possible from the cost standpoint. On the other hand, suction- and discharge-line pressure drops cause a loss of compressor capacity, and excessive liquid-line pressure drops may cause flashing of the liquid refrigerant with consequent faulty expansion-valve operation.

Refrigerant piping systems, to operate successfully, should satisfy the following:

- Proper refrigerant feed to the evaporators should be ensured.

- Refrigerant lines should be of sufficient size to prevent an excessive pressure drop.

- An excessive amount of lubricating oil should be prevented from being trapped in any part of the system.

- Liquid refrigerant should be prevented from entering the compressor at all times.

Pressure-Drop Considerations

Pressure drop in liquid lines is not as critical as it is in the suction and discharge lines. The important thing to remember is that the pressure drop should not be so great as to cause gas formation in the liquid line and/or insufficient liquid pressure at the liquid-feed device. A system should normally be designed so that the pressure drop due to friction in the liquid line is not greater than that corresponding to 1 to 2°F (0.5555 to 1.111°C) change in saturation temperature. Friction pressure drops in the liquid line include the drop in accessories, such as the solenoid valve, strainer-drier, and hand valves, as well as in the actual pipe and fittings from the receiver outlet to the refrigerant-feed device at the evaporator.

Friction pressure drop in the suction line means a loss in system capacity because it forces the compressor to operate at a lower suction pressure to maintain the desired evaporating temperature in the coil. It is usually standard practice to size the suction line to have a pressure drop due to friction not any greater than the equivalent of a 1 to 2°F (0.5555 to 1.111°C) change in saturation temperature.

Liquid Refrigerant Lines

The liquid lines do not generally present any design problems. Refrigeration oil is sufficiently miscible with commonly used refrigerants in the liquid form to ensure adequate mixture and positive oil return. The following factors should be considered when designating liquid lines:

- The liquid lines, including the interconnected valves and accessories, must be of sufficient size to prevent excessive pressure drops.

- When interconnecting condensing units with condenser receivers or evaporative condensers, the liquid lines from each unit should be brought into a common liquid line.

- Each unit should join the common liquid line as far below the receivers as possible, with a minimum of 2 ft preferred. The common liquid line should rise to the ceiling of the machine room. The added head of liquid is provided to prevent, as far as possible, hot gas blowing back from the receivers.

- All liquid lines from the receivers to the common line should have equal pressure drops in order to provide, as nearly as possible, equal liquid flow and prevent the blowing of gas.

- Remove all liquid-line filters from the condensing units, and install them in parallel in the common liquid line at the ceiling level.

- Hot gas blowing from the receivers can be condensed in reasonable quantities by liquid subcoolers, as specified for the regular condensing units, having a minimum lift of 60 ft at 80°F (26.7°C) condensing medium temperature.

- Interconnect all the liquid receivers of the evaporative condensers above the liquid level to equalize the gas pressure.

- The common and interconnecting liquid line should have an area equal to the sum of the areas of the individual lines. Install a hand shutoff valve in the liquid line from each receiver. Where a reduction in pipe size is necessary in order to provide sufficient gas velocity to entrain oil up the vertical risers at partial loads; greater pressure drops will be imposed at full load. These can usually be compensated for by over sizing the horizontal and down comer lines to keep the total pressure drop within the desired limits.

Interconnection of Suction Lines

When designing suction lines, the following important considerations should be observed:

- The lines should be of sufficient capacity to prevent any considerable pressure drop at full load.

- In multiple-unit installations, all suction lines should be brought to a common manifold at the compressor.

- The pressure drop between each compressor and main suction line should be the same in order to ensure a proportionate amount of refrigerant gas to each compressor, as well as a proper return of oil to each compressor.
- Equal pipe lengths, sizes, and spacing should be provided.
- All manifolds should be level.
- The inlet and outlet pipes should be staggered.
- Never connect branch lines at a cross or tee.
- A common manifold should have an area equal to the sum of the areas of the individual suction lines.
- The suction lines should be designed so as to prevent liquid from draining into the compressor during shutdown of the refrigeration system.

Discharge Lines

The hot-gas loop accomplishes two functions in that it prevents gas that may condense in the hot-gas line from draining back into the heads of the compressor during the *off* cycles, and prevents oil leaving one compressor from draining down into the head of an idle machine. It is important to reduce the pressure loss in hot-gas lines because losses in these lines increase the required compressor horsepower per ton of refrigeration and decrease the compressor capacity. The pressure drop is kept to a minimum by sizing the lines generously to avoid friction losses but still making sure that refrigerant line velocities are sufficient to entrain and carry along oil at all load conditions. In addition, the following pointers should be observed:

- The compressor hot-gas discharge lines should be connected as shown in Fig. 15-7.
- The maximum length of the risers to the horizontal manifold should not exceed 6 ft.

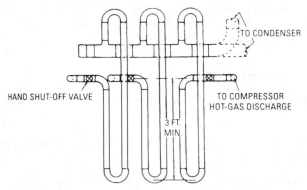

Figure 15-7 One way to connect hot-gas discharge lines.

- The manifold size should be at least equal to the size of the common hot-gas line to the evaporative condenser.

- If water-cooled condensers are interconnected, the hot-gas manifolds should be at least equal to the size of the discharge of the largest compressor.

- If evaporative condensers are interconnected, a single gas line should be run to the evaporative condensers, and the same type of manifold provided at the compressors should be installed.

- Always stagger and install the piping at the condensers.

- When the condensers are above the compressors, install a loop having a minimum depth of 3 ft in the hot-gas main line.

- Install a hand shutoff valve in the hot-gas line at each compressor.

Water Valves

The water-regulating valve is the control used with water-cooled condensers. When installing water valves, the following should be observed:

- The condenser water for interconnected compressor condensers should be applied from a common water line.

- Single automatic water valves or multiple valves in parallel (Fig. 15-8) should be installed in the common water line.

- Pressure-control tubing from the water valves should be connected to a common line, which, in turn, should be connected to one of the receivers or to the common liquid line.

Multiple-Unit Installation

Multiple compressors operating in parallel must be carefully piped to ensure proper operation. The suction piping at parallel compressors should be designed so that all compressors run at the same suction pressure and

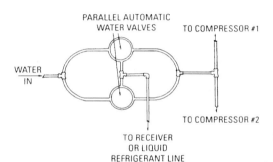

Figure 15-8 A method of interconnecting water valves.

oil is returned in equal proportions to the running compressors. All suction lines should be brought into a common suction header in order to return the oil to each crankcase as uniformly as possible.

The suction header should be run above the level of the compressor suction inlets so that oil can drain into the compressors by gravity. The header should not be below the compressor suction inlets because it can become an oil trap. Branch suction lines to the compressors should be taken off from the side of the header. Care should be taken to make sure that the return mains from the evaporators are not connected into the suction header so as to form crosses with the branch suction lines to the compressors. The suction header should be run full size along its entire length. The horizontal takeoffs to the various compressors should be the same size as the suction header. No reduction should be made in the branch suction lines to the compressors until the vertical drop is reached.

Figure 15-9 shows the suction and hot-gas header arrangements for two compressors operating in parallel. Takeoffs to each compressor from the common suction header should be horizontal and from the side to ensure equal distribution of oil and prevent accumulating liquid refrigerant in an idle compressor in case of slop-over.

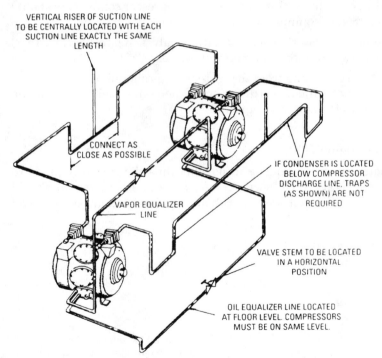

Figure 15-9 Connections for the suction and hot-gas headers in a multiple-compressor installation.

Piping insulation

Insulation is required for refrigeration piping to prevent moisture condensation and prevent heat gain from the surrounding air. The desirable properties of insulation are that it should have a low coefficient of heat transmission, be easy to apply, have a high degree of permanency, and provide protection against air and moisture infiltration. Finally, it should have a reasonable installation cost.

The type and thickness of insulation used depends on the temperature difference between the surface of the pipe and the surrounding air and also on the relative humidity of the air. It should be clearly understood that although a system is designed to operate at a high suction temperature, it is quite difficult to prevent colder temperatures occurring from time to time. This may be due to a carrying over of some liquid from the evaporator or the operation of an evaporator pressure valve. Interchangers are preferable to insulation, in this case.

One of the safest pipe insulations available is molded cork or rock cork of the proper thickness. Hair-felt insulation may be used, but great care must be taken to have it properly sealed. For temperatures above 40°F (4.4°C), wool felt or a similar insulation may be used, but here again, success depends on the proper seal against air and moisture infiltration.

Liquid-refrigerant lines carry much higher temperature refrigerant than suction lines, and if this temperature is above the temperature of the space through which they pass, no insulation is usually necessary. However, if there is danger of the liquid lines going below the surrounding air temperatures and causing condensation, they should be insulated when condensation will be objectionable. If they must unavoidably pass through highly heated spaces, such as those adjacent to steam pipes, through boiler rooms, then the liquid lines should also be insulated to ensure a solid column of liquid to the expansion valve.

There were four types of insulation in use before the discovery of modern insulation materials. Those you may encounter and were in general use for refrigerator piping, are namely:

1. Cork

2. Rock cork

3. Wool felt with waterproof jacket

4. Hair felt with waterproof jacket

Cork insulation

Cork pipe covering is prepared by pressing dried and granulated cork in metal molds. The natural resins in the cork bind the entire mass into its new shape. In the case of the cheaper cork, an artificial binder is used.

The cork may be molded to fit pipe and fittings, or it may be made into flat boards of varying sizes and thickness. Cork has a low thermal conductivity. The natural binder in the material itself makes cork highly water-resistant, and its structure ensures a low capillarity. It can be made practically impervious to water by surfacing with odorless asphalt.

All fittings in the piping, as well as the pipe itself, should be thoroughly insulated to prevent heat gain to protect the pipe insulation from moisture infiltration and deterioration and eliminate condensation problems. Molded cork covering made especially for this purpose is available for all common types of fittings. Each covering should be the same in every respect as the pipe insulation, with the exception of the shape, and should be formed so that it joins to the pipe insulation with a break. Typical cork fitting covers are furnished in three standard thicknesses for ice water, brine, and special brine.

To secure maximum efficiency and long life from cork covering, it must be correctly applied and serviced, as well as properly selected. Hence, it is essential that the manufacturer's recommendations and instructions be followed in detail. The following general information is a summary of the data that are of general interest.

All pipelines should be thoroughly cleaned, dried, and free from all leaks. It is also advisable to paint the piping with waterproof paint before applying the insulation, although this is not recommended by all manufacturers. All joints should be sealed with waterproof cement when applied. Fitting insulation should be applied in substantially the same manner, with the addition of a mixture of hot crude paraffin and granulated cork used to fill the space between the fittings, as shown in Fig. 15-4.

Rock-cork insulation

Rock-cork insulation is manufactured commercially by molding a mixture of rock wool and a waterproof binder into any shape or thickness desired. The rock wool is made from limestone melted at about 3000°F (1666.6°C) and then blown into fibers by high-pressure steam. It is mixed with an asphalt binder and molded into various commercial forms. The heat conductivity is about the same as cork, and the installed price may be less. Because of its mineral composition, it is odorless, vermin-proof, and free from decay. Like cork, it can be made completely waterproof by surfacing with odorless asphaltum. The pipe covering fabricated from rock wool and a binder is premolded in single-layer sections 36 in. long to fit all standard pipe sizes and is usually furnished with a factory-applied waterproof jacket.

When pipelines are insulated with rock-cork covering, the fittings are generally insulated with built-up rock wool impregnated with asphalt. This material is generally supplied in felted form, having a nominal thickness of about 1 in. and a width of about 18 in. It can be readily adapted to any type of fitting and is efficient as an insulator when properly applied.

Before applying the formed rock-cork insulation, it is first necessary to thoroughly clean and dry the piping and then paint it with waterproof asphalt paint. The straight lengths of piping are next covered with the insulation, which has the two longitudinal joints and one end joint of each section coated with plastic cement. The sections are butted tightly together with the longitudinal joints at the top and bottom and temporarily held in place by staples. The plastic cement should coat that part of the exterior area of each section to be covered by the waterproof lap and the lap pressed smoothly into it. The end joints should be sealed with a waterproof fabric embedded in a coat of the plastic cement. Each section should then be secured permanently in place with three to six loops of copper-plated annealed steel wire.

Wool-felt insulation

Wool felt is a relatively inexpensive type of pipe insulation and is made up of successive layers of waterproof wool felt that are indented in the manufacturing process to form air spaces. The inner layer is a waterproof asphalt-saturated felt, while the outside layer is an integral waterproof jacket. This insulating material is satisfactory when it can be kept air- and moisture-tight. If air is allowed to penetrate, condensation will take place in the wool felt, and it will quickly deteriorate. Thus, it is advisable to use it only where temperatures above 40°F (4.4°C) are encountered and when it is perfectly sealed. Under all conditions, it should carry the manufacturer's guarantee for the duty that it is to perform.

After all the piping is thoroughly cleaned and dried, the sectional covering is usually applied directly to the pipe with the outer layer slipped back and turned so that all joints are staggered. The joints should be sealed with plastic cement, and the flap of the waterproof jacket should be sealed in place with the same material. Staples and copper-clad steel wire should be provided to permanently hold the insulation in place, and then the circular joints should be covered with at least two layers of waterproof tape to which plastic cement is applied.

Pipe fittings should be insulated with at least two layers of hair felt (Fig. 15-6) built up to the thickness of the pipe covering, but before the felt is placed around the fittings, the exposed ends of the pipe insulation should be coated with plastic cement.

After the felt is in place, two layers of waterproof tape and plastic cement should be applied for protection from moisture infiltration.

Insulation of this type is designed for installation in buildings where it is normally protected against outside weather conditions. When outside pipes are to be insulated, one of the better types of pipe covering should be used. In all cases, the manufacturer's recommendations should be followed as to the application.

Hair-felt insulation

Hair-felt insulation is usually made from pure cattle hair that has been especially prepared and cleaned. It is a very good insulator against heat, having a low thermal conductivity. Its installed cost is somewhat lower than cork; but it is more difficult to install and seal properly, and hence its use must be considered a hazard with the average type of workmanship. Prior to installation, the piping should be cleaned and dried and then prepared by applying a thickness of waterproof paper or tape wound spirally, over which the hair felt of approximately 1-in thickness is spirally wound for the desired length of pipe. It is then tightly bound with jute twine, wrapped with a sealing tape to make it entirely airtight, and finally painted with waterproof paint. If more than one thickness of hair felt is desired, it should be built up in layers with tarpaper between. When it is necessary to make joints around fittings, the termination of the hair felt should be tapered down with sealing tape and the insulation applied to the fittings should overlap this taper, thus ensuring a permanently tight fit.

The important point to remember is that this type of insulation must be carefully sealed against any air or moisture infiltration, and even then difficulty may occur after it has been installed. At any point where air infiltration (or "breathing," as it is called) is permitted to occur, condensation will start and travel great distances along the pipe, even undermining the insulation that is properly sealed.

There are several other types of pipe insulation available, but they are not used extensively. These include various types of wrapped and felt insulation, but they are seldom applied with success. Whatever insulation is used, it should be critically examined to see whether it will provide the protection and permanency required of it, otherwise it should never be considered. Although all refrigerant piping, joints, and fittings should be covered, it is not advisable to do so until the system has been thoroughly leak tested and operated for a time.

Pressure drop in the various parts of commercial refrigeration systems due to pipe friction and the proper dimensioning to obtain the best operating results are important items when installation of equipment is made.

By careful observation of the foregoing detailed description of refrigeration piping and methods of installation, the piping problem will be greatly simplified and result in proper system operation.

16

Electrical Safety

Electrical Safety Devices

Air-conditioning, refrigeration, and heating systems utilize the convenience of electrical controls. These devices are made with a particular application in mind. Each is designed for a specific purpose. By using low voltage in most instances to control higher voltages, the devices are suited for remote operation since the size of the wire is rather small and inexpensive to install.

Some control devices are designed to protect the technician while repairing or maintaining equipment. Certain safety procedures are necessary to make sure you do not receive a fatal electrical shock. The following should be helpful in making it safe to work on electrically controlled and operated equipment.

Safety Precautions

It takes very little current to cause physical damage to the human body. In some cases death may result from as little a one-tenth (0.1) of an ampere (see Table 16-1). You may react to a slight shock and then move quickly and come in contact with operating machinery. Involuntary actions caused by electrical shock are more harmful in most instances than the actual mild shock. It is very easy to become careless. Just keep in mind that the human body has a skin resistance of between 400,000 and 800,000 Ω. This can be used in an Ohm's law formula to determine the current:

$$I = E/R$$

$$I = 120/400,000 = 0.3 \text{ mA}$$

TABLE 16-1   **Physiological Effects of Electric Currents**

Reading (mA)		Effects
Safe current values	1 or less	Causes no sensation—not felt.
	1–7	Sensation of shock, not painful; individual can let go at will since muscular control is not lost.
Unsafe current values	8–15	Painful shock; individual can let go at will since muscular control is not lost.
	15–20	Painful shock; control of adjacent muscles lost; victim cannot let go.
	20–50	Painful, severe muscular contractions; breathing difficult.
	50–100	Ventricular fibrillation, a heart condition that can result in instant death is *possible*.
	100–200	Ventricular fibrillation occurs.
	200 and over	Severe burns, severe muscular contractions, so severe that chest muscles clamp the heart and stop it for the duration of the shock. (This prevents ventricular fibrillation.)

This is not enough to cause you to feel it. However, if you have wet hands or make contact with part of the body that is not dry, you may have a body resistance as low as 50,000 Ω and receive a current of 2.4 mA. That is enough to cause you to jump back from a slight tingle. The backward movement on a roof may be enough to cause you to fall a great distance. You may also drop the equipment or tools you are working with and cause further damage.

If you are working around 240 V, the danger is greater. At 240 V a contact resistance of 50,000 Ω produces 4.8 mA or enough to cause a heavy shock with tightening of the muscles. A low body resistance of say 200 Ω can be fatal. This can happen whenever the skin is wet or you are standing on a wet surface. With 200 Ω resistance and 120 V, you will receive 600 mA, enough to cause death.

Main switches

If you are working on equipment with the power on, be sure you are standing on a dry surface and your hands do not make contact with anything other than the probes being used to make measurements. It is always good practice when working around live circuits to keep one hand in your pocket so that you will not complete the path across your chest. It is, of course, always better to have the power off when working on equipment. A good procedure is to turn off the main circuit breakers and put a sign on them so that no one turns them on without your knowledge. If there is a control box or distribution panel with a lock on it, make sure the lock is in place and locked with the lever in the down or off position (see Figs. 16-1 and 16-6). Fusing of the circuits is shown in Fig. 16-1.

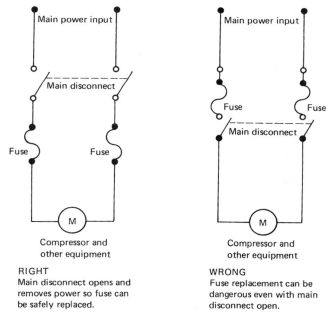

RIGHT
Main disconnect opens and
removes power so fuse can
be safely replaced.

WRONG
Fuse replacement can be
dangerous even with main
disconnect open.

Figure 16-1 Main disconnect should be located before working on equipment.

Portable electrical tools

Grounding portable tools aids in preventing shocks and damage to equipment under certain conditions. A grounding conductor has a white-colored jacket in a two- or three-wire cable (neutral wire). It is terminated to the white- or silver-colored terminal in a plug cap or connector. And it is terminated at the neutral bar in the distribution box. Keep in mind that gray is also used to color-code wires in industrial and commercial installations. Gray wires are also used for grounding.

When there is an electrical fault that allows the hot line to contact the metal housing of electrical equipment, in a typical two-wire system, or some other ungrounded conductors, any person who touches that equipment or conductor will receive a shock. The person completes the circuit from the hot line to the ground and current passes through the body. Because a body is not a good conductor, the current is not high enough to blow the fuse. Thus, the current continues to pass through the body as long as the body remains in contact with the equipment (see Fig. 16-2).

A grounding conductor, or equipment ground, is a wire attached to the housing or other conductive parts of electrical equipment that are not normally energized to carry current from them to the ground. Thus, if a person touches a part that is accidentally energized, there will be no shock, because the grounding line furnishes a much lower resistance

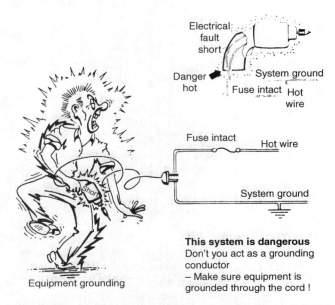

Electrical fault short

Danger hot

System ground

Fuse intact   Hot wire

Fuse intact   Hot wire

System ground

This system is dangerous
Don't you act as a grounding conductor
— Make sure equipment is grounded through the cord !

Equipment grounding

Figure 16-2   A person who contacts the charged housing of a drill or piece of equipment becomes the conductor in a short circuit to ground. (*Courtesy of National Safety Council.*)

path to the ground (see Fig. 16-3). The high current passes through the wire conductor and blows the fuses and stops the current. In normal operation, a grounding conductor does not carry current.

The grounding conductor in a three-wire conductor cable has a green jacket. The grounding conductor is always terminated at the green-colored hex-head screw on the cap or connector. It utilizes either a green-colored conductor or a metallic conductor as its path to ground. In Canada, this conductor is referred to as the earthing conductor, which is somewhat more descriptive and helpful in distinguishing between grounding conductors and neutral wires, or grounded conductors.

Ground-fault circuit interrupters

A safety device that should be used by all technicians in the field who do not have time to test out each circuit before it is used is the ground-fault circuit interrupter (GFCI). It is designed to protect you from shock.

The differential ground-fault interrupter, available in various modifications, has current-carrying conductors passing through the circular iron core of a doughnut-shaped differential transformer. As long as all the electricity passes through the transformer, the differential transformer is not affected and will not trigger the sensing circuit. If a portion of the current flows to ground and through the fault-detector line, however,

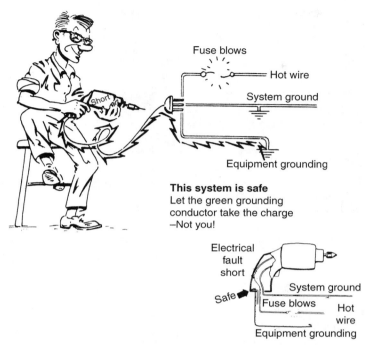

Fuse blows
Hot wire
System ground
Equipment grounding

This system is safe
Let the green grounding
conductor take the charge
—Not you!

Electrical
fault
short
Safe
System ground
Fuse blows
Hot wire
Equipment grounding

Figure 16-3 Properly wired circuit causes the shorted equipment to be shorted to ground instead of through the person to ground. (*Courtesy of National Safety Council.*)

the flow of electricity through the sensing windings of the differential transformer causes the sensing circuit to open the circuit breaker. These devices can be arranged to interrupt a circuit for currents of as little as 5 mA flowing to ground (see Fig. 16-4).

Another design is the isolation-type ground-fault interrupter. This unit combines the safety of an isolation system with the response of an electronic-sensing circuit. In this setup, an isolating transformer provides an inductive coupling between load and line. Both the hot and neutral wires are connected to the isolating transformer. There is no continuous wire connected between.

In the latter type of interrupter, a ground fault must pass through the electronic-sensing circuit, which has sufficient resistance to limit current flow to as low as 2 mA, well below the level of human perception.

Types of circuit protectors

There are two types of circuit protectors made in the circuit-breaker configuration. They are circuit breakers that work on the heating principle. A bimetallic strip is heated by having circuit current pass through it.

Figure 16-4 Ground-fault circuit interrupter (GFCI) used for portable tools.

When too much current flows, the strip is overheated. The expansion of the strip causes the breaker to trip, thereby causing the circuit to be opened (see Fig. 16-5). This type is slower to cool down than the magnetic type. It takes a little longer for it to cool to a point where it can be reset. If the overload still exists, the breaker will trip again.

The magnetic type uses a coil to operate. The circuit current is drawn through the coil. If too much current flows, the magnetic properties of the coil cause the circuit breaker to trip and open the circuit. This type is quicker in terms of being able to reset it. The overload must, of course, be removed before it will remain in the reset position.

In most cases where the power is 240 V single-phase, the circuit breakers will be locked together with a pin. If one side of the power circuit is opened, the other circuit is also tripped since they are tied together physically. In three-phase circuits, the three circuit breakers are tied together so that when one trips all three circuits are disconnected from the power line.

If air-conditioning or refrigeration equipment is served by a separate box, make sure the lock is in place and the handle is in the down or off position before making an inspection or working on the equipment (see Fig. 16-6).

One of the most important parts of working around air-conditioning and refrigeration equipment is that of doing the job safely. The possibility of incorrect procedures being followed can make it very painful both physically and mentally. Some of the suggestions that follow should aid in your understanding of careful work habits and use of the proper tool for the job.

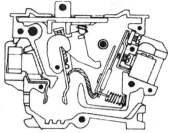

Tripped position
- Contacts open — no current flow
- Handle stationary when tripped

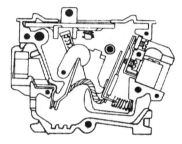

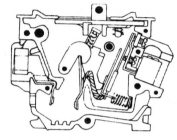

"On" position
- Contacts closed — current on
- Handle in "On" position (shows "On")

"Off" position
- Contacts open — no current flow
- Handle in "Off" position (shows "Off")

To restore service when fault is cleared you simply move operating handle to "Off" position and then to "On."

Figure 16-5 Cutaway view of a circuit breaker in various positions. (*Courtesy of Wadsworth.*)

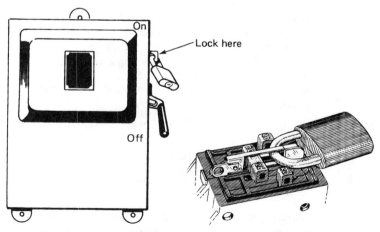

Lock here

On

Off

Figure 16-6 Lock the box by using a lock and keep the key until you need to unlock it.

Electrical Safety

Many Tecumseh single-phase compressors are installed in systems requiring off-cycle crankcase heating. This is designed to prevent refrigerant accumulation in the compressor housing. The power is on at all times. Even if the compressor is not running, power is applied to the compressor housing where the heating element is located.

Another popular system uses a run-capacitor that is always connected to the compressor motor windings, even when the compressor is not running. Other devices are energized when the compressor is not running. That means there is electrical power applied to the unit even when the compressor is not running. This calls for an awareness of the situation and the proper safety procedures.

Be safe. Before you attempt to service any refrigeration system, make sure that the main circuit breaker is open and all power is off.

Safe Practices

Safe practices are important in servicing refrigeration units. Such practices are common sense, but must be reinforced to make one aware of the problems that can result when a job is done incorrectly.

Handling cylinders

Refrigeration and air-conditioning servicepersons must be able to handle compressed gases. Accidents occur when compressed uses are not handled properly.

Oxygen or acetylene must never be used to pressurize a refrigeration system. Oxygen will explode when it comes into contact with oil. Acetylene will explode under pressure, except when properly dissolved in acetone as used in commercial acetylene cylinders.

Dry nitrogen or dry carbon dioxide are suitable gases for pressurizing refrigeration or air-conditioning systems for leak tests or system cleaning. However, the following specific restrictions must be observed:

- Commercial nitrogen (N_2) cylinders contain pressures in excess of 2000 psi at normal room temperature.

- Commercial carbon dioxide (CO_2) cylinders contain pressures in excess of 800 psi at normal room temperature.

- Cylinders should be handled carefully. Do not drop them or bump them.

- Keep cylinders in a vertical position and securely fastened to prevent them from tipping over.

- Do not heat the cylinder with a torch, or other open flame. If heat is necessary to withdraw gas from the cylinder, apply heat by immersing the lower portion of the cylinder in warm water. Never heat a cylinder to a temperature over 110°F (43°C).

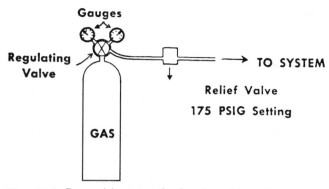

Figure 16-7   Pressurizing set-up for charging refrigeration systems.

Pressurizing

Pressure testing or cleaning refrigeration and air-conditioning systems can be dangerous! Extreme caution must be used in the selection and use of pressurizing equipment. Follow these procedures:

- Never attempt to pressurize a system without first installing an appropriate pressure-regulating valve on the nitrogen or carbon dioxide cylinder discharge. This regulating valve should be equipped with two functioning pressure gages. One gage indicates cylinder pressure. The other gage indicates discharge or downstream pressure.

- Always install a pressure relief valve or frangible-disc type pressure relief device in the pressure supply line. This device should have a discharge port of at least 1/2 in. NPT size. This valve or frangible-disc device should be set to release at 175 psig.

- A system can be pressurized up to a *maximum* of 150 psig for leak testing or purging (see Fig. 16-7).

Tecumseh hermetic-type compressors are low-pressure housing compressors. The compressor housings (cans or domes) are not normally subjected to discharge pressures. They operate instead at relatively low suction pressures. These Tecumseh compressors are generally installed on equipment where it is impractical to disconnect or isolate the compressor from the system during pressure testing. Therefore, do not exceed 150 psig when pressurizing such a complete system.

When flushing or purging a contaminated system, care must be taken to protect the eyes and skin from contact with acid-saturated refrigerant or oil mists. The eyes should be protected with goggles. All parts of the body should be protected by clothing to prevent injury by refrigerant. If contact with either skin or eyes occurs, flush the exposed area with cold water. Apply an ice pack if the burn is severe, and see a physician at once.

Working with refrigerants

R-12 have effectively been replaced in modern air-conditioning equipment with R-134a or any of the approved substitutes and R-22 has some acceptable substitutes also. They are considered to be nontoxic and noninflammable. However, any gas under pressure can be hazardous. The latent energy in the pressure alone can cause damage. In working with R-12 and R-22 (or their substitutes), observe the same precautions that apply when working with other pressurized gases.

Never completely fill any refrigerant gas cylinder with liquid. Never fill more than 80 percent with liquid. This will allow for expansion under normal conditions.

Make sure an area is properly ventilated before purging or evacuating a system that uses R-12, R-22, or their equivalents. In certain concentrations and in the presence of an open flame, such as a gas range or a gas water heater, R-12 and R-22 may break down and form a small amount of harmful phosgene gas. This gas was used in World War I as the poison gas designed for warfare.

Lifting

Lifting heavy objects can cause serious problems. Strains and sprains are often caused by improper lifting methods. Figure 16-8 indicates the right and the wrong way to lift heavy objects. In this case, a compressor is shown.

To avoid injury, learn to lift the safe way. Bend your knees, keep your back erect, and lift gradually with your leg muscles.

The material you are lifting may slip from your hands and injure your feet. To prevent foot injuries, wear proper shoes.

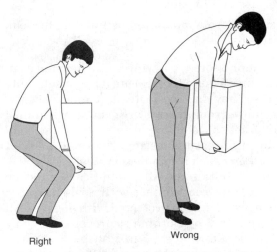

Right   Wrong

Figure 16-8 Safety first. Lift with the legs not the back.

17

Temperatures, Thermometers, and Psychrometrics

Transferring Heat

Heat always passes from a warmer to a colder object or space. The action of refrigeration depends upon this natural law. The three methods by which heat can be transferred are convection, conduction, and radiation.

- *Convection* is heat transfer that takes place in liquids and gases. In convection, the molecules carry the heat from one point to another.

- *Conduction* is heat transfer that takes place chiefly in solids. In conduction, the heat passes from one molecule to another without any noticeable movement of the molecules.

- *Radiation* is heat transfer in waveform, such as light or radio waves. It takes place through a transparent medium such as air, without affecting that medium's temperature, volume, and pressure. Radiant heat is not apparent until it strikes an opaque surface, where it is absorbed. The presence of radiant heat is felt when it is absorbed by a substance or by your body.

Convection can be used to remove heat from an area. Then it can be used to cool. Air or water can be cooled in one plan and circulated through pipes of radiators in another location. It this way the cool water or air is used to remove heat.

Air Temperature and Comfort Conditions

The surface temperature of the average adult's skin is 80°F (26.7°C). The body can either gain or lose heat according to the surrounding air. If the surrounding air is hotter than the skin temperature, the body gains heat and the person may become uncomfortable. If the surrounding air is cooler than the skin temperature, then the body loses heat. Again, the person may become uncomfortable. If the temperature is much higher than the skin temperature or much cooler than the body temperature, then the person becomes uncomfortable. If the air is about 70°F (21.1°C) then the body feels comfortable. Skin temperature fluctuates with the temperature of the surface air. The total range of skin temperature is between 40 and 105°F (4.4 and 40.6°C). However, if the temperature rises 10°F (5.5°C), the skin temperature rises only 3°F (1.7°C). Most of the time the normal temperature of the body ranges from 75 to 100°F (23.9 to 37.8°C). Both humidity and temperature affect the comfort of the human body. However, they are not the only factors that cause a person to be comfortable or uncomfortable. In heating or cooling a room, the air velocity, noise level, and temperature variation caused by the treated air must also be considered.

Velocity

When checking for room comfort, it is best to measure the velocity of the air at the distance of 4 to 72 in. from floor level. Velocity is measured with a velometer (see Fig. 17-1). Following is a range of air velocities and their characteristics.

- Slower than 15 ft/min (fpm): stagnant air
- 20 fpm to 50 fpm: acceptable air velocities
- 25 fpm to 35 fpm: the best range for human comfort
- 35 fpm to 50 fpm: comfortable for cooling purposes

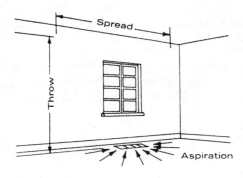

Figure 17-1   Aspiration, throw, and spread. (*Courtesy of Lima.*)

Velocities of 50 fpm or higher call for a very high speed for the air entering the room. A velocity of about 750 fpm or greater is needed to create a velocity of 50 fpm or more inside the room. When velocities greater than 750 fpm are introduced noise will also be present,

Sitting and standing levels must be considered when designing a cooling system for a room. People will tolerate cooler temperatures at the ankle level than at the sitting level, which is about 30 in. from the floor. Variations of 4°F (2.2°C) are acceptable between levels. This is also an acceptable level for temperature variations between rooms.

To make sure that the air is properly distributed for comfort, it is necessary to look at the methods used to accomplish the job.

Terminology

The following terms apply to the movement of air. They are frequently used in referring to air-conditioning systems:

- *Aspiration* is the induction of room air into the primary air stream. Aspiration helps eliminate stratification of air within the room. When outlets are properly located along exposed walls, aspiration also aids in absorbing undesirable currents from these walls and windows (see Fig. 17-1).

- *Feet per minute* (fpm) is the measure of the velocity of an air stream. This velocity can he measured with a velocity meter that is calibrated in feet per minute.

- *Cubic feet per minute* (cfm) is the measure of a volume of air. Air now in cubic feet per minute of a register or grille is computed by multiplying the face velocity times the free area in square feet. For example, a register with 144 in.^{2} (1 ft^2) of free area and a measured face velocity of 500 fpm would be delivering 500 cfm.

- *Decibels* (dB) are units of measure of sound level. It is important to keep this noise at a minimum. In most catalogs for outlets, there is a line dividing the noise level of the registers or diffusers. Lower total pressure loss provides a quieter system.

- *Drop* is generally associated with cooling where air is discharged horizontally from high sidewall outlets. Since cool air has a natural tendency to drop, it will fall progressively as the velocity decreases. Measured at the point of terminal velocity, drop is the distance in feet that the air has fallen below the level of the outlet (see Fig. 17-2).

- *Diffusers* are outlets that have a widespread, fan-shaped pattern of air.

- *Effective area* is the smallest net area of an outlet utilized by the air stream in passing through the outlet passages. It determines the

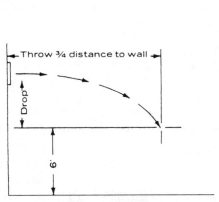

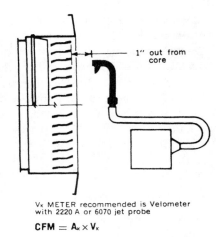

Figure 17-2 Drop. (*Courtesy of ARI.*)

Figure 17-3 Air measurement at the grille. (*Courtesy of Lima.*)

maximum, or jet, velocity of the air in the outlet. In many outlets, the effective area occurs at the velocity measuring point and is equal to the outlet area (see Fig. 17-3).

- *Face velocity* is the average velocity of air passing through the face of an outlet or a return.

- *Free area* is the total area of the openings in the outlet or inlet through which air can pass. With gravity systems, free area is of prime importance. With forced air systems, free area is secondary to total pressure loss, except in sizing return air grilles.

- *Noise criteria* (NC) is an outlet sound rating in pressure level at a given condition of operation, based on established criteria and a specific room acoustic absorption value.

- *Occupied zone* is that interior area of a conditioned space that extends to within 6 in. of all room walls and to a height of 6 ft above the floor.

- *Outlet area* is the area of an outlet utilized by the air stream at the point of the outlet velocity as measured with an appropriate meter. The point of measurement and type of meter must be defined to determine cfm accurately.

- Outlet *velocity* (V_k) is the measured velocity at the started point with a specific meter.

- *Perimeter systems* are heating and cooling installations in which the diffusers are installed to blanket the outside walls. Returns are usually located at one or more centrally located places. High sidewall or ceiling returns are preferred, especially for cooling. Low returns are

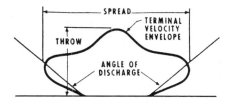

Figure 17-4 Typical airstream pattern. (*Courtesy of Tuttle & Bailey.*)

acceptable for heating. High sidewall or ceiling returns are highly recommended for combination heating and cooling installations.

- *Registers* are outlets that deliver air in a concentrated stream into the occupied zone.
- *Residual velocity* (V_R) is the average sustained velocity within the confines of the occupied zone, generally ranging from 20 to 70 fpm.
- *Sound power level* (L_w) is the total sound created by an outlet under a specified condition of operation.
- *Spread* is the measurement (in feet) of the maximum width of the air pattern at the point of terminal velocity (see Fig. 17-4).
- *Static pressure* (sp) is the outward force of air within a duct. This pressure is measured in inches of water. The static pressure within a duct is comparable to the air pressure within an automobile tire. A manometer measures static pressure.
- *Temperature differential* (ΔT) is the difference between primary supply and room air temperatures.
- *Terminal velocity is* the point at which the discharged air from an outlet decreases to a given speed, generally accepted as 50 fpm.
- *Throw* is the distance (measured in feet) that the air stream travels from the outlet to the point of terminal velocity. Throw is measured vertically from perimeter diffusers and horizontally from registers and ceiling diffusers (see Fig. 17-1).
- *Total pressure* (tp) is the sum of the static pressure and the velocity pressure. Total pressure is also known as impact pressure. This pressure is expressed in inches of water. The total pressure is directly associated with the sound level of an outlet. Therefore, any factor that increases the total pressure will also increase the sound level. The under sizing of outlets or increasing the speed of the blower will increase total pressure and the sound level.
- *Velocity pressure* (V_p) is the forward moving force of air within a duct. This pressure is measured in inches of water. The velocity pressure is comparable to the rush of air from a punctured tire. A velometer is used to measure air velocity (see Fig. 17-3).

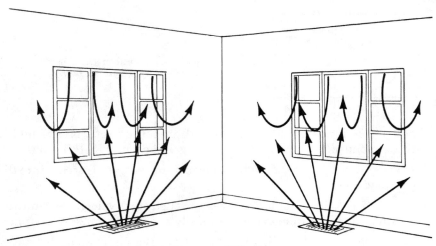

Figure 17-5   Location of an outlet. (*Courtesy of Lima.*)

Designing a Perimeter System

After the heat loss or heat gain has been calculated, the sum of these losses or gains will determine the size of the duct systems and the heating and cooling unit.

The three factors that ensure proper delivery and distribution of air within a room are location of outlet, type of outlet, and size of outlet. Supply outlets, if possible, should always be located to blanket every window and every outside wall (see Fig. 17-5). Thus, a register is recommended under each window.

The outlet selected should be a diffuser whose air pattern is fan shaped to blanket the exposed walls and windows.

The American Society of Heating, Refrigerating and Air-Conditioning Engineers (ASHRAE) furnishes a chart with the locations and load factors needed for the climate of each major city in the United States. The chart should be followed carefully. The type of house, the construction materials, house location, room sizes, and exposure to sun and wind are important factors. With such information, you can determine how much heat will be dissipated. You can also determine how much heat will be lost in a building. You can also determine how much cooling will be lost in a building. The *ASHRAE Handbook* of *Fundamentals* lists the information needed to compute the load factors.

Calculate the heat loss or heat gain of the room; divide this figure by the number of outlets to be installed. From this you can determine the Btu per hour required of each outlet. Refer to the performance data furnished by the manufacturer to determine the size the outlet should be. For residential application, the size selected should be large enough so that the Btu per hour capacity on the chart falls to the side where the quiet zone is indicated. There is still a minimum vertical throw of 6 ft where cooling is involved.

Locating and sizing returns

Properly locating and sizing return air grilles are important. It is generally recommended that the returns be installed in high sidewall or the ceiling. They should be in one or more centrally located places. This depends upon the size and floor plan of the structure. Although such a design is preferred, low returns are acceptable for heating.

To minimize noise, care must be taken to *size* correctly the return air grille. The blower in the equipment to be used is rated in cfm by the manufacturer. This rating can usually be found in the specification sheets. Select the grille or grilles necessary to handle this cfm.

The grille or grilles selected should deliver the necessary cfm for the air to be conditioned. Thus, the proper size must be selected. The throw should reach approximately three-quarters of the distance from the outlet to the opposite wall (see Fig. 17-2). The face velocity should not exceed the recommended velocity for the application (see Table 17-1). The drop should be such that the air stream will not drop into the occupied zone. The occupied zone is generally thought of as 6 ft above the floor level.

The sound caused by an air outlet in operation varies in direct proportion to the velocity of the air passing through it. Air velocity depends partially on outlet size. Table 17-2 lists recommendations for outlet velocities within safe sound limits for most applications.

Air-flow distribution

Bottom or side outlet openings in horizontal or vertical supply ducts should be equipped with adjustable flow-equalizing devices. Figure 17-6 indicates the pronounced one-sided flow effect from an outlet opening. This is before the corrective effect of air-turning devices. A control grid is added in Fig. 17-7 to equalize flow in the takeoff collar. A Vectrol is added in Fig. 17-8 to turn air into the branch duct and provide volume control. Air-turning devices are recommended for installation at all outlet collars and branch duct connections.

Square unvaned elbows are also a source of poor duct distribution and high-pressure loss. Nonuniform flow in a main duct, occurring after an unvaned ell, severely limits the distribution of air into branch ducts in the vicinity of the ell. One side of the duct may be void, thus starving a branch duct. Conversely, all flow may be stacked up on one side. This requires dampers to be excessively closed, resulting in higher sound levels.

TABLE 17-1 Register or Grille Size Related to Air Capacities

Register of grille size	Area in ft²	Air capacities in cfm										
		250 fpm	300 fpm	400 fpm	500 fpm	600 fpm	700 fpm	750 fpm	800 fpm	900 fpm	1000 fpm	1250 fpm
8 × 4	0.163	41	49	65	82	98	114	122	130	147	163	204
10 × 4	0.206	52	62	82	103	124	144	155	165	185	206	258
10 × 6	0.317	79	95	127	158	190	222	238	254	285	317	396
12 × 4	0.249	62	75	100	125	149	174	187	199	224	249	311
12 × 5	0.320	80	96	128	160	192	224	240	256	288	320	400
12 × 6	0.383	96	115	153	192	230	268	287	306	345	383	479
14 × 4	0.292	73	88	117	146	175	204	219	234	263	292	365
14 × 5	0.375	94	113	150	188	225	263	281	300	338	375	469
14 × 6	0.449	112	135	179	225	269	314	337	359	404	449	561
16 × 5	0.431	108	129	172	216	259	302	323	345	388	431	539
16 × 6	0.515	129	155	206	258	309	361	386	412	464	515	644
20 × 5	0.541	135	162	216	271	325	379	406	433	487	541	676
20 × 6	0.647	162	194	259	324	388	453	485	518	582	647	809
20 × 8	0.874	219	262	350	437	524	612	656	699	787	874	1093
24 × 5	0.652	162	195	261	326	391	456	489	522	587	652	815
24 × 6	0.779	195	234	312	390	467	545	584	623	701	779	974
24 × 8	1.053	263	316	421	527	632	737	790	842	948	1053	1316
24 × 10	1.326	332	398	530	663	796	928	995	1061	1193	1326	1658
24 × 12	1.595	399	479	638	798	951	1117	1196	1276	1436	1595	1993
30 × 6	0.978	245	293	391	489	587	685	734	782	880	978	1223
30 × 8	1.321	330	396	528	661	793	925	991	1057	1189	1371	1651
30 × 10	1.664	416	499	666	832	998	1165	1248	1331	1498	1664	2080
30 × 12	2.007	502	602	803	1004	1204	1405	1505	1606	1806	2007	2509
36 × 8	1.589	397	477	636	795	953	1112	1192	1271	1430	1589	1986
36 × 10	2.005	501	602	802	1003	1203	1404	1504	1604	1805	2005	2506
36 × 12	2.414	604	724	966	1207	1448	1690	1811	1931	2173	2414	3018

*Based on LIMA registers of the 100 Series.

TABLE 17-2　Outlet Velocity Ratings

Area	Rating (in fpm)
Broadcast studios	500
Residences	500–750
Apartments	500–750
Churches	500–750
Hotel bedrooms	500–750
Legitimate theatres	500–1000
Private offices, acoustically treated	500–1000
Motion picture theatres	1000–1250
Private offices, not acoustically treated	1000–1250
General offices	1250–1500
Stores	1500
Industrial buildings	1500–2000

Figure 17-6　This flow path diagram shows the pronounced one-sided flow effect from an outlet opening before corrective effect of air-turning devices. (*Courtesy of Tuttle & Bailey.*)

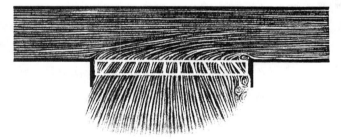

Figure 17-7　A control grid is added to equalize flow in the takeoff collar. (*Courtesy of Tuttle & Bailey.*)

Flow diagrams show the pronounced turbulence and piling up of airflow in an ell (see Fig. 17-9). Ducturns reduce the pressure loss in square elbows as much as 80 percent. Their corrective effect is shown in Fig. 17-10.

Selection of Diffusers and Grilles

The selection of a linear diffuser or grille involves job condition requirements, selection judgment, and performance data analysis. Diffusers and

Figure 17-8   A Vectrol is added to turn air into the branch duct and provide volume control. (*Courtesy of Tuttle & Bailey.*)

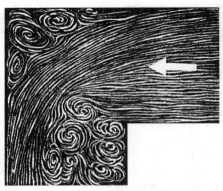

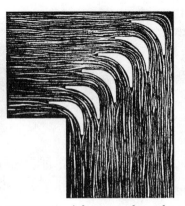

Figure 17-9   Note the turbulence and piling up of airflow in an ell. (*Courtesy of Tuttle & Bailey.*)

Figure 17-10   A ducturn reduces the pressure loss in square elbows by as much as 80 percent. (*Courtesy of Tuttle & Bailey.*)

grilles should be selected and sized according to the following characteristics:

- Type and style
- Function
- Air volume requirement
- Throw requirement
- Pressure requirement
- Sound requirement

Air volume requirement

The air volume per diffuser or grille is that which is necessary for the cooling, heating, or ventilation requirements of the area served by the unit. The air volume required, when related to throw, sound, or pressure-design limitations, determines the proper diffuser or grille size.

Generally, air volumes for internal zones of building spaces vary from 1 to 3 cfm/ft^2 of floor area. Exterior zones will require higher air volumes of 2.5 to 4 cfm/ft^2 of floor area. In some cases, only the heating or cooling load of the exterior wall panel or glass surface is to be carried by the distribution center. Then, the air volume per linear foot of diffuser or grille will vary from 20 to 200 cfm, depending on heat transfer coefficient, wall height, and infiltration rate.

Throw requirement

Throw and occupied area air location are closely related. Both could be considered in the analysis of specific area requirements. The minimum-maximum throw for a given condition of aeration is based upon a terminal velocity at that distance from the diffuser. The residual room velocity is a function of throw to terminal velocity. Throw values are based on terminal velocities ranging from 75 to 150 fpm with corresponding residual room velocities of 75 to 150 fpm. The diffuser or grille location together with the air pattern selected, should generally direct the air path above the occupied zone. The air path then induces room air along its throw as it expands in cross section. This equalizes temperature and velocity within the stream. With the throw terminating in a partition or wall surface, the mixed air path further dissipates energy.

Ceiling-mounted grilles and diffusers are recommended for vertical down pattern. Some locations in the room may need to be cooler than others. Also, some room locations may be harder to condition because of airflow problems. They are used in areas adjacent to perimeter wall locations that require localized spot conditioning. Ceiling heights of 12 ft or greater are needed. The throw for vertical projection is greatly affected by supply air temperature and proximity of wall surfaces.

Sidewall-mounted diffusers and grilles have horizontal values based on a ceiling height of 8 to 10 ft. The diffuser or grille is mounted approximately 1 ft below the ceiling. For a given listed throw, the room air motion will increase or decrease inversely with the ceiling height. For a given air pattern setting and room air motion the listed minimum-maximum throw value can be decreased by 1 ft for each 1 ft increase in ceiling height above 10 ft. Throw values are furnished by the manufacturer.

When sidewall grilles are installed remote from the ceiling (more than 3 ft away), it reduces rated throw values by 20 percent.

Sill mounted diffusers or grilles have throw values based on an 8 to 10 ft ceiling height. This is with the outlet installed in the top of a 30-in.-high sill. For a given listed throw, the room air motion will change with the ceiling height. For a given air pattern setting and room air motion, the listed minimum-maximum throw value can be decreased by 2 ft for each 1 ft increase in ceiling height above 10 ft. Decrease 1 ft for each 1 ft decrease in sill height.

The minimum throw results in a room air motion higher than that obtained when utilizing the maximum throw. Thus, 50 fpm, rather than 35 fpm, is the air motion. The listed minimum throw indicates the minimum distance recommended. The minimum distance is from the diffuser to a wall or major obstruction, such as a structural beam. The listed maximum throw is the recommended maximum distance to a wall or major obstruction. Throw values for sidewall grilles and ceiling diffusers and the occupied area velocity are based on flush ceiling construction providing an unobstructed air stream path. The listed maximum throw times 1.3 is the complete throw of the air stream. This is where the terminal velocity equals the room air velocity. Rated occupied area velocities range from 25 to 35 fpm for maximum listed throws and 35 to 50 fpm for minimum listed throw values.

Cooled-air drop or heated-air rise are of practical significance when supplying heated or cooled air from a sidewall grille. If the throw is such that the air stream prematurely enters the occupied zone, considerable draft may be experienced. This is due to incomplete mixing. The total airdrop must be considered when the wall grille is located at a distance from the ceiling. Cooled airdrop is controlled by spacing the wall grille from the ceiling and adjusting the grilles upward 15 in. Heated-air rise contributes significantly to temperature stratification in the upper part of the room.

The minimum separation between grille and ceiling must be 2 ft or more. The minimum mounting separation must be 2 ft or more. The minimum mounting height should be 7 ft.

Pressure requirement

The diffuser or grille minimum pressure for a given air volume reflects itself in ultimate system fan horsepower requirements.

A diffuser or grille with a lower pressure rating requires less total energy than a unit with a higher pressure rating for a given air volume and effective area. Diffusers and grilles of a given size having lower pressure ratings usually have a lower sound level rating at a specified air volume.

Sound requirement

Diffusers and grilles should be selected for the recommended noise criteria rating for a specific application. The data for each specific diffuser or grille type contains a noise criteria (NC) rating. Table 17-3 lists recommended NC and area of application.

TABLE 17-3 **Recommend NC Criteria**

NC curve	Communication environment	Typical occupancy
Below NC 25	Extremely quiet environment; suppressed speech is quite audible; suitable for acute pickup of all sounds.	Broadcasting studios, concert halls, music rooms.
NC 30	Very quiet office suitable for large conferences; telephone use satisfactory.	Residences, theatres, libraries, executive offices, directors' rooms.
NC 35	Quiet office; satisfactory for conference at a 15-ft table; normal voice 10 to 30 ft; telephone use satisfactory.	Private offices, schools, hotel rooms, courtrooms, churches, hospital rooms.
NC 40	Satisfactory for conferences at a 6- to 8-ft table; normal voice 6 to 12 ft; telephone use satisfactory.	General offices, labs, dining rooms.
NC 45	Satisfactory for conferences at a 4- to 5-ft table; normal voice 3 to 6 ft; raised voice 6 to 12 ft; telephone use occasionally difficult.	Retail stores, cafeterias, lobby areas large drafting and engineering offices, reception areas.
Above NC 50	Unsatisfactory for conferences of more than two or three persons; normal voice 1 to 2 ft; raised voice 3 to 6 ft; telephone use slightly difficult.	Photocopy rooms, stenographic pools, print machine rooms, process areas.

Air Noise

High velocities in the duct or diffuser typically generate air noise. The flow turbulence in the duct and the excessive pressure reductions in the duct and diffuser system also generate noise. Such noise is most apparent directly under the diffuser. Room background levels of NC 35 and less provide little masking effect. Any noise source stands out above the background level and is easily detected.

Typically, air noise can be minimized by the following procedures:

- Limiting branch duct velocities to 1200 fpm
- Limiting static pressure in branch ducts adjacent to outlets to 0.15 in. of water
- Sizing diffusers to operate at outlet jet velocities up to 1200 fpm (neck velocities limited to 500 to 900 fpm), and total pressures of 0.10 in. of water
- Using several small diffusers (and return grilles) instead of one or two large outlets or inlets that have a higher sound power
- Providing low-noise dampers in the branch duct where pressure drops of more than 0.20 in. of water must be taken
- Internally lining branch ducts near the fan to quiet this noise source
- Designing background sound levels in the room to be a minimum of NC 35 or NC 40

Casing Radiated Noise

Casing noise differs from air noise in the way it is generated. Volume controllers and pressure-reducing dampers generate casing noise. Inside terminal boxes are sound baffles, absorbing blankets, and orifice restrictions to eliminate line of sight through the box. All these work to reduce the generated noise before the air and air noise discharge from the box into the outlet duct. During this process, the box casing is vibrated by the internal noise. This causes the casing to radiate noise through the suspended ceiling into the room (see Fig. 17-11).

Locating terminal boxes

In the past, terminal boxes and ductwork were separated from the room by dense ceilings. These ceilings prevented the system noise from radiating into the room. Plaster and taped sheetrock ceilings are examples of dense ceilings. Current architectural practice is to utilize lightweight (and low-cost) decorative suspended ceilings. These ceilings are not dense. They have only one-half the resistance to noise transmission

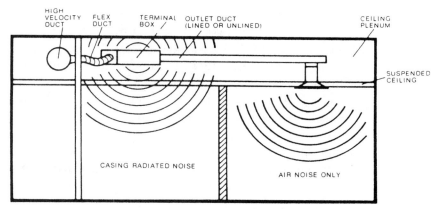

Figure 17-11   Casing noises. (*Courtesy of Tuttle & Bailey.*)

that plaster and sheetrock ceilings have. Exposed tee-bar grid ceilings with 2×4 glass-fiber pads, and perforated metal-pan ceilings are examples. The end result is readily apparent. Casing radiated noise in lightweight modern buildings is a problem.

Controlling casing noise

Terminal boxes can sometimes be located over noisy areas (corridors, toilet areas or machine equipment rooms), rather than over quiet areas. In quiet areas casing noise can penetrate the suspended ceiling and become objectionable. Enclosures built around the terminal box (such as sheetrock or sheet lead over a glass-fiber blanket wrapped around the box) can reduce the radiated noise to an acceptable level.

However, this method is cumbersome and limits access to the motor and volume controllers in the box. It depends upon field conditions for satisfactory performance, and is expensive. Limiting static pressure in the branch ducts minimizes casing noise. This technique, however, limits the flexibility of terminal box systems. It hardly classifies as a control.

Vortex shedding

Product research in controlling casing noise has developed a new method of reducing radiated noise. The technique is known as vortex shedding. When applied to terminal boxes, casing radiated noise is dramatically lowered. Casing radiation attenuation (CRA) vortex shedders can be installed in all single- or dual-duct boxes up to 7000 cfm, constant volume or variable volume, with or without reheat coils. CRA devices provide unique features and the following benefits:

- No change in terminal box size. Box is easier to install in tight ceiling plenums to ensure minimum casing noise under all conditions.

- Factory-fabricated box and casing noise eliminator, a one-piece assembly, reduces cost of installation. Only one box is hung. Only one duct connection is made.

- Quick-opening access door is provided in box. This ensures easy and convenient access to all operating parts without having to cut and patch field-fabricated enclosures.

- Equipment is laboratory tested and performance rated. Engineering measurements are made in accordance with industry standards. Thus, on-the-job performance is ensured. Quiet rooms result and owner satisfaction is assured.

Return Grilles

Performance

Return air grilles are usually selected for the required air volume at a given sound level or pressure value. The intake air velocity at the face of the grille depends mainly on the grille size and the air volume.

The grille style and damper setting have a small effect on this intake velocity. The grille style, however, has a very great effect on the pressure drop. This, in turn, directly influences the sound level.

The intake velocity is evident only in the immediate vicinity of the return grille. It cannot influence room air distribution. Recent ASHRAE research projects have developed a scientific computerized method of relating intake grille velocities, measured 1 in. out from the grille face, to air volume. Grille-measuring factors for straight, deflected bar, open, and partially closed dampers are in the engineering data furnished with the grille.

It still remains the function of the supply outlets to establish proper coverage, air motion, and thermal equilibrium. Because of this, the location of return grilles is not critical and their placement can be largely a matter of convenience. Specific locations in the ceiling may be desirable for local heat loads, or smoke exhaust, or a location in the perimeter sill or floor may be desirable for an exterior zone intake under a window wall section. It is not advisable to locate large centralized return grilles in an occupied area. The large mass of air moving through the grille can cause objectionable air motion for nearby occupants.

Return grille sound requirement

Return air grilles should be selected for static pressures. These pressures will provide the required NC rating and conform to the return system performance characteristics. Fan sound power is transmitted through the return air system as well as the supply system. Fan silencing may be necessary or desirable in the return side. This is particularly so if silencing is being considered on the supply side.

Transfer grilles venting into the ceiling plenum should be located remote from plenum noise source. The use of a lined sheet metal elbow can reduce transmitted sound. Lined elbows on vent grilles and lined common ducts on ducted return grilles can minimize "cross talk" between private offices.

Types of Registers and Grilles

The spread of an unrestricted air stream is determined by the grille bar deflection. Grilles with vertical face bars at 0° deflection will have a maximum throw value. As the deflection setting of vertical bars is increased, the air stream covers a wider area and the throw decreases.

Registers are available with adjustable valves. An air-leakage problem is eliminated if the register has a rubber gasket mounted around the grille. When it pulls up tightly against the wall, an airtight seal is made. This helps to eliminate noise. The damper has to be cam-operated so that it will stay open and not blow shut when the air comes through.

On some registers, a simple tool can be used to change the direction of the deflection bars. This means adjusting the bars in the register can have a number of deflection patterns.

Fire and Smoke Dampers

Ventilating, air-conditioning, and heating ducts provide a path for fire and smoke, which can travel throughout a building. The ordinary types of dampers that are often installed in these ducts depend on gravity-close action or spring-and-level mechanisms. When their releases are activated, they are freed to drop inside the duct.

A fusible link attachment to individual registers also helps control fire and smoke. Figure 17-12 shows a fusible-link-type register. The link is available with melting points of 160°F (71.1°C) or 212°F (100°C). When the link melts, it releases a spring that forces the damper to a fully closed position. The attachment does not interfere with damper operation.

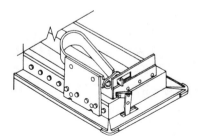

Figure 17-12   Register with fusible link for fire control. (*Courtesy of Lima.*)

Smoke dampers for high-rise buildings

Fire and smoke safety concepts in high-rise buildings are increasingly focusing on providing safety havens for personnel on each floor. This provision is to optimize air flow to or away from the fire floor or adjacent floors. Such systems require computer-actuated smoke dampers. Dampers are placed in supply and return ducts that are reliable. They must be tight closing, and offer minimum flow resistance when fully open.

Ceiling Supply Grilles and Registers

Some ceiling grilles and registers have individually adjustable vanes. They are arranged to provide a one-way ceiling air pattern. They are recommended for applications in ceiling and sidewall locations for heating and cooling systems. They work best where the system has 0.75 to 1.75 cfm/ft^2 of room area (see Fig. 17-13).

Some supply ceiling grilles and registers have individually adjustable curved vanes. They are arranged to provide a three-way ceiling air pattern. The vertical face vanes are a three-way diversion for air. A horizontal pattern with the face vanes also produces a three-way dispersion of air. These grilles and registers are recommended for applications in ceiling locations for heating and cooling systems handling 1.0 to 2.0 cfm/ft^2 of room area.

Figure 17-14 shows a grille with four-way vertical face vanes. Horizontal face vanes are also available. They, too, are adjustable individually for focusing an air stream in any direction. Both the three-way and four-way pattern grilles can be adjusted to a full or partial downblow position. The curved streamlined vanes are adjusted to a uniform partially closed position. This deflects the air path while retaining an effective area capacity of 35 percent of the neck area. In the full downblow position, grille effective area is increased by 75 percent.

Figure 17-13 Ceiling grille. (*Courtesy of Tuttle & Bailey.*)

Figure 17-14 Vertical face vanes in a four-way ceiling supply grille. (*Courtesy of Tuttle & Bailey.*)

Figure 17-15 Perforated face adjustable diffuser for full flow and a deflector for ceiling installation. (*Courtesy of Tuttle & Bailey.*)

Perforated adjustable diffusers for ceiling installation are recommended for heating and cooling (see Fig. 17-15). They are also recommended for jobs requiring on-the-job adjustment of air diffusion patterns.

Full-flow square or round necks have expanded metal air-pattern deflectors. They are adjustable for four-, three-, two-, or one-way horizontal diffusion patterns. This can be done without change in the air volume, pressure, or sound levels. This deflector and diffuser have high diffusion rates. The result is rapid temperature and velocity equalization of the mixed air mass well above the zone of occupancy. They diffuse efficiently with 6 to 18 air changes per hour.

Ceiling Diffusers

There are other designs in ceiling diffusers. The type shown in Fig. 17-16 is often used in a supermarket or other large store. Here, it is difficult to mount other means of air distribution. These round diffusers with a flush face and fixed pattern are for ceiling installation. They are used for heating, ventilating, and cooling. They are compact and simple flush diffusers. High induction rates result in rapid temperature and velocity equalization of the mixed air mass. Mixing is done above the zone of occupancy.

Figure 17-16 Round diffusers with flush face, fixed pattern for ceiling installation. (*Courtesy of Tuttle & Bailey.*)

Figure 17-17 Control grid with multi-blade devices to control airflow in a diffuser collar. (*Courtesy of Tuttle & Bailey.*)

Figure 17-18 Antismudge ring. (*Courtesy of Tuttle & Bailey.*)

Grids are used and sold as an accessory to these diffusers. The grid (see Fig. 17-17) is a multiblade device designed to ensure uniform airflow in a diffuser collar. It is individually adjustable. The blades can be moved to control the air stream precisely.

For maximum effect, the control grid should be installed with the blades perpendicular to the direction of approaching airflow. Where short collars are encountered, a double bank of control grids is recommended. The upper grid is placed perpendicular to the branch duct flow. The lower grid is placed parallel to the branch duct flow. The control grid is attached to the duct collar by means of mounting straps. It is commonly used with volume dampers.

Antismudge rings

The antismudge ring is designed to cause the diffuser discharge-air path to contact the ceiling in a thin-layered pattern. This minimizes local turbulence, the cause of distinct smudging (see Fig. 17-18).

For the best effect, the antismudge ring must fit evenly against the ceiling surface. It is held in position against the ceiling by the diffuser margin. This eliminates any exposed screws.

Air-channel diffusers

Air-channel supply diffusers are designed for use with integrated air-handling ceiling systems. They are adaptable to fit between open parallel tee bars.

They fit within perforated or slotted ceiling runners. The appearance of the integrated ceiling remains unchanged regardless of the size of the unit. They are painted-out to be invisible when viewing the ceiling.

Figure 17-19 High-capacity air channel diffuser with fixed pattern for suspended grid ceilings. (*Courtesy of Tuttle & Bailey.*)

Figure 17-20 Single-side diffuser with side inlet. (*Courtesy of Tuttle & Bailey.*)

These high-capacity diffusers provide a greater air-handling capability (see Fig. 17-19).

Luminaire diffusers

The luminaire is a complete lighting unit. The luminaire diffuser fits close to the fluorescent lamp fixtures in the ceiling. The single-side diffuser with side inlet is designed to provide single-side concealed air distribution (see Fig. 17-20). These diffusers are designed with oval-shaped side inlets and inlet dampers. They provide effective single point dampering.

Dual-side diffusers with side inlet are designed to provide concealed air distribution. Note the crossover from the oval-shaped side inlet to the other side of the diffuser. This type of unit handles more air and spreads it more evenly when used in large areas (see Fig. 17-21). This type of diffuser is also available with an insulation jacket when needed.

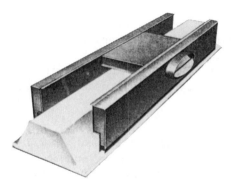

Figure 17-21 Dual-side diffuser with side inlet. (*Courtesy of Tuttle & Bailey.*)

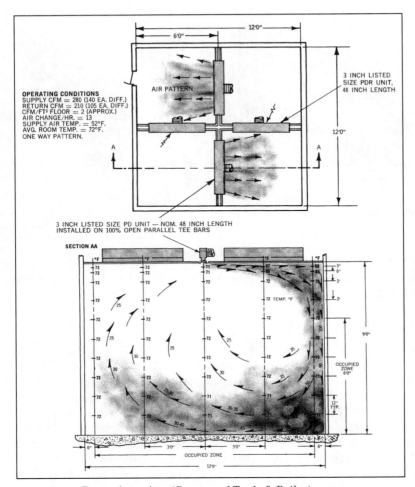

Figure 17-22   Room air motion. (*Courtesy of Tuttle & Bailey.*)

Room air motion

Figure 17-22 illustrates the airflow from ceiling diffusers. The top view illustrates the motion from the diffuser. The side view shows how the temperature differential is very low. Note that the temperature is 68°F near the ceiling and sidewall and 73°F on the opposite wall near the ceiling.

Linear Grilles

Linear grilles are designed for installation in the sidewall, sill, floor, and ceiling. They are recommended for supplying heated, ventilated, or cooled air and for returning or exhausting room air (see Fig. 17-23).

Figure 17-23 Linear grille with a hinged access door. (*Courtesy of Tuttle & Bailey.*)

Figure 17-24 Debris screen for linear grilles. (*Courtesy of Tuttle & Bailey.*)

When installed in the sidewall near the ceiling, the linear grilles provide a horizontal pattern above the occupied zone. Core deflections of 15° and 30° direct the air path upward to overcome the drop effect resulting from cool primary air.

When installed in the top of a sill or enclosure, the linear grilles provide a vertical up-pattern. This is effective in overcoming uncomfortable cold downdrafts. It also offsets the radiant effect of glass surfaces. Core deflections of 0° and 15° directed toward the glass surface provide upward airflow to the ceiling, and along the ceiling toward the interior zone.

When installed in the ceiling, linear grilles provide a vertically downward air pattern. This pattern is effective in projection heating and in cooling the building perimeter from ceiling heights above 13 to 15 ft. Application of down flow primary air should be limited to ensure against excessive drafts at the end of the throw. Core deflections of 0°, 15°, and 30° direct the air path angularly downward as required. Debris screens can be integrally attached (see Fig. 17-24).

Fans and Mechanical Ventilation

Mechanical ventilation differs from natural ventilation mainly in that the air circulation is performed by mechanical means (such as fans or blowers). In natural ventilation, the air is caused to move by natural forces. In mechanical ventilation, the required air changes are effected partly by diffusion, but chiefly by positive currents put in motion by electrically operated fans or blowers, as shown in Fig. 17-25. Fresh air is usually circulated through registers connected with the outside and warmed as it passes over and through the intervening radiators.

Air volume

The volume of air required is determined by the size of the space to be ventilated and the number of times per hour the air in that space is to

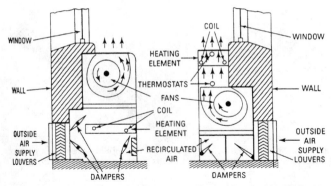

Figure 17-25 Typical mechanical ventilators for residential use. Note placement of fans and other details.

be changed. Table 17-4 shows the recommended rate of air change for various types of spaces. In many cases, existing local regulations or codes will govern the ventilating requirements. Some of these codes are based on a specified amount of air per person, and others on the air required per square foot of floor area. Table 17-4 should serve as a guide to average conditions. Where local codes or regulations are involved, they should be taken into consideration. If the number of persons occupying the space is larger than would be normal for such a space, the air should be changed more often than shown.

TABLE 17-4   Volume of Air Required

Space to be ventilated	Air changes per hour	Minutes per change
Auditoriums	6	10
Bakeries	20	3
Bowling alleys	12	5
Club rooms	12	5
Churches	6	10
Dining rooms (restaurants)	12	5
Factories	10	6
Foundries	20	3
Garages	12	5
Kitchens (restaurants)	30	2
Laundries	20	3
Machine shops	10	6
Offices	10	6
Projection booths	60	1
Recreation rooms	10	6
Sheet-metal shops	10	6
Ship holds	6	10
Stores	10	6
Toilets	20	3
Tunnels	6	10

Fans and blowers

The various devices used to supply air circulation in air-conditioning applications are known as fans, blowers, exhausts, or propellers. The different types of fans may be classified with respect to their construction as follows:

- Propeller
- Tube axial
- Vane axial
- Centrifugal

A *propeller fan* consists essentially of a propeller or disk-type wheel within a mounting ring or plate and includes the driving-mechanism supports for either belt or direct drive. A *tube axial fan* consists of a propeller or disk-type wheel within a cylinder and includes the driving-mechanism supports for either belt drive or direct connection. A *vane axial fan* consists of a disk-type wheel within a cylinder and a set of air guide vanes located before or after the wheel. It includes the driving-mechanism supports for either belt drive or direct connection. A *centrifugal fan* consists of a fan rotor or wheel within a scroll-type housing and includes the driving-mechanism supports for either belt drive or direct connection. Figure 17-26 shows the mounting arrangements.

Fan performance may be stated in various ways, with the air volume per unit time, total pressure, static pressure, speed, and power input being the most important. The terms, as defined by the National Association of Fan Manufacturers, are as follows:

- *Volume* handled by a fan is the number of cubic feet of air per minute expressed as fan-outlet conditions.

- *Total pressure* of a fan is the rise of pressure from fan inlet to fan outlet.

- *Velocity pressure* of a fan is the pressure corresponding to the average velocity determination from the volume of airflow at the fan outlet area.

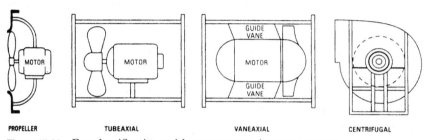

PROPELLER   TUBEAXIAL   VANEAXIAL   CENTRIFUGAL

Figure 17-26   Fan classifications with proper mounting arrangement.

- *Static pressure* of a fan is the total pressure diminished by the fan-velocity pressure.

- *Power output* of a fan is expressed in horsepower and is based on fan volume and the fan total pressure.

- *Power input* of a fan is expressed in horsepower and is measured as horsepower delivered to the fan shaft.

- *Mechanical efficiency* of a fan is the ratio of power output to power input.

- *Static efficiency* of a fan is the mechanical efficiency multiplied by the ratio of static pressure to the total pressure.

- *Fan-outlet area* is the inside area of the fan outlet.

- *Fan-inlet area* is the inside area of the inlet collar.

Horsepower requirements

The horsepower required for any fan or blower varies directly as the cube of the speed, provided that the area of the discharge orifice remains unchanged. The horsepower requirements of a centrifugal fan generally decrease with a decrease in the area of the discharge orifice if the speed remains unchanged. The horsepower requirements of a propeller fan increase as the area of the discharge orifice decreases if the speed remains unchanged.

Fan-driving methods

Whenever possible, the fan wheel should be directly connected to the motor shaft. This can usually be accomplished with small centrifugal fans and with propeller fans up to about 60 in. in diameter. The deflection and the critical speed of the shaft, however, should be investigated to determine whether or not it is safe.

When selecting a motor for fan operation, it is advisable to select a standard motor one size larger than the fan requirements. It should be kept in mind, however, that direct-connected fans do not require as great a safety factor as that of belt-driven units. It is desirable to employ a belt drive when the required fan speed or horsepower is in doubt, since a change in pulley size is relatively inexpensive if an error is made (see Fig. 17-27).

Directly connected small fans for single-phase AC motors of the split-phase, capacitor, or shaded-pole type usually drive various applications. The capacitor motor is more efficient electrically and is used in districts where there are current limitations. Such motors, however, are usually arranged to operate at one speed. With such a motor, if it is necessary to vary the air volume or pressure of the fan or blower, the throttling of air by a damper installation is usually made.

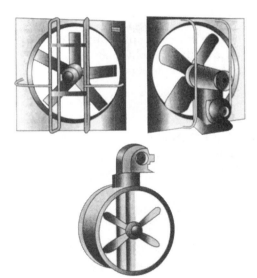

Figure 17-27 Various types of propeller fan drives and mounting arrangements.

In large installations (such as when mechanical draft fans are required), various drive methods are used:

- A slip-ring motor to vary the speed.

- A constant-speed, directly connected motor, which, by means of moveable guide vanes in the fan inlet, serves to regulate the pressure and air volume.

Selecting a fan

Most often, the service determines the type of fan to use. When operation occurs with little or no resistance, and particularly when no duct system is required, the propeller fan is commonly used because of its simplicity and economy in operation. When a duct system is involved, a centrifugal or axial type of fan is usually employed. In general, centrifugal and axial fans are comparable with respect to sound effect, but the axial fans are somewhat lighter and require considerably less space. The following information is usually required for proper fan selection:

- Capacity requirement in cfm
- Static pressure or system resistance
- Type of application or service
- Mounting arrangement of system
- Sound level or use of space to be served
- Nature of load and available drive

The various fan manufacturers generally supply tables or characteristic curves that ordinarily show a wide range of operating particulars for each fan size. The tabulated data usually include static pressure, outlet velocity, revolutions per minute, brake horsepower, tip or peripheral speed, and so on.

Applications for fans

The numerous applications of fans in the field of air-conditioning and ventilation are well known, particularly to engineers and air-conditioning repair and maintenance personnel. The various fan applications are as follows:

- Attic fans
- Circulating fans
- Cooling-tower fans
- Exhaust fans
- Kitchen fans

Exhaust fans are found in all types of applications, according to the American Society of Heating and Ventilating Engineers. Wall fans are predominantly of the propeller type, since they operate against little or no resistance. They are listed in capacities from 1000 to 75,000 cfm. They are sometimes incorporated in factory-built penthouses and roof caps or provided with matching automatic louvers. Hood exhaust fans involving duct work are predominantly centrifugal, especially in handling hot or corrosive fumes.

Spray-booth exhaust fans are frequently centrifugal, especially if built into self-contained booths. Tube axial fans lend themselves particularly well to this application where the case of cleaning and of suspension in a section of ductwork is advantageous. For such applications, built-in cleanout doors are desirable.

Circulating fans are invariably propeller or disk-type units and are made in a vast variety of blade shapes and arrangements. They are designed for appearance as well as utility. *Cooling-tower fans* are predominantly the propeller type. However, axial types are also used for packed towers, and occasionally a centrifugal fan is used to supply draft. *Kitchen fans* for domestic use are small propeller fans arranged for window or wall mounting and with various useful fixtures. They are listed in capacity ranges of from 300 to 800 cfm.

Attic fans are used during the summer to draw large volumes of outside air through the house or building whenever the outside temperature is lower than that of the inside. It is in this manner that the relatively cool evening or night air is utilized to cool the interior in one or several rooms, depending on the location of the air-cooling unit. It should be clearly understood, however, that the attic fan is not strictly a piece of air-conditioning equipment since it only moves air and does not cool, clean, or dehumidify. Attic fans are used primarily because of their low cost and economy of operation, combined with their ability to produce comfort cooling by circulating air rather than conditioning it.

Operation of fans

Fans may be centrally located in an attic or other suitable space (such as a hallway), and arranged to move air proportionately from several rooms. A local unit may be installed in a window to provide comfort cooling for one room only when desired. Attic fans are usually propeller types and should be selected for low velocities to prevent excessive noise. The fans should have sufficient capacity to provide at least 30 air changes per hour.

To decrease the noise associated with air-exchange equipment, the following rules should be observed:

- The equipment should be properly located to prevent noise from affecting the living area.
- The fans should be of the proper size and capacity to obtain reasonable operating speed.
- Equipment should be mounted on rubber or other resilient material to assist in preventing transmission of noise to the building.

If it is unavoidable to locate the attic air-exchange equipment above the bedrooms, it is essential that every precaution be taken to reduce the equipment noise to the lowest possible level. Since high-speed AC motors are usually quieter than low-speed ones, it is often preferable to use a high-speed motor connected to the fan by means of an endless V-belt, if the floor space available permits such an arrangement.

Installation of attic fans

Because of the low static pressures involved (usually less than 1/8 in. of water) disk or propeller fans are generally used instead of the blower or housed types. It is important that the fans have quiet operating characteristics and sufficient capacity to give at least 30 air changes per hour. For example, a house with 10,000 ft^3 content would require a fan with a capacity of 300,000 ft^3/h or 5000 cfm to provide 30 air changes per hour.

The two general types of attic fans in common use are *boxed-in fans* and *centrifugal fans*. The boxed-in fan is installed within the attic in a box or suitable housing located directly over a central ceiling grille or in a bulkhead enclosing an attic stair. This type of fan may also be connected by means of a direct system to individual room grilles. Outside

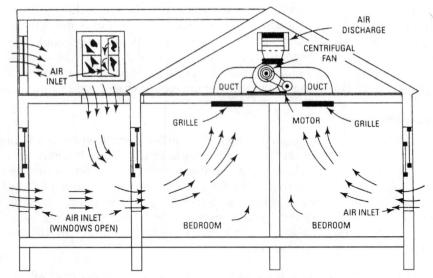

Figure 17-28   Installing a centrifugal fan in a one-family dwelling.

cool air entering through the windows in the downstairs room is discharged into the attic space and escapes to the outside through louvers, dormer windows, or screened openings under the eaves.

Although an air-exchange installation of this type is rather simple, the actual decision about where to install the fan and where to provide the grilles for the passage of air up through the house should be left to a ventilating engineer. The installation of a multiblade centrifugal fan is shown in Fig. 17-28. At the suction side, the fan is connected to exhaust ducts leading to grilles, which are placed in the ceiling of the two bedrooms. The air exchange is accomplished by admitting fresh air through open windows and up through the suction side of the fan; the air is finally discharged through louvers as shown.

Another installation is shown in Fig. 17-29. This fan is a centrifugal curved-blade type mounted on a light angle-iron frame, which supports the fan wheel, shaft, and bearings. The air inlet in this installation is placed close to a circular opening, which is cut in an airtight board partition that serves to divide the attic space into a suction and discharge chamber. The air is admitted through open windows and doors and is then drawn up the attic stairway through the fan into the discharge chamber.

Routine fan operation

The routine of operation to secure the best and most efficient results with an attic fan is important. A typical operating routine might require that, in the late afternoon when the outdoor temperature begins to fall, the

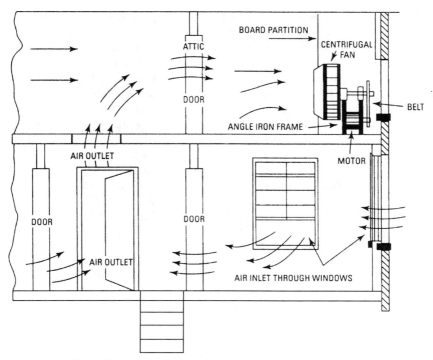

Figure 17-29 Typical attic installation of a belt-driven fan.

windows on the first floor and the grilles in the ceiling or the attic floor be opened and the second-floor windows kept closed. This will place the principal cooling effect in the living rooms. Shortly before bedtime, the first-floor windows may be closed and those on the second floor opened to transfer the cooling effect to the bedrooms. A suitable time clock may be used to shut the motor off before arising time.

Ventilation Methods

Ventilation is produced by two basic methods: natural and mechanical. Open windows, vents, or drafts obtain natural ventilation, whereas mechanical ventilation is produced by the use of fans.

Thermal effect is possibly better known as flue effect. Flue effect is the draft in a stack or chimney that is produced within a building when the outdoor temperature is lower than the indoor temperature. This is caused by the difference in weight of the warm column of air within the building and the cooler air outside.

Air may be filtered two ways: dry filtering and wet filtering. Various air-cleaning equipments (such as filtering, washing, or combined filtering and washing devices) are used to purify the air. When designing the duct network, ample filter area must be included so that the air velocity passing through the filters is sufficient. Accuracy in estimating the resistance to the flow of air through the duct system is important in the selection of blower motors. Resistance should be kept as low as possible in the interest of economy. Ducts should be installed as short as possible.

Competent medical authorities have properly emphasized the effect of dust on health. Air-conditioning apparatus removes these contaminants from the air. The apparatus also provides the correct amount of moisture so that the respiratory tracts are not dehydrated, but are kept properly moist. Dust is more than just dry dirt. It is a complex, variable mixture of materials and, as a whole, is rather uninviting, especially the type found in and around human habitation. Dust contains fine particles of sand, soot, earth, rust, fiber, animal and vegetable refuse, hair, and chemicals.

Troubleshooting

Air-conditioning and refrigeration units experience problems with the electrical and the mechanical aspects of their operation. In some cases they have been moved and damaged and in other cases they have been repaired with components that are incorrect or good components incorrectly wired.

The service technician must be aware of these problems. In this chapter an attempt has been made to show how certain troubleshooting procedures can aid in making the job faster and easier.

Safety

Safety first has its direct and implied meaning. You can work with air-conditioning and refrigeration equipment safely if a few common sense rules are followed.

Handling refrigerants

The proper use of gloves, eye protection, and clothing to protect the body is necessary. Freon or any refrigerant escaping from a refrigeration unit can cause permanent damage to the skin. It can also cause blindness if you are hit in the face with the gases under pressure. Make sure you wear the proper clothing at all times when troubleshooting or recharging a unit.

Testing precaution

Pressure testing or cleaning refrigeration and air-conditioning systems can be dangerous. Be careful not to exceed 150 psig when pressure testing a complete system.

Electrical safety

The main rule for electrical safety is to see that the main circuit breaker is in the off position and locked before starting to remove or check (with an ohmmeter) any refrigeration, air-conditioning, or heating equipment. Keep in mind that some compressors have power applied at all times to the off-cycle crankcase heater. Even if the compressor is not running, the power is applied to the crankcase heater. Some run capacitors are connected to the compressor motor windings even when the compressor is not running. Other devices are energized when the compressor is not running. Thus, electrical power is applied to the unit even when the compressor is not running.

Compressor Problems

There are a number of compressor problems that can be quickly identified from a table of problems, possible causes, and suggested repair (see Table 18-1). The compressor is the heart of the refrigeration system whether it is the air conditioner, refrigerator, or freezer. That makes it of primary concern in any troubleshooting procedure.

PSC compressors

The permanent split-capacitor compressor has some problems that should be uppermost in your mind so that you are aware of them. In Table 18-2 the low-voltage problems are listed and possible corrections given. The branch circuit fuse or circuit breaker trips and causes the unit to become disabled. The possible cause could be that the rating of the protection device is not high enough to handle the current. Some problems are within the scope of the work the technician can handle. Others, like low line voltage caused by factors beyond the control of the home-owner, call for the utility company to correct them.

A lot of troubleshooting is common sense. If the fuse is blown and you put in another and it blows, it tells you that there is something drawing too much current. Simply check for cause.

Air-conditioner compressors

Home air conditioners, those that fit into a window or those that are part of a central air-conditioning system, are subject to problems with their compressors and electrical controls. Most of those that operate in the window are designed for 120- or 240-V operation. In some cases, during the summer when line voltage is low, they may experience some low-voltage problems.

The units are equipped with some kind of filter to make sure the air is cleaned before it is forced into the room being cooled. These filters

TABLE 18-1 Compressor Troubleshooting and Service

Complaint	Possible cause	Repair
Compressor will not start. There is no hum.	1. Line disconnect switch open. 2. Fuse removed or blown. 3. Overload protector tripped. 4. Control stuck in open position. 5. Control off due to cold location. 6. Wiring improper or loose.	1. Close start or disconnect switch. 2. Replace fuse. 3. Refer to electrical section. 4. Repair or replace control. 5. Relocate control. 6. Check wiring against diagram.
Compressor will not start. It hums, but trips on overload protector.	1. Improperly wired. 2. Low voltage to unit. 3. Starting capacitor defective. 4. Relay failing to close. 5. Compressor motor has a winding open or shorted. 6. Internal mechanical trouble in compressor. 7. Liquid refrigerant in compressor.	1. Check wiring against diagram. 2. Determine reason and correct. 3. Determine reason and replace. 4. Determine reason and correct, replace if necessary. 5. Replace compressor. 6. Replace compressor. 7. Add crankcase heater and/or accumulator.
Compressor starts, but does not switch off of start winding.	1. Improperly wired. 2. Low voltage to unit. 3. Relay failing to open. 4. Run capacitor defective. 5. Excessively high discharge pressure. 6. Compressor motor has a winding open or shorted. 7. Internal mechanical trouble in compressor (tight).	1. Check wiring against diagram. 2. Determine reason and correct. 3. Determine reason and replace if necessary. 4. Determine reason and replace. 5. Check discharge shutoff valve, possible overcharge, or insufficient cooling of condenser. 6. Replace compressor. 7. Replace compressor.
Compressor starts and runs, but short cycles on overload protector.	1. Additional current passing through the overload protector. 2. Low voltage to unit (or unbalanced if three phase).	1. Check wiring against diagram; check added fan motors, pumps, etc., connected to wrong side of protector. 2. Determine reason and correct.

(Continued)

TABLE 18-1 Compressor Troubleshooting and Service (Continued)

Complaint	Possible cause	Repair
	3. Overload protector defective.	3. Check current, replace protector.
	4. Run capacitor defective.	4. Determine reason and replace.
	5. Excessive discharge pressure.	5. Check ventilation, restrictions in cooling medium, restrictions in refrigeration system.
	6. Suction pressure too high.	6. Check for possibility of misapplication; use stronger unit.
	7. Compressor too hot—return gas is hot.	7. Check refrigerant charge. (Repair leak.) Add refrigerant if necessary.
	8. Compressor motor has a winding shorted.	8. Replace compressor.
Unit runs, but short cycles on.	1. Overload protector.	1. Check current; replace protector.
	2. Thermostat.	2. Differential set too close; widen.
	3. High-pressure cutout due to insufficient air or water supply, overcharge, or air in system.	3. Check air or water supply to condenser; reduce refrigerant charge, or purge.
	4. Low-pressure cutout due to:	4.
	a. Liquid line solenoid leaking.	a. Replace.
	b. Compressor valve leak.	b. Replace.
	c. Undercharge.	c. Repair leak and add refrigerant.
	d. Restriction in expansion device.	d. Replace expansion device.
Unit operates long or continuously.	1. Shortage of refrigerant.	1. Repair leak; add charge.
	2. Control contacts stuck or frozen closed.	2. Clean contacts or replace control.
	3. Refrigerated or air-conditioned space has excessive load or poor insulation.	3. Determine fault and correct.
	4. System inadequate to handle load.	4. Replace with larger system.
	5. Evaporator coil iced.	5. Defrost.
	6. Restriction in refrigeration system.	6. Determine location and remove.
	7. Dirty condenser.	7. Clean condenser.
	8. Filter dirty.	8. Clean or replace.
Start capacitor open, shorted, or blown.	1. Relay contacts not operating properly.	1. Clean contacts or replace relay if necessary.
	2. Prolonged operation on start cycle due to:	2. a. Determine reason and correct.
	a. Low voltage to unit.	b. Replace.
	b. Improper relay.	c. Correct by using pump-down arrangement if necessary.
	c. Starting load too high.	

Complaint	Possible Cause	Remedy
	3. Excessive short cycling	3. Determine reason for short cycling as mentioned in previous complaint.
	4. Improper capacitor.	4. Determine correct size and replace.
Run capacitor open, shorted, or blown.	1. Improper capacitor.	1. Determine correct size and replace.
	2. Excessively high line voltage (110% of rated maximum).	2. Determine reason and correct.
Relay defective or burned out.	1. Incorrect relay.	1. Check and replace.
	2. Incorrect mounting angle.	2. Remount relay in correct position.
	3. Line voltage too high or too low.	3. Determine reason and correct.
	4. Excessive short cycling.	4. Determine reason and correct.
	5. Relay being influenced by loose vibrating mounting.	5. Remount rigidly.
	6. Incorrect run capacitor.	6. Replace with proper capacitor.
Space temperature too high.	1. Control setting too high.	1. Reset control.
	2. Expansion valve too small.	2. Use larger valve.
	3. Cooling coils too small.	3. Add surface or replace.
	4. Inadequate air circulation.	4. Improve air movement.
Suction line frosted or sweating.	1. Expansion valve oversized or passing excess refrigerant.	1. Readjust valve or replace with smaller valve.
	2. Expansion valve stuck open.	2. Clean valve of foreign particles. Replace if necessary.
	3. Evaporator fan not running.	3. Determine reason and correct.
	4. Overcharge of refrigerant.	4. Correct charge.
Liquid line frosted or sweating.	1. Restriction in dehydrator or strainer.	1. Replace part.
	2. Liquid shutoff (king valve) partially closed.	2. Open valve fully.
Unit noisy.	1. Loose parts or mountings.	1. Tighten.
	2. Tubing rattle.	2. Reform to be free of contact.
	3. Bent fan blade causing vibration.	3. Replace blade.
	4. Fan motor bearings worn.	4. Replace motor.

Courtesy of Kelvinator.

TABLE 18-2 PSC Compressor Motor Troubles and Corrections

Causes	Corrections
Low Voltage	
1. Inadequate wire size.	1. Increase wire size.
2. Watt-hour meter too small.	2. Call utility company.
3. Power transformer too small or feeding too many homes.	3. Call utility company.
4. Input voltage too low.	4. Call utility company.
Note: Starting torque varies as the square of the input voltage.	
Branch Circuit Fuse or Circuit Breaker Tripping	
1. Rating too low.	1. Increase size to a minimum or 175% of unit FLA (full load amperes) to a maximum of 225% of FLA.
System Pressure High or Not Equalized	
1. Pressures not equalizing within 3 minutes.	1. a. Check metering device (capillary tube or expansion valve).
	b. Check room thermostat for cycling rate; off cycle should be at least 5 minutes; also check for "chattering."
	c. Has some refrigerant dryer or some other possible restriction been added?
2. System pressure too high.	2. Make sure refrigerant charge is correct.
3. Excessive liquid in crankcase (split-system applications).	3. Add crankcase heater and suction line accumulator.
Miscellaneous	
1. Run capacitor open or shorted.	1. Replace with new, properly sized capacitor.
2. Internal overload open.	2. Allow 2 hours to reset before changing compressor.

usually require a change or cleaning at least once a year. Where dust is a problem, such maintenance should be more frequent. At this time, the condenser coil should be brushed with a soft brush and flushed with water. The filters should be vacuumed and then washed to remove dust. The outside of the case should be wiped clean with a soapy cloth. The cleaner the filter, the more efficient the unit. In some cases a clogged filter can cause compressor problems. Table 18-1 shows some of the problems associated with the compressor in these units. Note how most of the problems associated with air conditioners are electrical.

Low-Voltage Operation

Electrical apparatus designed to produce at full capacity at the voltage indicated on the rating plate can malfunction at lower than designated voltages. Motors operated at lower than rated voltage cannot provide full horsepower without shortening their service life. Relays and solenoids can also fail to operate if low voltage is present.

The Air-Conditioning and Refrigeration Institute (ARI) certifies the cooling units after testing them. The units are tested to make sure they will operate with ±10 percent of the rated voltage. This does not mean they will operate continuously without damage to the motor. Most air-conditioning compressor burnouts are caused by low-voltage operation. A hermetic compressor is entirely enclosed within the refrigerant cycle; it is very important that it not be abused by overloading or low voltage. Both conditions can occur during peak load periods. A national survey has shown that the most common cause of compressor low voltage is the use of undersized conductors between the utility lines and the condensing unit.

Using a System to Troubleshoot (Electrical)

Refrigeration, air-conditioning, and heating units all have various systems that cause them to operate and do the job they were designed for. First, you have to identify the system and the problems most commonly associated with the particular system. Then you have to be able to determine exactly what caused the symptom you are witnessing. Next, you have to be able to correct the problem and put the unit back into operation.

Electrical tests are the most common because the electrical problems are the most frequently encountered in all three types of equipment. A systematic procedure is necessary to obtain the needed results.

Motor testing

Testing a motor in a sealed condensing unit to determine why it does not operate becomes a very simple process if the correct procedure is followed. Each test made should be one of a series of eliminations to determine what part of the system is defective. By checking other parts of the wiring system before checking the unit itself, a great deal of time can be saved since, in most cases, the trouble will be in the wiring or controls rather than in the unit.

To make a complete electrical test on electrical outlets and on the unit itself, it is advisable to make a test cord (see Fig. 18-1). By connecting the black and white terminals together and placing a light bulb in the socket, the cord may be used to check the wall outlet into which the unit is connected. By connecting the white and red terminal clips, this same test may be made by depressing the push button. This will serve as a test to make certain the push button is in working order. This test cord should have a capacitor installed in the red lead to the push button if the compressor is a capacitor-start type.

When these tests have been completed and it is known that current is being supplied to the unit, the next step is to check the three wires on the base of the compressor unit. Pull the plug from the wall receptacle

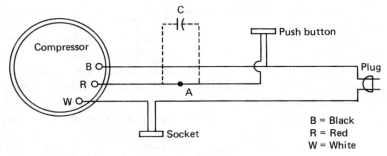

Figure 18-1 Test lamp for troubleshooting.

and carefully examine the nuts that hold the wires in position. Then try each wire to be sure it is held firmly in place, as a loose wire may keep the unit from operating. Test the thermostatic switch to determine whether or not contact is being made at that point. Turn the cold-control knob several times. If this fails to start the unit, then short across the thermostatic switch terminals on the switch. To do this, it will be necessary to remove the switch cover from the top of the switch. If the unit starts, it is an indication that the thermostatic switch is not operating properly and must be repaired or replaced. After the thermostatic switch has been checked and if the trouble is not located, it will be necessary to determine whether the trouble is in the motor, motor-protector relay, or capacitor.

Capacitor testing

The capacitor must be checked before testing the unit itself. This is done in the following manner:

1. Disconnect the capacitor wires from the motor-protector relay.
2. Connect these two wires to the black and white terminals of the test cord.
3. Put a 150-W light bulb in the receptacle on the test cord, and plug it into an outlet.

If the 150-W bulb does not light, it is an indication that the capacitor has an open circuit and must be replaced. However, if the bulb does light, it is not an indication that the capacitor is perfect. This must be checked further by shorting across the two terminals of the capacitor with a screwdriver that has an insulated handle. If the brilliance of the light changes, that is, if the light bulb burns brighter when the terminals of the capacitor are shorted, it indicates that the capacitor is in proper operating condition. A decided sparking of the terminals will also be noticed when the terminals are shorted. If the brilliance of the bulb does not change, it is an indication that the capacitor has an internal short and must be replaced.

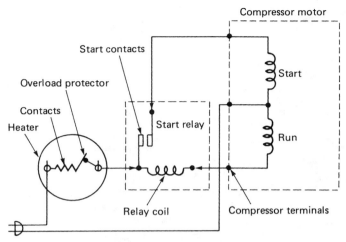

Figure 18-2 Location of overload protector in the circuit.

Motor-protector relay testing

If the overload protector is found defective during the preceding test, replace it (see Fig. 18-2). The motor overload protector is usually accessible for replacement and is located near the compressor (see Fig. 18-3). Remove the clips holding the overload in position, and remove the wire

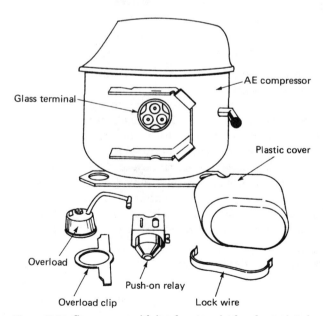

Figure 18-3 Compressor with its glass terminal and associated components (*Courtesy of Tecumseh.*)

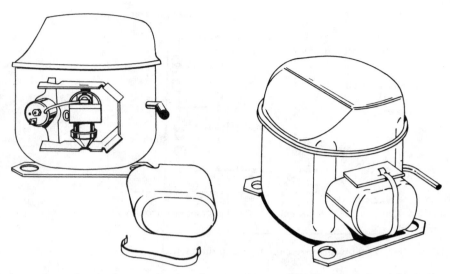

Figure 18-4 Overload protection location on the compressor. (*Courtesy of Tecumseh.*)

Figure 18-5 Compressor with cover. (*Courtesy of Tecumseh.*)

from the terminal by pulling it outward. Notice the position of the electrical lead before replacing it (see Fig. 18-4). Place the electrical leads on the replacement overload and check for correct connections against the wiring diagram. Close up the unit (see Fig. 18-5). Test for proper operation.

Using Meters to Check for Problems

The voltmeter and the ohmmeter can be used to isolate various problems. You should be able to read the schematic and make the proper voltage or resistance measurements. An incorrect reading will indicate the possibility of a problem. Troubleshooting charts will aid in isolating the problem to a given system. Once you have arrived at the proper system that may be causing the symptoms noticed, you will then need to use the ohmmeter with the power off to isolate a section of the system. Once you have zeroed in on the problem, you can locate it by knowing what the proper reading should be. Deviation from a stated reading of over 10 percent is usually indicative of a malfunction, and in most cases the component part must be replaced to ensure proper operation and no call backs.

Using a Volt-Ammeter for Troubleshooting Electric Motors

Most electrical equipment will work satisfactorily if the line voltage differs ±10 percent from the actual nameplate rating. In a few cases, however, a 10 percent voltage drop may result in a breakdown. Such may be the case with an induction motor that is being loaded to its fullest capacity both on start and run. A 10 percent loss in the line voltage will result in a 20 percent loss in torque.

The full-load current rating on the nameplate is an approximate value based on the average unit coming off the manufacturer's production line. The actual current for any one unit may vary as much as ±10 percent at rated output. However, a load current that exceeds the rated value by 20 percent or more will reduce the life of the motor due to higher operating temperatures, and the reason for excessive current should be determined. In many cases it may simply be an overloaded motor. The percentage of increase in load will not correspond with percentage of increase in load current. For example, in the case of a single-phase induction motor, a 35 percent increase in current may correspond to an 80 percent increase in torque output.

The operating conditions and behavior of electrical equipment can be analyzed only by actual measurement. A comparison of the measured terminal voltage and current will check whether the equipment is operating within electrical specifications.

A voltmeter and an ammeter are needed for the two basic measurements. To measure voltage, the test leads of the voltmeter are in contact with the terminals of the line under test. To measure current, the conventional ammeter must be connected in series with the line so that the current will flow through the ammeter.

To insert the ammeter, you must shut down the equipment, break open the line, connect the ammeter, and then start up the equipment to read the meter. And you have to do the same to remove the meter once it has been used. Other time-consuming tests may have to be made to locate the problem. However, all this can be eliminated by the use of a clamp-on volt-ammeter.

Clamp-on volt-ammeter

The pocket size volt-ammeter shown in Fig. 18-6 is the answer to most troubleshooting problems on the job. The line does not have to be disconnected to obtain a current reading. The meter works on the transformer principle; it picks up the magnetic lines surrounding a current-carrying conductor and presents this as a function of the entire amount flowing through the line. Remember, in transformers we discussed how the magnetic field strength in the core of the transformer determines the amount of current in the secondary. The same principle is used here to detect the flow of current and amount of current.

To get transformer action, the line to be tested is encircled with the split-type core by simply pressing the trigger button. Aside from measuring terminal voltages and load currents, the split-core ammeter-voltmeter can be used to track down electrical difficulties in electric motor repair.

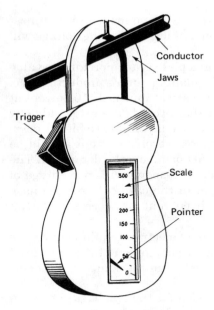

Conductor

Jaws

Trigger

Scale

Pointer

Figure 18-6 Clamp-on volt-ammeter. (*Courtesy of Amprobe.*)

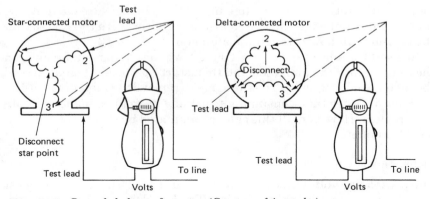

Figure 18-7   Grounded phase of a motor. (*Courtesy of Amprobe.*)

Looking for grounds

To determine whether a winding is grounded or has a very low value of insulation resistance, connect the unit and test leads as shown in Fig. 18-7. Assuming the available line voltage is approximately 120 V, use the unit's lowest voltage range. If the winding is grounded to the frame, the test will indicate full-line voltage.

A high-resistance ground is simply a case of low-insulation resistance. The indicated reading for a high-resistance ground will be a little less than line voltage. A winding that is not grounded will be evidenced

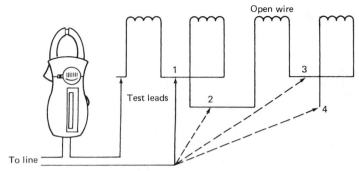

Figure 18-8   Isolating an open phase. (*Courtesy of Amprobe.*)

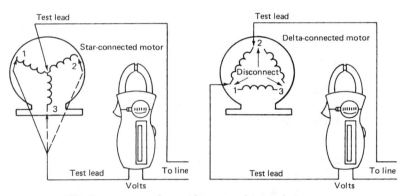

Figure 18-9   Finding an open phase. (*Courtesy of Amprobe.*)

by a small or negligible reading. This is due mainly to the capacitive effect between the windings and the steel lamination.

To locate the grounded portion of the windings, disconnect the necessary connection jumpers and test. Grounded sections will be detected by a full-line voltage indication.

Looking for opens

To determine whether a winding is open, connect test leads as shown in Figs. 18-8 and 18-9. If the winding is open, there will be no voltage indication. If the circuit is not open, the voltmeter indication will read full-line voltage.

Looking for shorts

Shorted turns in the winding of a motor behave like a shorted secondary of a transformer. A motor with a shorted winding will draw excessive current while running at no load. Measurement of the current can be made without disconnecting lines.

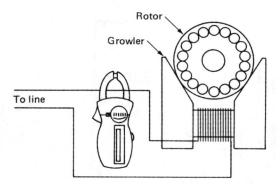

Figure 18-10   Using a growler to test a motor. (*Courtesy of Amprobe.*)

You engage one of the lines with the split-core transformer of the tester. If the ammeter reading is much higher than the full-load ampere rating on the nameplate, the motor is probably shorted.

In a two- or three-phase motor, a partially shorted winding produces a higher current reading in the shorted phase. This becomes evident when the current in each phase is measured.

Motors with Squirrel-Cage Rotors

Loss in output torque at rated speed in an induction motor may be due to opens in the squirrel-cage rotor. To test the rotor and determine which rotor bars are loose or open, place the rotor in a growler. Engage the split-core ammeter around the lines going to the growler, as shown in Fig. 18-10. Set the switch to the highest current range. Switch on the growler and then set the test unit to the approximate current range. Rotate the rotor in the growler and take note of the current indication whenever the growler is energized. The bars and end rings in the rotor behave similarly to a shorted secondary of a transformer. The growler windings act as the primary. A good rotor will produce approximately the same current indications for all positions of the rotor. A defective rotor will exhibit a drop in the current reading when the open bars move into the growler field.

Testing the Centrifugal Switch in a Single-Phase Motor

A defective centrifugal switch may not disconnect the start winding at the proper time. To determine conclusively that the start winding remains in the circuit, place the split-core ammeter around one of the start-winding leads. Set the instrument to the highest current range. Turn on the motor switch. Select the appropriate current range. Observe if there is any current in the start-winding circuit. A current indication signifies that the centrifugal switch did not open when the motor came up to speed (see Fig. 18-11).

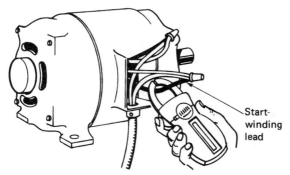

Figure 18-11 Checking the centrifugal switch with a clamp-on meter. (*Courtesy of Amprobe.*)

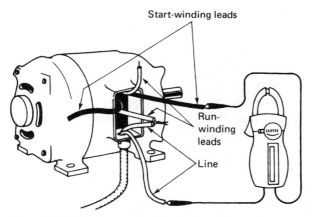

Figure 18-12 Finding a shorted winding using a clamp-on meter. (*Courtesy of Amprobe.*)

Testing for a Short Circuit between Run and Start Windings

A short between run and start windings may be determined by using the ammeter and line voltage to check for continuity between the two separate circuits. Disconnect the run and start-winding leads and connect the instrument as shown in Fig. 18-12. Set the meter on voltage. A full-line voltage reading will be obtained if the windings are shorted to one another.

Capacitor Testing

Defective capacitors are very often the cause of trouble in capacitor-type motors. Shorts, opens, grounds, and insufficient capacity in microfarads are conditions for which capacitors should be tested to determine whether they are good.

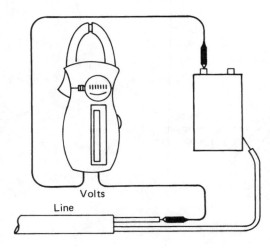

Figure 18-13 Finding a grounded capacitor with a clamp-on meter. (*Courtesy of Amprobe.*)

You can determine a grounded capacitor by setting the instrument on the proper voltage range and connecting it and the capacitor to the line as shown in Fig. 18-13. A full-line voltage indication on the meter signifies that the capacitor is grounded to the can. A high-resistance ground is evident from a voltage reading that is somewhat below the line voltage. A negligible reading or a reading of no voltage indicates that the capacitor is not grounded.

Measuring the capacity of a capacitor

To measure the capacity of the capacitor, set the test unit's switch to the proper voltage range and read the line-voltage indication. Then set to the appropriate current range and read the capacitor-current indication. During the test, keep the capacitor on the line for a very short period of time, because motor starting electrolytic capacitors are rated for intermittent duty (see Fig. 18-14). The capacity in microfarads is then computed by substituting the voltage and current readings in the following formula, assuming that a full 60-Hz line is used:

$$\text{Microfarads} = \frac{2650 \times \text{amperes}}{\text{volts}}$$

An open capacitor will be evident if there is no current indication in the test. A shorted capacitor is easily detected. It will blow the fuse when the line switch is turned on to measure the line voltage. Troubleshooting can be broken down into a simple procedure. Logical thinking is what it takes to be able to accurately diagnose a problem and then correct it. For instance, you can isolate a particular system

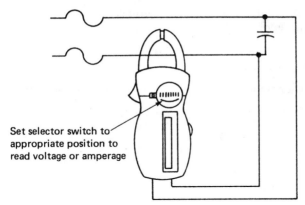

Figure 18-14 Finding the size of a capacitor with a clamp-on meter. (*Courtesy of Amprobe.*)

since you know how the system is supposed to operate. If it is not operating according to expectations, then you know something is causing the problem. By concentrating on a particular system that has been located, it is possible to concentrate your thinking on things that would cause the malfunction

Once you have determined which components in the system can cause the symptoms being experienced, it is possible to find the faulty component.

After the problem has been located and the component identified, it is then necessary to take the component out and repair it or replace it.

Once the problem has been located, the part removed and a new one inserted or the old one repaired it is necessary to check the operation of the system to make sure it is performing according to its design characteristics.

All this can be broken down into key words that will aid you in any troubleshooting situation.

Troubleshooting procedure

Isolate. Isolate the system giving the trouble. The system may be

The evaporator
The condenser
The compressor
The controls

Use the problem-probable cause-remedy charts to aid in isolating the problem.

Concentrate. Concentrate on the isolated system. Concentrate on how it works and why it would malfunction.

Differentiate. Determine which components could cause the identified symptoms, then locate each possible troublemaking part.

Eliminate. Eliminate the possible troublemakers one by one. Use the proper test instrument to do so.

Repair or replace. Either repair or replace the identified troublemaker.

Check. Check to see if the unit operates properly with the repaired or replaced parts.

19

Solid-State Controls

Semiconductors

The word semiconductor identifies a type of electronic device. Transistors and diodes are semiconductors. These units are part of the field of electronics. They have been used over the past years in every aspect of control circuits on every type of equipment. Electronic controls for air-conditioning, refrigeration, and heating equipment have been upgraded to use transistors, diodes, and computer chips as well, and other types of semiconductor devices (see Fig. 19-1).

Semiconductor principles

Semiconductor technology is usually called *solid state*. This means that the materials used are of one piece. This is in contrast with the vacuum tube, which consisted of a series of assembled parts. Of course, the substances from which semiconductors are made are not really solid. They have atomic structures consisting largely of empty space. The spaces are essential for the movement of electrons.

Many materials can be classified as semiconductors, but two have been used extensively for electronic circuits. They are silicon and germanium. Germanium was used originally for diodes and transistors, but has since been largely replaced by silicon.

Both silicon and germanium are hard, crystal-type materials that are very brittle. They can be made pure to 99.9 percent. Doping agents or impurities are added in controlled amounts. Impurities used may be boron, aluminum, indium, gallium, arsenic, and antimony.

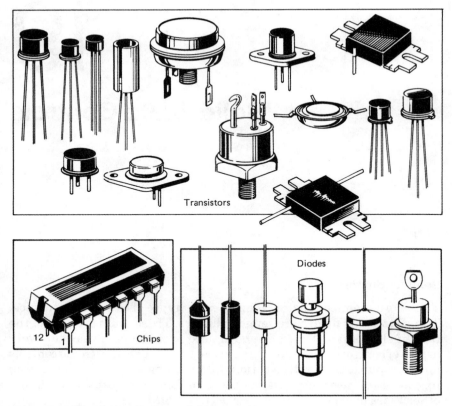

Figure 19-1 Transistors, diodes, and chips.

By doping silicon, it is possible to allow the electrons added by doping (in N-type material) to move easily. In P-type material the holes or places where an electron should be are moved along by the voltage applied. The addition of arsenic or antimony makes an N-type semiconductor material. The material will have an excess of electrons. Electrons have a negative charge.

When gallium or indium is used as the impurity, a P-type semiconductor material is produced. This means it has a positive charge, or is missing an electron.

Diode

The semiconductor diode is used to allow current to flow in only one direction. It can be used to change alternating current to direct current. It is made by using N-type and P-type semiconductor materials.

When N- and P-type materials are joined, they form a diode (see Fig. 19-2). The diode is also called a rectifier since it can change (rectify)

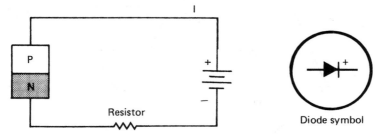

Figure 19-2   N-type diode in a circuit. Note the symbol for a diode.

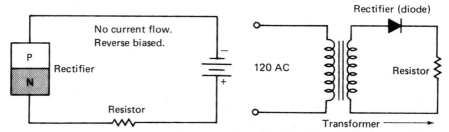

Figure 19-3   Reverse-bias diode in a circuit. Note how this differs from Fig. 19-2.

Figure 19-4   Circuit showing how a diode is used to rectify AC to produce DC.

AC and make it DC. The PN junction (diode) acts as a one-way valve to control the current flow. The forward or low-resistance direction through the junction allows current to flow through it. The high-resistance direction does not allow current to flow. Thus, only one-half of an alternating current hertz is allowed to flow in a circuit with a diode (see Fig. 19-2). The diode used in this circuit is forward biased and allows current to flow. Figure 19-3 indicates the arrangement in the reverse-bias configuration. No current is allowed to flow under these conditions. Also, note the polarity of the battery.

Diodes are used in isolating one circuit from another. A simple rectifier circuit is shown in Fig. 19-4. The output from the transformer is an AC voltage, as shown in Fig. 19-5. However, the rectifier action of the diode blocks current flow in one-half of the sine wave and produces a pulsating DC across the resistor (see Fig. 19-5b).

A number of factors affect the voltage and current ratings of a PN-junction diode. As you work with them, you will learn that they have ratings for peak and inverse AC currents. This information is available for both germanium and silicon diodes in transistor handbooks produced by their manufacturers.

Figure 19-6 shows a number of specialized diodes. This illustration demonstrates that diodes are designed for specific jobs. Both internal and external construction is determined by circuit requirements. For

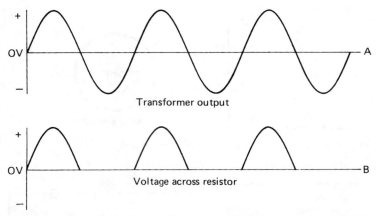

Figure 19-5 Results of the rectifier circuit. Transformer output (AC) is changed to DC.

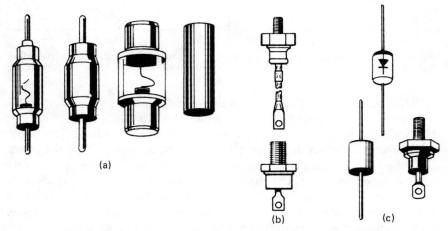

Figure 19-6 Diodes for special circuits: (*a*) radar and computer circuits; (*b*) large-current circuits; and (*c*) low-current circuits.

example, Fig. 19-6*a* shows diodes used for radar and computer circuits. The diodes in Fig. 19-6*b* are designed to carry large currents. So their cases are heat conductors. As the diodes become warm, the generated heat is transferred to the air. Figure 19-6*c* shows zener diodes. These units protect sensitive meter movements and regulate voltage.

Silicon-controlled rectifiers

The silicon-controlled rectifier (SCR) is a specialized type of semiconductor used for control of electrical circuits. This is a four-layer device. The structure can be either NPNP or PNPN.

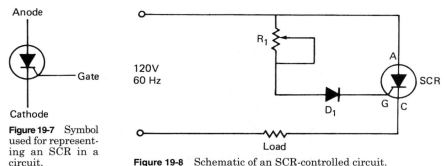

Anode

Gate

Cathode

Figure 19-7 Symbol used for representing an SCR in a circuit.

120V
60 Hz

R_1

A

SCR

D_1

G | C

Load

Figure 19-8 Schematic of an SCR-controlled circuit.

An SCR conducts current in a forward direction only. The symbol for an SCR is shown in Fig. 19-7. Current always flows through an SCR from the cathode (C) to the anode (A). The illustration indicates that the SCR also has a gate (G).

The function of an SCR is shown in the circuit diagram in Fig. 19-8. The typical use of an SCR is for a controlled circuit. Examples include a light dimmer or a speed control for a motor. This type of circuit is shown in Fig. 19-8. The resistor in the circuit, R_1, is a rheostat, or adjustable resistor. This is used to control the amount of voltage delivered to the gate of the SCR.

The more voltage delivered, the greater the flow. Thus, adjusting the rheostat will control the circuit. If the circuit illuminates a lamp, lowering the voltage to the rheostat dims the bulb. If the load is a motor, its speed is lowered. Figure 19-9 shows typical SCRs.

Transistors

Transistors are made from N- and P-type crystals. Once joined, the two different types of crystals produce junctions. Transistors are identified according to emitter junction and collector junction. Thus, they are either PNP- or NPN-types.

A PNP transistor is formed by a thin N-region between two P-regions (see Fig. 19-10). The center N-region is called the base. This base is usually 0.001 in. thick. A collector junction and an emitter junction

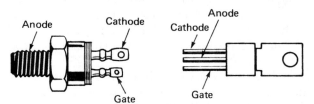

Anode

Cathode

Gate

Anode

Cathode

Gate

Figure 19-9 Two types of SCRs. Both types are used for control circuits, depending on the amount of current involved.

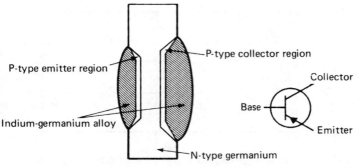

Figure 19-10   Transistor junction using germanium and indium.

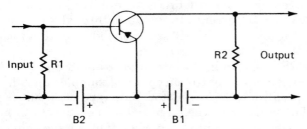

Figure 19-11   Schematic diagram of a common-emitter circuit using a PNP transistor.

are also formed. Note the emitter is represented by an arrow either going toward the vertical line or away from it. The collector is the line without an arrow. The third line, the one that has the other two contacting it, is the base. These are usually abbreviated C, B, and E, respectively.

Transistors can be used for amplification of signals or they can be used for switching. There are three common configurations for transistor circuits. They are the common base, common collector, and common emitter. The common-emitter circuit is the most often used (see Fig. 19-11). In this type of circuit, the current through the load flows between the emitter and collector. The input signal is applied between the emitter and the base. In normal operation, the collector junction is reverse biased by the supply voltage, B_1. The emitter junction is forward biased by the applied voltage, B_2. Electrons flow across the forward-biased emitter into the base. They diffuse through the base region and flow across the collector junction. Then they flow through the external collector circuit.

Battery B_2 voltage is applied in the forward direction. This means the voltage is positive to the emitter P-type crystal. It also means that voltage is negative to the N-type crystal. Thus, the emitter-base junction has low impedance.

The voltage of battery B_1 is applied in the reverse direction. This means the voltage is positive to the N-type crystal. It also means that voltage is negative to the P-type crystal. This, then, produces a collector-base junction with high impedance.

Transistor impedances.   The impedance of the emitter junction is low. Thus, electrons flow from the emitter region to the base region. At the junction, the electrons combine with the holes in the N-type base crystal. If the base is thin enough, almost all the holes are attracted to the negative terminal of the collector. They then flow through the load to B_1.

The collector current is stopped by applying a positive voltage to the base and a negative voltage to the emitter. In actual transistors, however, this cannot be done because of several basic limitations. Some of the electrons in the base region flow across the emitter junction. Some combine with the holes in the base region. For this reason, it is necessary to supply a current to the base. This makes up for these losses.

The ratio of the collector current to the base current is known as the current gain of the transistor. Current gain, called beta (β), is found by dividing base current into collector current. At high frequencies, the fundamental limitation is the time for carriers to diffuse across the base region. They move from the emitter to the collector. This is why the base region width or thickness is so important. The thinner the base region, the less time is required for the carriers to diffuse across it. This causes the transistor to operate faster.

Figure 19-12 shows a schematic for a common-base PNP circuit. The signal is injected into the emitter-base circuit. The output signal is taken from the collector-base circuit. An important advantage of the transistor is its ability to transfer impedances. This is where the transistor gets its name. The word transistor comes from *transfer resistor*.

The emitter circuit has low impedance. This low impedance allows current to flow. This current flow then creates a current through the collector circuit. The emitter has low impedance and low current. The collector has high impedance and even slightly less current than the

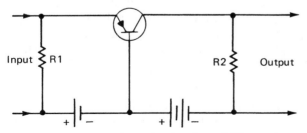

Figure 19-12   Common-base circuit using a PNP transistor.

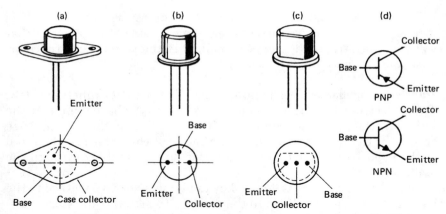

Figure 19-13   Three types of transistor packages: (*a*) power transistors; (*b*) TO-5 case; (*c*) small-signal transistors; and (*d*) transistor symbols.

emitter. However, more power is the result in the collector. This is because $P = I^2R$ or, in this case, $P = I^2Z$. Impedance (Z) is high and the current is squared. Thus, the collector circuit has more power than the emitter circuit with its low impedance.

A common-emitter circuit has about 1300 Ω of input impedance. This is compared to an output impedance of about 50,000 Ω. Thus, there is an increase in impedance from emitter to collector of about 39 times. The junction transistor amplifies in this way. It acts as a power amplifier. A small change in emitter voltage causes a large change in the collector circuit. The different impedances cause this reaction.

The power transistor is shown in Fig. 19-13. Along with the power transistor is the much smaller signal amplifying and switching transistor normally seen in electronic equipments.

Integrated Circuits

The semiconductor monolithic chip was developed in 1958 by J. S. Kilby. Active and passive circuit components were successively diffused and deposited on a single chip. Shortly after robert Noyce made a complete circuit on a single chip. This was the beginning of the modern, inexpensive integrated circuit (IC or chip). Resistors, capacitors, transistors, and diodes can be placed on a chip. These chips are available in three standard packages as follows:

Flat pack: The flat pack is hermetically sealed. This means it is vacuum packed. The ceramic flat pack has either 10 or 14 pins. This type of packaging is no longer used. It may be seen in some older-model equipment with chips (see Fig. 19-14).

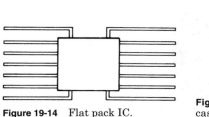

Figure 19-14 Flat pack IC.

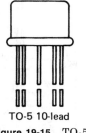

TO-5 10-lead

Figure 19-15 TO-5 case used for IC.

Figure 19-16 Integrated circuits in dual in-line packages (DIPs).

Multipin circular: This type of packaging is no longer in common use. It was originally used because it fit into the same type of container as a transistor. It had more than three leads coming from a TO-5 case originally made for transistors (see Fig. 19-15).

Dual in-line package (DIP): This is the most commonly used type of integrated circuit today. When you think of a "chip" you think of this type. It may have many leads from each side (see Fig. 19-16). This is an easily used type of IC since it can be placed into a socket and removed if it develops any problem. This type of package is standardized as to size. It can be placed into circuit boards by machines, which increases its usefulness in electronics equipment. Many thousands of ICs can be made at a time. This means the cost of manufacturing is very low. DIPs are used in computers, calculators, and control devices for air-conditioning and heating equipment. Amplifiers are fabricated as complete units. Everything seems to use ICs today. Toys, calculators, computers, and automobiles all use ICs or chips in one form or another. The possibilities are unlimited. ICs will play a more important role in air-conditioning, refrigeration, and heating devices in the future.

Solid-State Demand Defrost Control

An excellent example of a refrigeration control using semiconductor technology is the demand defrost control. The idea of defrosting only when needed saves money and energy and protects the quality of the refrigerated product, since extremes in temperature swings are minimized.

Thermistor sensing

A temperature difference concept can be used to automatically initiate the defrost cycle on vertically open frozen food display cases that uses electric defrost (see Fig. 19-17). If the cover is lifted, you can see the electronic package. There are a few transistors and the usual capacitors, resistors, and a transformer. The sensor that feeds in the information

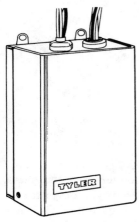

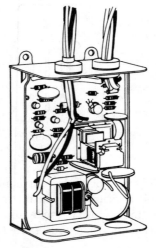

Figure 19-17 Solid-state demand defrost control for vertical open frozen food display cases. (*Courtesy of Tyler.*)

Figure 19-18 Cover removed from the DF-1 defrost control. (*Courtesy of Tyler.*)

Figure 19-19 Thermistors used to sense temperature differences. (*Courtesy of Tyler.*)

needed to operate properly as a control is shown in Fig. 19-18. These two sensors are simple in construction. They are simple thermistors that change resistance when temperature changes (see Fig. 19-19). Their operation is not affected by dirt, moisture, lint, food particles, or ice. They can sense the least temperature change precisely.

Figure 19-20 shows how frost buildup is sensed to cause the circuit to know what is happening. This tells when defrost should be started and terminated. The ever widening temperature difference is a sure

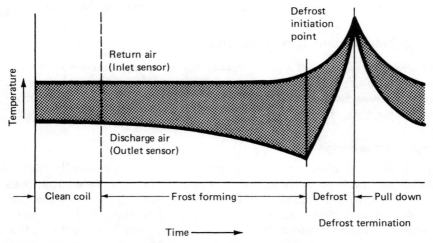

Figure 19-20 Temperature-difference chart. (*Courtesy of Tyler.*)

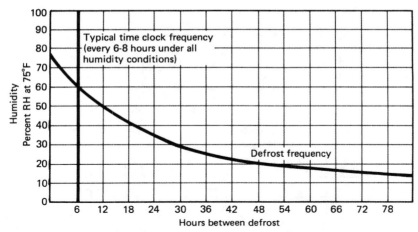

Figure 19-21 Effect of humidity on hours between defrosting. (*Courtesy of Tyler.*)

sign of frost buildup. The temperature difference of air flowing over a coil increases in direct proportion to frost buildup. This is what the thermistor monitors to trigger defrost only when needed. Refrigeration fixtures run at peak efficiency all the time. They use less energy and keep the product at a lower, steadier temperature.

Humidity sensing

Figure 19-21 shows how defrost frequency and humidity are related. Improvements in the design of humidity-sensing elements and the materials used in their construction have minimized many past limitations of humidity sensors. One type of humidity sensor used with electronic controls is a resistance CAB (cellulose acetate butyrate) element (see Fig. 19-22). This resistance element is an improvement over other resistance elements. It has greater contamination resistance, stability, and durability. The humidity CAB element is a multilayered,

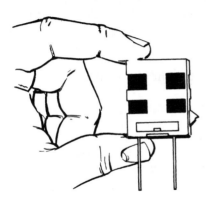

Figure 19-22 CAB resistive element.

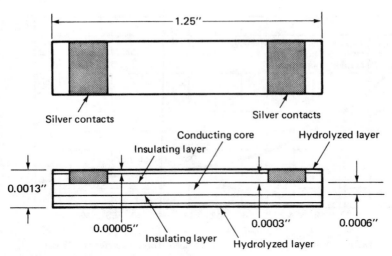

Figure 19-23   Hydrolyzed humidity element.

humidity-sensitive, polymeric film. It consists of an electrically conductive core and insulating outer layers. These layers are partially hydrolyzed. The element has a nominal resistance of 2500 Ω. It has a sensitivity of 2 Ω per 1 percent relative humidity (rh) at 50 percent rh. Its humidity-sensing range is rated at 0 to 100 percent rh.

The CAB element consists of conductive humidity-sensitive film, mounting components, and a protective cover (see Fig. 19-23). The principal component of this humidity sensor is the film. The film has five layers of CAB in the form of a ribbon strip. The CAB material is used because of its good chemical and mechanical stability and high sensitivity to humidity. It also has excellent film-forming characteristics.

The CAB resistance element is a carbon element having a resistance/humidity tolerance favorable to inclusion in a control circuit. With an increase in relative humidity, water is absorbed by the CAB, causing it to swell. This swelling of the polymer matrix causes the suspended carbon particles to move farther apart from each other. This results in an increased element resistance.

When relative humidity decreases, water is given up by the CAB. The contraction of the polymer causes the carbon particles to come closer together. This, in turn, makes the element more conductive or less resistive.

Bridge Circuit

A bridge circuit is a network of resistances and capacitive or inductive impedances. The bridge circuit is usually used to make precise

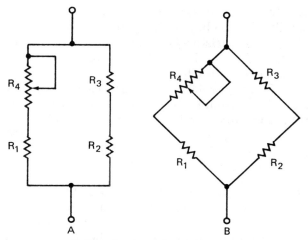

Figure 19-24 Bridge circuit configurations.

measurements. The most common bridge circuit is the Wheatstone bridge. This consists of variable and fixed resistances. Simply, it is a series-parallel circuit (see Fig. 19-24). The branches of the circuit forming the diamond shape are called legs.

If 10-V DC is applied to the bridge circuit shown in Fig. 19-25, one current will flow through R_1 and R_2 and another through R_3 and R_4. Since

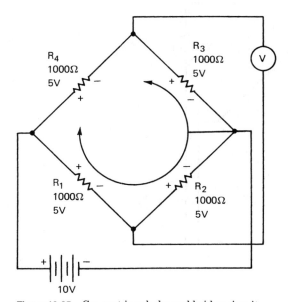

Figure 19-25 Current in a balanced bridge circuit.

R_1 and R_2 are both fixed 1000-Ω resistors, the current through them is constant. Each resistor will drop one-half of the battery voltage, or 5 V. Thus, 5 V is dropped across each resistor. The meter senses the sum of the voltage drops across R_2 and R_3. Both are 5 V. However, the R_2 voltage drop is a positive (+) to negative (−) drop. The R_3 drop is a negative to positive drop. They are opposite in polarity and cancel each other. This is called a *balanced bridge*. The actual resistance values are not important. What is important is that this ratio is maintained and the bridge is balanced.

Unbalanced bridge

In Fig. 19-26, the value, of the variable resistor R_4 is 950 Ω. The other resistors have the same value. Using Ohm's law, the voltage drop across R_4 is found to be 4.9 V. The remaining voltage, 5.1 V, is dropped across R_3 (see Fig. 19-26). The voltmeter measures the sum of the voltage drops across R_2 and R_3 as 5.0 V (+ to −) and 5.1 V (− to +). It registers a total of −0.1 V.

In Fig. 19-27, the converse is true. The value of R_4 is 1050 Ω. The voltage drop across R_3 is 4.9 V. The voltmeter senses the sum of 5 V (+ to −) and 4.9 V (− to +) or +0.1 V. When R_4 changes the same amount above or below the balanced-bridge resistance, the magnitude of the DC output measured by the voltmeter is the same. However, the polarity is reversed.

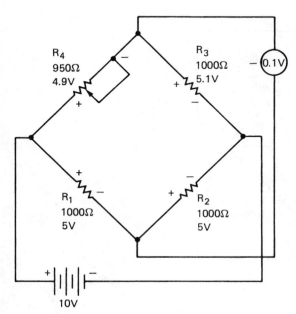

Figure 19-26 Operation of a bridge circuit.

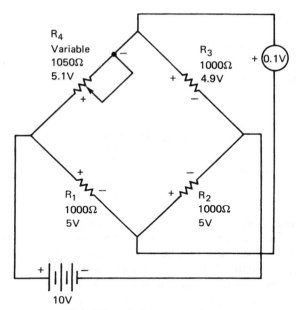

Figure 19-27 Operation of a bridge circuit.

Sensors

The *sensor* in a control system is a resistance element that varies in resistance value with charges in the variable it is measuring. This may be humidity or temperature. These resistance changes are converted into proportional amounts of voltage by a bridge circuit. The voltage is amplified and used to position actuators that regulate the controlled variable.

Controllers

The sensing bridge is the section of the controller circuit that contains the temperature-sensitive element or elements. The potentiometer for establishing the set point is also part of the control system. The bridges are energized with a DC voltage. This permits long-wire runs in sensing circuits without the need for compensating wires or for other capacitive compensating arrangements.

Both integral (room) and remote-sensing element controllers produce a proportional 0- to 16-V DC output signal in response to a measured temperature change. Controllers can be wired to provide direct or reverse action. Direct-acting operation provides an increasing output signal in response to an increase in temperature. Reverse-acting operation provides an increasing output signal in response to a decrease in temperature.

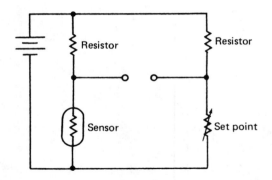

Figure 19-28   Bridge circuit with a sensor and a set point.

Electronic controllers

Electronic controllers have three basic elements: the bridge, the amplifier, and the output circuit. Two legs of the bridge are variable resistances in Fig. 19-28. The sensor and the set-point potentiometer are shown in the bridge configuration. If temperature changes or if the set-point is changed, the bridge is in an unbalanced state. This gives a corresponding output result. The output signal, however, lacks power to position actuators. Therefore, this signal has to be amplified to become useful in the control of devices associated with making sure the right amount of heat of cooled air gets to the room intended.

Differential Amplifiers

Controllers utilize direct-coupled DC differential amplifiers to increase the millivolt signal from the bridge to the necessary 0- to 16-V level for the actuators. There are two amplifiers, one for direct reading and the other for reversing signals. Each amplifier has two stages of amplification. This arrangement is shown in block form in Fig. 19-29.

The differential transistor circuit provides gain and good temperature stability. Figure 19-30 compares a single-transistor amplifier stage with a differential amplifier. Transistors are temperature sensitive. That is, the currents they allow to pass depend on the voltage at the transistor and its ambient temperature. An increase in the ambient temperature in the circuit shown in Fig. 19-30a causes the current through the transistor to increase. The output voltage decreases. The emitter resistor R_E reduces this temperature effect. It also reduces the available voltage gain in the circuit because the signal voltage across the resistor amounts to a negative feedback voltage. That is, it causes a decrease in the voltage difference that was originally produced by the change in temperature at the sensing element.

Since it is desirable for the output voltage of the controller to correspond only to the temperature of the sensing elements and not to the

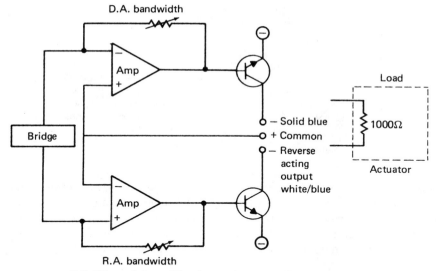

Figure 19-29 DC differential amplifier for use in a controller circuit.

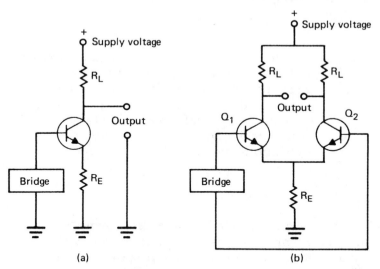

Figure 19-30 (a) Single-transistor amplifier stage. (b) Two-transistor amplifier stage.

ambient temperature of the amplifier, the circuit shown in Fig. 19-30b is used. Here any ambient temperature changes affect both transistors at the same time. The useful output is taken as the difference in output levels of each transistor and the effects of temperature changes are canceled. The voltage gain of the circuit shown in Fig. 19-30b is much

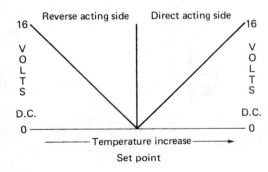

Figure 19-31 Result of sequentially varying DC signals in response to temperature change at the sensing element.

higher than that shown in Fig. 19-30a. This is because the current variations in the two transistors produced by the bridge signal are equal and opposite. An increase in current through Q_1 is accompanied by a decrease in current through Q_2. The sum of these currents through R_E is constant. No signal voltage appears at the emitters to cause negative feedback as in Fig. 19-30a.

The result of sequentially varying DC signals in response to temperature change at the sensing element is shown in Fig. 19-31.

Actuators

Cybertronic actuators perform the work in an electronic system. They accept a control signal and translate that signal into mechanical movement. This is used to position valves and dampers. The electrohydraulic actuators are so called because they convert electrical signals into a fluid movement or force. Damper actuators, equipped with linkage for connection to dampers, and value actuators, having a yoke and linkage to facilitate mounting on a valve body, are also available.

Other Devices

Low- and high-signal selectors accept several control signals. Such selectors then compare the signals and pass the lowest or highest. For example, a high-signal selector can be used on a multizone unit to control the cooling coil. The zone requiring the most cooling transmits the highest control signal. This, in turn, will be passed by the high-signal selector to energize the cooling.

The future of electronics in the control of heating, cooling, and refrigeration is unlimited. Only a few examples have been given. More and more mechanically operated thermostats and valves will be replaced and become electronically controlled. A computer has already been utilized in making a programmable controller that can process the information fed to it by many inputs. This will improve process control in industry and increase the comfort of large buildings by more accurately controlling the flow of heat or cooled air. The computer also aids in locating the source of trouble in a system, thereby eliminating a lot of troubleshooting.

Solid-State Compressor Motor Protection

Solid-state circuitry for air-conditioning units has been in use for some time. The following is an illustration of how some of the circuitry has been incorporated into the protection of compressor motors. This module is made by Robertshaw Controls Company.

Solid-state motor protection prevents motor damage caused by excessive temperature in the stator windings. These solid-state devices provide excellent phase-leg protection by means of separate sensors for each phase winding. The principal advantage of this solid-state system is its speed and sensitivity to motor temperature and its automatic reset provision.

There are two major components to the protection system:

1. The protector sensors are embedded in the motor windings at the time the motor is manufactured.

2. The control module is a sealed enclosure containing a transformer and a switch. Figure 19-32 shows two models.

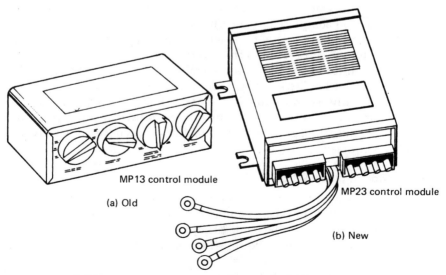

Figure 19-32   Solid-state control modules: (*a*) older unit and (*b*) newer unit. (*Courtesy of Robertshaw.*)

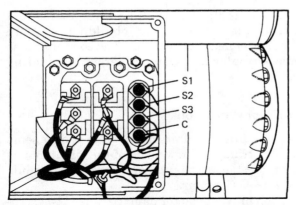

Figure 19-33   Compressor terminal board. (*Courtesy of Robertshaw.*)

Operation of the module

Leads from the internal motor sensors are connected to the compressor terminals, as shown in Fig. 19-33. Leads from the compressor terminals to the control module are connected, as shown in Fig. 19-34. Figure 19-34*a* shows the older model and Fig. 19-34*b* the newer model. While the exact internal circuitry is quite complicated, basically the modules sense resistance changes through the sensors as the result of motor temperature changes in the motor windings. This resistance change triggers the action of the control circuit relay at predetermined opening and closing settings, which causes the line-voltage circuit to the compressor to be broken and completed, respectively.

The modules are available for either 208/240- or 120-V circuits. The module is plainly marked as to the input voltage. The sensors operate at any of the stated voltages because an internal transformer provides the proper power for the solid-state components.

The two terminals on the module marked power supply (T1 and T2) are connected to a power source of the proper voltage; normally the line terminals on the compressor motor contactor or the control circuit transformer as required.

Troubleshooting the control

The solid-state module cannot be repaired in the field, and if the cover is opened or the module physically damaged, the warranty on the module is voided. No attempt should be made to adjust or repair this module, and if it becomes defective, it must be returned intact for replacement. This is the usual procedure for most solid-state units. However, if the unit becomes defective, you should be able to recognize that fact and replace it.

If the compressor motor is inoperable or is not operating properly, the solid-state control circuit may be checked as follows:

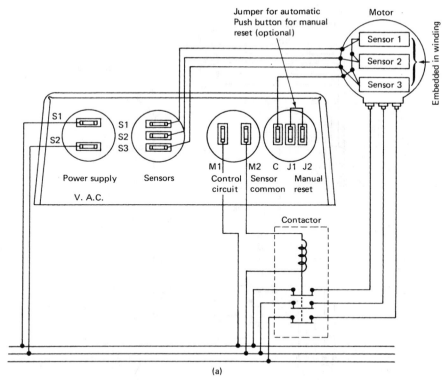

Figure 19-34 Solid-state control modules: (*a*) older unit wiring details and (*b*) newer unit wiring details.

1. If the compressor has been operating and has tripped on the protector, allow the compressor to cool for at least 1 hour before checking to allow time for the motor to cool and the control circuit to reset.

2. Connect a jumper wire across the control circuit terminals on the terminal board (see Fig. 19-34). This will bypass the relay in the module. If the compressor will not operate with the jumper installed, then the problem is external to the solid-state protection system. If the compressor operates with the module bypassed, but will not operate when the jumper wire is removed, then the control circuit relay is open.

3. If after allowing time for motor cooling, the protector still remains open, the motor sensors may be checked as follows:
   a. Remove the wiring connections from the sensor and common terminals on the compressor board (see Figs. 19-33 and 19-34).
   b. *Warning:* Use an ohmmeter with a 3-V maximum battery power supply. The sensors are sensitive and easily damaged, and no attempt should be made to check continuity through them. Any

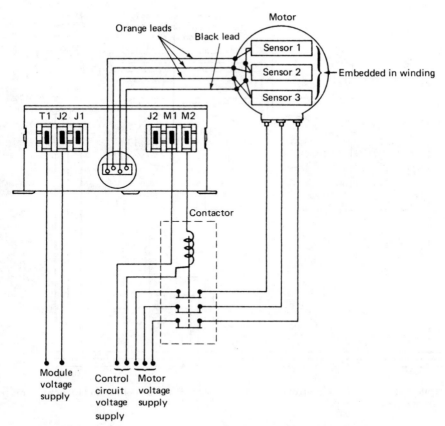

Note: Control is automatic reset when terminals J1 and J2 are not
included. The control is manual reset when terminals J1 and J2
are included.

(b)

Figure 19-34   (*Continued*)

external voltage or current applied to the sensors may cause
damage, necessitating compressor replacement.

c. Measure the resistance from each sensor terminal to the common terminal. The resistance should be in the range. 75 Ω (cold) to 125 Ω
(hot). Resistance readings in this range indicate the sensors are good.
A resistance approaching zero indicates a short. A resistance
approaching infinity indicates an open connection. If the sensors are
damaged, they cannot be repaired or replaced in the field, and the
compressor must be replaced to restore motor protection.

4. If the sensors have proper resistance and the compressor will run with
the control circuit bypassed but will not run when connected properly,

the solid-state module is defective and must be replaced. The replacement module must be the same voltage and made by the same manufacturer as the original module on the compressor.

Restoring service

In the unlikely event that one sensor is damaged and has an open circuit, the control module will prevent compressor operation even though the motor may be in perfect condition. If such a situation is encountered in the field, as an emergency means of operating the compressor until a replacement is made, a properly sized resistor can be added between the terminal of the open sensor and the common sensor terminal in the compressor terminal box (see Figs. 19-33 and 19-35). This then indicates to the control module an acceptable resistance in the damaged sensor circuit, and compressor operation can be restored. The emergency resistor should be a 2-W, 82-Ω wire wound with a tolerance of ±5 percent.

In effect, the compressor will continue operation with two-leg protection rather than three-leg protection. While this obviously does not provide the same high degree of protection, it does provide a means of continuing compressor operation with a reasonable degree of safety.

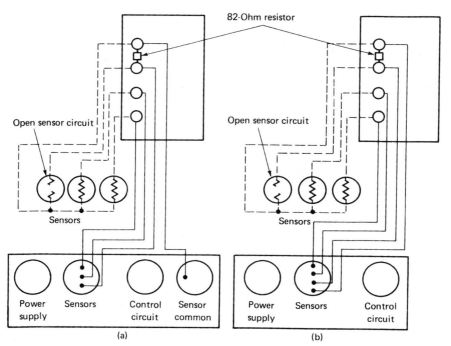

Figure 19-35 Adding a sensor to compensate for an open sensor. (*Courtesy of Robertshaw.*)

20

Electrical and Electronic Symbols Used in Schematics

Diagrams are more useful if you know what the symbols mean. The schematic diagram of an electrical circuit aids in being able to troubleshoot. They are also useful in making it possible to understand what happens in a given arrangement of symbols.

These symbols are part of ARI Standard 130-88. ARI is the abbreviation for the Air-Conditioning and Refrigeration Institute.

In some instances notes are added near the symbol for a special purpose. For instance, if IEC shows up near the symbol it means the symbol has been recommended by the International Electro-Technical Commission. The following symbols are not necessarily in alphabetical order.

Fundamental Items

Resistor

General

Tapped resistor

Variable resistor: with adjustable contact

Thermistor

General

> *Note:* The asterisk is not part of the symbol. It indicates an appropriate value.

With independent integral heater

Capacitor

General.   If it is necessary to identify the capacitor electrodes, the curved element shall represent the outside electrode.

Variable

Battery.   The long line is always positive, but polarity may be indicated in addition.

Multicell   Sensing link; fusible link, ambient temperature operated.

Temperature-measuring thermocouple

IEC

Thermopile for pumping heat

Spark gap, igniter gap

Transmission path

Transmission path: conductors, cables, wiring, factory wired

Field installed or sales option, if specified

Crossing of paths or conductors not connected.   The crossing is not necessarily at a 90° angle.

Junction of paths or conductors junction (connection)

IEC

•

Application: junction of connected paths, conductors, or wires

IEC

Terminal block

Assembled conductors: cable, shielded single conductor

IEC

Shielded cable (5-conductor shown)

Shielded cable with shield grounded (2-conductor shown)

Cable (2-conductor shown)

Ribbon cable

Twisted cable (pair, triple, and the like)

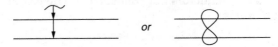

or

Circuit return

Ground. (1) A direct conducting connection to the earth or body of water that is a part thereof. (2) A conducting connection to a structure (chassis) that serves a function similar to that of an earth ground (that is, a structure such as a frame of an air, space, or land vehicle that is not conductively connected to earth).

Earth ground   Chassis ground

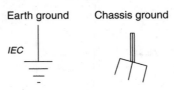

IEC

Normally closed contact (break)

Normally open contact (make)

Operating coil (relay coil)

Solenoid coil

Switch.   Fundamental symbols for contacts, mechanical connections, and so forth, may be used for switch symbols.

The standard method of showing switches is in a position with no operating force applied. For switches that may be in any one of two or more positions with no operating force applied and for switches actuated by some mechanical device (as in air-pressure, liquid-level, rate-of flow, and so forth, switches), a clarifying note may be necessary to explain the point at which the switch functions.

Pushbutton, momentary, or spring-return.   Normally open, circuit closing (make)

Normally closed, circuit opening (break)

Two-circuit (dual)

Two circuit, maintained or not spring-return

Maintained (locking) switch
Toggle switch single throw

Application: 3 disconnect switch

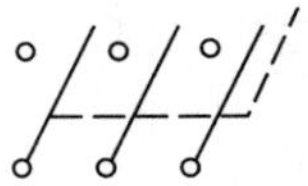

Transfer, 2-position—double throw

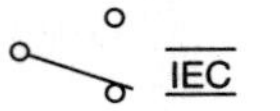

IEC

Transfer, 3-position

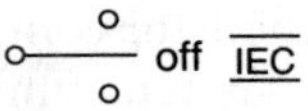

off IEC

Transfer, 2-position

Transfer, 3-position

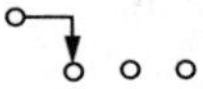

Selector or multiposition switch. The position in which the switch is shown may be indicated by a note or designation of switch position.

General (for power or control diagrams). Any number of transmission paths may be shown.

Segmental contact

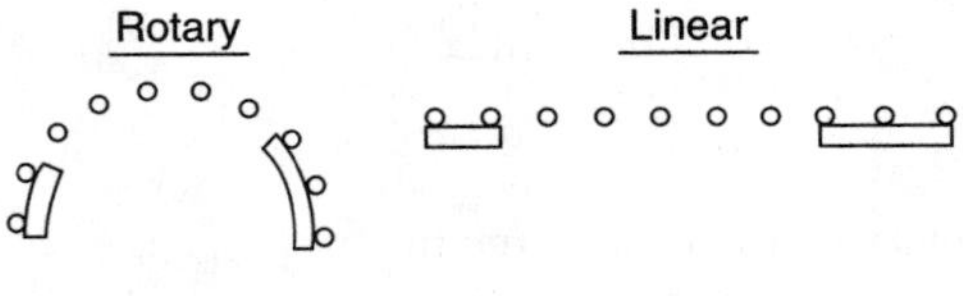

Rotary   Linear

Slide

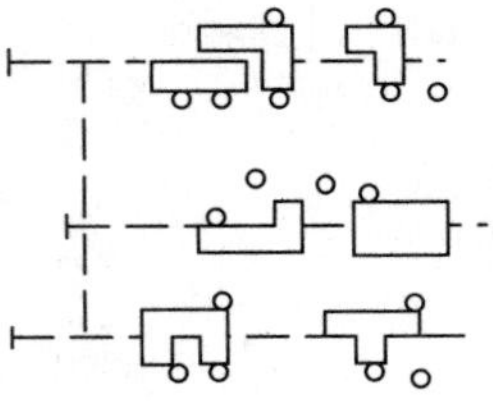

Master or control

Detached contacts
shown elsewhere
on diagram

Contact	Indicator position		
	A	B	C
1–2			X
3–4	X		
5–6			X
7–8	X		
X-indicates contacts closed			

Limit switch, directly actuated, spring returned normally open

Normally open—held closed

Normally open switch with time-delay closing (notc)

Normally closed switch with time-delay opening (ncto)

With single heater (single phase)

With heaters (three phase)

Humidity-actuated switch. Closes on rising humidity

Opens on rising humidity

Connectors

Connector, disconnecting device

The connector symbol is not an arrowhead. It is larger and the lines are drawn at a 90° angle.

Female contact

IEC —⟨

Male contact

IEC —▶

Separable connectors (engaged)

—▶⟩— IEC

Application: engaged 4-conductor

Connectors. The plug has one male and three female contacts.

Transformers, Inductors, and Windings

Transformer

Current transformer

IEC ⌢⌢

Magnetic core transformer (nonsaturating)

With taps—single phase

Autotransformer, single phase

Adjustable autotransformer

Semiconductor Devices

Semiconductor device, transistor, and diode

In general, the angle at which a lead is brought to a symbol element has no significance. $\overline{\text{IEC}}$

Orientation, including a mirror-image presentation, does not change the meaning of a symbol. $\overline{\text{IEC}}$

The elements of the symbol must be drawn in such an order as to show clearly the operating function of the device. $\overline{\text{IEC}}$

Element symbols

Rectifying junction or junctions, which influence a depletion layer: Arrowheads (→►—) shall be half the length of the arrow away from the semiconductor base region. $\overline{\text{IEC}}$

The equilateral (→►—) triangle shall be filled and shall touch the semiconductor base-region symbol. $\overline{\text{IEC}}$

The triangle points in the direction of the forward (easy) current as indicated by a direct current ammeter, unless otherwise noted adjacent to the symbol. Electron flow is in the opposite direction.

Special property indicators

If necessary, a special function or property essential for circuit operation shall be indicated by a supplementary symbol included as part of the symbol.

Typical applications: two-terminal devices

Semiconductor diode: semiconductor rectifier diode

$$\overline{\text{IEC}} \quad \longrightarrow\!\!\!|\!\!-$$

Breakdown diode: overvoltage absorber

Unidirectional diode; voltage regulator; or zener diode

$$\longrightarrow\!\!|\!\!\!\!-$$

Bidirectional diode

$$\overline{\text{IEC}} \quad \longrightarrow\!\!\blacktriangleright\!\!|\!\!\longleftarrow$$

Unidirectional negative-resistance breakdown diode; trigger diac
NPN-type

$$\overline{\text{IEC}} \quad \diagdown\!\!\diagup$$

PNP-type

$$\overline{\text{IEC}} \quad \diagdown\!\!\diagup$$

Bidirectional negative-resistance breakdown diode; trigger diac
NPN-type

$$\diagdown\!\!\diagup$$

PNP-type

$$\diagdown\!\!\diagup$$

Photodiode

Photosensitive type

$$\overline{\text{IEC}} \quad \longrightarrow\!\!\!|\!\!-$$

Photoemissive type

$$\overline{\text{IEC}} \quad \longrightarrow\!\!\!|\!\!-$$

Phototransistor (NPN-type) (without external base-region connection)

Typical appliations: three (or more) terminal devices

PNP transistor (also PNIP transistor, if omitting the intrinsic region will not result in ambiguity)

Application: PNP transistor with one electrode connected to envelope

NPN transistor (also NPIN transistor, if omitting the intrinsic region will not result in ambiguity)

Unijunction transistor with N-type base

IEC

Unijunction transistor with P-type base

IEC

Field-effect transistor with N-channel junction gate

Field-effect transistor with P-channel junction gate

Thyristor, reverse-blocking triode-type, N-type gate; semiconductor-controlled rectifier, N-type gate

Thyristor, reverse-blocking triode, P-type gate; semiconductor-controlled rectifier, P-type gate

Thyristor, reverse-blocking tetrode-type; semiconductor-controlled switch

Thyristor, bidirectional triode-type; triac; gated switch

Phototransistor (PNF-type)

Photon-coupled isolator

> *Note:* T is the transmitter; R is the receiver. The letters are for explanation and are not part of the symbol. Explanatory information should be added to explain circuit operation.

General

$$\boxed{T} \rightrightarrows \boxed{R}$$

Complete isolator (single-package type)

Application: incandescent lamp and symmetrical photoconductive transducer

Application: photoemissive diode and phototransistor

Field-effect transistor with N-channel mos gate

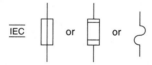

Field-effect transistor with P-channel mos gate

Thyristor, gate turn-off type

Circuit Protectors

Fuse

General

Circuit breaker

General

Application: three-pole circuit breaker with thermal-overload device in all three polesok

Application: three-pole circuit breaker with magnetic-overload device in all three poles

Acoustic Devices

Audible-signaling device

Bell, electrical

If specific identification is required, the abbreviation AC or DC may be added within the square.

Horn, electrical

Lamps and Visual-Signaling Devices

Indicating, pilot, signaling, or switchboard light

To indicate the characteristic, insert the specified letter or letters inside the symbol.

A   Amber

B   Blue

C   Clear

F   Fluorescent

G   Green

NE   Neon

O   Orange

OP   Opalescent

P   Purple

R   Red

W   White

Y   Yellow

Application: green signal light

Rotating Machinery

Rotating machine

Generator (general)

Single phase

Three phase

Motor (general)

Single phase

Three phase

Application: alternating-current motors

Two-lead type

External capacitor type

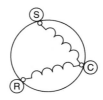

Polyphase type

Application: single phase with internal line break protector

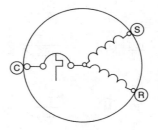

Application: three-phase with internal line break protector

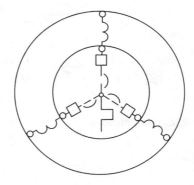

Overloads (current)

Thermal
  Service trip

Remote trip

Magnetic

Series trip

IEC

Remote trip

Overload coils

Thermal

Magnetic

Application: bimetallic (thermal)

No heater

Professional Organizations

Organization	Description	Contact information
Air Movement and Control Association International, Inc. (AMCA)	The Air Movement and Control Association International, Inc. is a non-profit international association of the world's manufacturers of related air system equipment—primarily, but not limited to fans, louvers, dampers, air curtains, airflow measurement stations, acoustic attenuators, and other air-system components for the industrial, commercial, and residential markets.	**Web site:** http://www.amca.org **Address:** 30 West University Drive, Arlington Heights, IL 60004-1893 **Phone:** 847-394-0150 **Fax:** 847-253-0088 **Email:** amca@amca.org
Air Conditioning Contractors of America (ACCA)		**Web site:** http://www.acca.org
American Institute of Architects (AIA)		**Web site:** http://www.aia.org **Address:** 1735 New York Ave. NW, Washington, DC 20006 **Phone:** 800-AIA-3837 **Fax:** 202-626-7547 **Email:** infocentral@aia.org
American National Standards Institute (ANSI)	ANSI is a private, nonprofit organization [501(c)3] that administers and coordinates the U.S. voluntary standardization and conformity assessment system. The Institute's mission is to enhance both the global competitiveness of U.S. business and the U.S. quality of life by promoting and facilitating voluntary consensus standards and conformity assessment systems, and safeguarding their integrity.	**Web site:** http://www.ansi.org **Address:** 25 West 43rd Street, 4th Floor, New York, NY 10036 **Phone:** 212-642-4900 **Fax:** 212-398-0023 **Email:** info@ansi.org

Organization	Description	Contact information
HVAC Excellence	HVAC Excellence, a nonprofit organization, establishes standards of excellence within the heating, ventilation, air-conditioning, and refrigeration (HVAC&R) industry. The two primary responsibilities of HVAC Excellence are Technician Competency Certification and Programmatic Accreditation. HVAC Excellence has achieved national acceptance and is involved in cooperative efforts with the National Skiffs Standard Board, the Manufactures Skills Standards Council, VTECHS, and various other industry and educational organizations.	**Web site:** http://www.hvacexcelle nce.org **Address:** Box 491, Mount Prospect, IL 60056-0491 **Phone:** 800-394-5268 **Fax:** 800-546-3726
Heating, Refrigeration, and Air Conditioning Institute of Canada (HRAI)	The Heating, Refrigeration and Air Conditioning Institute of Canada (HRAI), founded in 1968 is a nonprofit trade association of manufacturers, wholesalers, and contractors in the Canadian heating, ventilation, air-conditioning, and refrigeration (HVACR) industry. HRAI member companies provide products and services for indoor comfort and essential refrigeration processes.	**Web site:** http://www.hrai.ca **Address:** 5045 Orbitor Drive, Building 11, Suite 300, Mississauga, ON L4W 4Y4 Canada **Phone:** 800-267-2231; 905-602-4700 **Fax:** 905-602-1197 **Email:** hraimail@hrai.ca
International Institute of Ammonia Refrigeration (IIAR)	IIAR is an international association serving those who use ammonia refrigeration technology. Your membership in HAR is an investment that pays dividends for your company and for the ammonia refrigeration industry.	**Web site:** http://www.iiar.org **Address:** 1110 North Globe Road, Suite 250, Arlington, VA 22201 **Phone:** 703-312-4200 **Fax:** 703-312-0065 **Email:** iiar@naii.org
National Fire Protection Association (NFPA)	The mission of the International nonprofit NFPA is to reduce the worldwide burden of fire and other hazards on the quality of life by providing and advocating scientifically based consensus codes and standards, research, training, and education. NFPA membership totals more than 75,000 individuals from around the world and more than 80 national trade and professional organizations.	**Web site:** http://www.nfpa.org **Address:** 1 Batterymarch Park, Quincy, MA 02269-9101 **Phone:** 617-770-3000 **Fax:** 617-770-0700
BOCA International, Inc. See ICC Code	Founded in 1915, Building Officials and Code Administrators International, Inc. is a nonprofit membership association comprised of more than 16,000 members, who span the building community from code enforcement officials to materials manufacturers. They are dedicated to preserving the public health, safety, and	**Web site:** http://www.bocai.org **Address:** 4051 West Flossmoor Road, Country Club Hills, IL 60478 **Phone:** 708-799-2300 **Fax:** 708-799-4981

Organization	Description	Contact information
	welfare in the environment through the effective, efficient use and enforcement of model codes. BOCA provides a unique opportunity for any individual to join and derive the benefits of membership. Their members are professionals who are directly or indirectly engaged in the construction and regulatory process. BOCA is the original professional association representing the full spectrum of code enforcement disciplines and construction industry interests. They are the premier publishers of model codes. If you are interested in the development, maintenance, and enforcement of progressive and responsive building regulations, take a closer look at how membership in BOCA International can promote excellence in your profession.	
Cooling Tower Institute (CTI)	The CTI mission is to advocate and promote the use of environmentally responsible evaporative heat transfer systems (EHTS) for the benefit of the public by encouraging education, research, standards development and verification, government relations, and technical information exchange	**Web site:** http://www.cti.org **Address:** 2611 FM 1960 West, Suite H-200, Houston, TX 77068-3730 **Phone:** 281-583-4087 **Fax:** 281-537-1721 **Email:** vmanser@cti.org
Federation of European HVAC Associations	In was in the early 1960s that the idea of some form of cooperation between technical associations in the field of heating, ventilating, and air-conditioning would be extremely useful was recognized by leading professionals who met each other in international meetings and conferences. On September 27, 1963 representatives of technical associations of nine European countries met in The Hague, Netherlands, at the invitation of the Dutch association TWL. These associations were ATIC (Belgium), AICVF (France), VDI TGA (Germany) IHVE (United Kingdom), CARR (Italy), TWL (The Netherlands) Norsk WS (Norway), SWKI (Switzerland), and Swedish WS (Sweden). An impressive list of subjects about which further cooperation was essential was brought for discussion. Ways of exchanging technical knowledge was at the top of the list, together with the promotion of European standardization and regulations, education and coordination of congress and conferences.	**Web site:** http://www.rehva.com **Address:** Ravenstein 3 Brussels, 1000 BE **Phone:** 32-2-5141171

Industrial Associations

American National Standards Institute
1819 L Street NW
Washington, DC 20036
www.ansi.org

American Society for Testing and Materials
100 Barr Harbor Drive
W. Conshohocken, PA 19428-2959
www.astm.org

American Society of Heating, Refrigerating, and
Air-Conditioning Engineers, Inc.
1791 Tullie Circle NE
Atlanta, GA 30329-2305
www.ashrae.org

American Society of Safety Engineers
1800 East Oakton
Des Plaines, IL 60018
www.asse.org

Building and Fire Research Laboratory
National Institute of Standards and Technology
Building 226, Room 6216
Gaithersburg, MD 20899
www.bfrl.nist.gov

Building Research Establishment
Bucknalls Lane
Garston Watford WD25 9XX England
www.bre.co.uk

Building Officials and Code Administrators International
4051 W. Flossmoor Road
Country Club Hills, IL 50478-5794
www.bocai.org

International Organization for Standardization
1, rue de Varembe
Case postale 56 CH1211
Geneve 20 Switzerland
www.iso.ch

National Institute of Standards and Technology
100 Bureau Drive, Stop 3460
Gaithersburg, MD 20899-3460
www.nist.gov

National Research Council of Canada
1500 Montreal Road
Ottawa, ON K1A 0R6 Canada
www.nrc.ca

North American Technical Excellence
4100 North Fairfax Drive #210
Arlington, VA 22203
www.nate.org

Southern Building Code Congress International
900 Montclair Road
Birmingham, AL 35213-1206
www.sbcci.org

Underwriters Laboratories, Inc.
333 Pfingsten Road
Northbrook, IL 60062-2096
www.ul.com

International Code Council
5203 Leesburg Pike, Suite 600
Falls Church, VA 22041
www.intlcode.org

International Conference of Building Officials
5360 Workman Mill Road
Whittier, CA 90601-2298
www.icbo.org

New Refrigerants

Recommendations for R-12 Retrofit Products

Closest match/easiest

R-12 AC	R-500 AC	R-12 small equipment		R-12 large equipment	
		Higher T	Lower T	Higher T	Lower T
R-414B	R-409A	R-414B	R-409A	R-414B	R-409A
R-416A	R-401B	R-416A	R-401A	R-409A	R-401A
R-401A	R-401A	R-401A	R-414B	R-401A	R-414B
R-409A	R-414B	R-409A	R-416A	R-416A	R-416A
R-134a	R-134a				R-134a
	R-416A	R-134a	R-134a	R-134a	

Poorest match/most difficult

"R-12" Refrigerants: Property Comparison

Refrigerant	Components	Composition	Glide	Lube	Pressure match			
					−20	10	40	90P
R-12	(pure)	100	0	M	0.6	14.6	37	100
R-134a	(pure)	100	0	P	4"v	12	35	104
R-401A	22/152a/124	53/13/34	8	MAP	1	16	42	116
R-401B	22/152a/124	61/11/28	8	AP	2	19	46	124
R-409A	22/124/142b	60/25/15	13	MAP	0	16	40	115
R-414B	22/600a/124/142b	50/1.5/39/9.5	13	MAP	1	16	41	113
R-416A	134a/600/124	59/2/39	3	P	7.5"v	8	28	88
Freezone	134a/142b	80/20	4	P	6"v	15	31	93

Note: M—mineral oil; A—alkyl benzene; P—polyolester.

"R-502" Refrigerants: Property Comparison

Refrigerant	Components	Composition	Glide	Lube	Pressure match			
					−20	10	40	90P
R-502	22/115	49/51	0	MA	15	41	81	187
Retrofit blends								
R-402A	125/290/22	60/2/38	2.5	M+AP	19	48	93	215
R-402B	125/290/22	38/2/60	2.5	M+AP	15	42	83	198
R-408A	125/143a/22	7/46/47	1	M+AP	14	38	77	186
HFC blends								
R-404A	125/143a/134a	44/52/4	1.5	P	16	48	84	202
R-507	125/143a	50/50	0	P	18	46	89	210

Note: M—mineral oil; A—alkyl benzene; P—polyolester.

Thermodynamic Properties of R-401A

Temp (°F)	Pressure liquid (psia)	Pressure vapor (psia)	Density liquid (lb/ft³)	Density vapor (lb/ft³)	Enthalpy liquid (Btu/lb)	Enthalpy vapor (Btu/lb)	Entropy liquid (Btu/R-lb)	Entropy vapor (Btu/R-lb)
-60	6.5	4.7	88.18	0.1049	-5.371	94.93	-0.01309	0.2418
-55	7.5	5.5	87.71	0.1215	-4.035	95.60	-0.00977	0.2402
-50	8.7	6.4	87.24	0.1401	-2.694	96.26	-0.00648	0.2386
-45	9.9	7.4	86.77	0.1610	-1.350	96.93	-0.00323	0.2372
-40	11.4	8.6	86.29	0.1842	0.000	97.59	0.00000	0.2358
-35	12.9	9.9	85.82	0.2101	1.354	98.25	0.00320	0.2345
-30	14.7	11.3	85.33	0.2386	2.714	98.91	0.00637	0.2333
-25	16.6	12.9	84.85	0.2701	4.078	99.56	0.00952	0.2321
-20	18.7	14.7	84.36	0.3048	5.449	100.2	0.01265	0.2310
-15	21.0	16.6	83.86	0.3429	6.825	100.9	0.01575	0.2299
-10	23.6	18.8	83.37	0.3846	8.207	101.5	0.01882	0.2289
-5	26.4	21.2	82.86	0.4302	9.595	102.1	0.02188	0.2279
0	29.4	23.8	82.36	0.4799	10.99	102.8	0.02492	0.2269
5	32.7	26.6	81.84	0.5340	12.39	103.4	0.02793	0.2261
10	36.2	29.7	81.33	0.5927	13.80	104.0	0.03093	0.2252
15	40.1	33.1	80.80	0.6563	15.21	104.6	0.03391	0.2244
20	44.2	36.7	80.27	0.7251	16.64	105.2	0.03687	0.2236
25	48.7	40.7	79.74	0.7995	18.07	105.8	0.03982	0.2229
30	53.5	45.0	79.20	0.8798	19.51	106.4	0.04275	0.2221
35	58.6	49.6	78.65	0.9662	20.95	107.0	0.04566	0.2214
40	64.2	54.6	78.10	1.059	22.41	107.6	0.04857	0.2208
45	70.1	59.9	77.54	1.159	23.88	108.2	0.05145	0.2201
50	76.4	65.6	76.97	1.267	25.35	108.7	0.05433	0.2195
55	83.1	71.8	76.39	1.382	26.83	109.3	0.05720	0.2189
60	90.2	78.3	75.81	1.505	28.33	109.8	0.06005	0.2183

(*Continued*)

Thermodynamic Properties of R-401A (Continued)

Temp (°F)	Pressure liquid (psia)	Pressure vapor (psia)	Density liquid (lb/ft³)	Density vapor (lb/ft³)	Enthalpy liquid (Btu/lb)	Enthalpy vapor (Btu/lb)	Entropy liquid (Btu/R-lb)	Entropy vapor (Btu/R-lb)
65	97.8	85.3	75.21	1.637	29.83	110.4	0.06290	0.2178
70	105.9	92.8	74.61	1.779	31.35	110.9	0.06573	0.2172
75	114.5	100.7	74.00	1.930	32.87	111.4	0.06856	0.2167
80	123.5	109.2	73.37	2.092	34.41	111.9	0.07138	0.2162
85	133.1	118.1	72.74	2.265	35.96	112.4	0.07420	0.2156
90	143.2	127.6	72.09	2.449	37.52	112.8	0.07701	0.2151
95	153.9	137.7	71.43	2.647	39.10	113.3	0.07981	0.2146
100	165.2	148.3	70.76	2.858	40.69	113.7	0.08261	0.2141
105	177.0	159.6	70.08	3.083	42.30	114.1	0.08541	0.2136
110	189.5	171.4	69.38	3.324	43.92	114.5	0.08822	0.2131
115	202.6	183.9	68.66	3.581	45.56	114.9	0.09102	0.2126
120	216.3	197.1	67.93	3.857	47.21	115.2	0.09382	0.2120
125	230.7	211.0	67.17	4.152	48.89	115.6	0.09663	0.2115
130	245.8	225.6	66.40	4.468	50.58	115.9	0.09945	0.2110
135	261.7	240.9	65.60	4.807	52.30	116.2	0.1023	0.2104
140	278.2	257.1	64.77	5.171	54.04	116.4	0.1051	0.2098
145	295.5	274.0	63.92	5.564	55.81	116.6	0.1080	0.2092
150	313.6	291.7	63.04	5.987	57.61	116.8	0.1108	0.2085
155	332.6	310.3	62.12	6.444	59.43	116.9	0.1137	0.2078

Thermodynamic Properties of R-134a

Temp (°F)	Pressure (psia)	Density (L) (lb/ft³)	Density (V) (lb/ft³)	Enthalpy (L) (Btu/lb)	Enthalpy (V) (Btu/lb)	Entropy (L) (Btu/R-lb)	Entropy (V) (Btu/R-lb)
−60	4.0	90.49	0.09689	−5.957	94.13	−0.01452	0.2359
−55	4.7	90.00	0.1127	−4.476	94.89	−0.01085	0.2347
−50	5.5	89.50	0.1305	−2.989	95.65	−0.00720	0.2336
−45	6.4	89.00	0.1505	−1.498	96.41	−0.00358	0.2325
−40	7.4	88.50	0.1729	0.000	97.17	0.00000	0.2315
−35	8.6	88.00	0.1978	1.503	97.92	0.00356	0.2306
−30	9.9	87.49	0.2256	3.013	98.68	0.00708	0.2297
−25	11.3	86.98	0.2563	4.529	99.43	0.01058	0.2289
−20	12.9	86.47	0.2903	6.051	100.2	0.01406	0.2282
−15	15.3	85.95	0.3277	7.580	100.9	0.01751	0.2274
−10	16.6	85.43	0.3689	9.115	101.7	0.02093	0.2268
−5	18.8	84.90	0.4140	10.66	102.4	0.02433	0.2262
0	21.2	84.37	0.4634	12.21	103.2	0.02771	0.2256
5	23.8	83.83	0.5173	13.76	103.9	0.03107	0.2250
10	26.6	83.29	0.5761	15.33	104.6	0.03440	0.2245
15	29.7	82.74	0.6401	16.90	105.3	0.03772	0.2240
20	33.1	82.19	0.7095	18.48	106.1	0.04101	0.2236
25	36.8	81.63	0.7848	20.07	106.8	0.04429	0.2232
30	40.8	81.06	0.8663	21.67	107.5	0.04755	0.2228
35	45.1	80.49	0.9544	23.27	108.2	0.05079	0.2224
40	49.7	79.90	1.050	24.89	108.9	0.05402	0.2221
45	54.8	79.32	1.152	26.51	109.5	0.05724	0.2217
50	60.2	78.72	1.263	28.15	110.2	0.06044	0.2214
55	65.9	78.11	1.382	29.80	110.9	0.06362	0.2212
60	72.2	77.50	1.510	31.45	111.5	0.06680	0.2209
65	78.8	76.87	1.647	33.12	112.2	0.06996	0.2206

(Continued)

Thermodynamic Properties of R-134a (Continued)

Temp (°F)	Pressure (psia)	Density (L) (lb/ft³)	Density (V) (lb/ft³)	Enthalpy (L) (Btu/lb)	Enthalpy (V) (Btu/lb)	Entropy (L) (Btu/R·lb)	Entropy (V) (Btu/R·lb)
70	85.8	76.24	1.795	34.80	112.8	0.07311	0.2204
75	93.5	75.59	1.953	36.49	113.4	0.07626	0.2201
80	101.4	74.94	2.123	38.20	114.0	0.07939	0.2199
85	109.9	74.27	2.305	39.91	114.6	0.08252	0.2197
90	119.0	73.58	2.501	41.65	115.2	0.08565	0.2194
95	128.6	72.88	2.710	43.39	115.7	0.08877	0.2192
100	138.9	72.17	2.935	45.15	116.3	0.09188	0.2190
105	149.7	71.44	3.176	46.93	116.8	0.09500	0.2187
110	161.1	70.69	3.435	48.73	117.3	0.09811	0.2185
115	173.1	69.93	3.713	50.55	117.8	0.1012	0.2183
120	185.9	69.14	4.012	52.38	118.3	0.1044	0.2180
125	199.3	68.32	4.333	54.24	118.7	0.1075	0.2177
130	213.4	67.49	4.679	56.12	119.1	0.1106	0.2174
135	228.3	66.62	5.052	58.02	119.5	0.1138	0.2171
140	243.9	65.73	5.455	59.95	119.8	0.1169	0.2167
145	260.4	64.80	5.892	61.92	120.1	0.1201	0.2163
150	277.6	63.83	6.366	63.91	120.4	0.1233	0.2159
155	295.7	62.82	6.882	65.94	120.6	0.1265	0.2154
160	314.7	61.76	7.447	68.00	120.7	0.1298	0.2149

Physical properties of refrigerants	R-134a		
Environmental classification	HFC		
Molecular weight	102.3	Available in the following	
Boiling point (1 atm, °F)	−14.9	sizes:	
Critical pressure (psia)	588.3		
Critical temperature (°F)	213.8	R-134a	
Critical density (lb/ft^3)	32.0	012R134a	12-oz cans
Liquid density (70°F, lb/ft^3)	76.2	30R134a	30-lb cylinder
Vapor density (bp, lb/ft^3)	0.328	A30R134a	30-lb auto AC
Heat of vaporization (bp, BTU/lb)	93.3	50R134a	50-lb cylinder
Specific heat liquid (70 F, BTU/lb F)	0.3366	125R134a	125-lb cylinder*
Specific heat vapor (1 atm, 70 F, BTU/lb F)	0.2021	1000R134a	½-ton cylinder*
Ozone depletion potential (CFC 11 = 1.0)	0	2000R134a	ton cylinder*
Global warming potential (CO$_2$ = 1.0)	1320	* Deposit required	
ASHRAE Standard 34 safety rating	A1		

Pressure-temp chart

Temp (°F)	R-134a (psig)
−40	14.8
−35	12.5
−30	9.9
−25	6.9
−20	3.7
−15	0.6
−10	1.9
−5	4.0
0	6.5
5	9.1
10	11.9
15	15.0
20	18.4
25	22.1
30	26.1
35	30.4
40	35.0
45	40.1
50	45.5
55	51.3
60	57.5
65	64.1
70	71.2
75	78.8
80	86.8
85	95.4
90	104
95	114
100	124
105	135
110	147
115	159
120	171
125	185
130	199
135	214
140	229
145	246
150	263

R-401A and R-401B

Physical properties of refrigerants		R-401A	R-401B
Environmental classification	HCFC	HCFC	
Molecular weight	94.4	92.8	
Boiling point (1 atm, °F)	−29.9	−32.3	
Critical pressure (psia)	669	679.1	
Critical temperature (°F)	221	218.3	
Critical density (lb/ft^3)	30.9	31.1	
Liquid density (70°F, lb/ft^3)	74.6	74.6	
Vapor density (bp, lb/ft^3)	0.306	0.303	
Heat of vaporization (bp, BTU/lb F)	97.5	98.2	
Specific heat liquid (70°F, BTU/lb)	0.3037	0.3027	
Specific heat vapor (1 atm, 70°F, BTU/lb F)	0.1755	0.1725	
Ozone depletion potential (CFC 11 = 1.0)	0.037	0.039	
Global warming potential (CO$_2$ = 1.0)	1163	1267	
ASHRAE Standard 34 safety rating	A1	A1	

Available in the following sizes:

R-401A
30R401A   30-lb cylinder
125R401A   125-lb cylinder*
1700R401A   1-ton cylinder*

R-401B
30R401B   30-lb cylinder
125R401B   125-lb cylinder*

* Deposit required

Temperature glide (°F) (see section II)	8	8

Pressure-temp chart

	R-401A		R-401B	
Temp (°F)	Liquid (psig)	Vapor (psig)	Liquid (psig)	Vapor (psig)
−40	8.1	13.2	6.5	11.8
−35	5.1	10.7	3.3	9.1
−30	1.7	7.9	0.2	6.1
−25	1.0	4.8	2.1	2.8
−20	3.0	1.4	4.3	0.5
−15	5.2	1.2	6.6	2.5
−10	7.7	3.3	9.2	4.7
−5	10.3	5.5	12.0	7.1
0	13.2	8.0	15.1	9.7
5	16.3	10.7	18.4	12.6
10	19.7	13.7	22.0	15.8
15	23.4	16.9	25.9	19.2
20	27.4	20.4	30.1	23.0
25	31.7	24.2	34.6	27.0
30	36.4	28.3	39.5	31.4
35	41.3	32.8	44.8	36.1
40	46.6	37.6	50.4	41.1
45	52.4	42.7	56.4	46.6
50	58.5	48.2	62.8	52.4
55	65.0	54.1	69.6	58.7
60	71.9	60.4	76.9	65.4
65	79.3	67.2	84.7	72.5
70	87.1	74.4	92.9	80.1

(Continued)

Pressure-temp chart

Temp (°F)	R-401A		R-401B	
	Liquid (psig)	Vapor (psig)	Liquid (psig)	Vapor (psig)
75	95.4	82.1	102	88.2
80	104	90.2	111	96.8
85	114	98.9	121	106
90	123	108	131	116
95	134	118	142	126
100	145	128	153	137
105	156	139	166	148
110	169	151	178	160
115	181	163	192	173
120	195	176	206	187
125	209	189	220	201
130	224	203	236	216
135	239	218	252	231
140	255	234	269	248
145	272	250	287	265
150	290	267	305	283

R-402A and R-402B

Physical properties of refrigerants	R-402A	R-402B
Environmental classification	HCFC	HCFC
Molecular weight	101.6	94.7
Boiling point (1 atm, °F)	−56.5	−52.9
Critical pressure (psia)	600	645
Critical temperature (°F)	168	180.7
Critical density (lb/ft^3)	33.8	33.1
Liquid density (70°F, lb/ft^3)	72.61	72.81
Vapor density (bp, lb/ft^3)	0.356	0.328
Heat of vaporization (bp, BTU/lb)	83.58	90.42
Specific heat liquid (70 F, BTU/lb F)	0.3254	0.317
Specific heat vapor (1 atm, 70°F, BTU/lb F)	0.1811	0.1741
Ozone depletion potential (CFC 11 = 1.0)	0.019	0.03
Global warming potential (CO$_2$ = 1.0)	2746	2379
ASHRAE Standard 34 safety rating	A1	A1
Temperature glide (see section II)	2.5	2.5

Available in the following sizes:

R-402A
27R402A   27-lb cylinder
110R402A   110-lb cylinder*

R-402B
13R402B   13-lb cylinder

* Deposit required

Pressure-temp chart

Temp (°F)	R402A (psig)	R402B (psig)
−40	6.3	3.6
−35	9.1	6.0
−30	12.1	9.0
−25	15.4	12.0
−20	18.9	15.4
−15	22.9	18.6
−10	27.1	22.6
−5	31.7	27.0
0	36.7	31.0
5	42.1	36.0
10	48.0	42.0
15	54.2	47.0
20	60.9	54.0
25	68.1	60.0
30	75.8	67.0
35	84.0	75.0
40	92.8	83.4
45	102	91.6
50	112	100
55	123	110
60	134	120
65	146	133
70	158	143
75	171	155
80	185	170
85	200	183
90	215	198
95	232	213
100	249	230
105	267	247
110	286	262
115	305	283
120	326	303
125	347	323
130	370	345
135	393	—
140	418	—
145	443	—
150	470	—

Glossary

absolute humidity   The weight of water vapor per unit volume; grains per cubic foot; or grams per cubic meter.

absolute pressure   The sum of gage pressure and atmospheric pressure. Thus, for example, if the gage pressure is 154 psi (pounds per square inch), the absolute pressure will be 154 + 14.7 or 168.7 psi.

absolute zero   A temperature equal to −459.6°F or −273°C. At this temperature the volume of an ideal gas maintained at a constant pressure becomes zero.

absorption   The action of a material in extracting one or more substances present in the atmosphere or a mixture of gases or liquids accompanied by physical change, chemical change, or both.

acceleration   The time rate of change of velocity. It is the derivative of velocity with respect to time.

accumulator   A shell placed in a suction line for separating the liquid entrained in the suction gas. A storage tank at the evaporator exit or suction line used to prevent flood-backs to the compressor.

acrolein   A warning agent often used with methyl chloride to call attention to the escape of refrigerant. The material has a compelling, pungent odor and causes irritation of the throat and eyes. Acrolein reacts with sulfur dioxide to form a sludge.

ACR tube   A copper tube usually hard-drawn and sold to the trade cleaned and sealed with nitrogen inside to prevent oxidation. Identified by its actual outside diameter.

activated alumina   A form of aluminum oxide (Al_2O_3) that absorbs moisture readily and is used as a drying agent.

adiabatic   Referring to a change in gas conditions where no heat is added or removed except in the form of work.

adiabatic process   Any thermodynamic process taking place in a closed system without the addition or removal of heat.

adsorbent   A sorbent that changes physically, chemically, or both during the sorption process.

aeration   Exposing a substance or area to air circulation.

agitation   A condition in which a device causes circulation in a tank containing fluid.

air, ambient   Generally speaking, the air surrounding an object.

air changes   A method of expressing the amount of air leakage into or out of a building or room in terms of the number of building volumes or room volumes exchanged per unit of time.

air circulation   Natural or imparted motion of air.

air cleaner   A device designed for the purpose of removing airborne impurities, such as dust, gases, vapors, fumes, and smoke. An air cleaner includes air washers, air filters, electrostatic precipitators, and charcoal filters.

air conditioner   An assembly of equipment for the control of at least the first three items enumerated in the definition of *air-conditioning*.

air conditioner, room   A factory-made assembly designed as a unit for mounting in a window, through a wall, or as a console. It is designed for free delivery of conditioned air to an enclosed space without ducts.

air-conditioning   The simultaneous control of all, or at least the first three, of the following factors affecting the physical and chemical conditions of the atmosphere within a structure—temperature, humidity, motion, distribution, dust, bacteria, odors, toxic gases, and ionization—most of which affect human health or comfort.

air-conditioning system, central-fan   A mechanical indirect system of heating, ventilating, or air-conditioning in which the air is treated or handled by equipment located outside the rooms served, usually at a central location and conveyed to and from the rooms by means of a fan and a system of distributing ducts.

air-conditioning system, year-round   An air-conditioning system that ventilates, heats, and humidifies in winter, and cools and dehumidifies in summer to provide the desired degree of air motion and cleanliness.

air-conditioning unit   A piece of equipment designed as a specific air-treating combination, consisting of a means for ventilation, air circulation, air cleaning, and heat transfer, with a control means for maintaining temperature and humidity within prescribed limits.

air cooler   A factory-assembled unit including elements whereby the temperature of air passing through the unit is reduced.

air cooler, spray type   A forced-circulation air cooler wherein the coil surface capacity is augmented by a liquid spray during the period of operation.

air cooling A reduction in air temperature due to the removal of heat as a result of contact with a medium held at a temperature lower than that of the air.

air diffuser A circular, square, or rectangular air-distribution outlet, generally located in the ceiling, and comprising deflecting members discharging supply air in various directions and planes, arranged to promote mixing of primary air with secondary room air.

air, dry In psychrometry, air unmixed with or containing no water vapor.

air infiltration The in-leakage of air through cracks, crevices, doors, windows, or other openings caused by wind pressure or temperature difference.

air, recirculated Return air passed through the conditioner before being again supplied to the conditioned space.

air, return Air returned from conditioned or refrigerated space.

air, saturated Moist air in which the partial pressure of the water vapor is equal to the vapor pressure of water at the existing temperature. This occurs when dry air and saturated water vapor coexist at the same dry-bulb temperature.

air, standard Air with a density of 0.075 lb/ft^3 and an absolute viscosity of 1.22×10 lb mass/ft-s. This is substantially equivalent to dry air at 70°F and 29.92 in. Hg (inches of mercury).

air washer An enclosure in which air is forced through a spray of water in order to cleanse, humidify, or precool the air.

ambient temperature The temperature of the medium surrounding an object. In a domestic system having an air-cooled condenser, it is the temperature of the air entering the condenser.

ammonia machine An abbreviation for a compression-refrigerating machine using ammonia as a refrigerant. The term is also used for Freon machine, sulfur dioxide machine, and the like.

ampere Unit used to measure electrical current. It is equal to 1 C of electrons flowing past a point in 1 second. A coulomb is 6.28×10^{18} electrons.

analyzer A device used in the high side of an absorption system for increasing the concentration of vapor entering the rectifier or condenser.

anemometer An instrument for measuring the velocity of air in motion.

antifreeze, liquid A substance added to the refrigerant to prevent formation of ice crystals at the expansion valve. Antifreeze agents in general do not prevent corrosion due to moisture. The use of a liquid should be a temporary measure where large quantities of water are involved, unless a drier is used to reduce the moisture content. Ice crystals may form when moisture is present below the corrosion limits, and in such instances, a suitable noncorrosive antifreeze liquid is often of value. Materials such as alcohol are corrosive and, if used, should be allowed to remain in the machine for a limited time only.

atmospheric condenser   A condenser operated with water that is exposed to the atmosphere.

atmospheric pressure   The pressure exerted by the atmosphere in all directions as indicated by a barometer. Standard atmospheric pressure is considered to be 14.695 psi, which is equivalent to 29.92 in. Hg.

atomize   To reduce to a fine spray.

automatic air-conditioning   An air-conditioning system that regulates itself to maintain a definite set of conditions by means of automatic controls and valves usually responsive to temperature or pressure.

automatic expansion valve   A pressure-actuated device that regulates the flow of refrigerant from the liquid line into the evaporator to maintain a constant evaporator pressure.

baffle   A partition used to divert the flow of air or a fluid.

balanced pressure   The same pressure *in* a system or container that exists *outside* the system or container.

barometer   An instrument for measuring atmospheric pressure.

blast heater   A set of heat-transfer coils or sections used to heat air that is drawn or forced through it by a fan.

bleeder   A pipe sometimes attached to a condenser to lead off liquid refrigerant parallel to the main flow.

boiler   A closed vessel in which liquid is heated or vaporized.

boiler horsepower   The equivalent evaporation of 34.5 lb of water per hour from and at 212°F, which is equal to a heat output of $970.3 \times 34.5 = 33,475$ Btu.

boiling point   The temperature at which a liquid is vaporized upon the addition of heat, dependent on the refrigerant and the absolute pressure at the surface of the liquid and vapor.

bore   The inside diameter of a cylinder.

bourdon tube   Tube of elastic metal bent into circular shape that is found inside a pressure gage.

brine   Any liquid cooled by a refrigerant and used for transmission of heat without a change in its state.

brine system A system whereby brine cooled by a refrigerating system is circulated through pipes to the point where the refrigeration is needed.

British thermal unit (Btu) The amount of heat required to raise the temperature of 1 lb of water 1°F. It is also the measure of the amount of heat removed in cooling 1 lb of water 1°F and is so used as a measure of refrigerating effect.

butane A hydrocarbon, flammable refrigerant used to a limited extent in small units.

calcium chloride A chemical having the formula $CaCl_2$, which, in granular form, is used as a drier. This material is soluble in water, and in the presence of large quantities of moisture may dissolve and plug up the drier unit or even pass into the system beyond the drier.

calcium sulfate A solid chemical of the formula $CaSO_4$, which may be used as a drying agent.

calibration The process of dividing and numbering the scale of an instrument; also of correcting and determining the error of an existing scale.

calorie Heat required to raise the temperature of 1 g of water 1°C (actually, from 4 to 5°C). A mean calorie is one-hundredth of the heat required to raise 1 g of water from 0 to 100°C.

capacitor An electrical device that has the ability to store an electrical charge. It is used to start motors, among other purposes.

capacity, refrigerating The ability of a refrigerating system, or part thereof, to remove heat. Expressed as a rate of heat removal, it is usually measured in Btu's per hour or tons per 24 hours.

capacity reducer In a compressor, a device, such as a clearance pocket, movable cylinder head, or suction bypass, by which compressor capacity can be adjusted without otherwise changing the operating conditions.

capillarity The action by which the surface of a liquid in contact with a solid (as in a slender tube) is raised or lowered.

capillary tube In refrigeration practice, a tube of small internal diameter used as a liquid refrigerant-flow control or expansion device between high and low sides; also used to transmit pressure from the sensitive bulb of some temperature controls to the operating element.

carbon dioxide ice Compressed solid CO_2, also called dry ice.

Celsius A thermometric system in which the freezing point of water is called 0°C and its boiling point 100°C at normal pressure. This system is used in the scientific community for research work and also by most

European countries and Canada. This book has the Celsius value of each Fahrenheit temperature in parenthesis.

centrifugal compressor A compressor employing centrifugal force for compression.

centrifuge A device for separating liquids of different densities by centrifugal action.

change of air Introduction of new, cleansed, or recirculated air to a conditioned space, measured by the number of complete changes per unit time.

change of state A change from one state to another, as from a liquid to a solid, from a liquid to a gas, and so on.

charge The amount of refrigerant in a system.

chimney effect The tendency of air or gas in a duct or other vertical passage to rise when heated due to its lower density compared with that of the surrounding air or gas. In buildings, the tendency toward displacement, caused by the difference in temperature, of internal heated air by unheated outside air due to the difference in density of outside and inside air.

clearance Space in a cylinder not occupied by a piston at the end of the compression stroke or volume of gas remaining in a cylinder at the same point, measured in percentage of piston displacement.

coefficient of expansion The fractional increase in length or volume of a material per degree rise in temperature.

coefficient of performance (heat pump) Ratio of heating effect produced to the energy supplied, each expressed in the same thermal units.

coil Any heating or cooling element made of pipe or tubing connected in series.

cold storage A trade or process of preserving perishables on a large scale by refrigeration.

comfort chart A chart showing effective temperatures with dry-bulb temperatures and humidities (and sometimes air motion) by which the effects of various air conditions on human comfort maybe compared.

compression system A refrigerating system in which the pressure-imposing element is mechanically operated.

compressor That part of a mechanical refrigerating system which receives the refrigerant vapor at low pressure and compresses it into a smaller volume at higher pressure.

compressor, centrifugal   A nonpositive displacement compressor that depends on centrifugal effect, at least in part, for pressure rise.

compressor displacement   Compressor volume in cubic inches found by multiplying piston area by stroke by number of cylinders.

$$\text{Displacement in cubic feet per minute} = \frac{\pi \times r^2 \times L \times \text{rpm} \times n}{1728}$$

compressor, open-type   A compressor with a shaft or other moving part, extending through a casing, to be driven by an outside source of power, thus requiring a stuffing box, shaft seals, or equivalent rubbing contact between a fixed and moving part.

compressor, reciprocating   A positive displacement compressor with a piston or pistons moving in a straight line but alternately in opposite directions.

compressor, rotary   One in which compression is attained in a cylinder by rotation of a positive displacement member.

compressor booster   A compressor for very low pressures, usually discharging into the suction line of another compressor.

condenser   A heat-transfer device that receives high-pressure vapor at temperatures above that of the cooling medium, such as air or water, to which the condenser passes latent heat from the refrigerant, causing the refrigerant vapor to liquefy.

condensing   The process of giving up latent heat of vaporization in order to liquefy a vapor.

condensing unit   A specific refrigerating machine combination for a given refrigerant, consisting of one or more power-driven compressors, condensers, liquid receivers (when required), and the regularly furnished accessories.

condensing unit, sealed   A mechanical condensing unit in which the compressor and its motor are enclosed in the same housing, with no external shaft or shaft seal, the compressor motor operating in the refrigerant atmosphere.

conduction, thermal   Passage of heat from one point to another by transmission of molecular energy from particle to particle through a conductor.

conductivity, thermal   The ability of a material to pass heat from one point to another, generally expressed in terms of Btu per hour per square foot of material per inch of thickness per degree temperature difference.

conductor, electrical   A material that will pass an electric current as part of an electrical system.

connecting rod   A device connecting the piston to a crank and used to change rotating motion into reciprocating motion, or vice versa, as from a rotating crankshaft to a reciprocating piston.

constant-pressure valve   A valve of the throttling type, responsive to pressure, located in the suction line of an evaporator to maintain a desired constant pressure in the evaporator higher than the main suction-line pressure.

constant-temperature valve   A valve of the throttling type, responsive to the temperature of a thermostatic bulb. This valve is located in the suction line of an evaporator to reduce the refrigerating effect on the coil to just maintain a desired minimum temperature.

control   Any device for regulation of a system or component in normal operation either manual or automatic. If automatic, the implication is that it is responsive to changes of temperature, pressure, or any other property whose magnitude is to be regulated.

control, high-pressure   A pressure-responsive device (usually an electric switch) actuated directly by the refrigerant-vapor pressure on the high side of a refrigerating system (usually compressor-head pressure).

control, low-pressure   An electric switch, responsive to pressure, connected into the low-pressure part of a refrigerating system (usually closes at high pressure and opens at low pressure).

control, temperature   An electric switch or relay that is responsive to the temperature change of a thermostatic bulb or element.

convection   The circulatory motion that occurs in a fluid at a nonuniform temperature owing to the variation of its density and the action of gravity.

convection, forced   Convection resulting from forced circulation of a fluid as by a fan, jet, or pump.

cooling tower, water   An enclosed device for evaporative cooling water by contact with air.

cooling unit   A specific air-treating combination consisting of a means for air circulation and cooling within prescribed temperature limits.

cooling water   Water used for condensation of refrigerant; condenser water.

copper plating   Formation of a film of copper, usually on compressor walls, pistons, or discharge valves caused by moisture in a methyl chloride system.

corrosive Having a chemically destructive effect on metals (occasionally on other materials).

counter-flow In the heat exchange between two fluids, the opposite direction of flow, the coldest portion of one meeting the coldest portion of the other.

critical pressure The vapor pressure corresponding to the critical temperature.

critical temperature The temperature above which a vapor cannot be liquefied, regardless of pressure.

critical velocity The velocity above which fluid flow is turbulent.

crohydrate A eutectic brine mixture of water and any salt mixed in proportions to give the lowest freezing temperature.

cycle A complete course of operation of working fluid back to a starting point measured in thermodynamic terms. Also used in general for any repeated process in any system.

cycle, defrosting That portion of a refrigeration operation which permits the cooling unit to defrost.

cycle, refrigeration A complete course of operation of a refrigerant back to the starting point measured in thermodynamic terms. Also used in general for any repeated process for any system.

Dalton's law of partial pressure Each constituent of a mixture of gases behaves thermodynamically as if it alone occupied the space. The sum of the individual pressures of the constituents equals the total pressure of the mixture.

defrosting The removal of accumulated ice from a cooling unit.

degree-day A unit based on temperature difference and time used to specify the nominal heating load in winter. For one day there exist as many degree-days as there are degrees Fahrenheit difference in temperature between the average outside air temperature, taken over a 24-hour period, and a temperature of 65°F.

dehumidifier An air cooler used for lowering the moisture content of the air passing through it. An absorption or adsorption device for removing moisture from the air.

dehumidify A device used to remove water vapor from the atmosphere or water or liquid from stored goods.

dehydrator A device used to remove moisture from the refrigerant.

density The mass or weight per unit of volume.

dew point, air   The temperature at which a specified sample of air, with no moisture added or removed, is completely saturated. The temperature at which the air, on being cooled, gives up moisture, or dew.

differential (of a control)   The difference between the cutin and cutout temperature. A valve that opens at one pressure and closes at another. This allows a system to adjust itself with a minimum of overcorrection.

direct connected   Driver and driven, as motor and compressor, positively connected in line to operate at the same speed.

direct expansion   A system in which the evaporator is located in the material or space refrigerated or in the air-circulating passages communicating with such space.

discharge gas   Hot, high-pressure vapor refrigerant which has just left the compressor.

displacement, actual   The volume of gas at the compressor inlet actually moved in a given time.

displacement, theoretical   The total volume displaced by all the pistons of a compressor for every stroke during a definite interval (usually measured in cubic feet per minute).

drier   Synonymous with dehydrator.

dry-type evaporator   An evaporator of the continuous-tube type where the refrigerant from a pressure-reducing device is fed into one end and the suction line connected to the outlet end.

duct   A passageway made of sheet metal or other suitable material, not necessarily leak-tight, used for conveying air or other gas at low pressure.

dust   An air suspension (aerosol) of solid particles of earthy material, as differentiated from smoke.

economizer   A reservoir or chamber wherein energy or material from a process is reclaimed for further useful purpose.

efficiency, mechanical   The ratio of the output of a machine to the input in equivalent units.

efficiency, volumetric   The ratio of the volume of gas actually pumped by a compressor or pump to the theoretical displacement of the compressor.

ejector   A device that utilizes static pressure to build up a high fluid velocity in a restricted area to obtain a lower static pressure at that point so that fluid from another source maybe drawn in.

element, bimetallic   An element formed of two metals having different coefficients of thermal expansion, such as used in temperature-indicating and -controlling devices.

emulsion A relatively stable suspension of small, but not colloidal, particles of a substance in a liquid.

engine Prime mover; device for transforming fuel or heat energy into mechanical energy.

enthalpy The total heat content of a substance, compared to a standard value 32°F or 0°C for water vapor, or −40°F (−40°C) for refrigerant.

entropy The ratio of the heat added to a substance to the absolute temperature at which it is added.

equalizer A piping arrangement to maintain a common liquid level or pressure between two or more chambers.

eutectic solution A solution of such concentration as to have a constant freezing point at the lowest freezing temperature for the solution.

evaporative condenser A refrigerant condenser utilizing the evaporation of water by air at the condenser surface as a means of dissipating heat.

evaporative cooling The process of cooling by means of the evaporation of water in air.

evaporator A device in which the refrigerant evaporates while absorbing heat.

expansion valve, automatic A device that regulates the flow of refrigerant from the liquid line into the evaporator to maintain a constant evaporator pressure.

expansion valve, thermostatic A device that regulates the flow of refrigerant into an evaporator so as to maintain an evaporation temperature in a definite relationship to the temperature of a thermostatic bulb.

extended surface The evaporator or condenser surface that is not a primary surface. Fins or other surfaces that transmit heat from or to a primary surface, which is part of the refrigerant container.

external equalizer In a thermostatic expansion valve, a tube connection from the chamber containing the pressure-actuated element of the valve to the outlet of the evaporator coil. A device to compensate for excessive pressure drop through the coil.

Fahrenheit A thermometric system in which 32°F denotes the freezing point of water and 212°F the boiling point under normal pressure.

fan An air-moving device comprising a wheel or blade, and housing or orifice plate.

fan, centrifugal A fan rotor or wheel within a scroll-type housing and including driving-mechanism supports for either belt-drive or direct connection.

fan, propeller   A propeller or disk-type wheel within a mounting ring or plate and including driving-mechanism supports for either belt-drive or direct connection.

fan, tube-axial   A disk-type wheel within a cylinder, a set of air-guide vanes located either before or after the wheel, and driving-mechanism supports for either belt-drive or direct connection.

filter   A device to remove solid material from fluid by straining action.

flammability   The ability of a material to burn.

flare fitting   A type of connector for soft tubing that involves the flaring of the tube to provide a mechanical seal.

flash gas   The gas resulting from the instantaneous evaporation of the refrigerant in a pressure-reducing device to cool the refrigerant to the evaporation temperature obtained at the reduced pressure.

float valve   Valve actuated by a float immersed in a liquid container.

flooded system   A system in which the refrigerant enters into a header from a pressure-reducing valve and the evaporator maintains a liquid level. Opposed to dry evaporator.

fluid   A gas or liquid.

foaming   Formation of foam or froth of oil refrigerant due to rapid boiling out of the refrigerant dissolved in the oil when the pressure is suddenly reduced. This occurs when the compressor operates; and, if large quantities of refrigerant have been dissolved, large quantities of oil may "boil" out and be carried through the refrigerant lines.

freezeup   Failure of a refrigeration unit to operate normally due to formation of ice at the expansion valve. The valve maybe frozen closed or open, causing improper refrigeration in either case.

freezing point   The temperature at which a liquid will solidify upon the removal of heat.

Freon-12   The common name for dichlorodifluoromethane (CCl_2F_2).

frostback   The flooding of liquid from an evaporator into the suction line, accompanied by frost formation on the suction line in most cases.

furnace   That part of a boiler or warm-up heating plant in which combustion takes place. Also a complete heating unit for transferring heat from fuel being burned to the air supplied to a heating system.

fusible plug   A safety plug used in vessels containing refrigerant. The plug is designed to melt at high temperatures [usually about 165°F (73.8°C)] to prevent excessive pressure from bursting the vessel.

gage   An instrument used for measuring various pressures or liquid levels (sometimes spelled gauge).

gas   The vapor state of a material.

generator   A basic component of any absorption-refrigeration system.

gravity, specific   The density of a standard material usually compared to that of water or air.

grille   A perforated or louvered covering for an air passage, usually installed in a sidewall, ceiling, or floor.

halide torch   A leak tester generally using alcohol and burning with a blue flame; when the sampling tube draws in halocarbon refrigerant vapor, the color of flame changes to bright green. Gas given off by the burning halocarbon is phosgene, a deadly gas used in World War I in Europe against Allied troops (can be deadly if breathed in a closed or confined area).

halogen   An element from the halogen group that consists of chlorine, fluorine, bromine, and iodine. Two halogens may be present in chlorofluorocarbon refrigerants.

heat   Basic form of energy that may be partially converted into other forms and into which all other forms may be entirely converted.

heat of fusion   Latent heat involved in changing between the solid and the liquid states.

heat, sensible   Heat that is associated with a change in temperature; specific heat exchange of temperature, in contrast to a heat interchange in which a change of state (latent heat) occurs.

heat, specific   The ratio of the quantity of heat required to raise the temperature of a given mass of any substance by one degree to the quantity required to raise the temperature of an equal mass of a standard substance (usually water at 59°F) by one degree.

heat of vaporization   Latent heat involved in the change between liquid and vapor states.

heat pump   A refrigerating system employed to transfer heat into a space or substance. The condenser provides the heat, while the evaporator is arranged to pick up heat from air, water, and the like. By shifting the flow of the refrigerant, a heat-pump system may also be used to cool the space.

heating system   Any of several heating methods usually termed according to the method used in its generation, such as steam heating, warm-air heating, and so on.

heating system, electric   Heating produced by the rise of temperature caused by the passage of an electric current through a conductor having a high resistance to the current flow. Residence electric-heating systems generally consist of one or several resistance units installed in a frame or casing, the degree of heating being thermostatically controlled.

heating system, steam   A heating system in which heat is transferred from a boiler or other source to the heating units by steam at, above, or below atmospheric pressure.

heating system, vacuum   A two-pipe steam-heating system equipped with the necessary accessory apparatus to permit operating the system below atmospheric pressure.

heating system, warm-air   A warm-air heating plant consisting of a heating unit (fuel-burning furnace) enclosed in a casing from which the heated air is distributed to various rooms of the building through ducts.

hermetically sealed unit   A refrigerating unit containing the motor and compressor in a sealed container.

high-pressure cutout   A control device connected into the high-pressure part of a refrigerating system to stop the machine when the pressure becomes excessive.

high side   That part of the refrigerating system containing the high-pressure refrigerant. The term is also used to refer to the condensing unit, consisting of the motor, compressor, condenser, and receiver mounted on a single base.

high-side float valve   A float valve that floats in high-pressure liquid. Opens on an increase in liquid level.

hold over   In an evaporator, the ability to stay cold after heat removal from the evaporator stops.

horsepower   A unit of power. Work done at the rate of 33,000 lb-ft/min, or 550 lb-ft/s.

humidifier   A device to add moisture to the air.

humidify   To add water vapor to the atmosphere; to add water vapor or moisture to any material.

humidistat   A control device actuated by changes in humidity and used for automatic control of relative humidity.

humidity, absolute   The definite amount of water contained in a definite quantity of air (usually measured in grains of water per pound or per cubic foot of air).

humidity, relative   The ratio of the water-vapor pressure of air compared to the vapor pressure it would have if saturated at its dry-bulb temperature. Very nearly the ratio of the amount of moisture contained in air compared to what it could hold at the existing temperature.

humidity, specific   The weight of vapor associated with 1 lb of dry air; also termed *humidity ratio.*

hydrocarbons   A series of chemicals of similar chemical nature, ranging from methane (the main constituent of natural gas) through butane, octane, and so on, to heavy lubricating oils. All are more or less flammable. Butane and isobutane have been used to a limited extent as refrigerants.

hydrolysis   Reaction of a material, such as Freon-12 or methyl chloride, with water. Acid materials in general are formed.

hydrostatic pressure   The pressure due to liquid in a container that contains no gas space.

hygrometer   An instrument used to measure moisture in the air.

hygroscope   *See* humidistat.

ice-melting equivalent   The amount of heat (144 Btu) absorbed by 1 lb of ice at 32°F in liquefying to water at 32°F.

indirect cooling system   *See* brine system.

infiltration   The leakage of air into a building or space.

insulation   A material of low heat conductivity.

irritant refrigerant   Any refrigerant that has an irritating effect on the eyes, nose, throat, or lungs.

isobutane   A hydrocarbon refrigerant used to a limited extent. It is flammable.

kilowatt   Unit of electrical power equal to 1000 W, or 1.34 hp, approximately.

lag of temperature control   The delay in action of a temperature-responsive element due to the time required for the temperature of the element to reach the surrounding temperature.

latent heat   The quantity of heat that may be added to a substance during a change of state without causing a temperature change.

latent heat of evaporation   The quantity of heat required changing 1 lb of liquid into a vapor with no change in temperature (reversible).

leak detector   A device used to detect refrigerant leaks in a refrigerating system.

liquid   The state of a material in which its top surface in a vessel will become horizontal. Distinguished from solid or vapor forms.

liquid line   The tube or pipe that carries the refrigerant liquid from the condenser or receiver of a refrigerating system to a pressure-reducing device.

liquid receiver   That part of the condensing unit that stores the liquid refrigerant.

load   The required rate of heat removal.

low-pressure control   An electric switch and pressure-responsive element connected into the suction side of a refrigerating unit to control the operation of the system.

low side   That part of a refrigerating system that normally operates under low pressure, as opposed to the high side. Also used to refer to the evaporator.

low-side float   A valve operated by the low-pressure liquid, which opens at a low level and closes at a high level.

main   A pipe or duct for distributing to or collecting conditioned air from various branches.

manometer   A U-shaped liquid-filled tube for measuring pressure differences.

mechanical efficiency   The ratio of work done by a machine to the work done on it or energy used by it.

mechanical equivalent of heat   An energy-conversion ratio of 778.18 lb-ft = 1 Btu.

methyl chloride   A refrigerant having the chemical formula CH_3Cl.

micron (μ)   A unit of length; the thousandth part of 1 mm or the millionth part of 1 m.

mollier chart   A graphical representation of thermal properties of fluids, with total heat and entropy as coordinates.

motor   A device for transforming electrical energy into mechanical energy.

motor capacitor   A device designed to improve the starting ability of single-phase induction motors.

noncondensables   Foreign gases mixed with a refrigerant, which cannot be condensed into liquid form at the temperatures and pressures at which the refrigerant condenses.

oil trap   A device to separate oil from the high-pressure vapor from the compressor. Usually contains a float valve to return the oil to the compressor crankcase.

output   Net refrigeration produced by the system.

ozone   The O_3 form of oxygen, sometimes used in air-conditioning or cold-storage rooms to eliminate odors; can be toxic in concentrations of 0.5 ppm (parts per million) and over.

packing   The stuffing around a shaft to prevent fluid leakage between the shaft and parts around the shaft.

packless valve   A valve that does not use packing to prevent leaks around the valve stem. Flexible material is usually used to seal against leaks and still permit valve movement.

performance factor   The ratio of the heat moved by a refrigerating system to heat equivalent of the energy used. Varies with conditions.

phosphorous pentoxide   An efficient drier material that becomes gummy reacting with moisture and hence is not used alone as a drying agent.

pour point, oil   The temperature below which the oil surface will not change when the oil container is tilted.

power   The rate of doing work measured in horsepower, watts, kilowatts, and similar units.

power factor, electrical devices   The ratio of watts to volt-amperes in an AC circuit.

pressure   The force exerted per unit of area.

pressure drop   Loss in pressure, as from one end of a refrigerant line to the other, due to friction, static head, and so on.

pressure gage   *See* gage.

pressure-relief valve   A valve or rupture member designed to relieve excessive pressure automatically.

psychrometric chart   A chart used to determine the specific volume, heat content, dew point, relative humidity, absolute humidity, and wet- and dry-bulb temperatures, knowing any two independent items of those mentioned.

purging   The act of blowing out refrigerant gas from a refrigerant containing vessel usually for the purpose of removing noncondensables.

pyrometer   An instrument for the measurement of high temperatures.

radiation   The passage of heat from one object to another without warming the space between. The heat is passed by wave motion similar to light.

refrigerant   The medium of heat transfer in a refrigerating system that picks up heat by evaporating at a lower temperature and gives up heat by condensing at a higher temperature.

refrigerating system   A combination of parts in which a refrigerant is circulated for the purpose of extracting heat.

relative humidity   The ratio of the water-vapor pressure of air compared to the vapor pressure it would have if saturated at its dry-bulb temperature. Very nearly the ratio of the amount of moisture contained in air compared to what it could hold at the existing temperature.

relief valve   A valve designed to open at excessively high pressures to allow the refrigerant to escape.

resistance, electrical   The opposition to electric current flow, measured in ohms.

resistance, thermal   The reciprocal of thermal conductivity.

room cooler   A cooling element for a room. In air-conditioning, a device for conditioning small volumes of air for comfort.

rotary compressor   A compressor in which compression is attained in a cylinder by rotation of a semiradial member.

running time   Usually indicates percent of time a refrigerant compressor operates.

saturated vapor   Vapor not superheated but of 100 percent quality, that is, containing no unvaporized liquid.

seal, shaft   A mechanical system of parts for preventing gas leakage between a rotating shaft and a stationary crankcase.

sealed unit   *See* hermetically sealed unit.

shell and tube   Pertaining to heat exchangers in which a coil of tubing or pipe is contained in a shell or container. The pipe is provided with openings to allow the passage of a fluid through it, while the shell is also provided with an inlet and outlet for a fluid flow.

silica gel   A drier material having the formula SiO_2.

sludge   A decomposition product formed in a refrigerant due to impurities in the oil or due to moisture. Sludges may be gummy or hard.

soda lime   A material used for removing moisture. Not recommended for refrigeration use.

solenoid valve   A valve opened by magnetic effect of an electric current through a solenoid coil.

solid   The state of matter in which force can be exerted in a downward direction only when not confined. As distinguished from fluids.

solubility   The ability of one material to enter into solution with another.

solution   The homogeneous mixture of two or more materials.

specific gravity   The weight of a volume of a material compared to the weight of the same volume of water.

specific heat   The quantity of heat required to raise the temperature of a definite mass of a material a definite amount compared to that required to raise the temperature of the same mass of water the same amount. May be expressed as Btu per pound per degrees Fahrenheit.

specific volume   The volume of a definite weight of a material. Usually expressed in cubic feet per pound. The reciprocal of density.

spray pond   An arrangement for lowering the temperature of water by evaporative cooling of the water in contact with outside air. The water to be cooled is sprayed by nozzles into the space above a body of previously cooled water and allowed to fall by gravity into it.

steam   Water in the vapor phase.

steam trap   A device for allowing the passage of condensate, or air and condensate, and preventing the passage of steam.

sub cooled   Cooled below the condensing temperature corresponding to the existing pressure.

sublimation   The change from a solid to a vapor state without an intermediate liquid state.

suction line   The tube or pipe that carries refrigerant vapor from the evaporator to the compressor inlet.

suction pressure   Pressure on the suction side of the compressor.

superheater   A heat exchanger used on flooded evaporators, wherein hot liquid on its way to enter the evaporator is cooled by supplying heat to dry and superheat the wet vapor leaving the evaporator.

sweating   Condensation of moisture from the air on surfaces below the dew-point temperature.

system   A heating or refrigerating scheme or machine, usually confined to those parts in contact with the heating or refrigerating medium.

temperature   Heat level or pressure. The thermal state of a body with respect to its ability to pick up heat from or pass heat to another body.

thermal conductivity   The ability of a material to conduct heat from one point to another. Indicated in terms of Btu per hour per square foot per inches of thickness per degrees Fahrenheit.

thermocouple   A device consisting of two electrical conductors having two junctions—one at a point whose temperature is to be measured, and the other at a known temperature. The temperature between the two junctions is determined by the material characteristics and the electrical potential setup.

thermodynamics   The science of the mechanics of heat.

thermometer   A device for indicating temperature.

thermostat   A temperature-actuated switch.

thermostatic expansion valve   A device to regulate the flow of refrigerant into an evaporator so as to maintain an evaporation temperature in a definite relationship to the temperature of a thermostatic bulb.

ton of refrigeration   Refrigeration equivalent to the melting of 1 ton of ice per 24 hours; 288,000 Btu/day, 12,000 Btu/h, or 200 Btu/min.

total heat   The total heat added to a refrigerant above an arbitrary starting point to bring it to a given set of conditions (usually expressed in Btu per pound). For instance, in a superheated gas, the combined heat added to the liquid necessary to raise its temperature from an arbitrary starting point to the evaporation temperature to complete evaporation, and to raise the temperature to the final temperature where the gas is superheated.

total pressure   In fluid flow, the sum of static pressure and velocity pressure.

turbulent flow   Fluid flow in which the fluid moves transversely as well as in the direction of the tube or pipe axis, as opposed to streamline or viscous flow.

unit heater   A direct-heating, factory-made, encased assembly including a heating element, fan, motor, and directional outlet.

unit system   A system that can be removed from the user's premises without disconnecting refrigerant containing parts, water connection, or fixed electrical connections.

unloader   A device in a compressor for equalizing high- and low-side pressures when the compressor stops and for a brief period after it starts so as to decrease the starting load on the motor.

vacuum   A pressure below atmospheric, usually measured in inches of mercury below atmospheric pressure.

valve   In refrigeration, a device for regulation of a liquid, air, or gas.

vapor   A gas, particularly one near to equilibrium with the liquid phase of the substance, which does not follow the gas laws. Frequently used instead of gas for a refrigerant and, in general, for any gas below the critical temperature.

viscosity   The property of a fluid to resist flow or change of shape.

water cooler   Evaporator for cooling water in an indirect refrigerating system.

wax   A material that may separate when oil/refrigerant mixtures are cooled. Wax may plug the expansion valve and reduce heat transfer of the coil.

wet-bulb depression   Difference between dry- and wet-bulb temperatures.

wet compression   A system of refrigeration in which some liquid refrigerant is mixed with vapor entering the compressor so as to cause discharge vapors from the compressor to tend to be saturated rather than superheated.

xylene   A flammable solvent, similar to kerosene, used for dissolving or loosening sludges and for cleaning compressors and lines.

zero, absolute, of pressure   The pressure existing in a vessel that is entirely empty. The lowest possible pressure. Perfect vacuum.

zero, absolute, of temperature   The temperature at which a body has no heat in it ($-459.6°F$ or $-273.1°C$).

zone, comfort (average)   The range of effective temperature during which the majority of adults feel comfortable.

Index